ENVIRONMENTAL BEHAVIOUR
OF CROP PROTECTION CHEMICALS

PROCEEDINGS SERIES

ENVIRONMENTAL BEHAVIOUR OF CROP PROTECTION CHEMICALS

PROCEEDINGS OF AN INTERNATIONAL SYMPOSIUM
ON THE USE OF NUCLEAR AND RELATED TECHNIQUES
FOR STUDYING ENVIRONMENTAL BEHAVIOUR
OF CROP PROTECTION CHEMICALS
JOINTLY ORGANIZED BY THE
INTERNATIONAL ATOMIC ENERGY AGENCY
AND THE
FOOD AND AGRICULTURE ORGANIZATION
OF THE UNITED NATIONS
AND HELD IN VIENNA, 1–5 JULY 1996

INTERNATIONAL ATOMIC ENERGY AGENCY
VIENNA, 1997

Permission to reproduce or translate the information contained in this publication may be obtained by writing to the International Atomic Energy Agency, Wagramerstrasse 5, P.O. Box 100, A-1400 Vienna, Austria.

© IAEA, 1997

VIC Library Cataloguing in Publication Data

International Symposium on the Use of Nuclear and Related Techniques for Studying Environmental Behaviour of Crop Protection Chemicals (1996 : Vienna, Austria)
    Environmental behaviour of crop protection chemicals : proceedings of an International Symposium on the Use of Nuclear and Related Techniques for Studying Environmental Behaviour of Crop Protection Chemicals jointly organized by the International Atomic Energy Agency and the Food and Agriculture Organization of the United Nations and held in Vienna, 1–5 July 1996. — Vienna : The Agency, 1997.
    p. ; 24 cm. — (Proceedings series, ISSN 0074–1884)
STI/PUB/1003
ISBN 92-0-104596-4
Includes bibliographical references.

    1. Agricultural chemicals — Environmental aspects — Congresses. 2. Radioisotopes in agriculture — Congresses. I. International Atomic Energy Agency. II. Food and Agriculture Organization of the United Nations. III. Title. IV. Series: Proceedings series (International Atomic Energy Agency).

VICL   96-00164

Printed by the IAEA in Austria
January 1997
STI/PUB/1003

FOREWORD

Pesticides are an integral component of agricultural systems throughout the world. It is generally accepted that this will continue for the foreseeable future if production of food of acceptable quality is to increase. However, pesticide use has costs as well as benefits. Concerning the environment, information must be provided before a product is registered in order to provide assurance that it can be used without unacceptable hazard to non-target organisms. In addition, post-registration surveillance and monitoring studies are necessary to check that the fate and environmental effects of pesticides under field conditions are consistent with predictions. Much of the data are generated using radioisotopes and other nuclear or related methods that can be applied in studies related to the fate and effects of pesticides in the various environmental compartments (soil, water and air), and in terrestrial (agricultural and non-agricultural) and marine ecosystems.

For a variety of reasons, developing countries often have to rely on data generated elsewhere in order to assess the acceptability of a compound, particularly if it is off-patent. This Symposium was organized to examine the circumstances under which extrapolation from one environment to another is valid on the basis of data generated under comparable conditions. It also considered ways in which relatively simple methods can be used to verify the field applicability of data obtained under sophisticated experimental conditions.

EDITORIAL NOTE

The Proceedings have been edited by the editorial staff of the IAEA to the extent considered necessary for the reader's assistance. The views expressed remain, however, the responsibility of the named authors or participants. In addition, the views are not necessarily those of the governments of the nominating Member States or of the nominating organizations.

Although great care has been taken to maintain the accuracy of information contained in this publication, neither the IAEA nor its Member States assume any responsibility for consequences which may arise from its use.

The use of particular designations of countries or territories does not imply any judgement by the publisher, the IAEA, as to the legal status of such countries or territories, of their authorities and institutions or of the delimitation of their boundaries.

The mention of names of specific companies or products (whether or not indicated as registered) does not imply any intention to infringe proprietary rights, nor should it be construed as an endorsement or recommendation on the part of the IAEA.

The authors are responsible for having obtained the necessary permission for the IAEA to reproduce, translate or use material from sources already protected by copyrights.

Material prepared by authors who are in contractual relation with governments is copyrighted by the IAEA, as publisher, only to the extent permitted by the appropriate national regulations.

CONTENTS

OPENING SESSION

Opening Statements

S. Machi 3
J.D. Dargie 5

PESTICIDES IN THE ENVIRONMENT: INTERNATIONAL PERSPECTIVE (Session 1)

Main provisions of the International Code of Conduct
  on the Distribution and Use of Pesticides
  for registration requirements (IAEA-SM-343/10) 11
  A. Ambrus
Pesticide residues in the marine environment
  and analytical quality assurance of the results (IAEA-SM-343/12) 35
  F.P. Carvalho, S.W. Fowler, J.P. Villeneuve, M. Horvat

PESTICIDES IN THE ENVIRONMENT: REGIONAL AND COUNTRY SITUATIONS (Session 2)

Control of pesticides in certain countries of South-East Asia
  (IAEA-SM-343/2) 61
  A. Balasubramaniam
Thiometon residues in cucumber (IAEA-SM-343/27) 73
  A. Sheikhigorgan, K. Talebi, H.R. Zolfagharieh, S. Mashayekhi
Pesticide residues in the soil of the Central Jordan Valley
  (IAEA-SM-343/29) 79
  T.M. Mustafa, A. Hatough, S. Khattari, K. Masha'al, A. Sa'adeh
Use of pesticides in Kazakstan (IAEA-SM-343/30) 91
  L. Pak
Fate and effects of pesticides under tropical field conditions:
  Implications for and research needs in a developing country
  (IAEA-SM-343/32) 93
  J. Espinosa-González

FATE AND BEHAVIOUR OF PESTICIDES IN THE TERRESTRIAL ENVIRONMENT (Session 3)

Extraction of pesticide residues from biological and
environmental samples (IAEA-SM-343/6) 111
S.U. Khan

Movement of ^{14}C-carbofuran in a silt clay soil: A laboratory study
(IAEA-SM-343/34) 127
I. Ghanem, S. Bali, F. Mohamad

Fate of ^{14}C-pirimicarb in Chinese cabbage and soil (IAEA-SM-343/18) 135
Jiarong Pan, Xianfang Wen

Distribution and fate of ^{14}C-acephate in tomato plants and soil
(IAEA-SM-343/26) 145
M.S. Tungguldihardjo, E. Anwar

Persistence in soil of endosulfan and lindane applied to
soyabean and maize pests in a field trial agrosystem in Zimbabwe
(IAEA-SM-343/41) 151
M.F. Zaranyika, P. Mugari

Persistence and fate of ^{14}C-lindane applied to soil
in a maize ecosystem (IAEA-SM-343/23) 163
P.O. Yeboah, K.G. Montford, F.E. Appoh, D.K. Dodoo

Degradation of chlorpyrifos in Turkish soil (IAEA-SM-343/35) 171
Ü. Yücel, M. Ilim, K. Gözek

Use of lysimeters for determining pesticide fate
in agroecosystems (IAEA-SM-343/5) 179
F. Führ, W. Mittelstaedt, T. Pütz, A. Stork, M. Dust

A microcosm system to evaluate the fate of ^{14}C labelled
pesticides in soils from different climate zones (IAEA-SM-343/21) 181
I. Scheunert, R. Schroll, G. Cao

Leaching and degradation of pesticides in groundwater layers
(IAEA-SM-343/22) 187
L. Vollner, D. Klotz

Model ecosystems for predicting the behaviour of pesticides
in the environment (IAEA-SM-343/24) 205
K. Raghu, N.B.K. Murthy, S.P. Kale, M.G. Kulkarni

Persistence of terbufos and its metabolites in soil and maize
(IAEA-SM-343/20) 215
E. Carazo, B.E. Valverde, O.M. Rodríguez, M. Barquero

Carbon-14-trifluralin residues in soil and carrots (IAEA-SM-343/36) 223
O. Tiryaki, K. Gözek

Radiation induced degradation of parathion in aqueous solution
(IAEA-SM-343/16) 235
L.C. Luchini, M.O. Oliveira de Rezende

FATE AND BEHAVIOUR OF PESTICIDES IN THE AQUATIC ENVIRONMENT (Session 4)

Agrochemical fate and effects in terrestrial, aquatic
and estuarine ecosystems (IAEA-SM-343/13) 247
  S.J. Klaine, P. Richards, D. Baker, R. Naddy, T. Brown, B. Joab,
  R. Casey, D. Fernández, J. Overmeyer, R. Benjamin
Fate of pesticides in a model rice paddy ecosystem
(IAEA-SM-343/33) 265
  A.W. Tejada, L.M. Varca, S.M.F. Calumpang,
  C.M. Bajet, M.J.B. Medina
Distribution and fate of ^{14}C-DDT in microcosm experiments
simulating the tropical marine environment of the Bay of Bengal
(IAEA-SM-343/15) 279
  M.A. Matin, E. Hoque, J. Khatoon, Y.S.A. Khan,
  M.M. Hossain, A.J. Mian
A tropical coastal lagoon affected by agricultural activities:
The importance of radiolabelled pesticide studies (IAEA-SM-343/31) 289
  F. González-Farias, F.P. Carvalho, S.W. Fowler, L.D. Mee
Fate of endosulfan in soil and in river and coastal waters of Jamaica
(IAEA-SM-343/28) 301
  D.E. Robinson, A. Mansingh, T.P. Dasgupta
Distribution and fate of ^{14}C-DDT in the estuarine environment
of the north of Viet Nam (IAEA-SM-343/40) 313
  Duc Nhan Dang, Van Thuan Vo, Manh Am Nguyen
Evolution of DDT residues in All Saints Bay, Brazil, 1985–1994
(IAEA-SM-343/17) 321
  T.M. Tavares, M. Beretta, M.A. Costa
Persistence of temephos and fenitrothion and their transformation
on products in rice field waters (IAEA-SM-343/1) 331
  D. Barceló
Organochlorine pesticides in sediment and biota in the coastal region
to the south of the Pinar del Río Province, Cuba (IAEA-SM-343/42) 343
  G. Dierksmeier, R. Hernández, P. Moreno, K. Martínez, C. Ricardo
Behaviour of BHC and DDT at the mouth of the Zhujiang River,
China (IAEA-SM-343/43) 349
  Fulong Cai, Zhifeng Lin, Ying Chen, Shumei Chen,
  Jiadong Yiang, Feng Cai, Lumin Qian

COMPARISON OF FATE AND BEHAVIOUR OF PESTICIDES IN DIFFERENT ENVIRONMENTS (Session 5)

A decision–support system for pesticide environmental
preregistration assessment (IAEA-SM-343/8) 361
P.H. Nicholls
Effective use of pesticide data for sustainable farming
in Australia and China (IAEA-SM-343/4) 371
I.G. Ferris, R.S. Kookana, Xianfang Wen, B.M. Haigh, Jiarong Pan
Environmental fate of herbicides in Hawaii, Peru and Panama
(IAEA-SM-343/7) 389
C.S. Helling
Polychlorinated biphenyls and cyclic pesticides in sediments and
macroinvertebrates from the coastal regions of different
climatological zones (IAEA-SM-343/45) 407
J.M. Everaarts, E.M. van Weerlee, C.V. Fischer, M.Th.J. Hillebrand

EFFECT OF PESTICIDES ON NON-TARGET SPECIES (Session 6)

Use of soil microcosms in assessing the effect of pesticides
on soil ecosystems (IAEA-SM-343/3) 435
C.A. Edwards, T. Knacker, A.A. Pokarzhevskij, S. Subler, R. Parmelee
Adverse effects on flora and fauna from use of organochlorine
pesticides in a maize agroecosystem (IAEA-SM-343/39) 453
E.M. Minja, R.A. Makusi, F. Tesha
Effect of organochlorine pesticides on birds
in the United Republic of Tanzania (IAEA-SM-343/38) 461
A.S.M. Ijani, J.M. Katondo, J.M. Malulu
Lindane and endosulfan residues in water and fish in the
Ashanti region of Ghana (IAEA-SM-343/44) 471
S. Osafo-Acquaah
Exposure of toad embryos and larvae to pesticides: Use of nuclear
techniques to determine their effect on the reproduction, survival
and potential risk to *Bufo arenarum* populations (IAEA-SM-343/14) 479
*A. Caballero de Castro, A. Venturino, V. Kirs, M. Loewy,
G. Carvajal, A.M. Pechen de D'Angelo*
Effects of methamidophos on soil microbial activity
(IAEA-SM-343/19) 489
Bujin Xu, Yongxi Zhang, Nanwen Zhu, Hong Ming, Meici Chen, Yuhua Zao
Extraction of beta-endosulfan and endosulfan sulphate in soils
with supercritical CO_2 (IAEA-SM-343/9) 495
R.M. López-Romero, S. Capella-Vizcaíno

MAXIMIZING THE USE OF ENVIRONMENTAL DATA
(Session 7)

Conclusions and recommendations 505

Chairpersons of Sessions and Secretariat of the Symposium 507
List of Participants 509
Author Index 517
Index of Papers by Number 519

OPENING SESSION

Chairperson

R.J. HANCE
FAO/IAEA

OPENING STATEMENTS

S. Machi
Deputy Director General,
Department of Research and Isotopes,
International Atomic Energy Agency,
Vienna

On behalf of the Directors General of the Food and Agriculture Organization of the United Nations and the International Atomic Energy Agency I have great pleasure in welcoming you to this FAO/IAEA Symposium on the Use of Nuclear and Related Techniques for Studying the Environmental Behaviour of Crop Protection Chemicals.

The two sponsoring organizations have a well established commitment to assisting Member States to combat and prevent environmental pollution of all types. They have both taken leading roles in the support of activities to implement the recommendations of Agenda 21 of the Report of the United Nations Conference on Environment and Development held in Rio de Janeiro in 1992. For example, within the IAEA, promotion of the safe development of nuclear energy makes an obvious contribution to the reduction of CO_2, SO_2 and NO_x emissions. In fact, throughout the world 437 nuclear power plants are in operation, generating 17% of the total electricity consumption. Technical and regulatory programmes dealing with radioactive waste management also clearly have environmental protection as an important objective.

In the Department of Research and Isotopes there are programmes that use isotope techniques to study groundwater pollution, water and pollutant dynamics in order to devise practices for the management of sustainable development of coastal and marine areas, and many other programmes concerned with the environment. Another example is the project to establish industrial technology to remove SO_2 and NO_x in flue gases from power stations that burn coal, using an electron accelerator.

In the agricultural context of the Joint FAO/IAEA Division of Nuclear Techniques in Food and Agriculture programme, the two major challenges are to increase or sustain food production, and to achieve this with minimum undesirable effects on the environment. This is being carried out against a background where population growth has outstripped food production in many developing countries, and where it is predicted that the area of arable land available worldwide to sustain each person will decline from 0.3 ha in 1981 to 0.13 ha in 2050. In addition to the work to be discussed at this Symposium, and in response to the above challenges, programmes include those that optimize fertilizer use, so minimizing losses that can lead to water pollution. Improved utilization of animal feed, although not primarily a pollution based programme, will also have an impact by reducing waste. Other programmes aim at developing disease and pest resistance in plant varieties and at promoting the sterile insect technique, both of which reduce pesticide use and, hence, diminish the pollution potential. For example, in March 1995 the Mediterranean fruit fly was successfully eradicated in Chile using the nuclear sterile insect technique with the support of the Joint FAO/IAEA Division.

Despite such developments in plant breeding and insect control, it is inevitable that in the foreseeable future the technology needed to produce a sustained increase in agricultural productivity will include the use of pesticides. Critics of pesticide technology are particularly concerned about the possible effects on the quality of food and the impact on the environment. This Symposium is addressed to the second of these concerns. It brings together not only scientists engaged in programmes of the Joint FAO/IAEA Division, but also those of the Agency's Marine Environment Laboratory in Monaco and of the FAO Plant Production and Protection Division in Rome. They share a common interest in the generation and interpretation of the data needed to provide reasonable confidence that a pesticide can be used without unacceptable risk to non-target organisms. This, of course, includes the monitoring needed when a product is in practical use in order to confirm that the field behaviour of the pesticide has been reasonably well predicted by the data provided for registration. Use of radioisotopes and other nuclear related methods is important for the success of such investigations.

This Symposium provides an opportunity for you to share the experience and data with which to synthesize views on the current situation and to develop recommendations on what needs to be done in future.

Once again, I would like to welcome you to this Symposium. May I wish you a successful meeting and hope that you are able to sample some of the sites of the beautiful city of Vienna.

OPENING STATEMENT

J.D. Dargie
Director,
Joint FAO/IAEA Division of Nuclear Techniques
in Food and Agriculture,
International Atomic Energy Agency,
Vienna

May I add my welcome to that of the Deputy Director General, Mr. S. Machi. The transfer of information and technology is a basic function of the Joint FAO/IAEA Division of Nuclear Techniques in Food and Agriculture. Symposia such as this are vital to this function, alongside training courses, fellowship training, expert services and Co-ordinated Research Programmes. It is, therefore, gratifying that, although the number of participants is restricted by financial considerations, over 40 Member States are represented here.

Mr. Machi has already spoken of the high priority placed on environmental issues at both the Food and Agriculture Organization of the United Nations and the International Atomic Energy Agency. This is particularly the case in the Joint FAO/IAEA Division, so much so that at the beginning of 1997 the Agrochemicals and Residues Section will be amalgamated with the Food Preservation Section to create the new Food and Environmental Protection Section, whose name emphasizes this commitment. Thus, it is particularly appropriate that we should be holding a Symposium devoted to an environmental issue in 1996.

Perhaps I may be permitted to digress a little at this point to consider one aspect of the programme of the new section that is closely related to the work that will be discussed this week.

Enforcement of pesticide control legislation and international agreements, in particular the FAO International Code of Conduct on the Distribution and Use of Pesticides, requires analytical facilities to monitor the quality of pesticides in trade and the level of pesticide residues in food and, as this audience is particularly aware, in the environment. Similarly, implementation of national legislation and international trade agreements to ensure the quality and safety of food, particularly the Uruguay Round Agreement on the Application of Sanitary and Phytosanitary Measures, also requires analytical laboratories and trained staff.

Many countries either do not have the facilities and personnel that are needed, or they have the facilities but do not operate them to the standards required to support legal and quasi-legal instruments. In order that the capability of such countries to control the quality and safety of food and the quality and use of pesticides is strengthened, we are in the process of establishing an FAO/IAEA Training and Reference Centre for the Control of Food and Pesticides in the IAEA Laboratories at Seibersdorf. The work of the centre will be relevant not only to the programme of the Joint FAO/IAEA Division in Vienna, but also to the Plant Production and Protection Division and the Food Policy and Nutrition Division of FAO in Rome.

The key activity of the centre will be training, by organizing fellowships, regional and interregional courses, and expert services. This will be carried out both at Seibersdorf and in co-operation with laboratories and institutions in different regions. Fellowship holders will be trained by working for a period of several months at Seibersdorf or in other laboratories, where they will obtain in-depth practical experience and widen their theoretical knowledge. Training courses will range in length from 1 to 6 weeks. The shorter courses will include topics such as the principles of analytical quality assurance and control, good laboratory practice, the obligations of signatories to World Trade Organization agreements and hazard analysis critical control point methodology. The longer courses will place more emphasis on laboratory work, and be mostly concerned with analytical methodologies.

The centre will also provide reference materials, organize interlaboratory comparisons and give advice on subjects such as analytical methods and the introduction of good laboratory practice and other relevant quality standards.

An additional important activity will be that of co-ordination. For example, it will be necessary to establish a network of laboratories, institutions and experts to support this work; the field is so extensive that we cannot hope to function without the assistance of colleagues around the world. We will also have to liaise with other relevant programmes and organizations to encourage harmonization and co-operation, while avoiding unnecessary duplication.

The range of analytical subjects included in the centre's programme will ultimately be very wide. In the immediate future, most attention will be given to pesticide residues, because the facilities and expertise are already available at Seibersdorf. This will support national monitoring programmes, which are vital to assess the effectiveness of regulations in protecting the health of consumers, and the environment, and to facilitate trade. To promote international trade, the Codex Alimentarius Committee on Pesticide Residues has established maximum residue limits for more than 150 pesticides in a wide range of foodstuffs, and national authorities are frequently required to certify that these standards are met, not only in food destined for export but also in food for domestic consumption. From pesticide residues it is but a short step to other organic contaminants of food and the environment, and by utilizing the expertise in heavy metal analyses and immunoassay procedures already available at Seibersdorf it should not take long before we are able to cope with almost the whole range of chemical contaminants. This will leave microbial food problems, an area in which we have little experience; clearly, urgent steps will have to be taken to make good this deficiency.

From what I have said so far you may have deduced that the main driving force behind this new development is the need to facilitate trade in foodstuffs, and we anticipate that most of the demand for the services of the centre will emanate from countries with an economic incentive for producing substantial food exports. Nevertheless, there are also Member States with powerful motives for monitoring environmental matrices, e.g. water, whether it be potable, for irrigation, or as a source for fish farming. Other interests such as ecotourism will also exert pressures in this direction.

We are now planning training programmes, and work will begin in 1997 at Seibersdorf to refurbish additional laboratory space to cope with the additional requirements.

I think you will agree that this is an important development. It is certainly one that we find exciting and challenging. I am sure that many of you here today will benefit from the programme in the not too distant future.

To return more specifically to the Symposium, we all know that pesticides will be essential to high yielding agriculture for many years to come. Equally, we all know that public perception of the value of pesticides tends to be negative, not least because of fears of their side effects on non-target species, both within and without agricultural ecosystems. Hence, pesticide control legislation now requires that manufacturers provide data on the fate and effect of their compounds in the environment.

The work required to generate the data for assessing environmental risk associated with the use of a pesticide is extensive, and such information expensive to obtain. Indeed, costs of research into the environmental toxicology of a new pesticide are of the same order as those for mammalian toxicology. It follows that manufacturers concentrate their research on environments that represent the most profitable markets and these, by and large, are in industrialized countries. Unfortunately, despite strenuous efforts, the International Group of National Associations of Manufacturers of Agrochemical Products was unable to send a speaker to give an industrial view to this Symposium, but it seems reasonable to assume that companies also take into account that people's perception and acceptance of risk differ according to their particular circumstances. People are risk takers when impoverished, and risk averse when affluent. So, when food is short, risks associated with increasing production are given less weight than when food supplies are adequate. This consideration further concentrates the manufacturers' minds on the concerns of affluent countries.

Thus, companies may be prepared to market newer compounds in less affluent countries, but not if they have to go to the expense of generating additional ecotoxicological data. The situation is similar with regard to the older compounds that are now off-patent, where clearly there is little incentive to spend money on ecotoxicological research.

This leaves the registration authorities in developing countries in a difficult position when having to decide whether or not to register a compound that may be badly needed but for which the only available ecotoxicological data were obtained in soil and climatic conditions that are very different to their own. An ideal situation, of course, would be for national laboratories to do the necessary work, contracted either by the government or the manufacturer; increasingly, Member States are seeking to do this. This is an important motivation for the establishment of our Training and Reference Centre, which will aim at assisting such laboratories to work to the appropriate standards; however, this solution is in most cases one for the longer term.

In practice, the authorities have to do their best to assess environmental risk with data that are largely generated elsewhere. An important objective of this Symposium is to review the extent to which this can be done and to consider ways in which simple experimental methods could provide bridging information that will allow extrapolations from one environment to another to be made with some confidence. The tendency in industrialized countries to reduce field and simulated field studies through a tiered approach to ecological risk assessment (tiers I and II are laboratory studies whose data are used in preliminary environmental fate modelling) is certainly helpful. Such an approach does, however, require information from the higher, field testing tiers for at least some compounds, because risk assessments based on only the lower tiers will inevitably be conservative, and useful compounds may be rejected. We hope that the results reported at this Symposium will provide some of the data needed to develop this approach in less affluent Member States.

A related issue is the need to avoid duplication of effort. Resources for ecotoxicological studies are limited in developing Member States, so it is important to share the available data. This Symposium will try to identify the opportunities and make recommendations in this regard. If recommendations cannot be agreed upon, then at least we hope that the relevant people will have been brought together, so that they can continue their discussion elsewhere.

Finally, I should like to draw attention to the fact that many of the delegates present are also participants in three of our Co-ordinated Research Programmes concerned with the fate and effect of pesticides in the environment. Two of these programmes are regional, one in Africa and the other in Central America, while the third is worldwide, concentrating on the marine environment. All three are financed by the Swedish International Development Authority, and it is a great pleasure to record our thanks for their generosity.

All that remains is for me to welcome you once again to the Symposium, and to wish you an interesting and productive week at the Vienna International Centre and an entertaining time in Vienna.

PESTICIDES IN THE ENVIRONMENT: INTERNATIONAL PERSPECTIVE

(Session 1)

Chairperson

R.J. HANCE
FAO/IAEA

IAEA-SM-343/10

MAIN PROVISIONS OF THE INTERNATIONAL CODE OF CONDUCT ON THE DISTRIBUTION AND USE OF PESTICIDES FOR REGISTRATION REQUIREMENTS

A. AMBRUS
Pesticide Management Group,
Plant Protection Service,
Food and Agriculture Organization
  of the United Nations,
Rome

Abstract

MAIN PROVISIONS OF THE INTERNATIONAL CODE OF CONDUCT ON THE DISTRIBUTION AND USE OF PESTICIDES FOR REGISTRATION REQUIREMENTS.
　　Pesticides are toxic and hazardous substances. They must be distributed and used properly to reduce unacceptable risk and to avoid harmful effects on humans and the environment. Judgement on the suitability, efficacy, safety or fate of the pesticide under particular conditions of use must be made by the responsible authority in the country. To promote the safe and efficient use of pesticides and fair practice in their sale and distribution, the Food and Agriculture Organization of the United Nations (FAO) developed the International Code of Conduct on the Distribution and Use of Pesticides. The objectives of the Code are to outline the responsibilities of and to establish the voluntary standards of conduct for all public and private entities engaged in or affecting the distribution and use of pesticides, particularly where there is no, or only an inadequate, national law to regulate pesticides. The Code defines the responsibilities of the pesticide manufacturers, governments of importing and exporting countries, international organizations and other bodies. Simultaneously with the development of the Code, FAO co-ordinated the elaboration and regular updating of several guidelines and manuals to assist Member Countries, manufacturers and distributors of pesticides in order to harmonize the data requirements and to ensure that the appropriate database is available for assessing the behaviour of pesticides and for regulating their use. Three of the guidelines especially relevant to the registration of pesticides, and the related recommendations of the FAO/WHO Joint Meeting on Pesticide Residues, are discussed in detail. It is emphasized that the guidelines and risk assessment schemes cannot be used on their own. A qualified and experienced expert's judgement is required for appropriate evaluation of the experimental data in order to perform the risk–benefit analysis, to determine the most suitable patterns of use and to introduce risk management measures. For all these considerations, detailed knowledge on local farming practices and the economic and social conditions is essential.

1. INTRODUCTION

Pests destroy up to one-third of the world's food crops during growth, harvesting and storage. In developing countries, crop losses are even higher. At present, use of pesticides is inevitable for producing and protecting food, animal feed and industrial crops. While pesticides are intended to effectively control organisms that destroy or endanger human food supply and health, under some circumstances and at concentrations above a certain threshold they, like virtually every chemical, may have physiological effects on other organisms living in the environment, including humans themselves. The effects depend on the dosage and the conditions of use.

As pesticides are toxic and hazardous substances, they must be distributed and used properly to reduce unacceptable risk and to avoid harmful effects on humans and the environment. Judgement on the suitability, efficacy, safety or fate of the pesticide under particular conditions of use must be made by the responsible authority in the country, in consultation with industry and other government authorities, on the basis of scientific evaluation and detailed knowledge of the conditions prevailing in the country of proposed use.

In the absence of an effective pesticide registration process and of a governmental infrastructure for controlling the availability of pesticides, some countries importing pesticides have to rely heavily on the assistance of the pesticide industry for the proper distribution and use of pesticides. Therefore, development of national regulatory programmes is the first priority of Food and Agriculture Organization of the United Nations (FAO) activities in this field.

To promote the safe and efficient use of pesticides and fair practice in their sale and distribution, FAO developed, through a series of consultations with appropriate United Nations agencies and other organizations, the International Code of Conduct on the Distribution and Use of Pesticides [1], which was adopted by the FAO Conference at its 20th Session in 1985.

Simultaneously, FAO co-ordinated the development and regular updating of several guidelines [2–9] and manuals [10, 11] to assist Member Countries, manufacturers and distributors of pesticides in the harmonization of the data requirements and to ensure that the appropriate database is available for assessing the behaviour of pesticides and for regulating their use.

The Joint Meeting of the Panel of FAO Experts on Pesticide Residues in Food and the Environment and a World Health Organization (WHO) Expert Group on Pesticide Residues (usually referred to in its abbreviated form, JMPR) evaluate a number of pesticides annually. In 1985, the WHO Environmental Core Assessment Group, whose role is to identify the risk to organisms in the environment, joined the JMPR, demonstrating that the intention of the two organizations is to give their full attention to the environmental impact of the proposed use of pesticides. The monographs and the appraisals of the results, published in the series Pesticide Residues in Food: Evaluations, Part I, Residues, by FAO, and, Part II, Toxicology, by WHO, give factual information on and comprehensive evaluation of the recommended patterns of use, and the fate and behaviour of pesticides. Complementing the above mentioned FAO guidelines, the JMPR reports also provide useful information on data requirements and interpretation of the results.

This paper outlines some of the recommendations included in the International Code of Conduct on the Distribution and Use of Pesticides [1], FAO guidelines and JMPR recommendations for assessing the fate and behaviour of pesticides for registration, and points out some particular aspects that require special attention in pre- or post-registration testing, reporting or evaluation of the results.

2. PROVISIONS OF THE CODE FOR REGISTRATION OF PESTICIDES

The objectives of the Code [1] are to outline the responsibilities of and to establish the voluntary standards of conduct for all public and private entities engaged in or affecting the distribution and use of pesticides, particularly where there is no, or only an inadequate, national law to regulate pesticides. One of its aims is to "promote practices which encourage the safe and efficient use of pesticides, including minimising adverse effects on humans and the environment and preventing accidental poisoning from improper handling" through appropriate pesticide management and testing of the product before and during its practical use.

The Code defines the responsibilities of the pesticide manufacturers, governments of importing and exporting countries, international organizations and other bodies. Those which are relevant to the characterization and testing of the properties of pesticides, and their registration, are summarized below.

The pesticide manufacturers are expected:

(1) To ensure that each pesticide and pesticide product is adequately and effectively tested by well recognized procedures and test methods so as to fully evaluate its safety, efficacy [2] and fate [3] with regard to various anticipated conditions in the regions or countries of use.
(2) To ensure that such tests are conducted in accordance with sound scientific procedures and good laboratory practice (GLP). The data produced by such tests, when evaluated by competent experts, must be capable of showing whether the product can be handled and used safely without unacceptable hazard to human health, plants, animals, wildlife and the environment.
(3) To provide an objective appraisal, together with the necessary supporting data, of each product.
(4) To make available copies or summaries of the original reports of such tests for assessment by the responsible government authorities in all countries in which the pesticide is to be offered for sale. Evaluation of the data should be referred to qualified experts.

(5) To ensure that the active ingredients and other ingredients of those pesticide preparations that are marketed correspond in identity, quality, purity and composition to the substances tested, evaluated and cleared for toxicological and environmental acceptability.
(6) To ensure that the active ingredients and formulated products for those pesticides for which international specifications have been developed conform with the specifications of FAO [12] where intended for use in agriculture, and with the specifications of WHO [13] where intended for use in public health.
(7) To verify the quality and purity of those pesticides that are offered for sale.
(8) To take care that the proposed pattern of use, label claims and directions, packages, technical literature and advertising truly reflect the outcome of these scientific tests and assessments.
(9) To use clear and concise labelling.
(10) To give, at the request of a country, advice on methods for analysing any active ingredient or formulation that they manufacture, and to provide the necessary analytical standards.

Governments should:

(a) Introduce the necessary legislation for the regulation, including registration, of pesticides and make provisions for its effective enforcement, including the establishment of appropriate educational, advisory, extension and health care services.
(b) Establish pesticide registration schemes and infrastructures under which products can be registered prior to domestic use. The FAO guidelines for the registration and control of pesticides [4, 5, 7] should be followed, as far as possible, taking full account of the local needs, the social and economic conditions, the levels of literacy, the climatic conditions and the availability of pesticide application equipment.

Exporting governments and international organizations must play an active role in assisting developing countries to train personnel in the interpretation and evaluation of test data. Industry and governments should:

(i) Collaborate in conducting post-registration surveillance or monitoring studies to determine the fate and environmental effect of pesticides under field conditions.
(ii) Reduce hazards by making provisions for the safe storage and disposal of pesticides and containers at both the warehouse and the farm level, and through proper siting and control of the wastes from formulating plants.

3. FAO GUIDELINES THAT ASSIST IN IMPLEMENTING THE CODE

The challenge of protecting crops and livestock from insects, diseases, weeds and other pests without hazard to people, animals or their environment requires the combined and sustained efforts of scientists, technicians and administrators; of producers, processors and distributors; of industry and government; and of nations working together to establish and administer sound, acceptable standards of food safety and environmental quality. The wholesomeness of any food supply depends in part on the quality of the total environment: the soil, water and air in which the food is grown, processed and consumed. Acute contamination of these basic natural resources by pesticide residues and other pollutants can affect not only the safety of food products but also other environmental values such as water supplies, wildlife preservation and outdoor recreation.

The risk to the environment of a pesticide or its formulated product depends on many factors, e.g. its toxic properties, the amount applied, the formulation, the method and time of application, and particularly the intensity of its use, mobility and persistence in the environment. In practice, information of environmental significance comes from three basic sources: application and pattern of use, the fate and possible occurrence of residues in relevant parts of the environment, and the effects of predicted exposures on non-target species.

Nations are actively seeking to protect and manage their natural resources in the interest of greater safety and human welfare by implementing or establishing legal systems to regulate the distribution, sale and use of pesticides.

FAO guidelines that deal with the major pertinent aspects of the Code in detail were prepared with a view to assisting governments strengthen their pesticide registration, to facilitating international harmonization of registration requirements and to producing data that supports registration.

3.1. Guidelines for the Registration and Control of Pesticides [4, 5]

These guidelines describe the time schedule and legal frame for the recommended regulatory scheme, the phases of registration and the data requirements for registration. The amount and depth of information to be provided depend on the phase of registration and the extent of target use. The main categories of data required and some specific information are discussed below.

(1) *Identity*

(2) *Physicochemical properties of the pure active ingredient:*

    (a) *Technical grade active ingredient:*

        — The minimum active ingredient content;
        — The identity and amount of isomers, impurities and other by-products, together with information on their possible range, expressed as g/kg.

(b) *General description (identity) of the formulated product.*

(c) *Composition and physicochemical properties of the formulated product:*
- Content of the technical grade active ingredient;
- Flammability; acidity; alkalinity; corrosiveness; wettability; persistent foam;
- Suspensibility; wet or dry sieve testing; emulsion stability.

Notes

The specified properties of the products should be checked regularly within the national pesticide quality control programme, since they may directly influence the applicability (biological efficacy), the acute toxicity to applicators and the environmental impact of the products.

It is emphasized that the toxicological evaluation, the estimated no-effect level and the acceptable daily intake (ADI) are valid only for a pesticide of a specified composition. Since some of the impurities (e.g. iso-malathion, ethylene thiourea dioxins and heavy metals) may be much more toxic than the active ingredient, different manufacturing processes, different sources of raw materials, and unfavourable and extended storage conditions may result in a substantial change in the composition, and consequently in the toxicity of the pesticide. It should be noted that analysis of the active ingredient content usually cannot reveal the increase in impurities, as they are present at a much lower concentration. The active ingredient content can be within the specified range, while a toxic impurity is present at an undesirably high level. For example, Table I shows the permitted ranges of active ingredients and the maximum concentrations of impurities for some pesticides given in FAO Specifications for Plant Protection Products [12]. The specifications for technical grade active ingredients and formulated products have been published since 1971 in booklets that have an FAO identification number.

The possible presence of highly toxic impurities (or persistent isomers such as alpha-benzene hexachloride (BHC)) has two practical consequences:

(i) The registration authorities should always request accurate characterization of the composition of technical active ingredients, especially for generic pesticides, and should carefully consider the extent of similarity between any active ingredient being considered for registration and the technical materials assessed by the JMPR [14]. They should not apply the JMPR toxicological evaluations, which primarily aim at supporting the recommended Codex maximum residue levels (MRLs) with scientific evidence, to products of unknown composition.

(ii) The toxic impurities should also be regularly checked as part of the pesticide quality control programme.

TABLE I. ACTIVE INGREDIENT CONTENT AND MAXIMUM CONCENTRATION OF IMPURITIES GIVEN FOR SOME PESTICIDES IN FAO SPECIFICATIONS [12]

Active ingredient	Formulation	Permitted concentration range (%)	Impurities	
			Name	Maximum concentration
Copper oxychloride	WP 500 (g/kg Cu)	±5	Arsenic	50 mg/kg
			Lead	250 mg/kg
			Cadmium	50 mg/kg
Malathion	WP 500 (g/kg)	±5	Iso-malathion iron	1.8% [a]
Mancozeb	Technical	±3	Ethylene thiourea	<0.5%
Maneb	Dust	±15	Ethylene thiourea	<0.5%
Zineb	DP 800 (g/kg)	±3	Ethylene thiourea	<0.5%
			Arsenic	200 mg/kg

[a] After a stability test at 54°C for 6 days.

(3) Data on efficacy

Notes

The efficacy evaluation should be based primarily on detailed data provided by the registration applicant, using internationally harmonized methods such as European and Mediterannean Plant Protection Organisation (EPPO) Guidelines for the Efficacy Evaluation of Plant Protection Products [15, 16].

The effective minimum dosage should be checked with local trials, since it can vary substantially from region to region. Overdosing should be avoided in order to reduce environmental exposure and cost.

FAO Guidelines on the Efficacy Data for the Registration of Pesticides for Plant Protection [2] can be of valuable help in the planning, performing and reporting of efficacy trials.

(4) Data on toxicity for assessing the human health hazard

Note

In 1995, the JMPR prepared the list of minimum data required for estimating ADI [17]. The listed information may be used as the minimum data requirement also at the national level.

(5) Data on residues in agricultural produce

(6) Data for prediction of the environmental effects

Note

FAO Guidelines on Producing Pesticide Residue Data from Supervised Trials [3] and the Revised Guidelines on Environmental Criteria for the Registration of Pesticides [7], as well as the recommendations of JMPR on data requirements, are discussed in detail in Section 3.3.

(7) Recommendation for the pattern of use and instructions on the label

Labels are the most important source of information on the approved or recommended use of pesticides. They should bear clearly legible directions for use. Warnings and warning symbols should be prepared in the language(s) of the country. It is essential that all restrictions, including the pre-harvest intervals, withholding periods, limitations for use of the treated crop for animal feed, etc., are indicated.

Note

A recent example also underlines the necessity of careful and comprehensive assessment of the possible uses of by-products. Helix, which contains chlorfluazuron active ingredient, obtained a special clearance in Australia for use in cotton. Owing to drought in 1994, farmers were forced to feed cattle with cotton trash as an alternative feed. According to recent reports, new born calves are still being contaminated with chlorfluazuron residues [18], and export of meat to several countries has been suspended.

(8) Proposed packing

Note

The size of packing should preferably be in concord with the average quantities purchased at one time within the country in order to avoid the need for repackaging and sale from open containers.

3.2. Guidelines on Producing Pesticide Residue Data from Supervised Trials [3]

These guidelines supersede the previous FAO Guidelines on Pesticide Residue Trials to Provide Data for the Registration of Pesticides and the Establishment of Maximum Residue Limits published in 1986. Use of a pesticide on crops or commodities can lead to, and occasionally aims at, a residue remaining at harvest or at other appropriate stages. Additionally, a pesticide may move from the site of application and remain for some time elsewhere in the environment.

Uptake of residues by animals, leading to residues in foods of animal origin, can occur following either direct application of the pesticide to the animal or ingestion of feed containing the pesticide residues.

Assessment of hazards to humans and the environment arising from pesticide residues has become an important part of the overall risk–benefit evaluation, and is essential before the pesticide can be introduced. One of the basic prerequisites of such an assessment is the availability of reliable data on the fate of pesticides in plants and animals and in the environment, and on pesticide residues in food, feed, water and soil. Data on residues are mostly obtained from supervised trials.

The OECD Member Countries require that both the field and the analytical part of current supervised field trials should be carried out according to OECD Good Laboratory Practice Principles [19–28]. The number of countries setting the same requirement is increasing continuously. Institutes and laboratories intending to perform supervised field trials are therefore urged to introduce and implement the GLP Principles in their regular operation in order to be able to provide internationally acceptable results.

The purpose of these FAO guidelines is:

(1) To indicate acceptable techniques that may be followed in order to secure valid experimental data appropriate to the above objectives;
(2) To promote harmonized procedures in order to facilitate international acceptance of the data obtained.

The information provided in these guidelines is outlined in the following subsections.

3.2.1. Definition of residues

(1) Metabolites as components of the total residue

Notes

The decision on which of the metabolites and degradation products should be analysed depends basically on their toxicity and concentration. To make the right decision, the metabolic pattern of the pesticide and the relative concentration of the metabolites should be known from metabolism studies.

To enable a comprehensive evaluation of residues to be made, all the toxicologically significant and major residue components (metabolites, degradation products and impurities of technical grade product) should be determined and reported individually as far as is technically possible.

Taking into account the limitations in analytical methodology, the guidelines specify the cases when the residues may be expressed as a parent compound, a single metabolite or an alteration product.

The 1995 JMPR report [17] further elaborated on the expression of residues and named the two basic requirements for defining residues. The first is to be the most suitable for monitoring compliance with good agricultural practice (GAP), i.e. the set of registered/recommended uses, interpreted by the 1987 and 1988 JMPR reports [29, 30]. The second is to include compounds of toxicological importance for dietary intake estimations and risk assessment. These two requirements are sometimes not compatible, and in the compromise various definitions of the residues are possible [17].

For the purpose of monitoring compliance with GAP it is sufficient to include in the definition a single compound, the concentration of which reflects the applied dosage, and the time that has elapsed between application and sampling. The selected compound should preferably be stable during the analytical process, and recoverable with multiresidue procedures.

(2) Non-extracted or 'bound' residues: conjugates

Notes

Conjugated residues in plants should be released and determined, except in cases where it is known from metabolism studies that the conjugates formed in plants pass unchanged through animals.

The metabolism studies and characterization of non-extracted and conjugated residues are generally performed with radiolabelled compounds. Guidance has not yet been completed for these types of study.

3.2.2. Metabolism/degradation studies in crops and plant products

Notes

This section is to be completed in the current guidelines. However, on the basis of previous FAO guidelines, JMPR statements and national registration guidelines, the basic principles can be summarized.

The objectives of these studies are to provide information on:

— The qualitative identification and quantitative estimation of metabolites and degradation products and their form of occurrence in various tissues (e.g. free, conjugated or bound);
— The metabolism pathway;
— The changes in composition of residues with time;
— The uptake of residues from soil;
— The translocation of residues within plants.

Usually, these studies are carried out with radiolabelled compounds.

The number of plant species on which the metabolism is to be studied depends on the extent of target use. Metabolism studies on representatives of major crop classes are generally sufficient (e.g. wheat, lettuce and tomato, if the pesticide is intended to protect cereals and vegetables).

The information obtained from metabolism studies is used to decide on the significance of residue components, and on the need for their determination in supervised field trials.

3.2.3. Supervised residue trials in crops and plant products

(1) Design of residue trials

— Selection of size, location and number of sites; replicate applications;
— The type and variety of the crop/commodity/cropping system;
— Application of the pesticide (formulation, method, dosage rate, number and timing);
— Additional pesticides applied on the experimental plot.

Note

The minimum database (number of trials and locations) required for supporting the petition for registration depends on many factors, e.g. the type and extent of target use, the nature of the residues and the geographical/climatic diversity of the country. Consequently, national requirements show great variation. There is no international guideline on minimum data. The JMPR describes the typical issues and considerations that are taken into account in judging the adequacy of available information [31].

(2) Studies on the degradation/disappearance of residues and on the magnitude of residues

(3) Principles of sampling

— Sampling for decline studies and at normal harvest time at field sites;
— Sampling of processed commodities, stored products and soil;
— Sampling from bulk;
— Packing, reception and handling, and storage of samples.

Notes

It should always be borne in mind that the analytical results cannot be better or more reliable than the samples themselves. Sampling error is usually the major source of uncertainty in residue results.

In the guidelines, detailed information is given on the size of samples (minimum weight and minimum number of natural units) to be taken from the groups of plants and crops specified in the Codex Commodity Classification [32]. It is essential that the size of the samples taken be reported, since it directly affects the residue range and the maximum residue level estimated on the basis of the results of the trials. The effect of sample size on the residue distribution is illustrated in Fig. 1 [33].

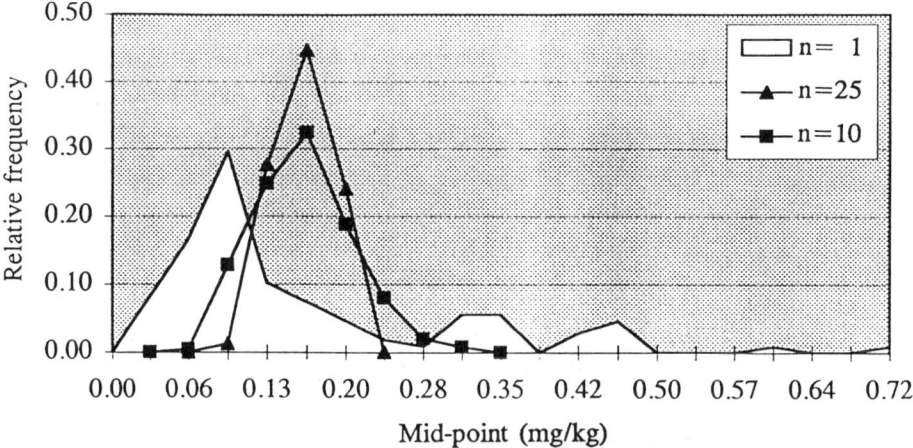

FIG. 1. Relative frequency of phosphamidon residues in primary and composite samples of apples. Relative frequency of residues in primary apple samples (sample size: n = 1) was obtained from analysis of 108 samples. Relative frequency of sample populations of sizes 10 and 25 were calculated on the basis of repeated random sampling with replacements from the primary sample population [33]. The figure indicates the residue ranges that may be found in samples of different size taken from the same field.

(4) Reporting on the field aspects of residue trials

(5) Analytical methods for the residues in crops and plant products

Notes

There are some aspects of residue analysis that are often overlooked and, therefore, need special attention:

(a) *Preparation of representative analytical samples*

As a consequence of the effect of sample size on the spread of residues, the total amount of the sample taken must be processed in the laboratory to obtain representative portions for analysis; any deviation from the recommended sampling [3] or sample preparation procedures [34] should be explained in the report.

(b) *Portion of commodity to be analysed*

The portion of commodity to be analysed in order to provide data for establishing the MRLs is specified in the Codex Guide [34].

In addition, residues in the edible portion should be analysed in order to provide data for estimating the dietary intake [35, 36].

The portion of plant commodity that may be used as animal feed should be analysed separately (e.g. maize fodder, straw).

There may be other, special objectives of the studies that require analysis of the specified part of the treated plants or crops. In this case, the analysed part and the method of preparation must be unambiguously described in the report.

(c) *Stability of residues during storage*

There are several examples indicating that degradation and dissipation can occur even under cold ($\leq -20\,°C$) storage conditions. When samples are held in storage for extended periods prior to analysis it must be demonstrated that the analysis will reflect the residues in terms of the nature and level that existed at the time of sampling. The JMPR considers it necessary that representative storage stability studies should be provided for all pesticides for which MRLs are proposed. Storage stability studies should include separate analysis of all the residue components of concern, not only the total residues.

The results of storage stability tests also provide information to regulatory laboratories on how long they may store monitoring, enforcement or other analytical samples before analysis [35].

The storage stability tests are usually carried out with cold materials, but using labelled compounds can be especially advantageous when the stability of several residue components and their possible transformation during storage are studied.

(d) Analysis of samples

Analysis of samples should be carried out with validated methods under properly controlled conditions. The Analytical Working Group of the Codex Committee on Pesticide Residues (CCPR) prepared guidelines for residue analysis [37] that provide useful technical help for analysts, and can also be used as a starting point for the introduction of the OECD GLP Principles.

Demonstrating the efficiency of extraction is one of the most difficult parts of the method validation procedure. The most convenient way of determining this is to use labelled material. Even in the latter case, there are several drawbacks that must be carefully avoided in order to obtain valid results:

— The labelled material to be applied to the test plants must be mixed in a commercial formulation, preferably in a manner that closely resembles practical use conditions and pre-harvest intervals.
— The selected plant should have properties that are similar to those to be extracted.
— Studies carried out with plant samples spiked with labelled materials shortly before extraction do not reveal more information than those with unlabelled analytical standards. Such studies are considered to be a waste of expensive labelled materials, and should not be carried out.
— Many pesticides bind to soil rapidly, therefore it is usually sufficient to mix well the labelled compound to the soil, to evaporate the solvent, and to store the sample in an airtight bottle in a dark place for 1 day. When metabolites are also to be analysed, a longer period is required before the efficiency of the extraction can be studied.
— The material balance should be established in all extraction efficiency studies.

It is important to provide a realistic picture of the residue distribution from the proposed pesticide use. On the other hand, increasing the sensitivity of the methods to detect levels that are 2–3 magnitudes lower than the average residue unnecessarily increases the difficulties and cost of analysis. The limit of determination (LOD) of residues should aim to be at least 1/10 of the established or expected MRL, if technically possible.

The latest EU Directive [38] on analytical methodology specifies the following minimum criteria:

MRL (mg/kg)	LOD (mg/kg)
>0.5	0.1
0.5–0.05	0.1–0.02
<0.05	MRL × 0.5

Exceptions are cases where the MRL is set at the LOD, and in analysis of residues in soil and water where there is no MRL. In the latter case, the analyst may consider the results obtained with other pesticides of similar chemical structure, and take into consideration other requirements such as the LOD of 0.1 µg/L in groundwater set by the European Union. The same directive specifies the requirements on the repeatability and reproducibility of analytical methods.

The repeatability and reproducibility must be less than:

Residue level (mg/kg)	Repeatability difference (%)	Reproducibility difference (%)
0.01	50	100
0.1	25	50
≥1	12.5	25

(6) Reporting the results of residue analysis

Note

The guidelines give clear instructions on the content of the reports. The appropriate expression of residues needs special consideration in the following cases:

— *Expression of the levels of multicomponent residues present at or about the LOD.* The 1995 JMPR report prepared a proposal for the expression of such residues [17].
— *Expression of the pesticide residues that are the metabolites of another pesticide.* The 1987 and 1989 JMPR reports contain guidance on this issue [29, 39].
— *Expression of the residues in samples of animal origin.* The fat solubility of residues should be considered. JMPR suggested [40] that a pesticide is fat soluble when the logarithm of the octanol–water partition coefficient of a

pesticide (log P_{ow}) is >4, and fat insoluble when it is <3. In cases where $3 \leq \log P_{ow} \leq 4$, the fat solubility of residues has to be decided specifically for each case, taking into consideration the distribution of residues. It should be pointed out that the parent compound and the major metabolites may not fall into the same category. The 1994 JMPR report summarizes the current MRLs for animal products according to the method of expression of residues [14].

— *Expression of residues in milk and milk products.* The CCPR could not elaborate a generally acceptable approach, and the matter is currently under discussion.

— *Expression of residues of forage crops with a changing water content.* In this case, the residues should be expressed on a dry weight basis.

3.2.4. Metabolism studies and supervised residue trials in animals

(1) Design of studies

Notes

It is customary to use radiolabelled compounds in metabolism studies on animals in order to identify the nature of the residues in the animal products. Quantitative measurements of these residues are then made using unlabelled compounds.

The study should indicate the distribution of residues in tissues, eggs and milk, and whether the residues are accumulated or excreted readily. The plateau of residues should be established for the regular intake of residues containing feed. (The plateau in milk and eggs is an indication that the levels in tissues may also be close to a plateau level.) As part of the study, the efficiency of extraction of the various components of the residue should be determined.

The position of radiolabelling must be chosen so that the label is not easily lost by metabolic transformation. For example, ring labelling is preferred for aromatic and cyclic compounds. Where the molecule is complex and more than one large fragment may be formed in animals it may be important to carry out more than one study, with pesticides labelled at different parts of the molecule.

The most important studies are those involving ruminants (lactating cows or goats) and poultry (chickens). For each position of the label, at least one ruminant and three chickens should be dosed. If metabolism in the rat is different from that in the cow, goat and chicken, pig (monogastric) metabolism studies may be necessary.

Animals should be slaughtered 24 hours after the final dose. For compounds remaining in the tissues — in studies with unlabelled compounds — animals from the treated groups should be slaughtered at predetermined intervals after the last dose, depending on prior information on the rate of elimination of residues, in order to determine the necessary withholding period.

Animal transfer studies are performed to provide data on residues in food of animal origin deriving from residues in feedstuff. The JMPR prepared guidelines on the need for animal transfer studies in estimating MRLs [41, 42].

(2) Studies with radiolabelled compounds

    (a) Oral administration studies

The dose should be at least equivalent to the daily intake from feed (MRL). If the intake is too low to identify the residues, a higher dose (e.g. 10 mg/kg of feed) is recommended.

Animals should be suitably housed in the same place for a few days for acclimatization before dosing commences. During the acclimatization period, milk and eggs can be taken to establish the background level of radioactivity.

The dose should be administered in the least stressful manner.

Notes

The JMPR uses parts per million to distinguish the residues in feed from the feeding level of animals, expressed as mg/kg body weight.

In extreme cases, a much higher dose (200 mg/kg of feed, or 50 mg active ingredient/kg body weight per day) can be applied (see, for example, buprofezin and bentazone in Ref. [14]).

    (b) Dermal treatment studies

If an oral administration study has been carried out, dermal treatment using the radiolabelled chemical may not be necessary, because oral ingestion of a dermally applied chemical, as a result of the animal grooming itself, is the major route of uptake, and metabolites may be satisfactorily characterized in the oral study.

The dermal treatment should simulate the proposed pattern of use. An area equivalent to 25% of the surface of the cow can be painted with the formulation at four times the normal rate. In chicken, the formulated product can be applied with a brush to the skin and to the base of the feathers on the breast, the abdominal area and the upper leg.

Following treatment, the animal should be allowed to lick itself or to groom normally.

(c) Sampling and analysis

Samples of excreta, milk or eggs should be collected for analysis during the dosing and withholding period.

The following tissues are essential: *ruminants*: meat (leg), liver and kidney; *chickens*: meat (leg and breast) with overlying skin and associated fat, liver and kidney.

The following analyses of samples are required: the levels of radioactivity in the excreta; identification of the metabolites in the urine and/or faeces; the levels of radioactivity in the tissues, milk and eggs; and identification of the residues in the tissues, milk and eggs.

For the compounds that remain in the tissues, animals from the treated groups should be slaughtered at predetermined intervals after the last dose, depending on prior information on the rate of elimination of residues, in order to determine the necessary withholding period.

(3) Studies with unlabelled compounds

If the crop to be treated with the pesticide is an important part of the animal diet, feeding studies or dermal treatment should be carried out with technical grade pesticide. These studies provide quantitative data for establishing the MRLs.

The feeding study should include a control group, a group dosed with the expected maximum daily intake ($1\times$) and a group with an exaggerated level ($3-10\times$).

Studies can be performed for oral administration, which includes sampling and analysis, and for dermal treatment.

(4) Storage of samples

The samples should be unambiguously labelled and stored in airtight containers at $\leq 20°C$. Milk samples should be stored in partly filled glass bottles. Prolonged storage of milk samples is undesirable, since the emulsion breaks on thawing and removal of the representative portion is very difficult, or virtually impossible.

(5) Reporting of results

3.3. Revised Guidelines on Environmental Criteria for the Registration of Pesticides [7]

Assessment of the effects on the environment is an integral part of the process of pesticide development and registration. The nature and amount of data required for registration depends on the properties and use of each substance. A stepwise sequence allows efficient selection of the tests essential to each individual risk analysis. The steps are:

(1) Standard laboratory tests on the physicochemical properties; and the primary fate of the compound, and the acute and short term biological effects.
(2) Supplementary laboratory tests on environmental distribution; and additional toxicity tests, including the sublethal and chronic effects.
(3) Simulated field and field studies.
(4) Post-registration monitoring.

The guidelines present the recommended stepwise approach and include the considerations listed below.

3.3.1. Part I. Principles

(1) Exposure

— Environmental concentrations, including residues in the diet of wildlife;
— Bioavailability;
— Biology of organisms;
— Stepwise sequence of data production.

(2) Effects

— On mammals, birds, honey bees, fish, daphnia, soil organisms, predators and/or parasites;
— On bird and daphnia reproduction studies;
— On phytotoxicity.

(3) Hazard

— Hazard assessment.

(4) Risk

— Risk management.

3.3.2. Part II. Appropriate test procedures

(1) Physicochemical properties

— Vapour pressure, water solubility and octanol–water partition coefficient.

(2)   Fate and mobility in the environment

— Degradation in mammals and plants;
— Degradation in soil;
— Degradation in the aquatic environment: hydrolysis, aqueous photolysis and biodegradation studies;
— Mobility in the environment: absorption/desorption and leaching.

(3)   Effects on the environment

— Vertebrate wildlife: mammals and birds;
— Non-target aquatic organisms;
— Soil non-target organisms and earthworms;
— Honey bees.

Notes

The guidelines give the criteria and test sequence for each test.

Very intensive research and harmonization activities are also carried out in this field by national and international organizations.

EPPO and the Council of Europe have elaborated a detailed Decision-making Scheme for the Environmental Risk Assessment of Plant Production Products [43, 44]. Applicability of the scheme is currently being tested in European countries. The chapters on the risk assessment scheme for higher plants and air are being finalized.

The most critical point in environmental risk assessment is prediction of the environmental concentration. The computer models used for this purposes are not standardized internationally. The WHO Environmental Core Assessment Group [17] used three models in their evaluations in 1995: the EPPO model; USES, from the Netherlands; and GENEEC, employed by the Environmental Protection Agency. All these approaches are broadly similar and differ only in details. The results obtained with the models must be used with great care because their validity is restricted to specified conditions and depend on the availability of reliable input data, which are often difficult to gather.

4.   CONCLUSIONS

The documents mentioned above give useful guidance on the establishment of a registration scheme, on the range and depth of information required for assessing the behaviour of pesticides, and on the risk associated with their application according to the registered patterns of use. They effectively contribute to international harmonization and acceptance of the registration requirements and to the development of national registration systems.

However, the guidelines and risk assessment schemes cannot be used on their own. As emphasized in most of the documents, a qualified and experienced expert's judgement is required for appropriate evaluation of the experimental data in order to perform the risk–benefit analysis, to determine the most suitable patterns of use and to introduce risk management measures. For all these considerations, detailed knowledge on local farming practices and the economic and social conditions is essential.

REFERENCES

[1] FOOD AND AGRICULTURE ORGANIZATION OF THE UNITED NATIONS, International Code of Conduct on the Distribution and Use of Pesticides (amended version), FAO, Rome (1990).
[2] FOOD AND AGRICULTURE ORGANIZATION OF THE UNITED NATIONS, Guidelines on the Efficacy Data for the Registration of Pesticides for Plant Protection, FAO, Rome (1985).
[3] FOOD AND AGRICULTURE ORGANIZATION OF THE UNITED NATIONS, Guidelines on Producing Pesticide Residue Data from Supervised Trials, FAO, Rome (1990).
[4] FOOD AND AGRICULTURE ORGANIZATION OF THE UNITED NATIONS, Guidelines for the Registration and Control of Pesticides, FAO, Rome (1985).
[5] FOOD AND AGRICULTURE ORGANIZATION OF THE UNITED NATIONS, Addenda to Guidelines for the Registration and Control of Pesticides, FAO, Rome (1988).
[6] FOOD AND AGRICULTURE ORGANIZATION OF THE UNITED NATIONS, Guidelines on Environmental Criteria for the Registration of Pesticides, FAO, Rome (1985).
[7] FOOD AND AGRICULTURE ORGANIZATION OF THE UNITED NATIONS, Revised Guidelines on Environmental Criteria for the Registration of Pesticides, FAO, Rome (1989).
[8] FOOD AND AGRICULTURE ORGANIZATION OF THE UNITED NATIONS, Guidelines on Good Labelling Practice for Pesticides, FAO, Rome (1995).
[9] FOOD AND AGRICULTURE ORGANIZATION OF THE UNITED NATIONS, Provisional Guidelines on Prevention of Accumulation of Obsolete Pesticide Stocks, FAO, Rome (1995).
[10] FOOD AND AGRICULTURE ORGANIZATION OF THE UNITED NATIONS, Pesticide Storage and Stock Control Manual, FAO, Rome (1996).
[11] FOOD AND AGRICULTURE ORGANIZATION OF THE UNITED NATIONS, Manual on the Development and Use of FAO Specifications for Plant Protection Products, 4th edn, FAO Plant Production and Protection Paper No. 128, FAO, Rome (1995).

[12] FOOD AND AGRICULTURE ORGANIZATION OF THE UNITED NATIONS, FAO Specifications for Plant Protection Products, technical papers prepared by the FAO Panel of Experts on Pesticide Specifications, Registration Requirements, Application Standards and Prior Inform Consent, FAO, Rome (continuous).

[13] WORLD HEALTH ORGANIZATION, Specification for Pesticides Used in Public Health, 6th edn, WHO, Geneva (1985).

[14] FOOD AND AGRICULTURE ORGANIZATION OF THE UNITED NATIONS/ WORLD HEALTH ORGANIZATION, Pesticide Residues in Food, Report of the 1994 Joint Meeting on Pesticide Residues, FAO Plant Production and Protection Paper No. 127, FAO, Rome (1994).

[15] EUROPEAN AND MEDITERRANEAN PLANT PROTECTION ORGANISATION, EPPO Guidelines for the Efficacy Evaluation of Plant Protection Products Nos 182–188, EPPO Bull. **24** (1994) 241–314.

[16] EUROPEAN AND MEDITERRANEAN PLANT PROTECTION ORGANISATION, EPPO Guidelines for the Efficacy Evaluation of Plant Protection Products Nos 189–198, EPPO Bull. **25** (1995) 473–593.

[17] FOOD AND AGRICULTURE ORGANIZATION OF THE UNITED NATIONS/ WORLD HEALTH ORGANIZATION, Pesticide Residues in Food, Report of the 1995 Joint Meeting on Pesticide Residues, FAO Plant Production and Protection Paper No. 133, FAO, Rome (1995).

[18] PESTICIDE ACTION NETWORK NORTH AMERICA, Information Release, May 21 1996 (1996).

[19] ORGANIZATION FOR ECONOMIC CO-OPERATION AND DEVELOPMENT, Good Laboratory Practice Guideline No. 1, The OECD Principles of Good Laboratory Practice, Environment Monograph No. 45, OECD, Paris (1992).

[20] ORGANIZATION FOR ECONOMIC CO-OPERATION AND DEVELOPMENT, Good Laboratory Practice Guideline No. 2 (revised), Guidance for GLP Monitoring Authorities: Revised Guides for Compliance Monitoring Procedures for Good Laboratory Practice, Environment Monograph No. 110, OECD, Paris (1995).

[21] ORGANIZATION FOR ECONOMIC CO-OPERATION AND DEVELOPMENT, Good Laboratory Practice Guideline No. 3 (revised), Guidance for GLP Monitoring Authorities: Revised Guidance for the Conduct of Laboratory Inspections and Study Audits, Environment Monograph No. 111, OECD, Paris (1995).

[22] ORGANIZATION FOR ECONOMIC CO-OPERATION AND DEVELOPMENT, Good Laboratory Practice Guideline No. 4, GLP Consensus Document: Quality Assurance and GLP, Environment Monograph No. 48, OECD, Paris (1992).

[23] ORGANIZATION FOR ECONOMIC CO-OPERATION AND DEVELOPMENT, Good Laboratory Practice Guideline No. 5, GLP Consensus Document: Compliance of Laboratory Suppliers with GLP Principles, Environment Monograph No. 49, OECD, Paris (1992).

[24] ORGANIZATION FOR ECONOMIC CO-OPERATION AND DEVELOPMENT, Good Laboratory Practice Guideline No. 6, GLP Consensus Document: The Application of the GLP Principles to Field Studies, Environment Monograph No. 50, OECD, Paris (1992).

[25] ORGANIZATION FOR ECONOMIC CO-OPERATION AND DEVELOPMENT, Good Laboratory Practice Guideline No. 7, GLP Consensus Document: The Application of the GLP Principles to Short-term Studies, Environment Monograph No. 73, OECD, Paris (1993).

[26] ORGANIZATION FOR ECONOMIC CO-OPERATION AND DEVELOPMENT, Good Laboratory Practice Guideline No. 8, GLP Consensus Document: The Role and Responsibilities of the Study Director in GLP Studies, Environment Monograph No. 74, OECD, Paris (1993).

[27] ORGANIZATION FOR ECONOMIC CO-OPERATION AND DEVELOPMENT, Good Laboratory Practice Guideline No. 9, Guidance for GLP Monitoring Authorities, Guidance for the Preparation of GLP Inspection Reports, Environment Monograph No. 115, OECD, Paris (1995).

[28] ORGANIZATION FOR ECONOMIC CO-OPERATION AND DEVELOPMENT, Good Laboratory Practice Guideline No. 10, GLP Consensus Document: The Application of the Principles of GLP to Computerised Systems, Environment Monograph No. 116, OECD, Paris (1995).

[29] FOOD AND AGRICULTURE ORGANIZATION OF THE UNITED NATIONS/ WORLD HEALTH ORGANIZATION, Pesticide Residues in Food, Report of the 1987 Joint Meeting on Pesticide Residues, FAO Plant Production and Protection Paper No. 84, FAO, Rome (1987).

[30] FOOD AND AGRICULTURE ORGANIZATION OF THE UNITED NATIONS/ WORLD HEALTH ORGANIZATION, Pesticide Residues in Food, Report of the 1988 Joint Meeting on Pesticide Residues, FAO Plant Production and Protection Paper No. 92, FAO, Rome (1988).

[31] FOOD AND AGRICULTURE ORGANIZATION OF THE UNITED NATIONS/ WORLD HEALTH ORGANIZATION, Pesticide Residues in Food, Report of the 1992 Joint Meeting on Pesticide Residues, FAO Plant Production and Protection Paper No. 116, FAO, Rome (1992).

[32] FOOD AND AGRICULTURE ORGANIZATION OF THE UNITED NATIONS/ WORLD HEALTH ORGANIZATION, Codex Alimentarius, Vol. 2, Pesticide Residues in Food, 2nd edn, Section 2, Codex Classification of Foods and Animal Feeds, FAO, Rome (1993).

[33] AMBRUS, A., Estimation of uncertainty of sampling for analysis of pesticide residues, J. Environ. Sci. Health **B31.3** 435 (1996).

[34] FOOD AND AGRICULTURE ORGANIZATION OF THE UNITED NATIONS/ WORLD HEALTH ORGANIZATION, Codex Alimentarius, Vol. 2, Pesticide Residues in Food, 2nd edn, Section 4.1, Portion of commodities to which Codex MRLs apply and which is analysed, FAO, Rome (1993).

[35] FOOD AND AGRICULTURE ORGANIZATION OF THE UNITED NATIONS/ WORLD HEALTH ORGANIZATION, Pesticide Residues in Food, Report of the 1990 Joint Meeting on Pesticide Residues, FAO Plant Production and Protection Paper No. 102, FAO, Rome (1990).

[36] WORLD HEALTH ORGANIZATION/FOOD AND AGRICULTURE ORGANIZATION OF THE UNITED NATIONS, Recommendations for the Revision of the Guidelines for Predicting Dietary Intake of Pesticide Residues, Report of a FAO/WHO Consultation, York 1995, WHO/FNU/FOS/95.11, WHO, Geneva (1995).

[37] FOOD AND AGRICULTURE ORGANIZATION OF THE UNITED NATIONS/ WORLD HEALTH ORGANIZATION, Codex Alimentarius, Vol. 2, Pesticide Residues in Food, 2nd edn, Section 4.2, Guidelines on Good Laboratory Practice in Pesticide Residue Analysis, FAO, Rome (1993).

[38] EUROPEAN UNION, Official Journal of the European Union, No. L 227, 1.9 (1994) 31.

[39] FOOD AND AGRICULTURE ORGANIZATION OF THE UNITED NATIONS/ WORLD HEALTH ORGANIZATION, Pesticide Residues in Food, Report of the 1989 Joint Meeting on Pesticide Residues, FAO Plant Production and Protection Paper No. 99, FAO, Rome (1989).

[40] FOOD AND AGRICULTURE ORGANIZATION OF THE UNITED NATIONS/ WORLD HEALTH ORGANIZATION, Pesticide Residues in Food, Report of the 1991 Joint Meeting on Pesticide Residues, FAO Plant Production and Protection Paper No. 111, FAO, Rome (1991).

[41] FOOD AND AGRICULTURE ORGANIZATION OF THE UNITED NATIONS/ WORLD HEALTH ORGANIZATION, Pesticide Residues in Food, Report of the 1986 Joint Meeting on Pesticide Residues, FAO Plant Production and Protection Paper No. 77, FAO, Rome (1986).

[42] FOOD AND AGRICULTURE ORGANIZATION OF THE UNITED NATIONS/ WORLD HEALTH ORGANIZATION, Pesticide Residues in Food, Report of the the 1993 Joint Meeting on Pesticide Residues, FAO Plant Production and Protection Paper No. 122, FAO, Rome (1993).

[43] EUROPEAN AND MEDITERRANEAN PLANT PROTECTION ORGANISATION, Decision-making Scheme for the Environmental Risk Assessment of Plant Protection Products, EPPO Bull. **23** (1993) 1–165.

[44] EUROPEAN AND MEDITERRANEAN PLANT PROTECTION ORGANISATION, Decision-making Scheme for the Environmental Risk Assessment of Plant Protection Products, EPPO Bull. **24** (1994) 1–87.

IAEA-SM-343/12

PESTICIDE RESIDUES IN THE MARINE ENVIRONMENT AND ANALYTICAL QUALITY ASSURANCE OF THE RESULTS

F.P. CARVALHO, S.W. FOWLER,
J.P. VILLENEUVE, M. HORVAT
Marine Environment Laboratory,
International Atomic Energy Agency,
Monaco

Abstract

PESTICIDE RESIDUES IN THE MARINE ENVIRONMENT AND ANALYTICAL QUALITY ASSURANCE OF THE RESULTS.

A brief review is given of the information that is available on the distribution and levels of pesticide residues in marine systems. Residues detected in coastal waters largely reflect the regional use of pesticides (e.g. DDTs, atrazine), although for more volatile and environmentally persistent compounds (e.g. hexachlorocyclohexane, lindane) long range atmospheric transport also contributes to their far field dispersal in the oceans. Despite the increasing number of pesticide reports in the scientific literature, data on residues are still very scarce for extensive coastal areas in regions of intensive pesticide usage such as the tropics. Therefore, the aim of IAEA Co-ordinated Research Programmes is to assist with the implementation of pesticide monitoring in tropical coastal ecosystems and with experimental research on pesticide cycling and its effects on the marine environment. The results of worldwide laboratory intercomparison exercises organized by the IAEA for analyses of the organochlorine pesticides in marine samples highlight the need to further improve quality control of the analytical results. Although research on marine contamination by pesticide residues is progressing, in view of the high number of compounds and formulations in use it is unlikely that all the data required for environmental risk assessment of agrochemical residues in marine ecosystems will be generated with sufficient rapidity. Therefore, enhanced development through experimental research with model compounds and subsequent modelling is required. Nevertheless, from current knowledge it is clear that environmental management programmes for coastal ecosystems should urgently adopt measures to prevent or reduce the impact of agrochemical residues on biological resources such as fisheries and aquaculture.

1. INTRODUCTION

Agrochemicals, and in particular pesticides, have become an integral part of modern agriculture systems, contributing significantly to improved crop yields and enhanced food production [1]. The continued need to produce more food for the rapidly growing world population renders it unlikely that agrochemicals will be replaced in the foreseeable future [2].

Although agrochemicals contribute to significant improvements in the production of food, the spread of agrochemical residues in the environment has created several current or potential future problems. Besides direct exposure of humans to residues in food and drinking water and to health related injuries [3, 4], it has been reported that pesticide residues introduced into aquatic ecosystems cause massive fish and shrimp kills [5], reduce the reproductive success of species [6], contribute to the death of coral reefs [7], and ultimately may have a major impact on fishery resources, biological diversity and the functional equilibrium of ecosystems.

In many countries, e.g. Indonesia, Japan and the Philippines, sea food represents 30–50% of the protein in the diet [8]. Of the 101.4 Mt of fisheries and aquaculture production worldwide in 1993, 80% were fished in marine waters, with aquaculture contributing 15.6% (15.9 Mt) [9]. If, on the one hand, the marine fisheries production has stabilized at around 80 Mt/a, aquaculture in inland and coastal areas is rapidly expanding, with an average 40% annual growth. Therefore, it is expected that aquaculture will contribute to a higher percentage of the food supply in coming years. However, the development of aquaculture may be compromised by pollution of coastal waters containing a large number of chemicals, including pesticide residues. Recent reports from Central America indicate that fungicides from banana plantations may be implicated in the devastating disease of farmed shrimp known as the Taura syndrome [10]. Ecuador alone has suffered US $50 million annual losses in shrimp exports as a result of this disease. There are, therefore, a number of reasons for concern about the effects of agrochemical residue discharges into aquatic environments and, in particular, into the sea.

Assessment of the ecological risk posed by pesticide residues to marine ecosystems has, for the most part, yet to be undertaken. Estimation of the environmental risk and introduction of measures to manage or counteract the risk of pesticide residues require expanded knowledge of the environmental behaviour and effects of pesticides. To this end, enhanced laboratory capacities are needed to implement ample marine monitoring programmes. Furthermore, experimental research is required to generate the necessary data on the cycling, fate and effects of pesticides in marine ecosystems.

2. TRENDS IN THE WORLD USE OF AGROCHEMICALS

Assessment of the distribution and effects of pesticide residues in the environment requires consideration of the past, present and future usage of agrochemicals. Unfortunately, the information available on pesticide usage is either very incomplete or not recorded in terms of quantities of the active agrochemical ingredients. Information on the agrochemical market values is more readily available. Although such information does not relate in a simple manner with the amount of active ingredients used, it does allow approximate expression of the growth in agrochemical usage and the relative use by region (Figs 1 and 2).

Since 1960, world use of pesticides has seen a steady increase, averaging an 11% annual growth and reaching about 5×10^6 t in 1993 (Fig. 1). During the same period, major changes occurred in the use of crop protection chemicals. First generation pesticides, mainly organochlorine (OC) compounds, were gradually replaced in the 1960s and 1970s by a second generation of pesticides (organophosphorus (OP) and carbamate compounds), whereas a third generation of pesticides (pyrethroids, sterilants, pherormones, chitin inhibitors) has appeared on the market more recently.

The ecotoxicological effects of intensive use of persistent OCs such as DDT, toxaphene, heptachlor and 'drins' were noticed in the early 1960s and, gradually, use of these compounds has been discontinued in many countries. Within the insecticide group, OP compounds and pyrethroids have undergone increased use, partially compensating for the declining usage of OCs. Persistent OCs (e.g. lindane, hexachlorocyclohexane (HCH) and DDT) are still in use in many regions in South and East Asia, partly because of the low cost of these compounds and less strict regulations and control, but also as a result of sanitary campaigns against malaria vectors.

Noticeable trends in the agrochemical market have also been the increased use of herbicides (carbamate, triazines) and the relative decrease in fungicides (Fig. 1). The growth in the market value of herbicides has resulted from increased herbicide usage in the rice growing regions of India and East Asia, combined with restrictions on older products (atrazines) and their replacement by higher value compounds (triazines, sulphonyl ureas). Use of the remaining agrochemicals (plant growth regulators, nematicides, fumigants, etc.) has been in decline [11, 12].

At present, the largest amounts of herbicide are used in North America, Western Europe and Asia, whereas insecticides are mostly employed in East Asia, North America and Latin America. It is estimated that about 50% of the world's use of insecticides are applied in developing countries, in particular in tropical regions, to control insect pests in cash crops and household insects, and to use as seed dressing [2]. In the coming years, a general increase in pesticide use is expected to occur in the rapidly developing countries of Asia, whereas a reduction will continue in Western Europe [13].

National reports on pesticide usage, when more detailed, provide more specific information that illustrates and confirms the changes in the worldwide sales of agrochemicals. For instance, China, a major world producer of rice and wheat, mainly used DDT and HCH in the 1950s and 1960s to control insect pests. Use of these chemicals was discontinued in the mid-1980s (except for export and the production of lindane), and national pesticide production and use have gradually shifted to OP compounds. In 1990, the total consumption of pesticides in China reached 2×10^6 t of active ingredient, of which 70% were insecticides and 14% fungicides. Of the insecticides used, 40% were OPs, with methamidophos account-

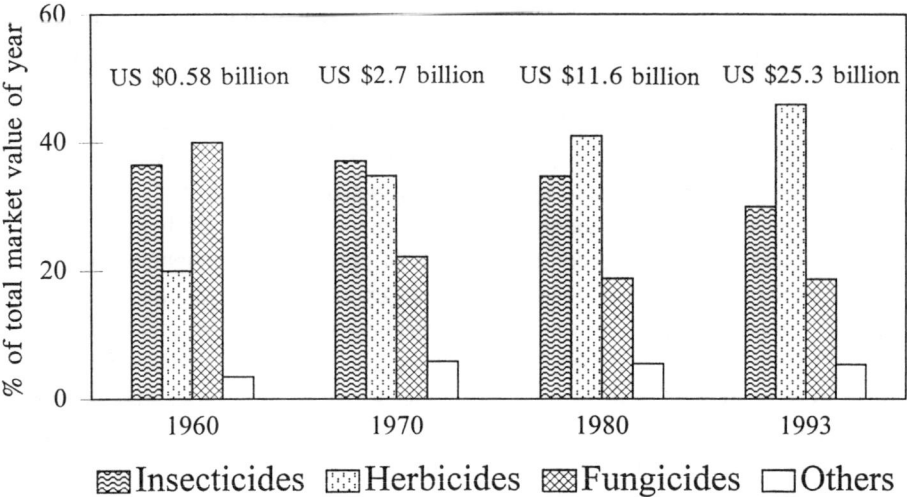

FIG. 1. Trends in the world use of pesticides. Pesticide groups are displayed in percentage of the total market value of the year (1 billion = 10^9) [11, 12].

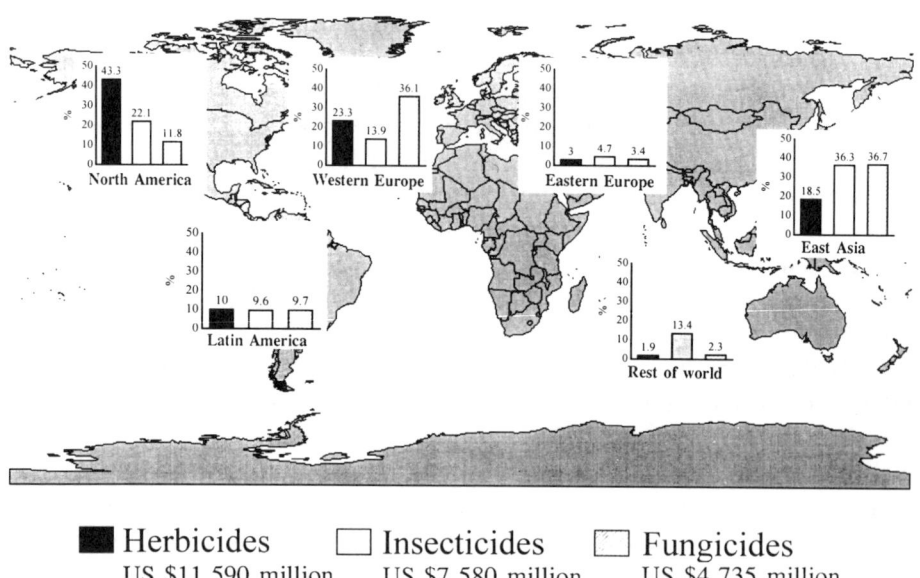

FIG. 2. Use of agrochemicals by region in 1993, based on the world market value.

TABLE I. ESTIMATES OF THE QUANTITY (t) OF PERSISTENT OC PESTICIDES USED WORLDWIDE, AND THE MAIN USERS (AFTER Ref. [17])

	DDT	Lindane (gamma-HCH)	HCH (technical)	Toxaphene
World use: 1950–1992	2 600 000			1 330 000
World use: 1970–1992	990 000			670 000
Use accounted in the search	1 500 000	720 000	550 000	450 000

Quantity | | | *Main users* | | |

Quantity	DDT	Lindane (gamma-HCH)	HCH (technical)	Toxaphene
>10^5 t	India, USA	India, USA	China	USA
10^4–10^5 t	Brazil, Canada, China, Egypt, Germany, Italy, Japan, Mexico, Poland, Spain, Turkey, Syrian Arab Republic, former USSR	Italy, Japan, Peru, Spain	India	Brazil, Colombia, Egypt, Germany, Mexico
10^3–10^4 t	Most countries in Latin America, Eastern Europe and the Middle East; Australia, Bangladesh, Cambodia, Indonesia, New Zealand, Thailand, Viet Nam	Algeria, Argentina, Austria, Côte d'Ivoire, Egypt, France, Germany, Ghana, Mexico, Sweden, Thailand, Turkey	Former USSR	China, Ecuador, Guatemala, Hungary, India, Nicaragua, Peru, Poland

ing for nearly half of these. Use of pyrethroids (mainly fenvalerate), herbicides (nitrofen, butachlor) and carbamate compounds is also increasing [14].

In the countries of Central America, very large quantities of pesticides were applied for decades to protect tropical fruits and cash crops (coffee, cotton) from insect pests. From 1977 to 1985 in Costa Rica, OCs decreased from 16 to 1% of the total pesticides used. Although some persistent OCs are still used, usage of OPs and carbamate compounds gradually increased during the same period [15]. A similar shift occurred in Nicaragua, where toxaphene was applied to cotton crops at a rate of 31 kg/ha in 1985, but had decreased to 7.3 kg/ha by 1989. Also DDT, which had been in regular use in Nicaragua to control malaria vectors, gradually decreased from 22 t in 1981 to 2 t in 1989 [16].

Estimates of the amounts used worldwide for several common compounds are given in Table I [17]. Although incomplete, from this information it can be concluded that at all times DDT, which had been applied in nearly every country, was the most heavily used pesticide. Use of DDT was restricted or banned in North America in the 1960s and 1970s, and in the early 1970s in Europe; it was banned in 1984 in China, but is still used in several Asian (India (20 000 t/a), Bangladesh, Pakistan) and African countries. Technical HCH (a mixture of 50-80% alpha-HCH, 8-15% gamma-HCH, 5-14% beta-HCH and 3-5% delta-HCH) is mainly employed in India (47 000 t/a) and China, whereas an alternate formulation made of pure gamma-HCH (lindane) has been used in Europe and America. Lindane is also employed in India and Malaysia, but its usage is restricted in China and the former USSR. Toxaphene, another persistent OC, was mainly used on the American continent on cotton crops until it was banned in the USA in 1986. Toxaphene was banned or restricted earlier in China, the former USSR and several countries in Western Europe [17].

2. PESTICIDE RESIDUES IN THE MARINE ENVIRONMENT

It has been estimated that less than 1% of the pesticides applied to crops attain the target pest species. Excess pesticide moves throughout the environment and may enter marine ecosystems by a number of pathways, provided the environmental persistence of the compound is sufficiently long. The potential of pesticides to have an impact on the marine environment can be inferred using the information available on pesticide half-lives in soils [18] (Fig. 3). Compounds with long environmental half-lives (e.g. most OCs) are likely to be transported to and found in marine ecosystems, whereas short lived compounds are likely to degrade closer to the area of application (e.g. most carbamates) (Fig. 4). This information, coupled with data on the partial vapour pressure of compounds, may indicate which compounds are likely to be more easily volatilized and transported as atmospheric vapour; in a similar way, information on the persistence of pesticides in soils, combined with data on

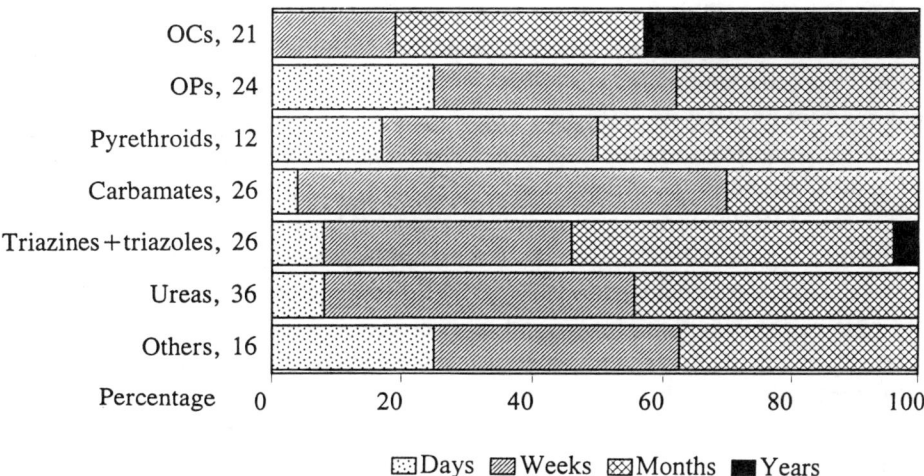

FIG. 3. Pesticide half-lives in soils, based on data compiled from the literature [18]. (Numerals indicate the number of compounds used to obtain the data.)

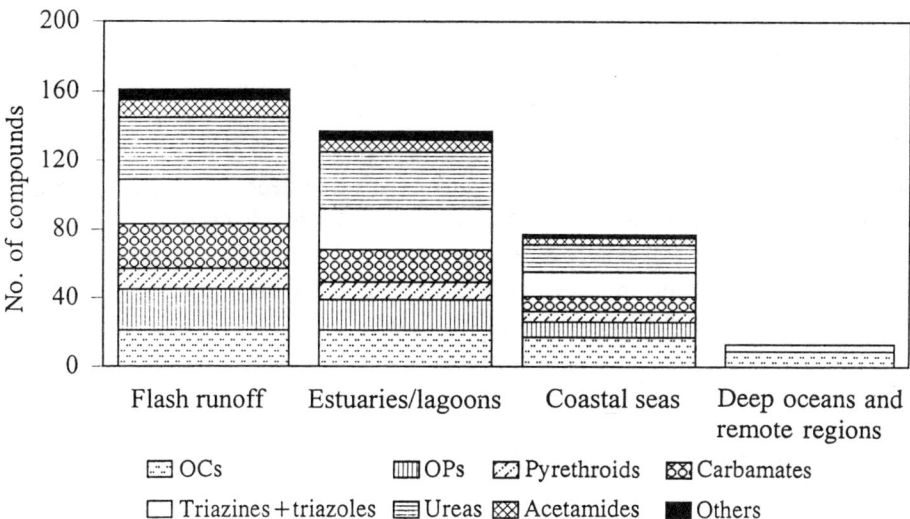

FIG. 4. Predicted distribution of pesticide residues in the environment, based on their persistence half-lives in soils. A large number of pesticides in different chemical groups may potentially have an impact on estuarine and coastal areas. (Data relate to compounds in commercial production for which the soil half-life is available.)

water solubility, will indicate those compounds that are likely to be primarily transported in solution by surface runoff. Compounds with low water solubility and low vapour pressure tend to remain sorbed on soil particles, and their transport in the environment probably is associated primarily with the transport of soil dust.

The atmospheric pathway may therefore account for the long range transport of pesticide residues and their introduction into marine ecosystems. This is the case for traces of DDT, aldrin, endrin, dieldrin and toxaphene measured in the Antarctic and Arctic regions [19–21]. Nevertheless, recent reports show that because of their low volatility, these compounds mostly remain in the region of application and, thus, their global redistribution has no significant correlation with latitude [22, 23]. In contrast, more volatile compounds such as lindane and technical HCH are transported from the continents to the oceans in the vapour phase [24, 25]. Volatilization of these compounds is more intense in tropical areas (e.g. South Asia) and, following atmospheric transport, condensation occurs at the low temperatures in high latitudes [22–24]. Concentration of HCH in surface sea water of the Pacific Ocean shows a significant positive correlation with latitude, with the concentrations at high latitudes (e.g. the Chuckchi Sea, 1600 pg/L) being more elevated than those at mid-latitudes (e.g. the South China Sea, 480 pg/L) [24, 25]. This global distillation–condensation effect was also recently demonstrated in the terrestrial environment through analyses of bark and leaves from trees [22, 23]. OC residues deposited in the bottom sediments of Arctic lakes in Canada confirm the long range atmospheric transport of these persistent compounds, and the differences in the environmental cycling of DDTs and HCHs [21].

Most of the world's agricultural land is located on coastal plains and in the valleys of major rivers. Therefore, it is not surprising that pesticide residues carried by surface runoff and residual waters enter fresh water streams and are discharged into coastal seas. For example, in 1989 the Mississippi River may have transported some 430 t of atrazine and a similar amount of atrazine degradation products from maize and soybean plantations in the Midwest of the United States of America to the Gulf of Mexico [26]. Triazine and derivatives carried by the Elbe River (Germany) into the German Bight and the North Sea result in concentrations in sea water ranging from 1 to 1100 ng/L. This coastal area is, therefore, considered to be highly contaminated and toxic effects on marine biota are likely to occur [27]. A wide variety of herbicides, including atrazine, simazine, alachlor, metolachlor and molinate at concentrations generally lower than 1500 ng/L, have also been reported for several estuaries in the Mediterranean Sea [28]. These results offer extensive evidence that some herbicides persist in marine systems, despite the generally accepted belief that they degrade rapidly. Less water soluble compounds such as OCs are mainly transported in association with suspended matter and sediments in river discharges [29]. Following restriction on the use of persistent pesticides and/or the introduction of better agricultural management practices, transport of persistent OC residues by rivers discharging into the Mediterranean Sea was observed to decline in comparison with past years [29]. At present, the OCs discharged by these rivers likely contribute much less than the atmospheric deposition to the input of OC residues into the Mediterranean Basin [29–31]. Nevertheless, owing to long persistence in soils, DDT from past usage is still remobilized and transported to estuaries and coastal areas. As a consequence, DDT concentrations measured in estuarine areas (e.g. the Seine estuary, Arcachon Bay, France) remain significantly enhanced in comparison to open coast areas [32].

The relative importance of atmospheric and surface water pathways in the delivery of pesticide residues into coastal ecosystems will depend upon a number of factors, namely, the physicochemical properties of the compounds, the pesticide application technologies and the weather conditions. A study on the fate of HCH applied in rice paddies in the Vellar River of southern India concluded that, under the tropical conditions of this region, 99.6% of the applied pesticide are volatilized and only 0.4% is drained into the estuary [33]. This low discharge of HCH with river water flowing into the Bay of Bengal would explain the relatively low concentrations in the adjacent coastal areas. Less volatile compounds such as DDTs are less dispersible through air on a wide geographical scale. Thus, DDT concentrations in coastal marine environments mainly reflect usage of pesticides in the nearby agricultural regions. Furthermore, the cumulative use and strength of the DDT sources are also expected to be reflected in the marine environment. This is evident in sea water concentrations of DDT, which are much higher in seas around Asia (5.6–16 pg/L) than in the North Pacific and North Atlantic coastal areas (0.8–1.2 pg/L) [24].

Accumulation of OC residues in marine biota has been monitored through national programmes frequently inspired by the Mussel Watch Program [34, 35]. Data are available for OC compounds in molluscs, fish, seabirds and marine mammals [36–38]. The results reported for different regions can be compared in order to assess the relative contamination of coastal seas (Table II [32, 34–36, 39–53]). However, it should be borne in mind that sampling sites, sample type and size, year of survey, season of year, etc. may introduce large variations in the results. For example, molluscs living in estuaries and coastal lagoons may display OC concentrations that are 1–3 orders of magnitude higher than the concentrations measured in molluscs sampled from the open coast (Table II).

The concentration of DDTs in marine biota are generally higher than the non-DDT compounds, which is a result of the more intensive worldwide use of DDT than of any other pesticide, combined with the long environmental persistence and high bioaccumulation of these compounds. The total DDTs (i.e. DDT plus DDT metabolites) in molluscs from coastal areas around Asia are generally higher than those in molluscs from the coastal areas of the American and European continents. In North America, results indicate that the DDTs in bivalves of the US coast display concentrations that average 4 ng/g (wet weight), whereas in most of the sites around the Gulf of Mexico, bivalves (oyster and mussels) display concentrations that generally vary between 1.6 and 16 ng/g (wet weight); in tropical and subtropical Latin

TABLE II. AVERAGE CONCENTRATION AND RANGE (ng/g WET WEIGHT) OF OC PESTICIDES IN BIVALVE MOLLUSCS (MUSSELS AND OYSTERS) FROM OPEN COAST SITES (C), ESTUARIES (E) OR COASTAL LAGOONS (L) [32, 34–36, 39–53][a]

Region	Year of survey	ΣDDT	ΣHCH	Gamma-HCH	Refs
USA					
Atlantic and Pacific coasts	1993 (C)	4	—	—	[39]
Cheasapeake Bay	1980 (E)	6	2	—	[40]
South Carolina coast	1989 (E)	10.1 (<5–23)	5 (<5–7.0)	—	[41]
Gulf of Mexico	1993 (C)	1.6–16 (1–154)*	—	—	[35]
Gulf of Mexico	1987 (C)	10.8 (0.48–571)	—	0.28 (0.04–1.4)	[34]
Mexico					
Sinaloa	1992 (L)	9.3	—	0.5	[42]
Campeche	1991 (L)	0.96	—	0.14	[43]
Central America	1993 (C)	1–32	—	—	[35]
South America	1993 (C)	1–34	—	—	[35]
Denmark					
Baltic proper	1985 (C)	5 (2.4–67)	>3.0	1.4 (0.6–3.4)	[44]
France					
Seine estuary	1979–1988 (E)	16 (3.1–46)	—	—	[32]
Atlantic coast	1979–1988 (C)	4–16	—	—	[32]
Mediterranean coast	1979–1988 (E, C)	24 (1.2–117)	—	—	[32]

Location	Year				Ref.
Spain					
Atlantic coast	1990–1991 (E)	0.4 (0.3–85)	ND	7 (ND-49)	[45]
Nigeria	1987 (C)	114 (91–138)	>2.6	1.28 (1.13–1.43)	[46]
Oman	1980–1986 (C)	0.06–0.78	—	—	[36]
India					
Arabian Sea	1988–1989 (C)	37 (6.0–39)	5.6 (4.9–9.7)	1.2 (0.7–2.0)	[47]
Bay of Bengal	1988–1989 (C)	11 (2.8–33)	6.5 (4.3–16)	1.1 (0.5–2.5)	[47]
Thailand					
Gulf of Thailand	1991 (C)	4 (0.74–5.38)	ND–0.09	ND–0.04	[48]
East Java Sea	1984 (C)	100–520	ND–30	—	[49]
Hong Kong	1983 (C)	<14–320	<4.8–34	—	[50]
Taiwan (China)	1980–1982 (C)	2.5–160	0.6–16	—	[51]
South China	1995 (E, C)	37.5	8.6	—	[52]
Japan	1984 (C)	ND–10	1–18	—	[53]

[a] ΣDDT = DDT + DDE + DDD + DDMU; ΣHCH = sum of the alpha, beta, gamma and delta isomers; ND = not detected; — = no data; numerals in parentheses with an asterisk = extreme values.

America, a higher percentage of sites was found with above 16 ng/g (wet weight) [35]. These geographical trends of DDTs in molluscs are related to the earlier banning of DDT in Canada and the USA (in 1972), compared with restrictions adopted later in South America (e.g. DDT was banned in Brazil in 1980). Furthermore, DDT is still used to control malaria vectors in tropical and subtropical areas of the American continent [16, 35]. In Asia, the DDT results in various regions range from relatively low concentrations, < 10 ng/g (wet weight) in Japan, up to 320 ng/g in Hong Kong (Table II). Furthermore, in samples of bivalves from the American continent, DDT metabolites largely predominate over DDT, with DDT:DDE ratios of less than 0.2. In samples from southern Asia, the DDT:DDE ratios are generally closer to and frequently much higher than unity, which indicates exposure to 'fresh' DDT [35, 47, 48]. The persistence of DDT and DDT metabolites in the marine environment is well demonstrated in the very slow decrease of DDT concentrations in molluscs, which remain easily measurable more than 20 years after the banning of DDT in the USA [39]. The decreasing trend observed between 1986 and 1993 indicates that ΣDDT would have a persistence half-life of about 7 years in the coastal environment.

HCH concentrations in mollusc samples around Asia are generally higher than elswhere, reflecting the heavy use of this compound in South and East Asian countries (Table II). Owing to the global distillation effect on the distribution of HCH, it would not be surprising if even higher concentrations were to be found in marine biota at higher latitudes, but this has not yet been documented. Analysis of HCH isomers in mussels from the coast of India [47] showed average composition, with 49% alpha-HCH, 34% beta-HCH and 12% gamma-HCH, demonstrating that the main source of these residues is the technical HCH used in the region. The results for other regions, e.g. the Baltic Sea [44], show similar concentrations of alpha and gamma isomers, which may be due to past use of technical HCH in Nordic countries before the shift to lindane. Furthermore, samples from the American continent, where technical HCH was not used, also display some percentage alpha-HCH. Isomerization of gamma-HCH to alpha-HCH has been hypothesized [22], which could help explain the concentrations of both HCH isomers measured in the northern regions of Canada [21]. Therefore, use of the ratio of alpha and gamma isomers, which was suggested to identify the sources of the HCH compounds to the oceans, may be problematic because of isomerization as well as the mixed use of technical HCH and lindane [22].

Although fewer in number, concentrations of other persistent OC pesticides, including dieldrin, endosulfan and chlordane compounds, have also been reported in marine biota from several regions [35, 36, 47, 48]. Concentrations of these compounds are generally lower than those of DDTs.

Data on OP residues in estuarine and coastal environments are very sparse. Nevertheless, evidence has been provided on the persistence and bioaccumulation of these compounds in tropical coastal lagoons [54] and estuaries in the Mediterranean

Sea [55]. Moreover, experimental results confirmed that OP compounds, although generally less persistent than OC compounds, may persist in marine systems [56, 57]. Therefore, further attention should be paid to some of these compounds which, because of their high lipophilic characteristics and high toxicity, have the potential to severely impact marine ecosystems [42, 54, 55]. Data on pyrethroid and carbamate compounds in marine ecosystems are entirely lacking.

3. FATE AND EFFECTS OF RESIDUES IN MARINE ECOSYSTEMS

There are probably more than 1000 compounds in use as crop production chemicals, and new compounds are being introduced every year. For the most part, environmental risk assessment has not yet been made and there is a lack of sufficient knowledge about their environmental cycling and fate. For example, of the pesticides in use in the USA, only 15 have established use criteria, with defined water quality criteria for fresh water and marine species, despite the extensive pesticide degradation and toxicity tests carried out to date [58]. The very high number of agrochemicals makes it nearly impossible to test the behaviour of each chemical, on each species, in each particular environment. Therefore, some degree of generalization and the development of predictive models are of the utmost importance to encompass the wide variety of agrochemical residues entering the aquatic environment.

Prediction of the environmental fate of agrochemical residues requires the collection of three sets of data. First, information on the chemical and physical properties of the compound; second, information on the rates of transfer through environmental compartments; and, third, the resulting distribution of the compound in the various environmental phases. For most of the pesticides in use, data of these three categories are frequently unavailable [18]. Prediction of the toxic effects of pesticides on marine biota requires appropriate knowledge about effective concentrations at the individual level and chronic effects on descendant generations; frequently, data on the breakdown or transformation of compounds are also needed. These data are generally obtained through experimental exposure of organisms to pesticides [59, 60].

Chemical structure–activity relationships (SARs) offer a promising approach to the development of predictive tools and may enhance our understanding of the environmental cycling of pesticides in aquatic ecosystems [61]. Use of molecular based properties in developing SARs has the advantage of ease of computation and provides a sensitive link to the cycling of compounds in the environment. For instance, the aqueous solubility of pesticides relates to the dispersion and transport processes, and hydrophobicity relates to partitioning between the water and the organisms/particles [62]. For example, Fig. 5 shows the correlation between the water solubility and the octanol–water partition coefficient (K_{ow}) for these pesticide

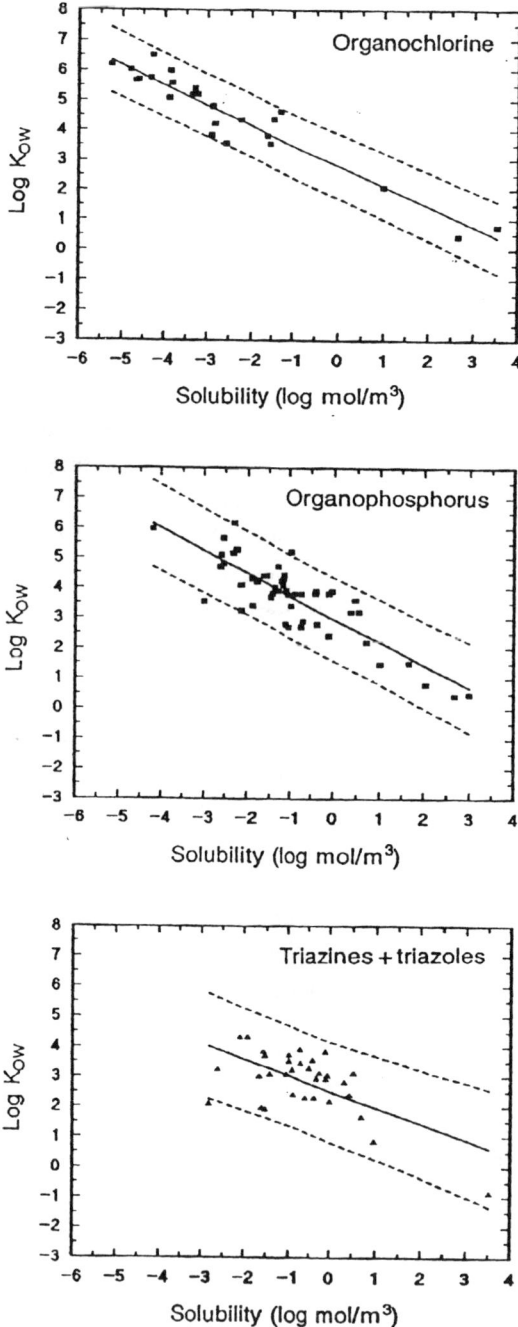

FIG. 5. *Correlation between the octanol–water partition coefficient (log K_{ow}) and the water solubility (log mol/m^3) for three pesticide groups (P < 0.001). The best fit lines have identical slopes.*

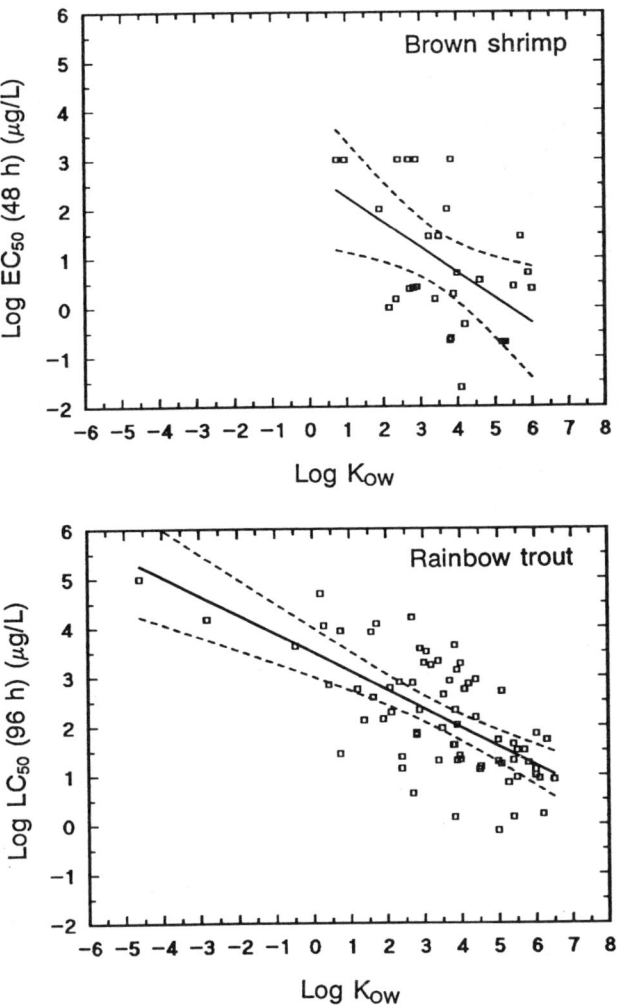

FIG. 6. *Correlation between the pesticide toxicity (LC_{50} and EC_{50}) for two fresh water organisms and the pesticide octanol–water partition coefficient (log K_{ow}) ($P < 0.001$). A wide range of pesticides of various chemical groups was tested on the same organism. Based on data in Ref. [63].*

groups. This correlation may permit estimation of the K_{ow} for compounds of known solubility.

The final aim of research on the partitioning, cycling and fate of pesticide residues is to assess their effects (toxic insult) on biota and ecosystems. Again, general relationships may be of great help in predicting toxicity. The toxic effects of lipophilic compounds have been related to their partitioning into cell membrane lipids and adverse effects on membrane proteins. Since a large number of pesticides are lipophilic, non-specific toxicity (narcosis) can be described on the basis of a simple relationship of log K_{ow} [61]. This non-specific toxicity (in addition, some compounds may display specific toxic effects) appears to be a universal biological response, and provides a baseline on which other chemicals can be assessed (Fig. 6) [63]. Moreover, this non-specific toxicity SAR allows prediction of the toxicity of a compound to a species [64].

The available database of acute toxicity of pesticides to fresh water species includes about 450 chemicals tested on 70 species, whereas the corresponding database to marine species includes results of nearly 220 chemicals tested on 55 species [63, 65]. The advantage in developing toxicity SARs would be to permit the use of the wider database for fresh water biota and extrapolation to marine species, thus avoiding repetition of a large number of tests.

Despite the progress being made in developing these predictive tools, there is still a need to take into account the microbial degradation, the hydrolysis of compounds, the effects of environmental conditions on the degradation/persistence of compounds, and the metabolic transformation of xenobiotics in vivo. Since no relationship seems to exist between the biodegradation of organic compounds and their chemical structure [66], the results of experimental research on these aspects probably cannot be anticipated by model results.

Development of models to predict the fate and toxic effects of residues in marine systems may greatly help in guiding field research (what compounds to look for and where) and in improving the design of experimental bioaccumulation and toxicological tests (compounds, what range of concentrations, on what species). However, they should not be regarded as a replacement to the actual toxicology tests required for pesticide registration and for ecological risk assessment. Tier protocols, which mainly include chronic toxicity tests on algae, crustacea and fish and a multi-generation test at sublethal concentrations, are currently being used to generate toxicity data for registration purposes and for chemical hazard assessment [60]. Ecological risk assessment of residues in aquatic environments is, however, better investigated through use of model ecosystems (outdoor tanks, mesocosms) that permit examination of the direct and indirect effects of pesticides on biota and linkages among ecosystem components [59, 67].

4. IAEA SUPPORT TO RESEARCH ON PESTICIDE RESIDUES IN THE MARINE ENVIRONMENT: QUALITY ASSURANCE OF THE ANALYTICAL RESULTS

The large number of pesticides and their widespread use require that efforts be made to enhance the analytical capacities in nations in order to monitor the residues in their food, water and environment. Several international programmes are aimed at contributing to this end [68]. This is the case, for example, of the IAEA Co-ordinated Research Programme on the Distribution, Fate and Effects of Pesticides on Biota in the Tropical Marine Environment: Use of Radiotracers, which is based on an extraordinary contribution made by the Government of Sweden (Swedish International Development Authority) to the IAEA. Laboratories in 17 IAEA Member States are participating in this research programme, which aims at enhancing the monitoring of residues in coastal areas and at promoting experimental research on the behaviour of selected pesticides using radiolabelled compounds and model ecosystems [69].

To assess contamination of the marine environment by pesticide residues it is deemed important to obtain more data. However, at the same time it is necessary to obtain data of good quality and, therefore, one essential component of every environmental monitoring programme is quality assurance of the analytical results. This is achieved through good laboratory practice and good performance in the analysis of reference materials and intercomparison samples.

For many years, the IAEA Analytical Quality Control Programme has been providing continued support to laboratories around the world through certified reference materials and samples for intercomparison exercises. Intercomparison samples are distributed free of charge to all interested laboratories. A code number is given to each laboratory and disclosed only to that laboratory. The results reported by these

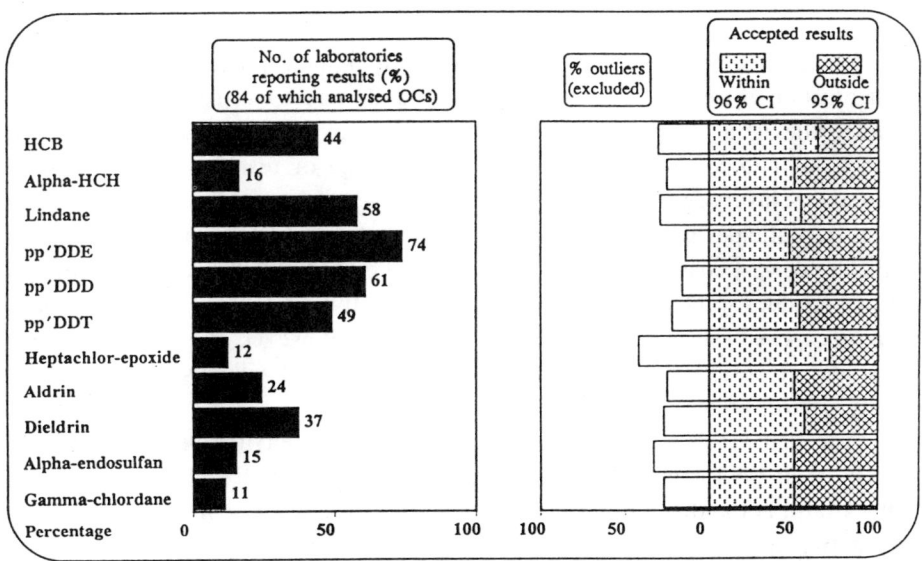

FIG. 7. Results obtained in 1995 by laboratories worldwide for chlorinated compounds in the intercomparison sample IAEA-142 (mussel flesh homogenate) (CI = confidence interval).

laboratories are statistically treated and compared with the consensus values for the concentration of analytes in the sample. The results obtained enable each laboratory to check the analytical performance and, through modifications and a tune up of the techniques, to improve the accuracy of the analytical results.

The laboratory results in the most recent intercomparison exercise (1995) are given in Fig. 7. In this exercise, a mussel tissue homogenate (IAEA-142) was distributed to 245 laboratories worldwide for analysis of the OC compounds and petroleum hydrocarbons [70]. A large number of laboratories reported their results, although the number of compounds that could be analysed by each laboratory varied. A first screening of these results was performed through appropriate statistical tests; outliers, in a variable number per analyte, were excluded. The results considered acceptable were then plotted and a consensus mean computed for each analyte. The final result showed that a number of laboratories had performed well and obtained acceptable values for several OC compounds. Nevertheless, it should be noted that a meaningful percentage of the results falls outside the acceptable range of accuracy (95% confidence interval). Moreover, good performance in the analysis of one compound does not imply identical performance in the analysis of other OC compounds.

The final results obtained by laboratories in this intercomparison exercise indicated that an important worldwide effort has still to be made to improve the quality of analytical results for chlorinated compounds.

5. CONCLUSIONS

Future improvements in the application of pesticides and the abatement of contamination in residual waters and agricultural runoff will certainly contribute towards alleviating contamination of the marine environment. Nevertheless, the diversity of agrochemicals in use and the discharge of residues into the marine environment justify appropriate monitoring and research programmes on the fate and effects of pesticides in marine systems. Data on persistent OC pesticides are not yet available for vast coastal regions. Contamination by OP, pyrethroid, carbamate and triazine residues still has to be investigated practically everywhere and, in particular, in more exposed estuarine and coastal lagoon areas. However, substantive progress in monitoring pesticide residues in the marine environment will depend on improved quality control of the analytical results.

The behaviour of many pesticides under marine environmental conditions, namely, their persistence, degradation rates, bioaccumulation and toxic effects on biota, is also not known. Taking into account the large number of agrochemicals in use, extrapolation of the data available from laboratory studies and from other regions, as well as use of predictive models, may permit economization of resources and time. However, carefully designed experiments using model compounds and model ecosystems have to be carried out in order to validate such extrapolations.

Radiolabelled compounds and nuclear techniques are particularly suited to experimental research on the cycling of pesticides in marine sytems.

In many cases, the knowledge currently available on the behaviour of residues should permit the establishment of integrated environmental management procedures that could reduce the discharges and the impact of pesticide residues in coastal ecosystems [26, 27, 42, 55, 71]. Implementation of these procedures is probably the best way of harmonizing the interests of agriculture, fisheries and aquaculture.

REFERENCES

[1] World Agriculture Toward 2000 (ALEXANDRATOS, N., Ed.), Food and Agriculture Organization of the United Nations, Rome, and Belhaven Press, London (1988).

[2] KLASSEN, W., "World food security up to 2010 and the global pesticide situation", Pesticide Chemistry (Proc. 8th Int. Congr. Washington, DC, 1995), American Chemical Society, Washington, DC (1995) 1–32.

[3] COLBORN, T., SAAL, F.S., SOTO, A.M., Developmental effects of endocrine-disrupting chemicals in wildlife and humans, Environ. Health Persp. **101** (1993) 378–384.

[4] WOLFF, M.S., TONIOLO, P.G., Environmental organochlorine exposure as a potential etiologic factor in breast cancer, Environ. Health Persp. **103** (1995) 141–145.

[5] CAPEL, P.D., GIGER, W., REICHERT, P., WANNER, O., Accidental input of pesticides into the Rhine River, Environ. Sci. Technol. **22** (1988) 992–997.

[6] LIEBERG-CLARK, P., BACON, C.E., BURNS, S.A., JARMAN, W.M., LE BOEUF, B.J., DDT in California sea-lions: A follow-up study after twenty years, Mar. Pollut. Bull. **30** (1995) 744–745.

[7] ACEVEDO, R., Preliminary observations on effects of pesticides carbaryl, naphthol, and chlorpyrifos on Planulae of the hermatypic coral *Pocillopora damicornis*, Pacific Sci. **45** (1991) 287–289.

[8] LAURETI, E., Fish and fishery products: World apparent consumption statistics based on food balance sheets (1961–1989), FAO Fish. Circ. **821** 1 (1991).

[9] FOOD AND AGRICULTURE ORGANIZATION OF THE UNITED NATIONS, La situation mondiale de l'alimentation et de l'agriculture 1995, Collection FAO: Agriculture 28, FAO, Rome (1995).

[10] JIMENEZ, R., Síndrome de Taura, Revista de la Cámara de Productores de Camarón, Quito, Ecuador (1992) 10–12.

[11] FOOD AND AGRICULTURE ORGANIZATION OF THE UNITED NATIONS, FAO Yearbook: Trade, Vol. 46, FAO, Rome (1992).

[12] Agrochemical Service, Wood Mackenzie Consultants Limited, Edinburgh (1994).

[13] WILLIAMS, A.J., The Common Agricultural Policy and the general environmental policies concerned with agriculture in the European Community and their implications for fertilizer consumption, Mar. Pollut. Bull. **29** (1994) 500–507.

[14] LI, S., "Pesticides: Environmental pollution and human health in China", Chemistry, Agriculture and the Environment (RICHARDSON, M.L., Ed.), The Royal Society of Chemistry, London (1991) 389–409.

[15] DUSZELN, J., "Pesticide contamination and pesticide control in developing countries: Costa Rica, Central America", ibid., pp. 410–428.
[16] APPEL, J., et al., Uso, manejo y riesgos associados a plaguicidas en Nicaragua, Proyecto Regional de Plaguicidas, Confederación Universitaria Centroamericana, Managua, Nicaragua (1991) 143 pp.
[17] VOLDNER, E.C., YI-FAN, L., Global usage of selected persistent organochlorines, Sci. Total Environ. **160/161** (1995) 201–210.
[18] The Agrochemicals Handbook, 3rd edn, The Royal Society of Chemistry, Cambridge (1991).
[19] LARSSON, P., JARNMARK, C., SODERGREN, A., PCBs and chlorinated pesticides in the atmosphere and aquatic organisms of Ross Island, Antarctica, Mar. Pollut. Bull. **23** (1992) 281–287.
[20] BIDLEMAN, T.F., WALLA, M.D., ROURA, R., CARR, E., SCHMIDT, S., Organochlorine pesticides in the atmosphere of the Southern Ocean and Antarctica, January–March, 1990, Mar. Pollut. Bull. **26** (1993) 258–262.
[21] MUIR, D.C.G., et al., Spatial trends and historical profiles of organochlorine pesticides in Arctic lake sediments, Sci. Total Environ. **160/161** (1995) 447–457.
[22] CALAMARI, D., et al., Role of plant biomass in the global environmental partitioning of chlorinated hydrocarbons, Environ. Sci. Technol. **25** (1991) 1489–1495.
[23] SIMONICH, S., HITES, R.A., Global distribution of persistent organochlorine compounds, Science **269** (1995) 1851–1854.
[24] IWATA, H., TANABE, S., SAKAL, N., TATSUKAWA, R., Distribution of persistent organochlorines in the oceanic air and surface seawater and the role of ocean in their global transport and fate, Environ. Sci. Technol. **27** (1993) 1080–1098.
[25] CHERNYAK, S.M., McCONNELL, L.L., RICE, C.P., Fate of some chlorinated hydrocarbons in Arctic and Far Eastern ecosystems in the Russian Federation, Sci. Total Environ. **160/161** (1995) 75–85.
[26] PEREIRA, W.E., ROSTAD, C.E., Occurrence, distributions, and transport of herbicides and their degradation products in the lower Mississippi River and its tributaries, Environ. Sci. Technol. **24** (1990) 1400–1406.
[27] BESTER, K., HUHNERFUSS, H., Triazines in the Baltic and North Sea, Mar. Pollut. Bull. **26** (1993) 423–427.
[28] READMAN, J.W., et al., Herbicide contamination of Mediterranean estuarine waters: Results from a MED POL pilot survey, Mar. Pollut. Bull. **26** (1993) 613–619.
[29] MONTANES, J.F.C., RISEBROUGH, R.W., LAPPE, B.W., MARINO, M.G., ALBAIGES, J., Estimated inputs of organochlorines from the River Ebro into the northwestern Mediterranean, Mar. Pollut. Bull. **21** (1990) 518–523.
[30] FOWLER, S.W., VILLENEUVE, J.P., BURNS, K.A., "Vertical flux of hexachlorobenzene in coastal waters of the north-west Mediterranean Sea", Hexachlorobenzene (Proc. Symp. Lyons, 1986) (MORRIS, C.R., CABRAL, J.R.P., Eds), International Agency for Research on Cancer, Lyons (1986) 67–73.
[31] VILLENEUVE, J.P., CATTINI, C., Input of chlorinated hydrocarbons through dry and wet deposition to the Western Mediterranean, Chemosphere **15** (1986) 115–120.
[32] CLAISSE, D., Chemical contamination of French coasts, Mar. Pollut. Bull. **20** (1989) 523–528.

[33] TAKEOKA, H., et al., Fate of the insecticide HCH in the tropical coastal area of South India, Mar. Pollut. Bull. **22** (1991) 290–297.

[34] SERICANO, J.L., ATLAS, E.L., WADE, T.L., BROOKS, J.M., NOAA's Status and Trends Mussel Watch Program: Chlorinated pesticides and PCBs in oysters (*Crassostrea virginica*) and sediments from the Gulf of Mexico, 1986–1987, Mar. Environ. Res. **29** (1990) 161–203.

[35] SERICANO, J.L., et al., Trace organic contamination in the Americas: An overview of the US National Status and Trends and the International "Mussel Watch" Programmes, Mar. Pollut. Bull. **31** (1995) 214–225.

[36] FOWLER, S.W., Critical review of selected heavy metal and chlorinated hydrocarbon concentrations in the marine environment, Mar. Environ. Res. **29** (1990) 1–64.

[37] TANABE, S., et al., Persistent organochlorine residues in dolphins from the Bay of Bengal, South India, Mar. Pollut. Bull. **26** (1993) 311–316.

[38] KELLY, A.G., CAMPBELL, L.A., Organochlorine contaminants in liver of cod (*Gadus morhua*) and muscle of herring (*Clupea harengus*) from Scottish waters, Mar. Pollut. Bull. **28** (1994) 103–108.

[39] O'CONNOR, T.P., Trends in chemical concentrations in mussels and oysters collected along the US coast from 1986 to 1993, Mar. Environ. Res. **41** (1996) 183–200.

[40] EISENBERG, M., TOPPING, J.J., Organochlorine residues in shellfish from Maryland waters, 1976–1980, J. Environ. Sci. Health **B19** (1984) 673–678.

[41] MARCUS, J.M., RENFROW, R.T., Pesticides and PCBs in South Carolina estuaries, Mar. Pollut. Bull. **21** (1990) 96–99.

[42] CARVALHO, F.P., FOWLER, S.W., GONZALEZ-FARIAS, F., MEE, L.D., READMAN, J.W., Agrochemical residues in the Altata-Ensenada del Pabellón coastal lagoon (Sinaloa, Mexico): A need for integral coastal zone management, Int. J. Environ. Health Res. (in press).

[43] GOLD-BOUCHOT, G., SILVA-HERRERA, T., ZAPATA-PEREZ, O., Chlorinated pesticides in the Rio Palizada, Campeche, Mexico, Mar. Pollut. Bull. **26** (1993) 648–650.

[44] GRANBY, K., SPLIID, N.H., Hydrocarbons and organochlorines in common mussels from the Kattegat and the belts and their relation to condition indices, Mar. Pollut. Bull. **30** (1995) 74–82.

[45] PINEIRO, M.E.A., LOZANO, J.S., YUSTY, M.A.L., Organochlorine compounds in mussels of the estuarine bays of Galicia (North-West Spain), Mar. Pollut. Bull. **30** (1995) 484–487.

[46] OSIBANJO, O., BAMGBOSE, O., Chlorinated hydrocarbons in marine fish and shellfish of Nigeria, Mar. Pollut. Bull. **21** (1990) 581–586.

[47] RAMESH, A., et al., Persistent organochlorine residues in green mussels from coastal waters of South India, Mar. Pollut. Bull. **21** (1990) 587–590.

[48] RUANGWISES, S., RUANGWISES, N., TABUCANON, M.S., Persistent organochlorine pesticide residues in green mussels (*Perna viridis*) from the Gulf of Thailand, Mar. Pollut. Bull. **28** (1994) 351–355.

[49] BOON, J.P., et al., Cyclic organochlorines in epibenthic organisms from coastal waters around East Java, Neth. J. Sea Res. **23** (1989) 427–439.

[50] PHILLIPS, D.J.H., Organochlorines and trace metals in greenlipped mussels, *Perna veridis*, from Hong Kong waters: A test of indicator ability, Mar. Ecol. Prog. Ser. **21** (1985) 251–258.

[51] LIN, Y.S., et al., "Mussel watch in Taiwan, Republic of China", Assimilative Capacity of the Oceans for Man's Wastes (Proc. Conf. Taipei, 1982), SCOPE/ICSU, Academia Sinica, Taipei (1982) 142–193.

[52] CAI, Fulong, et al., Paper IAEA-SM-343/43, these Proceedings.

[53] ENVIRONMENT AGENCY OF JAPAN, Chemicals in the Environment, Environmental Health Department, EAJ, Tokyo (1987).

[54] READMAN, J.W., et al., Persistent organophosphorus pesticides in tropical marine environments, Mar. Pollut. Bull. **24** (1992) 398–402.

[55] BARCELO, D., SOLE, M., DURAND, G., ALBAIGES, J., Analysis and behaviour of organophosphorus pesticides in a rice crop field, Fresenius' J. Anal. Chem. **339** (1991) 676–683.

[56] CARVALHO, F.P., FOWLER, S.W., READMAN, J.W., MEE, L.D., "Pesticide residues in tropical coastal lagoons", Applications of Isotopes and Radiation in Conservation of the Environment (Proc. Symp. Karlsruhe, 1992), IAEA, Vienna (1992) 637–653.

[57] LACORTE, S., LARTIGES, S.B., GARRIGUES, P., BARCELO, D., Degradation of organophosphorus pesticides and their transformation products in estuarine waters, Environ. Sci. Technol. **29** (1995) 431–438.

[58] ENVIRONMENTAL PROTECTION AGENCY, Water Quality Criteria Summary, Office of Science and Technology, Health and Ecological Criteria Division, EPA, Washinghton, DC (1991).

[59] LA POINT, T.W., FAIRCHILD, J.F., LITTLE, E.E., FINGER, S.E., "Laboratory and field techniques in ecotoxicological research: Strengths and limitations", Aquatic Ecotoxicology: Fundamental Concepts and Methodologies (BOUDOU, A., RIBEYRE, F., Eds), CRC Press, Boca Raton, FL (1989) 239–255.

[60] Progress in Standardization of Aquatic Toxicity Tests (SOARES, A.M.V.M., CALOW, P., Eds), Lewis Publishers, Boca Raton, FL (1993).

[61] CALAMARI, D., VIGHI, M., Quantitative structure–activity relationships in ecotoxicology: Value and limitations, Rev. Environ. Toxicol. **4** (1990) 1–112.

[62] BANERJEE, S., BAUGHMAN, G.L., Bioconcentration factors and lipid solubility, Environ. Sci. Technol. **25** (1991) 536–539.

[63] MAYER, F.L., Jr., ELLERSIECK, M.R., Manual of Acute Toxicity: Interpretation and Data Base for 410 Chemicals and 66 Species of Freshwater Animals, Resource Publication 160, Fish and Wild Life Service, United States Department of the Interior, Washington, DC (1986).

[64] CRONIN, M.T.D., DEARDEN, J.C., DOBBS, A.J., QSAR studies of comparative toxicity in aquatic organisms, Sci. Total Environ. **109/110** (1991) 431–439.

[65] MAYER, F.L., Jr., Acute Toxicity Handbook of Chemicals to Estuarine Organisms, Rep. EPA/600/8-87/017, Environmental Protection Agency, Gulf Breeze, FL (1987).

[66] PARSONS, J.R., GOVERS, H.A.J., Quantitative structure–activity relationships for biodegradation, Ecotoxicol. Environ. Saf. **19** (1990) 212–227.

[67] LA POINT, T.W., FAIRCHILD, J.F., "Use of mesocosm data to predict effects in aquatic ecosystems", Limits to Interpretation in Aquatic Mesocosm Studies in Ecological Risk Assessment (GRANEY, R., KENNEDY, J.H., RODGERS, J.H., Eds), Lewis Publishers, Boca Raton, FL (1994) 241–255.

[68] UNITED NATIONS ENVIRONMENT PROGRAMME, Global Programme of Action for the Protection of the Marine Environment from Land-Based Activities, Resolution adopted by the Intergovernmental Conference, Washington, DC, 23 Oct.–3 Nov. 1995, Rep. UNEP (OCA) LBA/IG. 2/7, UNEP, Nairobi (1995).

[69] CARVALHO, F.P., HANCE, R.J., Pesticides in tropical marine environments: Assessing their fate, IAEA Bull. **35** 2 (1993) 14–19.

[70] VILLENEUVE, J.P., HORVAT, M., CATTINI, C., World-wide and regional intercomparison for the determination of organochlorine compounds and petroleum hydrocarbons in mussel sample IAEA-142, Rep. No. 59, IAEA Marine Environment Laboratory, Monaco (1996).

[71] KLAINE, S.J., et al., Characterization of agricultural nonpoint pollution: Pesticide migration in a West Tennessee watershed, Environ. Toxicol. Chem. **7** (1988) 609–614.

PESTICIDES IN THE ENVIRONMENT: REGIONAL AND COUNTRY SITUATIONS

(Session 2)

Chairperson

F. GONZALES-FARIAS
Mexico

Rapporteur

M. HUSSAIN
FAO/IAEA

CONTROL OF PESTICIDES IN CERTAIN COUNTRIES OF SOUTH-EAST ASIA

A. BALASUBRAMANIAM
Food and Agriculture Organization
   of the United Nations,
Rome

Abstract

CONTROL OF PESTICIDES IN CERTAIN COUNTRIES OF SOUTH-EAST ASIA.
   Control of pesticides in some countries of South-East Asia, namely, Indonesia, Malaysia, the Philippines and Thailand, is discussed. All these countries have implemented pesticide registration and control schemes. The information required to register pesticides and the details of pesticide labels in all the countries are somewhat similar to those provided in Food and Agriculture Organization of the United Nations (FAO) guidelines. Certain pesticides are not registered because of their toxic properties, while others may be placed under severe handling and application restrictions. Advertising of pesticides is also controlled in these countries. The importance of licensing pesticide companies, and their post-registration activities, including enforcement of the law, monitoring of the environmental effects, control of the residue levels in food, and determination of the quality of the pesticides on the market, are highlighted. Wearing protective clothing that is suitable for applying pesticides in a temperate climate is not practical under tropical conditions. Records from hospital cases and surveys demonstrate that pesticide poisoning is a serious problem. Recognizing the problem, FAO issued a circular advising small farmers against using pesticides that fall into Class Ia or Ib of the World Health Organization hazard classification. FAO and the United Nations Environment Programme are implementing the prior informed consent (PIC) procedure. All the countries under discussion are participating in PIC and have prohibited the import of most of the pesticides currently subject to this procedure. The level of pesticide residues, particularly in vegetables, is of concern. It was found that a rather high proportion of pesticide products in Malaysia and Thailand did not comply with FAO specifications. Most countries do not appear to have the necessary resources to carry out analyses of an adequate number of formulation and residue samples, and representatives of FAO Member Countries in the Asian–Pacific region have unanimously expressed the need for establishing regional or subregional pesticide analytical facilities.

1. INTRODUCTION

In South-East Asia[1], pesticides have been used for several decades in agriculture, forestry, public health and households, and for timber treatment. The quantity of pesticides used is increasing from year to year. In Thailand, the amount of pesticide active ingredients imported during 1980 was 9855 t which, within a decade, rose to 17 812 t. The average amount of pesticides used per hectare of agricultural land also increased from 1.6 kg in 1978 to 3.3 kg in 1988. On rice, the quantity used increased from 2.7 kg in 1978 to 9.2 kg in 1990, i.e. more than three times in 12 years [1].

Similarly, in Malaysia it is estimated that the total end user value of agricultural pesticides in 1993 was US $105 million, with a 5% increase in 1994 to US $110 million, followed by another 5% increase in 1995 to US $115 million. The value of public health and household pesticides was estimated to be about US $74 million in 1995 [2].

In Indonesia, a total of 523 pesticide products was registered for use in 1993. A total of 71 products was registered for use on rice, comprising nine active ingredients of insecticides, 15 fungicides and 13 herbicides [3]. Although there are no reliable statistics on pesticide use, it is estimated that the pesticide market for rice is the largest of all. In plantation crops, herbicides are the most widely used [4].

Most countries in South-East Asia have laws that control various aspects of pesticides, namely, the import, manufacture, formulation, packaging, labelling, sale, licensing of imports, storage and sale, residues remaining on food or feed, quality of the pesticide, and manner of use. These laws are implemented by committees or boards comprising representatives of government agencies, such as agriculture, health, environment, trade, customs, universities and relevant research agencies.

2. REGISTRATION

The International Code of Conduct on the Distribution and Use of Pesticides defines registration as the process whereby the responsible national government authority approves the sale and use of a pesticide following the evaluation of comprehensive scientific data demonstrating that the product is effective for the purposes intended and not unduly hazardous to human or animal health, or the environment

---

[1] The pesticide control schemes operating primarily in Malaysia and Indonesia are discussed. Certain aspects of the situation in Thailand and the Phillipines are also included.

[5]. In countries where pesticide legislation has been implemented, registration of pesticides is mandatory. The information required to be submitted by an applicant for registration in Malaysia, Indonesia, Thailand and the Philippines is somewhat similar to that provided in Food and Agriculture Organization of the United Natiions (FAO) guidelines on registration. The application must include information on the physical and chemical properties of the product, toxicological data, efficacy data, residue data on those crops on which the product is intended to be used, environmental effects and storage stability, and a declaration as to whether the pesticide has been approved for use in other countries that have a sound registration procedure, as well as a proposed label that is based largely on FAO Guidelines on Good Labelling Practices for Pesticides [6].

In Malaysia, registration is implemented under the Pesticide (Registration) Regulations 1976 [7]. The application for registration is considered by a Technical Committee that consists of members from certain government agencies represented on the Pesticides Board. If the information is satisfactory, and if the product is of acceptable quality and, in the opinion of the Technical Committee, its handling and use under local conditions are not hazardous, the Technical Committee recommends that the product be approved by the Board. If the Board agrees to the recommendation, the product is registered for a period of 5 years.

In Indonesia, registration is primarily under the Decree of the Minister of Agriculture of 1973 [8]. The application is considered by the Pesticide Commission that consists of members from various government departments. Under the Pesticide Commission, there are four subcommissions that deal specifically with evaluation of the bioefficacy, evaluation of the health and environmental safety, monitoring of the health and environmental impact, and enforcement. Members of the subcommission are also members of the Pesticide Commission. The decision on an application may be: (1) permanent approval, valid for a period of 5 years; (2) temporary approval, effective for 1 year; (3) approval for experimentation or trial, effective for 1 year; and (4) rejection of application.

In the Philippines, pesticides are registered by the Fertilizer and Pesticide Authority (FPA). The registration requirements are basically similar to those of other countries in the region. However, one important feature of the system is the phased registration scheme:

(a) Conditional registration: This is granted to a product that has met all the requirements for full registration, except for some data that need to be submitted within an agreed period, normally 6 months. Products with this type of registration may be marketed subject to agreed conditions.

(b) Full registration: This is granted for products that fully meet the FPA requirements on the specifications, efficacy, toxicology, residue and environmental fate, and proper labelling [9].

3. LABELLING

In Indonesia, the Decree of the Minister of Agriculture of 1973 [8] includes provisions for the labelling of pesticides. In Malaysia, the Pesticide (Labelling) Regulations [7] are being implemented as part of the registration procedure. These regulations prohibit the sale of a pesticide unless it bears a label that complies with the labelling requirements specified in the regulations. The labelling requirements in the countries under discussion, except for minor modifications, are based on those stated in FAO Guidelines on Good Labelling Practices for Pesticides [6].

4. RESTRICTED OR BANNED PESTICIDES

In Malaysia, highly toxic pesticides, namely, paraquat, monocrotophos and calcium cyanide, are placed under the specific handling restrictions of the Pesticide (Highly Toxic Pesticides) Regulations 1996. According to these regulations, employers are expected to maintain records of those workers employed in the handling of these pesticides, the quantity of pesticides used, the method of application, and the number of hours worked; to store such pesticides in a locked and well ventilated enclosure accessible only to authorized personnel; to provide workers with protective clothing and a first-aid kit; to carry out yearly medical examinations of workers; and to provide training. Employers should permit work for a maximum of only 8 h/d and only for those persons who are above the age of 18, medically fit, trained and, in the case of female workers, those who are not pregnant or lactating. Workers who handle these pesticides are required to co-operate and wear the protective clothing provided, to wash the equipment and protective clothing every day after work and to follow the instructions provided. Workers and employers have to ensure that all the empty containers are safely disposed of, to abide by the label instructions, to ensure that the pesticides are not used in residential or recreational areas, to minimize the possibility of water contamination and to display a notice at the site where the pesticide has been used [10].

In Indonesia, through a Presidential Instruction of 1986, 57 of the 66 insecticides that had been in use were banned for rice, and adoption of integrated pest management practices was required. The 57 products comprise 30 active ingredients and their mixtures [11]. In addition, 33 active ingredients of pesticides were not allowed to be registered because of their acute toxicity and persistence in the environment [8].

4.1. Prior informed consent (PIC)

FAO and the United Nations Environment Programme (UNEP) are implementing the PIC procedure under which the exporting country will also be held responsible for taking steps, within its authority, to ensure that the export does not take place contrary to the decisions taken by the importing country [12]. Indonesia, Malaysia, Thailand and the Philippines are participating in the PIC procedure, and have prohibited the import of most of the pesticides subject to the procedure [13].

5. PESTICIDE QUALITY

To be consistent with the registration policies, the product marketed should conform with the specifications submitted during registration. Monitoring should include the quality of the technical material, the imported formulated products, the local formulations and the repacked products [14]. In all the countries under discussion there are laboratories responsible for analysis of the pesticide formulations. The number of samples taken from the market and analysed in Malaysia between 1987 and 1991 is given in Table I. The results show that 15–56% of the samples did not comply with the specifications [15].

TABLE I. SAMPLES TAKEN BY ENFORCEMENT OFFICERS AND ANALYSED IN MALAYSIA (1987–1991)

	1987	1988	1989	1990	1991
Total samples taken	96	78	65	39	118
Samples not in compliance	32	37	22	22	18
% samples not in compliance	33	47	34	56	15
Unregistered pesticides	37	36	65	16	52
Total samples analysed	133	114	130	55	170

In Thailand, analysis of pesticide products obtained from retail stores and from farmers between 1985 and 1989 showed that a large number of samples contained less than the declared active ingredient content, i.e. they were below the allowable variation established in FAO specifications. The quality was rather poor, and certain samples were found to contain no active ingredient at all. In another study, 20% of the wettable powder formulations and 60% of the emulsifiable concentrate formulations did not pass the test for suspensibility and emulsion stability, respectively [16].

6. PESTICIDE RESIDUES IN FOOD

Monitoring the residues on food commodities can provide useful information on assessing the safety to consumers of treated food, detecting residues from the improper use of pesticides and protecting the credibility of exporters with their customers [14]. In Indonesia, control of the residue levels in food is the responsibility of the Director-General of Medicine and Food Control [4]. In Malaysia, residues in food are controlled by the Ministry of Health under the Food Act 1983. Maximum residue levels (MRLs) have been established largely on the basis of those of the Codex Alimentarius Commission. The Ministry of Health analysed 1506 samples of fruits and vegetables between January and August 1991. Of these, 76 samples (5.1%) were found to contain pesticide residues in excess of the established MRLs. Monitoring for pesticide residues by the Department of Agriculture of Malaysia indicated that pesticide residues are problematic on leaf and fruit vegetables. In addition, the residues of pesticides not approved for certain crops have also been found in crops, indicating that farmers were not following recommended usage [15].

7. LICENSING OF PREMISES

Licensing pesticide companies, including formulators, provide the government with the assurance that their representatives are aware of the hazardous nature of the products with which they are dealing, and take the responsibility for complying with regulations on pesticide registration, trade, production and use [14]. FAO has formulated guidelines [17] that spell out the various conditions with which retail stores have to abide.

In Malaysia, the Pesticide (Licensing for Sale and Storage for Sale) Rules 1988 are enforced with the objective that premises which sell and/or store pesticides for sale conform to certain conditions for the safe and proper storage/display of registered pesticides. Up to May 1994, a total of 4000 premises had been licensed in Malaysia [18].

8. ADVERTISING

Article 11 of the Code [5] spells out the basic criteria for advertising pesticides. In Malaysia, control of pesticide advertisement came into effect on 1 April 1996 under the Pesticide (Advertisement) Regulations 1996. All advertising of pesticides is required to have the prior approval of the Pesticides Board before it can be relayed on electronic media such as radio, television, films and video casettes, or in any other form of mass media, including newspapers, magazines, periodicals, posters, leaflets and billboards [19]. In Indonesia, all advertising of pesticides has to be approved by the Pesticide Commission [4].

In Thailand, there is no specific law or guidelines for controlling pesticide advertisements. However, advertisers are encouraged to comply with the provisions of the Code. In the Philippines, the FPA Rules and Regulations, Article V, Section 7, states that pesticides approved and registered by the authority shall not be advertised in a manner that is false, misleading and deceptive, and not justified by the conditions of their registration [20].

9. ENFORCEMENT

The effectiveness of pesticide regulations depends on practical implementation and enforcement of the law. The registration authorities should have the power to confiscate unregistered products and hold the distributor, trader, retailer and user responsible for any violation of the law [14].

Between 1982 and 1991, enforcement officers in Malaysia made 738 seizures of pesticides, valued at US $236 000. Two hundred and forty-three offenders have been charged and found guilty of selling pesticides that were not registered, or were below specifications, adulterated/imitations, sold without a licence or labelled improperly [15]. Between 1993 and 1995 in Indonesia, 146 formulations of pesticides had their registration withdrawn by the applicant, or cancelled because of an offence committed against the law [18].

10. ENVIRONMENTAL MONITORING

Data submitted for registration allow prediction of the impact of the pesticide on the environment. After a pesticide has been used for some time, it is desirable to confirm that predictions on the environmental effects, made at the time of registration, are still valid [14]. However, in Malaysia [16], as in the other countries under discussion, data on the environmental fate of pesticides are rather scanty.

11. POISONING

Monitoring of poisoning cases will provide information on the hazards posed by the pesticide under conditions of field use, and must be accompanied by the necessary training on pesticide poisoning symptoms and an information campaign to instil awareness of the hazards posed by pesticide use. Proper and accurate monitoring of pesticide poisoning cases can form a strong basis for appropriate policy decisions and control measures on pesticide use [14].

The World Health Organization (WHO) has estimated that, on average, 3% of farmers and agricultural workers in developing countries suffer from acute pesticide poisoning each year [21]. In Malaysia, 40% of the human poisoning cases are due to pesticides, of which paraquat accounted for about 70% of all the cases reported over the period 1979–1988. Although pesticide poisoning is often associated with suicides, accidental exposure at work accounted for about 11% of the 6554 pesticide poisoning cases reported to government hospitals between 1989 and 1994 [22].

In Indonesia, where the situation is similar to that in most developing countries, no reliable statistics are maintained on pesticide poisoning cases. Therefore, one has to depend on the results of studies by workers such as Kishi et al. [23], who found in a survey of Indonesian farmers that 21% of the respondents had three or more neurological, intestinal or respiratory signs and symptoms per spray operation, and concluded that the risk to Indonesian farmers of acute pesticide poisoning is considerable. In a survey conducted by the Department of Health [24] in Central Java, Indonesia, it was shown that, of the 85 paddy farmers interviewed, 78% had experienced pesticide poisoning, with symptoms that included headaches, weakening of the body and paralysis.

In 1985, paraquat, followed by ethyl-parathion and methyl-parathion, caused the largest number of hospitalized or hospital treated cases of acute poisoning in Thailand. Only 2% of the farmers suffering from any degree of pesticide poisoning received treatment at a hospital [25].

12. DISCUSSION AND CONCLUSIONS

All the countries under discussion have implemented pesticide registration and control schemes. However, the monitoring activities so far carried out reveal that pesticides continue to be a problem in these countries.

Statistics available in Malaysia [22], surveys carried out in Indonesia [23, 24] and findings in Thailand [25] on poisoning cases clearly demonstrate the seriousness of the situation. It must be recognized that wearing protective clothing that is suitable for applying pesticides in a temperate climate is not practical under the hot and humid conditions of the tropics.

FAO, recognizing the problem with hazardous pesticides in developing countries, issued a Field Program Circular [26] pointing out that pesticide formulations which fall into Class Ia or Ib of the WHO hazard classification usually have severe restrictions on use in developed countries; in general, they can only be used by specially trained and certified applicators. Such pesticides should not be used by small farmers or untrained and unprotected workers in developing countries. Even pesticide formulations in WHO Class II should only be provided if it can be demonstrated that users adhere to the necessary precautionary measures.

FAO and UNEP are also jointly executing the PIC procedure, including those formulations which, under the conditions of use in developing countries, pose a hazard, particularly to those who handle them [27].

Analyses of the pesticide formulations available in the countries under discussion were also found to be of rather poor quality. In Malaysia, non-compliance with FAO specifications can be attributed to adulteration of the registered products such that in a number of cases the active ingredient content was less than half the declared value, and in other cases the declared active ingredient had been replaced by other chemicals. Another reason for non-compliance is that pesticide importers do not check the quality of the product prior to marketing. In Malaysia, 5.1% of the fruit and vegetable samples analysed were found to contain residues in excess of the established MRLs. Residues of pesticides not approved for certain crops have also been found in crops, indicating that farmers were not following recommended usage or complying with label directions [15]. It is known that certain vegetables exported from Indonesia and Malaysia have been rejected on account of their excessive level of pesticide residues.

It appears that most countries do not have the necessary resources to carry out analyses of an adequate number of formulation and residue samples. In several formal discussions held among representatives of FAO Member Countries in the Asia–Pacific region [28–30] it was unanimously agreed that regional or subregional pesticide analytical facilities should be established. A Regional Workshop [30] further agreed that regional laboratory facilities could also serve as a reference centre for the validation of critical samples.

REFERENCES

[1] GRANDSTAFF, S., Pesticide Policy in Thailand, Thailand Development Research Institute Foundation, Bangkok (1992) 86 pp.

[2] MALAYSIAN AGRICULTURAL CHEMICALS ASSOCIATION, Annual Report 1995/1996, Petaling Jaya (1996).

[3] PESTISIDA UNTUK PERTANIAN DAN KEHUTANAN, Pesticides for Agriculture and Forestry, Ministry of Agriculture, Jakarta (1993) 286 pp.

[4] DARYONTO, Secretary, Pesticide Commission Jakarta, personal communication, 1995.

[5] FOOD AND AGRICULTURE ORGANIZATION OF THE UNITED NATIONS, International Code of Conduct on the Distribution and Use of Pesticides, FAO, Rome (1986) 34 pp.

[6] FOOD AND AGRICULTURE ORGANIZATION OF THE UNITED NATIONS, Guidelines on Good Labelling Practices for Pesticides, FAO, Rome (1995) 51 pp.

[7] PESTICIDES BOARD MALAYSIA, Guidelines on Registration, Labelling and Classification of Pesticides, Department of Agriculture, Kuala Lumpur (1991) 35 pp.

[8] PEDOMAN PENDAFTARAN PESTISIDA, KOMISI PESTISIDA, DEPARTEMEN PERTANIAN, Directive for Registration of Pesticides, Pesticide Commission, Department of Agriculture, Jakarta (1992) 83 pp.

[9] DEEN, N.R., "Pesticide registration schemes and procedures: Country report", Regional Workshop on the Harmonization of Pesticide Registration Requirements, Beijing, 1990 (1990) 5 pp.

[10] PESTICIDES BOARD MALAYSIA, Guidelines on Pesticide (Highly Toxic Pesticides) Regulations 1996, Department of Agriculture, Kuala Lumpur (1966) 5 pp.

[11] PROSIDING FORUM DIALOG, PESTISIDA DAN PERMASALAHANNYA, LEMBAGA PEMBINAAN DAN PERLINDUNGAN KONSUMEN, Pesticides and Problems, Consumer Development and Protection Board, Semarang, Indonesia (1993) 25 pp.

[12] FOOD AND AGRICULTURE ORGANIZATION OF THE UNITED NATIONS, Guidance for Governments, Joint FAO/UNEP Programme for the Operation of Prior Informed Consent, FAO/UNEP, Rome (1991) 52 pp.

[13] FOOD AND AGRICULTURE ORGANIZATION OF THE UNITED NATIONS, Import Decisions from Participating Countries, Circular V, Joint FAO/UNEP Programme for the Operation of Prior Informed Consent, FAO/UNEP, Rome (1995) 86 pp.

[14] FOOD AND AGRICULTURE ORGANIZATION OF THE UNITED NATIONS, Guidelines on Post-registration Surveillance and Other Activities in the Field of Pesticides, FAO, Rome (1988) 11 pp.

[15] TAN, S.H., YEOH, H.F., YUNUS, M.F., "The Pesticides Act 1974 — Implementation and monitoring of compliance", Pesticides in Perspective (Proc. Int. Conf. Kuala Lumpur, 1992) (1992) 8.

[16] LEUTRAKUL, C., SRIPHALKIT, C., "Quality test of pesticide formulations", Pesticide Policy in Thailand, Thailand Development Research Institute Foundation, Bangkok (1992).

[17] FOOD AND AGRICULTURE ORGANIZATION OF THE UNITED NATIONS, Guidelines on Retail Distribution of Pesticides with Particular Reference to Storage and Handling at the Point of Supply to Users in Developing Countries, FAO, Rome (1988) 9 pp.

[18] TAN, S.H., NURSIAH, M.T., THIAGARAJAN, R., "Control and management of pesticides in Malaysia — The challenges ahead", Plant Protection (Proc. Seminar Serdang, 1994) (1994) 10.

[19] PESTICIDES BOARD MALAYSIA, Guidelines on Pesticide Advertisement 1996, Department of Agriculture, Kuala Lumpur (1966) 6 pp.

[20] FOOD AND AGRICULTURE ORGANIZATION OF THE UNITED NATIONS, Regional Workshop on Pesticide Labelling and Advertising, Chiangmai, 1990, Department of Agriculture, Jakarta/FAO, Rome (1990) 45 pp.

[21] JEYARATNAM, J., Acute pesticide poisoning: A major global health problem, World Health Statis. Quart. **43** 3 (1990) 139–144.

[22] MOKHTAR, A.M., "Safety in the use of pesticides at work", Pesticides (Proc. Int. Conf. Kuala Lumpur, 1996) (1996) 145–147.

[23] KISHI, M., et al., Relationship of pesticide spraying to signs and symptoms in Indonesian farmers, Scand. J. Work Environ. Health **21** 2 (1995) 124–133.

[24] SAID MAKSUDI, 78% of Farmers in Temanggung Poisoned by Pesticides, Suara Merdeka (1995) (in Indonesian).

[25] WONGPARICH, M., et al., "Pesticide poisoning among agricultural workers", Pesticide Policy in Thailand, Thailand Development Research Institute Foundation, Bangkok (1992) 86 pp.

[26] DE VEGA, J.P., Pesticides Selection and Use in Field Projects, Field Program Circular No. 8/92, Food and Agriculture Organization of the United Nations, Rome (1992) 4 pp.

[27] FOOD AND AGRICULTURE ORGANIZATION OF THE UNITED NATIONS, Report of the 8th Joint FAO/UNEP Meeting on Prior Informed Consent, FAO/UNEP, Rome (1995) 24 pp.

[28] Report on the Strengthening of Pesticide Regulations (Workshop and Symposium), Food and Agriculture Organization of the United Nations, Rome, Environmental Protection Agency, Washington, DC, and Deutsche Gesellschaft für Technische Zusammenarbeit (1989) 100 pp.

[29] FOOD AND AGRICULTURE ORGANIZATION OF THE UNITED NATIONS, Regional Experts Meeting on Pesticide Specifications and Quality Control, Ministry of Agriculture, Jakarta/FAO, Rome (1989) 50 pp.

[30] FOOD AND AGRICULTURE ORGANIZATION OF THE UNITED NATIONS, Report on the Regional Workshop on Harmonization of Pesticide Registration Requirements, Institute for the Control of Agrochemicals, Ministry of Agriculture, Beijing/FAO, Rome (1990) 50 pp.

IAEA-SM-343/27

THIOMETON RESIDUES IN CUCUMBER

A. SHEIKHIGORGAN, K. TALEBI
Plant Protection Department,
Agricultural College,
Tehran University,
Tehran

H.R. ZOLFAGHARIEH, S. MASHAYEKHI
Nuclear Research Centre
  for Agriculture and Medicine,
Atomic Energy Organization of Iran,
Karadj

Islamic Republic of Iran

Abstract

THIOMETON RESIDUES IN CUCUMBER.
   In the Islamic Republic of Iran, vegetables are treated repeatedly with pesticides to control pests and diseases. Crops harvested shortly after pesticide application are probably contaminated by pesticides that are toxic to humans. Pesticide residues in vegetables are especially important, since these crops are ingested directly by humans. In recent years, farmers have been using more persistent pesticides to protect their crops because of increasing pesticide prices. Field trials were carried out to determine the residues of thiometon, a systemic insecticide used on cucumber. In 1993 and 1994, Ekatin (a 25% thiometon emulsifiable concentrate) was applied with a knapsack sprayer to cucumber (var. Daminus) at late flowering at dilutions of 1:1000 and 2:1000. Plots were 25 m^2 in three replicates and were sampled 3, 6, 10, 15, 19 and 24 days after application. The samples were extracted with acetonitrile and cleanup was carried out using thin layer chromatography. Thiometon and two metabolites (thiometon sulphide and thiometon sulphone) were analysed by gas chromatography. In 1994, the amount of thiometon sulphide determined 24 days after treatment was 2 ppm in the peeled cucumber, which exceeded the maximum residue limit of 0.5 ppm. The total thiometon residues were higher in the peeled cucumber than those in the cucumber peel. Thus, peeling is ineffective for reducing the systemic residues of thiometon. Likewise, in this experiment lower dilution caused higher rather than lower thiometon residues in cucumber. The rapid disappearance of thiometon residues 2 weeks after treatment suggests that storing cucumbers at room temperature may be a better strategy for reducing excessive thiometon residues. Alternatively, a contact insecticide could be used. Thus, the bulk of the surface residues could be removed by washing or peeling.

1. INTRODUCTION

Pesticides are used widely in the Islamic Republic of Iran to control insects, diseases and weeds. The pesticide residues in vegetables, fruits and commercial crops may have undesirable effects on consumers. Sometimes these residues can cause acute and chronic toxicity to humans through consumption of meat, milk, vegetable and fruits. Leaching of pesticides through soil into rivers and groundwater may also result in direct or indirect toxicity to humans. The effects of long term exposure to pesticide residues are difficult to quantify, but could include mutations and some health disorders.

In some parts of Iran, excessive amounts of pesticides are used by farmers because of lack of information. This increases the risk of unnecessary pesticide applications and leads to the accumulation of pesticide residues in treated crops. Currently, several thousand tonnes of pesticide are used annually in the country. The Environment Protection Organization, the Ministry of Agriculture and the Ministry of Health are co-operating in an effort to reduce unnecessary pesticide use.

2. MATERIALS AND METHODS

2.1. Field activities

Field trials were established at Karadj and consisted of 25 m^2 plots in three replicates. The Daminus cucumber was transplanted in spring according to local agricultural practice. At late flowering, the vines were sprayed with Ekatin (a 25% thiometon emulsifiable concentrate) at dilutions of 1:1000 and 2:1000. The treatments were applied when there were only a few small cucumbers on the vines.

2.2. Sampling

Random sampling was carried out 3, 6, 10, 15, 19 and 24 days after application. At each sampling period, 5 kg of fruit were picked. In the laboratory, 0.5 subsamples (with skin) were blended and placed in the freezer until analysis.

2.3. Solvents and standard solutions

The solvents used for thiometon analyses were analytical grade acetonitrile, dichloromethane, petroleum ether, acetone, chloroform and hexane. To rinse and extract the pesticides from the extract solutions, acetonitrile and a mixture of petroleum ether and dichloromethane (9:1) were used. Acetone was used as a mobile phase for preparing the thiometon standard solutions, as well as a mixture of hexane/acetone/chloroform (10:15:75).

2.4. Extraction

Blended cucumber samples (25 g each) were weighed, together with sodium sulphate (0.5 g per gram of sample) and acetonitrile (60 mL). The samples were extracted for 30 min and then filtered through a Buchner funnel. The cucumber residue remaining on the filter paper was re-extracted with acetonitrile (50 mL) until the volume was 110 mL. This solution was concentrated in a rotary evaporator at 30°C and then transferred to a separating funnel with 30 mL of sodium sulphate solution (5% wt/vol.). The aqueous acetonitrile was extracted with 10, 20 and again 20 mL of petroleum ether and dichloromethane (9:1); 30 mL of this combined extract were concentrated with a rotary evaporator and made up to 2 mL.

2.5. Clean up

Thin layer chromatography was used to separate the thiometon sulphide from the cucumber impurities. The final sample extract (100 μL) was applied to a silica gel chromatographic plate (0.25 mm layer thickness). Two samples and two standards (thiometon sulphide and thiometon sulphone) were spotted on each plate. The plates were then developed with chloroform/acetone/hexane (75:15:5). After the solvent had evaporated, the standards were sprayed with a palladium chloride solution (0.2% wt/vol.). The spots, corresponding to the thiometon sulphide and thiometon sulphone standards, were scraped from the plates and washed with acetone. The acetone extract was evaporated with a stream of air and made up to 100 μL of acetone. This cleaned up sample was now ready to inject into the gas chromatographic (GC) facility.

2.6. GC injection and calibration curve

The GC was equilibrated for 1 day because of variations in the electricity voltage and the carrier gas flows. At the start of each run, the columns were conditioned with 30 successive injections of a standard solution. The concentration of thiometon and metabolites was then determined by referring to a calibration curve and by alternating the sample between the standard injections.

3. DISCUSSION AND RESULTS

Most of the thiometon residues in the cucumber peaked 2 weeks after treatment at 9.8 and 8.8 ppm in 1993 and 1994, respectively (Figs 1 and 2). An overall 70% reduction in thiometon residues was observed for the 1:1000 and 2:1000 dilutions between 2 and 3 weeks after treatment. Nevertheless, in both 1993 and 1994 the thiometon residues exceeded the maximum residue limit (0.5 ppm), therefore a safe withholding period was not reached during the course of the experiment.

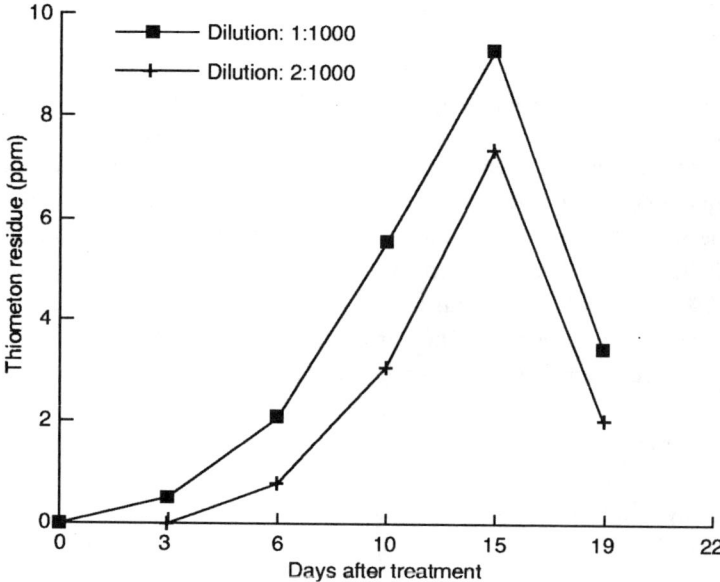

FIG. 1. Thiometon residue variations with time in cucumber (1993).

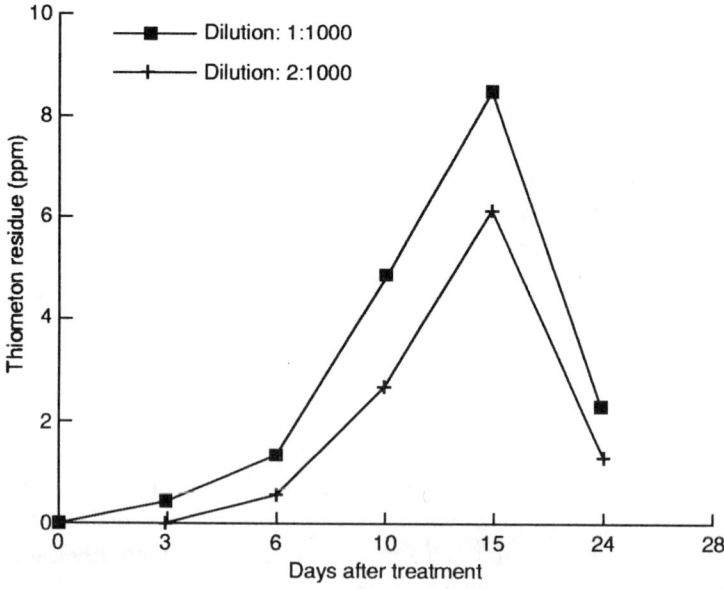

FIG. 2. Thiometon residue variations with time in cucumber (1994).

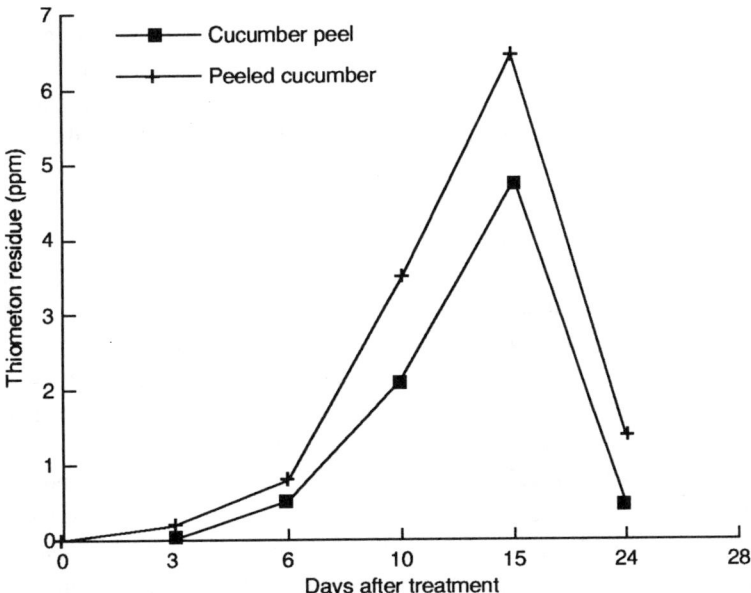

FIG. 3. Thiometon residue variations with time in peeled cucumber and cucumber peel at a spray dilution of 2:1000 (1994).

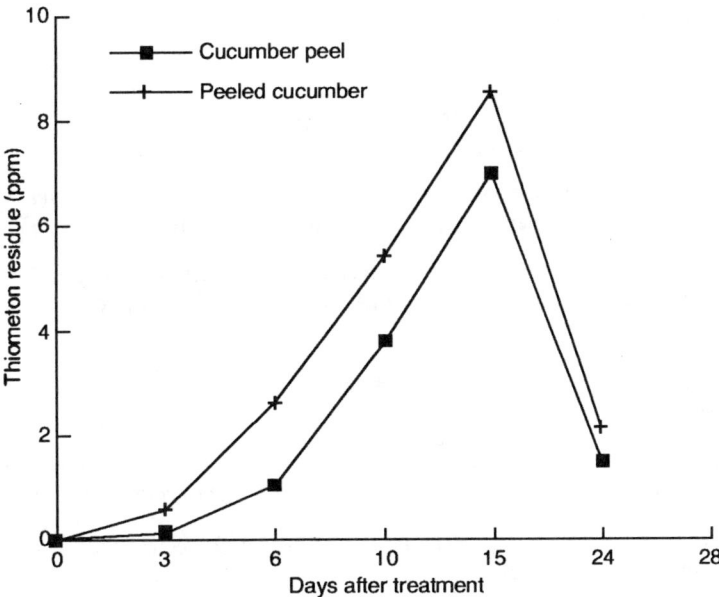

FIG. 4. Thiometon residue variations with time in peeled cucumber and cucumber peel at a spray dilution of 1:1000 (1994).

The withholding period depends on many factors: the crop morphology and physiology; the toxicological, chemical and physical properties of the pesticide; the rate and method of application; and the environmental factors, e.g. temperature, humidity and sunlight. This experiment examined the effect of two dilutions on the thiometon residues. Interestingly, uptake of thiometon by cucumber was higher at the lower dilution (Figs 1 and 2). Thus, lower application rates applied at flowering may enhance rather than reduce the residue levels.

Our findings revealed that the thiometon residues were higher in the peeled cucumber than in the cucumber peel (Figs 3 and 4). The increase in thiometon residues up to 2 weeks after application is consistent with the behaviour of a systemic pesticide. Unfortunately, such behaviour is undesirable from a residue point of view. Post-harvest storage was found to be effective for reducing residues. This finding was confirmed by the rapid thiometon degradation shown between 2 and 3 weeks after application. Alternatively, a contact insecticide could be used rather than a systemic insecticide against vegetable pests. In this way, the surface residues could be separated from the edible portion of the fruit. Prior washing of the cucumber is also recommended for sanitary reasons and to reduce the surface residues, including agrochemical adjuvants such as surfactants.

In summary, consumption of raw fruits, especially cucumber, requires that special attention be paid to the maximum pesticide residue limits because of potential direct ingestion of the pesticide. Knowledge of pesticide behaviour may assist in the design of agricultural strategies to minimize the residue levels and any adverse environmental impacts.

BIBLIOGRAPHY

KLISENKO, M.A., Identification and determination of organophosphorus and organochlorine pesticide residues in tomatoes, Agrokhimiya **6** (1990) 98–103.

LACO, W., JOE, T., Multiresidue screening method for fresh fruits and vegetables with mass chromatographic–mass spectrometric detection, J. Assoc. Off. Anal. Chem. **76** 3 (1991) 554–565.

POWELL, K.A., JUTSUMA, A.R., Technical and commercial aspects of biocontrol products, Pestic. Sci. **37** (1993) 315–321.

STAN, H.J., CHRISTAL, B., Residue analysis of onion and other food stuffs with complex matrix using two dimensional capillary GC, Anal. Chem. **339** 6 (1991) 395–398.

WORTHING, C.R., HANCE, R.J., The Pesticide Manual, 9th edn, British Crop Protection Council, Farnham, Surrey (1991).

IAEA-SM-343/29

PESTICIDE RESIDUES IN THE SOIL OF THE CENTRAL JORDAN VALLEY

T.M. MUSTAFA*, A. HATOUGH**, S. KHATTARI*,
K. MASHA'AL***, A. SA'ADEH***

*Faculty of Agriculture

**Faculty of Sciences

***Center for Strategic Studies

University of Jordan,
Amman, Jordan

Abstract

PESTICIDE RESIDUES IN THE SOIL OF THE CENTRAL JORDAN VALLEY.
    Soil samples were taken from three zones in the Central Jordan Valley to evaluate the range of contamination with pesticides. Several pesticides belonging to different groups, particularly chlorinated hydrocarbon insecticides such as the DDT family and cyclodienes, were detected. In zone I, most of the chlorinated hydrocarbon insecticides occurred in open fields cultivated with tomatoes at a depth of 30–60 cm, at 0–15 and 15–30 cm in zone II, and at 0–15 and 30–60 cm in zone III. When considering pesticide residues under plastic house conditions cultivated with tomatoes, most of the chlorinated hydrocarbon insecticide residues were detected at a depth of 0–15 and 15–30 cm in zone I, at 0–15, 15–30 and 30–60 cm in zone II, and at 0–15 and 15–30 cm in zone III of the Central Jordan Valley.

1. INTRODUCTION

    Most organic pesticides were developed after 1945. Broad spectrum pesticides have been very successfully used by farmers to control pesticides. Chlorinated hydrocarbon insecticides were used for many years to control agricultural and public health insects before their application was banned in many countries. These insecticides, which break down only slowly in soil, may reach fresh water and concentrate in the bodies of some terrestrial and aquatic invertebrates and plants that provide food for vertebrates [1]. There have been numerous surveys of soil insecticide residues in the United States of America [2, 3], Canada [4] and the United Kingdom [1]. Several studies on pesticide residues in Jordan have been made, most of which concentrated on pesticide residues in fruits and vegetables [5–9], whereas few studies have been conducted on pesticide residues in Jordanian soil [10, 11].
    This study was conducted to evaluate the range of contamination of soil in open fields and under plastic house conditions at three depths in the Central Jordan Valley.

2. MATERIALS AND METHODS

2.1. Soil sampling

Composite soil samples (1.5 kg) from three depths (0–15, 15–30 and 30–60 cm) were collected randomly, using a trowel, during the period January–June 1995 and covered the area of the Central Jordan Valley. The area under investigation was divided into three zones that extended from Wadi Buleiwq to Kuremeh (20 km length and 8 km wide). Zone I extended from Wadi Buleiwq to Deir Alla; zone II from Deir Alla to Wadi Rajeb; zone III from Wadi Rajeb to Kuremeh. Samples were taken from open fields cultivated with tomatoes, cucumbers and squash, from a plastic house cultivated with tomatoes and from the uncultivated soil bordering the streets. These samples were placed in polyethylene plastic bags, labelled and transferred for air drying to a mulch greenhouse on the university campus; they were then ground and sieved through a 2 mm screen. The samples were maintained in a deep frozen condition ($-30°C$) until analysis. Subsamples of 50 g were transferred to the Residue Laboratory in Al-Baqa'a for pesticide residue analysis.

2.2. Analytical work

Becker's [12] micromethod was used for the determination of pesticides. Soil samples (25 g) were placed in a glass bottle and 25 mL of distilled water and 100 mL of acetone were added. The glass bottle was shaken for 2 hours on a medium speed shaker and the mixture was filtered through a Buchner funnel. The samples were then transferred to a 250 mL separating funnel with 30 mL of dichloromethane. The organic layer was placed in 100 mL beakers, where it passed through a glass funnel packed with anhydrous sodium sulphate to absorb the water into a 25 mL round bottom flask. Using a rotary evaporator, the extract was concentrated to 0.5 mL at a bath temperature of 30°C. For further removal of the impurities, the soil samples were passed through a clean-up column, which consisted of 3 g of silica gel and 1 g of anhydrous sodium sulphate. The elution mixture consisted of dichloromethane, benzene and acetone (5:1:1).

Recovery tests were conducted to measure the efficiency of the procedure followed to determine the pesticide residues in the samples. A blank test was conducted to check for the presence of any contamination in the solvents and for any equipment with residues of the pesticide.

Organophosphate residues were determined by gas–liquid chromatography using a nitrogen phosphorus detector [8]. The chlorinated hydrocarbon and pyrethroid residues were determined by gas–liquid chromatography using an electron capture detector.

The peak area of each injection was recorded. The pesticide residues in each sample were calculated according to the following formula:

$$C_s = \frac{A_s \times C_{st} \times 200 \times 5 \times F}{A_{st} \times Wt \times 20} = \frac{A_s \times C_{st} \times 50 \times F}{A_{st} \times Wt}$$

where

C_s is the concentration of residues in samples (mg/kg);.
Wt is the weight of the analysed sample (g);
A_s is the peak area obtained for the sample;
A_{st} is the peak area obtained for the standard solution;
C_{st} is the concentration of standard solution injected (mg/L);
200 is the final volume of the aqueous solution (mL);
20 is the volume of the aqueous solution used in the analysis;
F is the recovery factor from $\frac{100}{\% \text{ recovery}}$;
5 is the final volume of the analysed solution (mL).

3. RESULTS

3.1. Pesticides detected

Table I shows the pesticides detected in the soil of three zones in the Central Jordan Valley in 1995. Chlorinated hydrocarbon pesticides were mainly detected in the soil of the three zones. Aldrin, beta-HCH, cis-chlordene, endrin, P,P-DDT, O,P-DDE, P,P-DDE, PCNB and chlorothalonil were found in the soil of zone I cultivated with cucumbers in open fields. The fungicide procymidone from the dicarboximide group was also detected. Aldrin, HCB, beta-HCH, O,P-DDE and P,P-DDT were found in the soil of zones cultivated with tomatoes in open fields. Cis-heptachlor, P,P-DDT, O,P-DDE, P,P-DDE, gamma-HCH, beta-HCH and transchlordene were detected in the soil of zone I cultivated with tomatoes under plastic house conditions. The vinclozolin fungicide from the dicarboximide group was also found. Cis-heptachlor, P,P-DDT, P,P-DDE and endrin were detected in the soil of zone I cultivated with squash in open fields. Procymidone was also detected. Gamma-HCH, PCNB, cis-heptachlor, dieldrin, O,P-DDE, O,P-DDT and P,P-DDT were detected in the uncultivated soil bordering the streets in zone I.

Gamma-HCH, P,P-DDE, O,P-DDE, P,P-DDT, aldrin and chlorothalonil were found in the soil of zone II cultivated with cucumbers in open fields.

TABLE I. PESTICIDES DETECTED IN THE SOIL OF THREE ZONES IN THE CENTRAL JORDAN VALLEY (1995)

Zone	Crop	Pesticide	Type	Group
I	Cucumbers in open fields	Aldrin	Insecticide	Chlorinated hydrocarbon
		Beta-HCH	Insecticide	Chlorinated hydrocarbon
		Cis-chlordene	Insecticide	Chlorinated hydrocarbon
		Endrin	Insecticide	Chlorinated hydrocarbon
		P,P-DDT	Insecticide	Chlorinated hydrocarbon
		PCNB	Fungicide	Chlorinated hydrocarbon
		P,P-DDE	Insecticide	Chlorinated hydrocarbon
		Chlorothalonil	Fungicide	Chlorinated hydrocarbon
		O,P-DDE	Insecticide	Chlorinated hydrocarbon
		Procymidone	Fungicide	Dicarboximide
	Tomatoes in open fields	Aldrin	Insecticide	Chlorinated hydrocarbon
		HCB	Insecticide	Chlorinated hydrocarbon
		Beta-HCH	Insecticide	Chlorinated hydrocarbon
		O,P-DDE	Insecticide	Chlorinated hydrocarbon
		P,P-DDT	Insecticide	Chlorinated hydrocarbon
	Tomatoes in plastic houses	Cis-heptachlor	Insecticide	Chlorinated hydrocarbon
		P,P-DDT	Insecticide	Chlorinated hydrocarbon
		O,P-DDE	Insecticide	Chlorinated hydrocarbon
		Gamma-HCH	Insecticide	Chlorinated hydrocarbon
		Beta-HCH	Insecticide	Chlorinated hydrocarbon
		Vinclozolin	Fungicide	Chlorinated hydrocarbon
		Trans-chlordene	Insecticide	Chlorinated hydrocarbon
		P,P-DDE	Insecticide	Chlorinated hydrocarbon
	Squash in open fields	Cis-heptachlor	Insecticide	Chlorinated hydrocarbon
		P,P-DDT	Insecticide	Chlorinated hydrocarbon
		Procymidone	Fungicide	Dicarboximide

TABLE I (cont.)

Zone	Crop	Pesticide	Type	Group
I (cont.)		Endrin	Insecticide	Chlorinated hydrocarbon
		P,P-DDE	Insecticide	Chlorinated hydrocarbon
	Uncultivated soil bordering the streets	Gamma-HCH	Insecticide	Chlorinated hydrocarbon
		PCNB	Fungicide	Chlorinated hydrocarbon
		Cis-heptachlor	Insecticide	Chlorinated hydrocarbon
		Dieldrin	Insecticide	Chlorinated hydrocarbon
		O,P-DDE	Insecticide	Chlorinated hydrocarbon
		O,P-DDT	Insecticide	Chlorinated hydrocarbon
		P,P-DDT	Insecticide	Chlorinated hydrocarbon
II	Cucumbers in open fields	Gamma-HCH	Insecticide	Chlorinated hydrocarbon
		P,P-DDE	Insecticide	Chlorinated hydrocarbon
		P,P-DDT	Insecticide	Chlorinated hydrocarbon
		Phosphamidon	Insecticide	Organophosphates
		Diazinon	Insecticide	Organophosphates
		Parathion	Insecticide	Organophosphates
		Chlorothalonil	Fungicide	Chlorinated hydrocarbon
		Procymidone	Fungicide	Dicarboximide
		Aldrin	Insecticide	Chlorinated hydrocarbon
		O,P-DDE	Insecticide	Chlorinated hydrocarbon
		Cypermethrin	Insecticide	Pyrethroids
	Tomatoes in open fields	Diazinon	Insecticide	Organophosphates
		Parathion	Insecticide	Organophosphates
		PCNB	Fungicide	Chlorinated hydrocarbon
		Phosphamidon	Insecticide	Organophosphates
		P,P-DDT	Insecticide	Chlorinated hydrocarbon
		Alpha-HCH	Insecticide	Chlorinated hydrocarbon

TABLE I (cont.)

Zone	Crop	Pesticide	Type	Group
II (cont.)		Chlorothalonil	Fungicide	Chlorinated hydrocarbon
		P,P-DDE	Insecticide	Chlorinated hydrocarbon
		O,P-DDT	Insecticide	Chlorinated hydrocarbon
		O,P-DDE	Insecticide	Chlorinated hydrocarbon
		Procymidone	Fungicide	Dicarboximide
	Tomatoes in plastic houses	Phosphamidon	Insecticide	Organophosphates
		Procymidone	Fungicide	Dicarboximide
		Endrin	Insecticide	Chlorinated hydrocarbon
	Uncultivated soil bordering the streets	O,P-DDT	Insecticide	Chlorinated hydrocarbon
III	Cucumbers in open fields	Parathion	Insecticide	Organophosphates
		Diazinon	Insecticide	Organophosphates
		Alpha-HCH	Insecticide	Chlorinated hydrocarbon
		HCB	Insecticide	Chlorinated hydrocarbon
		Beta-HCH	Insecticide	Chlorinated hydrocarbon
		Cis-chlordene	Insecticide	Chlorinated hydrocarbon
		P,P-DDE	Insecticide	Chlorinated hydrocarbon
		O,P-DDT	Insecticide	Chlorinated hydrocarbon
	Tomatoes in open fields	O,P-DDT	Insecticide	Chlorinated hydrocarbon
		Diazinon	Insecticide	Organophosphates
		Parathion	Insecticide	Organophosphates
		HCB	Insecticide	Chlorinated hydrocarbon
		Beta-HCH	Insecticide	Chlorinated hydrocarbon
		Aldrin	Insecticide	Chlorinated hydrocarbon
		Cis-chlordene	Insecticide	Chlorinated hydrocarbon

TABLE I (cont.)

Zone	Crop	Pesticide	Type	Group
III (cont.)	Tomatoes in plastic houses	HCB	Insecticide	Chlorinated hydrocarbon
		HCH	Insecticide	Chlorinated hydrocarbon
		Aldrin	Insecticide	Chlorinated hydrocarbon
		Cis-chlordene	Insecticide	Chlorinated hydrocarbon
		O,P-DDT	Insecticide	Chlorinated hydrocarbon

Phosphamidon and diazinon from the organophosphate group were also detected, in addition to procymidone and the cypermethrin insecticide from the pyrethroid group. P,P-DDT, P,P-DDE, O,P-DDT, O,P-DDE, alpha-HCH and the fungicides chlorothalonil and PCNB were detected in the soil of zone II cultivated with tomatoes in open fields. Diazinon, parathion and phosphamidon from the organophosphate group and procymidone from the dicarboximide group were also detected. Endrin from the chlorinated hydrocarbon group and phosphamidon from the organophosphate group were detected in the soil of zone II cultivated with tomatoes under plastic house conditions, in addition to procymidone from the dicarboximide group. O,P-DDT was detected in the uncultivated soil bordering the streets in zone II.

Alpha-HCH, HCB, beta-HCH, cis-chlordene, P,P-DDE and O,P-DDT were found in the soil of zone III cultivated with cucumbers in open fields, in addition to parathion and diazinon. O,P-DDT, HCB, beta-HCH, aldrin and cis-chlordene were found in the soil of zone III cultivated with tomatoes in open fields, in addition to diazinon and parathion.

3.2. Residues at the three depths

Table II shows the chlorinated hydrocarbon pesticide residues at three depths of soil cultivated with tomatoes in open fields in three zones in the Central Jordan Valley in 1995. A relatively small amount of aldrin and a high amount of P,P-DDT were detected at the 30–60 cm depth in zone I. A small amount of O,P-DDT and HCH and a high amount of P,P-DDT were found at the 15–30 cm depth in zone II, in addition to a small amount of P,P-DDE, O,P-DDT, HCB, HCH, O,P-DDE and a high amount of P,P-DDT at the 0–15 cm depth of the same zone. A small amount of O,P-DDT was detected at the 30–60 and 0–15 cm depths in zone III.

Table III shows the chlorinated hydrocarbon pesticide residues at three depths of soil cultivated with tomatoes under plastic house conditions in three zones in the

TABLE II. CHLORINATED HYDROCARBON PESTICIDE RESIDUES AT THREE DEPTHS IN SOIL CULTIVATED WITH TOMATOES IN OPEN FIELDS IN THREE ZONES IN THE CENTRAL JORDAN VALLEY (1995)[a]

Sample No.	Zone	Pesticide detected	Pesticide residues (mg/kg) at depth (cm)		
			0–15	15–30	30–60
7	I	Aldrin	ND	ND	0.0023
11		Aldrin	ND	ND	0.0015
30		P,P-DDT	ND	ND	3.2315
29	II	HCB	0.0007	ND	ND
		HCH	0.0203	ND	ND
		O,P-DDE	0.0157	ND	ND
		P,P-DDT	0.7559	ND	ND
36		P,P-DDT	ND	4.6675	ND
40		O,P-DDT	ND	0.0038	ND
		P,P-DDT	ND	4.4684	ND
47		P,P-DDE	0.0047	ND	ND
		O,P-DDT	0.0118	ND	ND
54		HCH	ND	0.0198	ND
59		O,P-DDE	0.0012	ND	ND
		P,P-DDT	0.0597	ND	ND
67		P,P-DDT	5.4696	ND	ND
93	III	O,P-DDT	ND	ND	0.0027
96		O,P-DDT	0.0043	ND	ND

[a] ND = not detected.

Central Jordan Valley in 1995. A relatively small amount of P,P-DDE and a high amount of P,P-DDT were found at the 15–30 cm depth in zone I. A small amount of HCH, cis-heptachlor, trans-chlordene and P,P-DDE and a high amount of P,P-DDT were detected at the 0–15 cm depth of zone I. A small amount of O,P-DDE and a high amount of P,P-DDT were detected at 0–15 cm, a high amount of

TABLE III. CHLORINATED HYDROCARBON PESTICIDE RESIDUES AT THREE DEPTHS OF SOIL CULTIVATED WITH TOMATOES UNDER PLASTIC HOUSE CONDITIONS IN THREE ZONES IN THE CENTRAL JORDAN VALLEY (1995)[a]

Sample No.	Zone	Pesticide detected	Pesticide residues (mg/kg) at depth (cm)		
			0–15	15–30	30–60
24	I	P,P-DDE	ND	0.0018	ND
		P,P-DDT	ND	3.6981	ND
25		HCH	0.0493	ND	ND
		Cis-heptachlor	0.0011	ND	ND
		Trans-chlordene	0.0015	ND	ND
		P,P-DDT	4.2270	ND	ND
26		P,P-DDE	0.0018	ND	ND
		P,P-DDT	0.5246	ND	ND
27		P,P-DDE	0.0037	ND	ND
		P,P-DDT	0.5373	ND	ND
22	II	O,P-DDE	0.0087	ND	ND
		P,P-DDT	0.5391	ND	ND
26		P,P-DDT	ND	4.0649	ND
48		O,P-DDT	ND	ND	0.0048
51		P,P-DDT	0.5593	ND	ND
91	III	HCB	0.0051	ND	ND
		HCH	0.0575	ND	ND
		Aldrin	0.0118	ND	ND
		Cis-chlordene	0.0052	ND	ND
		O,P-DDT	0.0369	ND	ND
95		O,P-DDT	ND	0.0100	ND
96		O,P-DDT	0.0043	ND	ND

[a] ND = not detected.

P,P-DDT at 15–30 cm, and a small amount of O,P-DDT at 30–60 cm in zone II. A small amount of HCB, HCH, aldrin, cis-chlordene and O,P-DDT were detected at 0–15 cm and a small amount of O,P-DDT at 15–30 cm in zone III.

4. DISCUSSION

The survey clearly showed that the residues most likely to occur in agricultural soil in the Central Jordan Valley are chlorinated hydrocarbon insecticides. Many samples contained DDT, DDE and DDD, because these insecticides had been used in Jordan for several years (from the 1950s up to the end of 1994) to control the malaria mosquito, *Anopheles* sp., in the Central Jordan Valley, as well as dicofol, which contains a small amount of DDT, to control the spider mite. DDT metabolized in soil to several degradation products, mainly DDE and DDD. This is in agreement with a report by Edwards [13] and research carried out by the Royal Scientific Society of Jordan [11]. Some samples contained beta-HCH, gamma-HCH, alpha-HCH and HCB, which are isomers of lindane, except for HCB, which is similar in structure. Lindane was commonly used to control animal ectoparasites and, illegally, agricultural insects until its use was banned in Jordan in 1994. Cyclodiene compounds such as aldrin, chlordene, endrin, heptachlor and dieldrin were also detected in the soil samples. However, DDT was found to be the most persistent insecticide in the soil, followed by its degradation products. Application of DDT and other chlorinated hydrocarbon components resulted in the buildup of residues in the soil, particularly where the spray was applied to the soil. Some parent pesticides may have converted in the soil to other products, e.g. DDT to DDE and DDD, aldrin to dieldrin, and heptachlor to chlordene. Fungicides similar in structure to chlorinated hydrocarbon insecticides were detected. These were PCNB, chlorothalolin, procymidone and vinclozolin, which were commonly used by farmers to control soil fungi in the Central Jordan Valley. Phosphamidon, diazinon and parathion from the organophosphate group and cypermethrin from the pyrethroid group were detected in the soil, particularly at the 0–15 cm depth. These insecticides were most probably used to control insects on crops in open fields or under plastic houses a few weeks before soil sampling. Use of parathion was illegal, since it had been banned for 15 years in Jordan.

Chlorinated hydrocarbons were detected at the 30–60 cm depth in the soil cultivated with tomatoes in open fields in zone I. Few aldrin residues from cyclodienes, banned for about 25 years, were found, while a relatively high amount of DDT was detected. Most of the chlorinated hydrocarbons were detected at the 0–15 and 15–30 cm depths in zone I. Few HCH and HCB residues were detected, since they were banned a few years ago, while a high amount of DDT and a low amount of DDT derivatives, namely, P,P-DDE and O,P-DDE, were found. Few DDT residues were detected at the 0–15 and 30–60 cm depths in zone I.

Chlorinated hydrocarbons were detected at the 0–15 and 15–30 cm depths in the soil cultivated with tomatoes under plastic house conditions in zone I. Few heptachlor and chlordene residues from cyclodienes and HCH were detected, while a high amount of DDT and a low amount of DDT derivatives, e.g. P,P-DDE, were found at the above depths. Soil under the above conditions showed a high amount of DDT residues at the 15–30 cm depth in zone II, while few DDT and DDE residues were detected at the 0–15 and 30–60 cm depths. Few HCB and HCH from lindane, aldrin and chlordene from cyclodienes, and DDT were detected at the 0–15 cm depth in the soil cultivated with tomatoes in open fields in zone III.

In the Central Jordan Valley, DDT was applied for many years to control mosquitoes. This DDT may have converted to DDT derivatives, particularly DDE, at the 0–15 cm depth. High amounts of DDT were used in the Central Jordan Valley; on average, 1 t/a. The total amount of DDT used might reach 160 t [11]. Several soil samples at the three depths contained DDT, lindane and cyclodiene compounds in open fields and under plastic house conditions, indicating that the soil of the Central Jordan Valley is still relatively contaminated with these persistent compounds, despite the high temperatures in the summer months, crop irrigation and compound evaporation. These results are not unique, since scientists in different parts of the UK [1] and the USA have also reported such data.

ACKNOWLEDGEMENTS

This work was organized and supported by the Environment Unit at the Center of Strategic Studies of the University of Jordan. The authors thank the director of the center, M. Hamareneh, for his support and encouragement.

REFERENCES

[1] EDWARDS, C.A., Problem of insecticide residues in agricultural soils, Pans **16** (1970) 271–274.
[2] DECKER, G.C., BRUCE, W.N., BIGGER, J.H., The accumulation and dissipation of residues resulting from the use of aldrin in soils, J. Econ. Entomol. **58** (1965) 266–271.
[3] FAHEY, J.E., BUTCHER, J.W., MURPHY, R.T., Chlorinated hydrocarbon insecticide residues in soil of urban areas, Battle Creek, Michigan, J. Econ. Entomol. **58** (1965) 1026–1027.
[4] HARRIS, C.R., SAN, W.W., MILES, J.R.W., Exploratory studies on occurrence of organochlorine insecticide residues in agricultural soils in Southwestern Ontario, J. Agric. Food Chem. **14** (1966) 398–406.
[5] ISHAQ, D., OMAR, J., Study Project of Insecticides and Herbicides in Vegetables and Fruits in Jordan, Final Report, Royal Scientific Society of Jordan, Amman (1985) (in Arabic).

[6] Al-SHURAIQI, Y.T., Evaluation of the residue situation of the most frequently used pesticides on and in the economically important fruits and vegetables in Jordan, PhD Thesis, University of Bonn, Bonn (1987).

[7] MUSTAFA, T.M., Al-RIFAE, J., Al-SHURAIQI, Y.T., Residues of pyrazophos in cucumber in the Central Highlands of Jordan, Dirasat **21B** (1994) 7–14.

[8] MUSTAFA, T.M., Al-SURAKHY, R.A., Al-SHURAIQI, Y.T., Removal of dimethoate residues from cucumber fruits, Dirasat **21B** (1994) 137–149.

[9] YAQUB, I.H.F., Persistence of formothion and fenarimol pesticide in leaves and fruits of grapes, MSc Thesis, University of Jordan, Amman (1995).

[10] Al-SURAKHY, R.A., Persistence of dimethoate on cucumber under plastic houses in the Central Highlands of Jordan, MSc Thesis, University of Jordan, Amman (1992).

[11] ROYAL SCIENTIFIC SOCIETY OF JORDAN, Pesticide Residues in Water, Soil, Sediments and Fish in the Jordanian Environment, Final Report, RSSJ, Amman (1995) (in Arabic).

[12] BECKER, V.G., Eine Multimethode zur gleichzeitigen Erfassung von 75 Pflanzenbehandlungsmitteln auf pflanzlichem Material, Dtsch. Lebensm.-Rundsch. **75** 5 (1979) 148–151.

[13] EDWARDS, C.A., Insecticide residues in soils, Res. Rev. **13** (1966) 83–132.

USE OF PESTICIDES IN KAZAKSTAN

L. PAK
Control and Toxicological Laboratory,
Republic Station for Plant Protection,
Almaty, Kazakstan

Abstract

USE OF PESTICIDES IN KAZAKSTAN.
   The Republic of Kazakstan has 210 regional stations, 19 provincial stations and one co-ordinating government research and production station, all known as Republic Stations for Plant Protection. These stations have the following functions: they control pesticide use and development, and monitor pest populations; they inform and teach farmers the methods for controlling insect pests, diseases and weeds, and the rules for safe pesticide use; and they continually monitor the presence or spread of any adverse effects in order to make short and long term prognoses, which are then used to determine the pesticide requirements. The major functions of these government stations is to control the timely distribution of pesticides and their safe and timely application, so that regulations for their use can be monitored and control over technology maintained. The co-ordinating station evaluates all projects, provides documentation on the use and storage of pesticides, controls field trials carried out by foreign companies on the Republic's territory and regulates pesticide usage during campaigns against insect pests, diseases and weeds. The regional and district subdivisions of these government stations promote integrated pest management for protecting harvests.

The Republic Station for Plant Protection in Almaty is in charge of and co-ordinates the activities of all government departments.

From 1992 onwards, the Republic of Kazakstan bought pesticides from developed countries such as France, Germany, Japan, Netherlands, Switzerland, the United Kingdom and the United States of America. Unfortunately, some firms only offered older pesticides because of the difficult economic situation in the country.

These pesticides have to be registered in a list of the pesticides that are permitted in the Republic. The State Commission on the Registration and Testing of Pesticides was specially created in 1992 for the purpose of protecting plants at the request of specialists at the Republic Station for Plant Protection in Almaty. Importation of pesticides is controlled by the custom services, the Ministry of Environment and Biology and the Ministry of Agriculture.

Previously, the pesticides used in Kazakstan covered a land area of about 18–20 million hectares. However, over the past few years, this was reduced to 10 million hectares because of the inaccessibility of pesticides to the farmers. Up to

1993, the government financed all the campaigns against serious insects pests such as the locust and the Colorado beetle, as well as an array of exotic pests. Now, farmers have to finance the purchase of pesticides themselves and also have to economize on their use because of the higher prices. These are the reasons why pesticide loading applications on 1 ha of land decreased 2.5 times between 1987 (0.57 kg active ingredient/ha) and 1994 (0.23 kg active ingredient/ha). In 1995, average application was $0.21 \text{ kg} \cdot \text{ha}^{-1} \cdot \text{a}^{-1}$. From these facts it can be seen that the adverse effects on the health of the population and the environment have decreased.

Thirty control and toxicological laboratories of these stations carry out toxicological control over the use of pesticides. Quality control of the pesticides received as well as control of the remaining pesticides in plant products are performed by specialists, i.e. these laboratories act as the control department for the Ministry of Agriculture in that they oversee all pesticide use in the country.

The Control and Toxicological Laboratory of the Republic Station for Plant Protection in Almaty carries out and adapts all methods for assessing active ingredients, i.e. it determines the standards for active ingredients and sends these to the regional laboratories. It controls the quality of the pesticides received, their storage, and the remaining quantities in agricultural products, and makes arbitrary analyses. It also co-ordinates all the work carried out with other departments on the regulations for pesticide use and studies the behaviour of pesticides in the environment and on agricultural products.

Since 1988, toxicological certificates on agricultural products have been issued by control and toxicological laboratories; annually, about 12 000–13 000 certificates on plant and food products are distributed. To control the above, these laboratories only have equipment that was adapted to the methods used by the former USSR, which were based on gas–liquid chromatography. Thus, control of pesticide use globally is not possible.

Despite this fact, the laboratories have controlled 2.5×10^3 t of received and stored pesticides. During control checks it was found that 79 t of pesticides had been spoiled because of the very bad conditions under which they had been stored. Also, 113×10^3 t of agricultural products were controlled.

FATE AND EFFECTS OF PESTICIDES UNDER TROPICAL FIELD CONDITIONS
Implications for and research needs in a developing country*

J. ESPINOSA-GONZALEZ
Laboratorio de Agroecotoxicología
 y Plaguicidas,
Instituto de Investigación Agropecuaria
 de Panamá (IDIAP),
Panama City, Panama

Abstract

FATE AND EFFECTS OF PESTICIDES UNDER TROPICAL FIELD CONDITIONS: IMPLICATIONS FOR AND RESEARCH NEEDS IN A DEVELOPING COUNTRY.

Public concern over the impact of pesticides on public health and on the quality of the environment is of importance in Central America and Panama. The main problems of pesticide use are the direct effects on health, mainly of applicators, and the side effects on non-target organisms and the environment. Important among the data that are needed to reduce the environmental risks potentially associated with pesticide use is a clear understanding of the environmental processes. Much of this information has been generated in the ecosystems of developed countries. Therefore, prediction of the environmental fate of a pesticide can be made using mathematical models and the physical and chemical properties of a given chemical; for example, volatilization and pesticide losses can be predicted in terms of the vapour pressure of the chemical and the nature of the surface. However, the agricultural system and the method of application are also important. These are complex parameters, which are sometimes specific to a region. The knowledge gained in laboratory and outdoor experiments carried out in temperate zones on the behaviour and effects of a pesticide cannot be transferred with certainty to a complex, dynamic, natural tropical ecosystem. Regional conditions may therefore be covered by selecting key pesticides, soils and research techniques, including simple methods and analytical techniques. Environmental monitoring plays an essential role in the evaluation and management of agrochemicals, and leaching has been recognized as a problem in certain cases. Between 1990 and 1995, the behaviour of various herbicides and insecticides was analysed under tropical field conditions. Taking into account the local practices and the environmental conditions, an evaluation is being made of widely used key pesticides using microecosystems, lysimeters, and radioisotopes and related techniques. The studies have yielded additional, useful data on the dissipation of pesticides (propanil, atrazine, maneb and carbofuran) in agricultural soils under tropical field conditions.

---

* Research carried out in co-operation with the IAEA under Technical Co-operation Project No. PAN/5/008.

1. INTRODUCTION

Use of pesticides in agriculture has improved the availability of food for the people of Panama. Nevertheless, increasing concern is being shown for possible and excessive pesticide residues in food and environmental components. Furthermore, data are lacking on the degradation and mobility of these agricultural inputs in local ecosystems. In theory, dissipation and degradation of pesticides in the tropics should occur at higher rates than in temperate zones, and therefore persistence of a given pesticide in the tropics should be of less magnitude and significance. On the other hand, farmers are using pesticides without adequate knowledge about the significance of residue levels in agricultural products and the new international market rules.

This paper focuses on current problems of pesticide use and the lack of data generated in situ in the tropics. The results of field studies and the research needs are presented, with special significance given to regulatory decisions on environmental and food safety.

2. LOCAL PRACTICES IN THE MANAGEMENT OF PESTICIDES AND THEIR POTENTIAL RISKS

As a tropical country with a hot and humid climate, Panama presents a favourable environment for the development of pests, plant pathogens and weeds. As fields can be cultivated at any time of the year, pests and plant pathogens are continuously present. Many pests cause serious damage to and loss of entire crops.

For these reasons, agriculture, as a primary social and economic activity, requires agrochemicals. Of those registered in 1995 for agriculture, 85 active ingredients in approximately 250 commercial formulations were available [1]. Six thousand tonnes of formulated pesticides were introduced into the country in 1990, of which 2.5×10^3 tonnes were active ingredients (Table I).

Crops that account for the greatest consumption and use of agrochemicals are the basic grains, sugar cane, tropical fruits and horticultural products. Crop activities occur in approximately 200×10^3 hectares in the highlands of Chiriquí and Coclé and in the lowlands of Barú, Changuinola, the central provices and Panama City.

Use and control of pesticides have been difficult because of geographical distribution and management practices, both of which are poor, although a notable improvement has been made in legislation on the use of agrochemicals and the support facilities for regulating their use and management. In the agricultural sector, reduction, minimization and optimization of chemical control have been initiated through continuous training on safety, biological monitoring and the introduction of integrated crop management systems. In addition, there are strategic plans for the transfer of technological information to producers in order to increase the quality and

TABLE I. PESTICIDES IMPORTED INTO PANAMA (1990)

Pesticides	Quantity (kg active ingredient)	(%)
Herbicides		
Molinate	30 620	1.5
Benthiocarb	11 050	0.6
Diuron	35 480	1.7
Paraquat	120 294	5.8
Propanil	430 153	20.8
Pendimethaline	10 731	0.5
Glyphosate	24 809	1.2
Picloram	10 143	0.5
2,4-D	575 575	27.8
Others	818 466	39.6
Subtotal herbicides	2 067 321 (56.7%)	
Insecticides		
Terbufos	17 628	9.9
Carbofuran	5 541	3.1
Mirex	27	0.1
Etoprophos	7 741	4.3
Methamidophos	1 495	0.8
Fenitrothion	124 020	69.7
Others	21 589	12.1
Subtotal insecticides	1 294 041 (35.5%)	
Fungicides		
Chlorothalonil	37 768	13.4
Tridemorph	29 432	10.4
Edifenphos	15 815	5.6
EBDCs	71 376	25.2
Metalaxyl	16 788	5.9
Others	111 788	39.5
Subtotal fungicides	282 736 (7.8%)	
Grand total active ingredients	3 644 329 (100%)	
Total formulated material	6 077 911	

efficiency of production and crop management. Farmers that produce for the export market generally utilize pesticides according to the instructions and/or the recommendations of the food seller regarding the time, type and dosage of the agrochemicals to be applied.

Tropical conditions[1] are characterized by heavy rains, which cause soil erosion in fields; the pesticides adsorbed on soil particles finally end up in coastal ecosystems, in lowlands near the coast, and in the Parita and Barú areas, where intensive agriculture has been developed for many years. Fishermen in these areas have indicated a significant decrease in the yield of marine fauna as a result of the possible excessive use of pesticides [2].

The potential risk of using agrochemicals varies significantly between chemicals. Thus, there are some compounds that are highly toxic and others that are corrosive and flammable. It is evident that the risk is greatest during the handling of concentrates by field personnel. After being applied in the field, pesticides can be absorbed by the soil particles (mainly by organic matter and clay minerals).

In addition, a portion of the applied chemical could be volatilized into the atmosphere immediately after spraying, or be leached by rains or irrigation water. At sites where chemicals are applied frequently and intensively, the existing problem becomes even greater. Waste water, resulting from cleaning the application equipment, also presents a risk to the environment. In addition, abandoned chemicals and containers require special attention because of the vast amount that lie around the fields. One estimation indicates that at least 500×10^3 containers/a (bottles and hard/soft plastic bags) containing pesticides constitute the waste on agricultural fields [4].

Pertaining to the disposal of such waste, although recommendations for its disposal appear on the labels, no generally acceptable practice exists. Some of the waste containers are burned and buried, or taken to municipal disposal sites; the rest are used for the transport and storage of fuels. Of special concern are the plastic insecticide bags (prepared with chlorpyrifos) used for protecting the banana fruit. Earlier, these bags were taken at the time of harvest and transported to isolated dumping sites, where heavy rains leached out the toxic material, carrying toxins to coastal areas, where they caused serious harm to aquatic fauna. Later, this waste was burned and buried, but more recently these waste bags have been considered for recycling.

The general problems of use and management of pesticides in Panama are incorrect selection or unrecommended use for treating crops; poor labelling; lack of

---

[1] The Panamanian tropical environment is characterized by global radiation: $7.8-27.7$ MJ/m^2; maximum temperature: $31.9-37\,°C$; mean temperature: $27.4\,°C$; humidity: 30–100% (depending on the site and season); annual precipitation: 1012–8274.4 mm (depending on the site); rainfall: 63–300 d/a (depending on the site); daily rainfall: 2.8–22.7 mm; evaporation: 10 mm/d (October) and 11.2 mm/d (March); and less precipitation during the months of February, March, July and August [3].

safety measures and personnel protection; lack of a generally accepted good practice in their use; inappropriate storage (at home and in inappropiate containers); lack of personnel training on the safety and handling of chemicals; poor supervision, control and monitoring of those inputs authorized for crops [5].

3. PROBLEMS PERTAINING TO HUMANS AND THE ENVIRONMENT

Use of pesticides is of increasing public concern. This can be seen in reports on the effects on health of people working with these agrochemicals, the claims made by hundred of workers and a doubling in the illness rate of agricultural workers (2.4%) [6].

During 1990–1992 in a western agricultural zone, acute poisoning was identified as a working hazard in 626 males between the ages of 10 and 40 (53%). The pesticides concerned were paraquat (31.8%) and cholinesterase inhibiting insecticides (43.3%) (organophosphates and carbamates) [7, 8]. During this time, the rate of poisoning varied between 0.5 and 31.1 per 100 000 inhabitants. A total of 58 persons were killed, demonstrating a significant risk for the people who handle the pesticides in Panama, even though this number is 20 times less than that of the persons killed in automobile accidents. In addition, dermal and allergic problems occur. Chronic as well as other effects also exist, but these are not as significant.

The potential risk is high as a result of weak legislation and technical controls, and the lack of effective safety measures and training. The people most affected are the workers of low educational level who handle the pesticides. The most exposed population is a group (thousands), primarily Indians, that works in the banana, horticulture and coffee fields in the highlands, although those working with export crops and vegetables in the lowlands of the central or eastern parts of the country are also potentially exposed to pesticides. The impact of pesticides on the total population potentially exposed has been estimated to be in the range of 210×10^3 persons [9].

High levels (12 mg/kg) of DDT residues in human fat were found in 1985; these are significant even in the range reported for Central America. The exposure level of pesticides to potato workers and the chlorpyrifos residues in the plasma of 'Dursbag' operators are indicative of the effects of these chemicals on human health.

The community is concerned about hazardous materials, including pesticides, in the environment. This could have been influenced by the number of chemical spills that have taken place near the entrance to the Panama Canal; however, the number of pesticide spills is very low. Two important events have taken place over the past 5 years. In 1992, a major spill occurred when 2.5 tonnes of fungicide (chlorothalonil) were released into river water and dispersed into a coastal area of about 5 km, causing temporarily heavy damage to marine fauna [10, 11]. Another accident occurred when a ship loaded with methylbromide and other goods sank near the Caribbean coast. The effects were not measured, but the fumigant may have dissipated into the atmosphere [12].

The environmental effects of agrochemicals have not been studied in detail; from the available data, lethal effects on birds were found during the 1960s when grain seeds were treated with endrin. During the early 1980s, aquatic fauna were killed by pyrethroid insecticides, and soil insecticides as granules have killed birds in rice fields. Small fishermen from the Pacific coast have noticed a reduction in the fish catch and attribute this to pesticide use in agriculture [13]. Pesticide residues are spilled, drained or run off by rains from agricultural zones. Furthermore, aerial applications of phenoxic herbicides on rice fields and grasslands have frequently killed the flora on field sites. The decomposition of insecticides during the early 1980s killed a significant amount of cattle.

The impact of residues detected in livestock and agricultural products has shown that the residue levels are below those recommended by the Food and Agriculture Organization of the United Nations/World Health Organization in their Codex Alimentarius. On the other hand, no refuse is based on pesticide residues in food exports. Small amounts of refuse mainly result from decomposition and potential contamination with insects and pathogens.

The low levels of organochlorine residues reported in beef 10 years ago can no longer be detected [14]. This demonstrates that there has been an improvement in the production of meat as a result of organochlorines being replaced by other insecticides such as organophosphates and pyrethroids. In vegetables, organochlorine residues were very low (below 50 ppb) when these insecticides were used. At present, 80% of the vegetables sampled are free of pesticide residues [15]. The levels reported in some samples are below the recommended safety limits, and only in a few cases were there significant amounts of residue. The government has adopted measures for improving pesticide use, has increased control, and has improved the advice given to farmers in an effort to reduce the potential pesticide residues on agricultural products.

The effects of pesticides on water have been identified in organochlorine residues in sediment; the beta-BHC (2-4 ng/g), endrin (10-160 ng/g) and lindane (3-9 ng/g) found in sediments in Panama Bay in 1987 were not detected during 1995. Sediment samples from a river at the Atlantic coast, an area where bananas are cultivated, had shown up to 35 ng/g of organochlorine residues, and water samples from the central part of the country, Azuero, were found to contain atrazine residues (10 ng/L, and above) [16, 17]. In a feasibility study on the risk to rivers of the Atlantic zone, where bananas are grown, only 10% of the 125 samples collected were found to be positive for pesticide residues [18].

The impact of pesticides seems to be less on soil systems than on aquatic systems, as they are adsorbed and degraded in soil to a greater extent than other compounds. However, the soil microflora of a potato growing area was affected significantly in terms of its nitrification capacity and enzymatic content [19]. Ten years ago, a soil insecticide lost its effectiveness as a result of fast degradation by soil bacteria. The population of Collembolae in the highlands seems to have been significantly affected by the crop protection measures taken [20].

Studies of pesticide residues on soil after intensive and long term applications of a Burdeaux mixture, lead arseniate and fertilizers revealed that soils in a banana area still have, at some sites, levels of 1000 mg/g of copper, 11 mg/g of lead, 973 mg/g of manganese, 145 mg/g of zinc and 0.2 mg/g of cadmium [21].

Normal application of pesticides at the recommended rates (1–2 kg active ingredient/ha) deposits concentrations of 1–3 μg/g on to the soil; in a number of cases, degradation (microbial or photolytic) rapidly reduces these residue levels. However, in the case of paraquat, only about 10% per year dissipates. Not very much information exists on what happens to the bound residues or the degradation products in soil, and their potential for contamination of the water sheds, or on bound residues and their bioavailability in food.

On fields where pesticide mixtures are prepared and applied, part of the waste water can damage the river fauna, while runoff due to heavy rainfall can potentially kill the river fauna. This is a critical factor, especially in areas where erosion is common, and soil loss can be up to about 300 $t \cdot ha^{-1} \cdot a^{-1}$. In this case, the pesticide residues adsorbed on to the soil are carried to the coast and thereby into the ocean [22].

A report of phenoxic herbicide residues on coral reefs at levels of 20 ng/g, and the presence of organochlorines on these organisms, can be correlated with the transport of residues by runoff [23]. Very little is known of the effects when foliage treated with a herbicide is burned, which is common practice in Panama.

4. DISSIPATION OF SOME PESTICIDES FROM AGRICULTURAL SOILS

The results of studies on the fate of agrochemicals in the environment allow us to make sustained judgements on their potential hazards and risks. The behaviour of a given pesticide depends on various factors, molecular structure being one of the most important; however, the dissipation rate and route can change, depending on the environment. It is generally believed that dissipation and degradation occur at higher rates under tropical conditions than in temperate zones. In the tropics, the temperature (25–40°C), humidity (70–100%), rainfall (2000–8000 mm/a) and solar radiation (6–12 h/d) are high, and consequently diverse microbial activity is also high, for example, variation in the dissipation rate of DDT. The half-life in soil of this insecticide can vary by a factor of 6 and occurs faster in tropical than in temperate zones [24].

Since relatively high amounts of some pesticides are used in Panama for the production of crops such as basic grains and horticulture, and because much of the applied chemicals falls on to the soil, studies were carried out to learn about the dissipation of propanil and maneb from the soils in rice and horticultural zones. Dissipation of atrazine and carbofuran from maize soils was also studied. The evaluations were conducted in accordance with FAO/IAEA Protocols and Guidelines for dissipation studies of pesticides under field conditions [25]. Lysimeters with soil cores of 1 kg were used in these studies. The soil was sieved to a particulate of 2 mm and conditioned in the field for at least 2 weeks before treatment, as described previously [26].

The soil properties are given in Table II. The volcanic soil from Cerro Punta drains well, shows rapid permeability, and has a granular fine, strong structure, with many tubular and very fine pores, whereas the marine sedimentary soil from El Coco drains poorly, and has slow permeability and a subangular block structure. The alluvial sedimentary soil from El Ejido drains well, has moderate permeability and a subangular to granular block structure, and has tubular, very fine to fine pores [27]. Organic carbon is relatively high in the soil from Cerro Punta, but the pH differs slightly between soils. The clay content in Cerro Punta soil was about 1/3, as in soils from El Coco and El Ejido.

The lysimeters were treated at that time of year and at that dose typically used by local farmers. The applied pesticides were ring labelled with the radioisotope ^{14}C; the fungicide maneb was labelled on the ethylene group. The applied specific activity was in the range of 7500 $dis \cdot min^{-1} \cdot g^{-1}$ of soil. Sampling of lysimeters in triplicate was carried out at 0, 8, 15, 30, 60, 90, 180, 300, 360 and/or 480 days after treatment. The soil cores were sectioned into an upper and a lower layer, and the weight was recorded and homogenized by good mechanical mixing; humidity was

TABLE II. SOIL PROPERTIES

Properties	Location		
	Cerro Punta	El Coco	El Ejido
Sand	72%	68%	55%
Lime	22%	16%	28%
Clay	6%	16%	17%
Texture	Sandy loam	Sandy loam	Clay
pH	6.1	5.5	5.7
Organic carbon	7.1%	1.5%	1.7%
Nutrients (P, K, Ca, Mg)	High	Medium	Medium

TABLE III. DISSIPATION OF ^{14}C FROM SOILS TREATED WITH ^{14}C-PROPANIL (0–15 cm)[a]

Days after treatment	Remaining ^{14}C			
	Cerro Punta		El Coco	
	(dis/min)	(%)	(dis/min)	%
0	3 537 875	95.2	3 649 780	98.2
15	1 530 672	41.2	3 326 146	88.7
30	1 579 408	42.5	2 143 207	57.1
60	1 423 326	38.8	1 877 365	50.0
90	1 378 730	37.1	2 010 429	53.6
132	1 326 632	35.7	1 828 306	49.2
180	1 412 987	38.0	1 687 095	45.4
250	1 183 440	31.9	1 571 615	41.9
300	1 031 350	27.5	1 400 958	37.7
360	774 152	20.9	1 243 412	33.4

[a] The initial concentration was 4.1 µg of propanil per gram of soil.

determined following the standard procedure (110°C/24 h); the total ^{14}C residues were analysed by combustion in an OX-600 Harvey biological oxidizer and the ^{14}C activity was measured in a Packard Tricarb 1000 liquid scintillation counter according to standard operational procedure.

4.1. Propanil

The dissipation rate of this important rice herbicide was high because of the volcanic soil of Cerro Punta; the half-life was less than 15 days, whereas that of propanil from El Coco soil varied between 60 and 132 days after treatment. Initial high dissipation was followed by a slow dissipation rate for up to 1 year after the original treatment; at the end of the experiment, the remaining ^{14}C residues were 20.9% for the soil from Cerro Punta and 33.4% for the soil from El Coco (Table III). Medium to slow mobility through the soil layers was observed; it was slightly lower in Cerro Punta soil (12.6%) than in El Coco soil (15.5%).

TABLE IV. DISSIPATION OF ^{14}C FROM SOILS TREATED WITH ^{14}C-MANEB (0–10 cm)[a]

Days after treatment	Remaining ^{14}C			
	Cerro Punta		El Coco	
	(dis/min)	(%)	(dis/min)	%
0	6 422 250	100.0	6 422 250	100.0
1	5 163 489	80.4	6 261 694	97.5
30	2 466 144	38.4	3 602 882	56.1
60	1 888 142	29.4	3 089 102	48.1
91	1 123 894	17.5	2 774 412	43.2
180	1 348 673	21.0	2 819 368	43.9
306	712 870	11.1	1 644 096	25.6
380	667 914	10.4	873 426	13.6
480	680 759	10.6	719 292	11.2

[a] The initial concentration was 0.642 μg of maneb per gram of soil.

4.2. Maneb

The results of the experiments with ^{14}C-maneb revealed high dissipation from both soils; however, the dissipation rate from the volcanic soil of Cerro Punta was higher than that of El Coco soil. The half-life in the volcanic soil was in the range of 30 days, whereas it was up to 60 days in the soil from El Coco. Dissipation of this fungicide occurs mainly during the rainy season; 1 year after original application to the soil, the remaining ^{14}C-maneb residues accounted for 11–13% of the initially applied amount. These residues are non-extractable, with hot organic solvents (bound residues) (Table IV). No downward movement into the soil layers was observed, therefore the potential for contamination of groundwater was low.

4.3. Atrazine

This important herbicide, used in the production of maize, dissipated slowly. The half-life in the clay soil was of the order of 50–60 days; 1 year after the initial application of atrazine, the remaining ^{14}C residues in the soil accounted for 21.8% of the original application (Table V). This herbicide is mobile to the underground layers of the soil and leaching can represent 21.6% of the original ^{14}C activity.

TABLE V. DISSIPATION OF ^{14}C FROM SOIL TREATED WITH
^{14}C-ATRAZINE (0–15 cm) (EL EJIDO/LOS SANTOS)[a]

Days after treatment	Remaining ^{14}C	
	(dis/min)	(%)
0	40 911 823	95.7
15	2 822 413	66.1
30	2 340 100	54.8
60	2 006 417	46.9
90	1 715 144	40.2
180	15 090 775	35.3
360	89 187 774	21.8

[a] The initial concentration was 1.41 µg of atrazine per gram of soil.

TABLE VI. DISSIPATION OF ^{14}C FROM SOIL TREATED WITH
^{14}C-CARBOFURAN (0–10 cm) (EL EJIDO/LOS SANTOS)[a]

Days after treatment	Remaining ^{14}C	
	(dis/min)	(%)
0	32 043 010	94.3
8	15 000 921	44.1
15	12 687 578	36.1
30	15 687 578	45.9
60	13 132 361	38.6
180	8 525 213	25.1

[a] The initial concentration was 1.96 µg of carbofuran per gram of soil.

4.4. Carbofuran

This insecticide is very mobile in soil, especially during heavy rains, when it is translocated to plants and leached into the soil layers. The dissipation rate is very high, and the half-life is in the range of 8–30 days. Six months after treatment at a dosage of 2.5 kg active ingredient/ha, the remaining ^{14}C residues accounted for 25.1% of the original amount added; these residues are not extractable with conventional extraction techniques (Table VI). The highest leaching rate in the 15 cm soil column occurred during the first 30 days after treatment, and leaching could have accounted for up to 50% of the original ^{14}C activity.

As all the experiments were conducted under the same climatic conditions, the great differences in the rates of dissipation during the initial phase have to be related to the soil properties. It seems highly likely that the structure, the microbiological complex and the organic matter content of the soils, as well as the environmental conditions, play an important role in the dissipation and degradation of these agrochemicals. Soil with a high organic matter content and a low amount of clay seems to greatly facilitate dissipation of the chemicals from the soil, whereas a high clay content extends the rate of dissipation. The soil has a profound influence on the microbial activity and consequently, in combination with other effects, contributes significantly to pesticide dissipation. In addition, leaching causes significant dissipation in the tropics. High and frequent precipitation has an important effect on the mobilization of pesticides from their original application site, which increases the potential of contamination of groundwater and surface water by runoff from heavy rains. The high rate of dissipation under field conditions, observed shortly after application, is due to volatilization into the atmosphere, water co-distillation and evapotranspiration from the soil; the change in temperature, humidity and atmospheric pressure at different field sites also seems to play a role.

5. RESEARCH NEEDS

On the basis of current problems and trends in the use, management and consumption of pesticides in Panama, there is a need to develop strategies for research and studies. The following research can be recommended for the successful development of pesticide studies in a developing country:

(1) Research should be strengthened and undertaken to optimize the safe and effective use of those pesticides that are employed in high amounts and those that are of potential risk to humans and the environment.
(2) The search must continue to find appropriate crop protection, where pesticides could be used as an important tool. This should include efforts to replace the

major hazardous chemicals that are currently available, especially from integrated pest management programmes and integrated crop production systems.

(3) Research should be undertaken to lower the pesticide hazard to humans and the environmental compartments.

(4) Research should be undertaken to evaluate the residues of important pesticides in local agroecosystem components, and their periphery. Special environmental attention should be given to residues in water and in coastal ecosystems.

(5) The search must continue to evaluate the bound residues of pesticides in food and soils, and their bioavailability.

(6) Since human health and the environment are of top priority, studies should be conducted to evaluate the additive or synergistic effects of multiple pesticides, following the local practices of farmers.

(7) Research should be expanded to include studies on the long term effects and the ecotoxicological impact of pesticides, especially on non-target organisms.

(8) Research should be undertaken to develop monitoring programmes for relevant pesticides in soil and food, especially taking into consideration the local practices of farmers and the prevailing climatic conditions.

(9) Research should be undertaken to develop or validate analytical techniques for pesticide residue analysis, especially those techniques that are simple, rapid and useful, in which reagent consumption and equipment maintenance are low; however, the effectiveness of such techniques will confirm their suitability for use in the monitoring and surveillance of pesticide residue programmes in food and water.

(10) A programme should be considered to minimize or safely dispose of pesticide wastes at the farmer's level and a programme should be undertaken to eliminate or safely dispose of all unused remaining pesticides from developing countries.

ACKNOWLEDGEMENTS

The author sincerely wishes to acknowledge the support of the IAEA, and is also grateful to V. García, J. Ceballos, F. Díaz and O. Pérez for their technical assistance.

REFERENCES

[1] MINISTERIO DE SALUD, Informe General de Registros, Categoría Agrícola, Año 1996, Panamá (1996).
[2] ORTEGA-LUNA, A., Misteriosa muerte de peces en ríos Mamoní y Bayano, El Panamá América de 12 de febrero de 1995, 2 pp.

[3] INSTITUTO NACIONAL DE RECURSOS RENOVABLES, Boletín Agrometeorológico 1990, Panamá (1991) 52.

[4] INSTITUTO DE RECURSOS NATURALES RENOVABLES/AGENCY FOR INTERNATIONAL DEVELOPMENT, Materiales y Desechos Peligrosos: Situación de Uso, Producción, Manejo, Almacenamiento, Transporte, Disposición Final y Potencial Efecto sobre la Salud y el Ambiente, Proyecto Marena, Informe de Consultoría, INRENARE, Panamá (1996) 141 pp.

[5] GARCES, H.A., Mercado de plaguicidas en la República de Panamá y constatación de formulaciones de pesticidas de uso en arroz, Institut für Tropentechnologie, Cologne **11** (1994) 34.

[6] CAJA DE SEGURO SOCIAL, Programa de Salud Ocupacional: Aspectos Conceptuales y Estadísticos de la Salud Ocupacional, Panamá (1994) 10 pp.

[7] MORA, E.L., Intoxicaciones Agudas por Plaguicidas en la Provincia de Chiriquí, Período 1990–1994, División de Epidemiología, Ministerio de Salud, Bol. Epidemiológ. **XIX** (1995).

[8] MINISTERIO DE SALUD, División Técnica de Epidemiología, Casos y Tasas de Intoxicación de Plaguicidas por Región de Salud, Año 1990–1994 (Feb. 1995) 2 pp.

[9] JENKINS-MOLIERI, J., "Plaguicidas: Salud y Desarrollo Sostenible en Centro América", Los Desafíos de la Salud Ambiental, Cuadernos de la Representación OPS/OMS en Panamá, Vol. 3, Organización Panamericana de Salud, Panamá (1995) 77.

[10] La contaminación en el río Chiriquí Viejo, La Prensa de 14 de julio de 1992, p. 7A.

[11] SANTIAGO, E., Bananera aclara caso de contaminación en río, El Panamá América de 19 de julio de 1992, p. 18A.

[12] DIAZ, J.M., Contenedores en alta mar podrían contener pesticidas, El Panamá América de 16 de junio de 1994.

[13] Fumigación aérea no sólo acaba con las plagas, El Siglo de 4 de mayo de 1992.

[14] TATIS, A., Ministerio de Desarrollo Agropecuario, Laboratorio de Residuos Tóxicos, Panama City, Personal communication, 1996.

[15] JENKINS-MOLIERI, J., Aproximación a la problemática sanitaria de la exposición a los plaguicidas en Centro América y Panamá (Actas II Cong. Nac. Sal. Publ. y Prim. Epidemiol., Panamá, 1995), Ministerio de Salud, Panamá (1995).

[16] MINISTERIO DE PLANIFICACION Y POLITICA ECONOMICA/COMISION NACIONAL DEL MEDIO AMBIENTE, Informe de Avance del Proyecto de Investigación de la Bahía de Panamá, MIPPE/CONAMA, Panamá (1986) 88.

[17] MINISTERIO DE PLANIFICACION Y POLITICA ECONOMICA/COMISION NACIONAL DEL MEDIO AMBIENTE, Informe del Estudio Preliminar sobre Niveles de Concentración de Residuos de Plaguicidas Organoclorados en Peces y Sedimentos del Río San San en la Provincia de Bocas del Toro, MIPPE/CONAMA, Panamá (1995) 22 pp.

[18] DUKE, V., Personal communication, Panama City, 1995.

[19] GARCIA-BLANDON, P.A., Aspectos Generales sobre Plaguicidas y su Efecto sobre las Personas y el Medio Ambiente, INCAP/OPS/OIT, Organización Panamerica de Salud, Panamá (1994).

[20] MÜLLER-STÖVER, D., Einfluss chemischer Pflanzenschutzmassnahmen im Kartoffelanbau Panamas auf mikrobielle Aktivitäten im Boden, Thesis, Universität Hohenheim (1994) 75 pp.
[21] SONDER, K., Instituto de Investigación Agropecuaria de Panamá, Personal communication, Panama City, 1994.
[22] MAHLBERG, A., Schwermetalle in Böden, Pflanzen, Fliessgewässern und ihren Sedimenten im Bananenanbaugebiet Barú, Panamá, Thesis, Ruprecht Karls-Universität, Heidelberg (1990) 145 pp.
[23] OSTER, R., "La Erosión y el Manejo de Suelos en las Tierras Altas de Chiriquí", Agonía de la Naturaleza (HECKADON-MORENO, S., ESPINOSA-GONZALEZ, S., Eds), Impretex, Panamá (1985) 87–102.
[24] GLYNN, P.W., HOWARD, L.S., CORCOVAN, E., FREAY, A.D., The occurrence and toxicity of herbicides in reef building corals, Mar. Pollut. Bull. **15** 10 (1984) 370–374.
[25] Special Issue on DDT in the Tropics, Pesticides, Food Contaminants, and Agricultural Wastes, J. Environ. Sci. Health, Part B **1** (1994) 231.
[26] ESPINOSA-GONZALEZ, J.V., GARCIA, J.C., Dissipation of ^{14}C-P'P' DDT in two Panamanian Soils, J. Environ. Sci. Health, Part B **1** (1994) 98.
[27] JARAMILLO, S., Pedones de Campo y Estaciones Experimentales del IDIAP Panamá, Instituto de Investigación Agroprecuaria de Panamá, Bol. Tec. **38** (1991) 28–49.

FATE AND BEHAVIOUR OF PESTICIDES IN THE TERRESTRIAL ENVIRONMENT

(Session 3)

Chairpersons

M.A. MATIN
Bangladesh

M.F. ZARANYIKA
Zimbabwe

Rapporteurs

C.S. HELLING
United States of America

P.H. NICHOLLS
United Kingdom

IAEA-SM-343/6

EXTRACTION OF PESTICIDE RESIDUES FROM BIOLOGICAL AND ENVIRONMENTAL SAMPLES*

S.U. KHAN
Centre for Land and Biological
 Resources Research,
Research Branch,
Agriculture and Agri-Food Canada,
Ottawa, Ontario, Canada

Abstract

EXTRACTION OF PESTICIDE RESIDUES FROM BIOLOGICAL AND ENVIRONMENTAL SAMPLES.

Various procedures that show potential in offering shorter extraction times with higher recoveries and low consumption of organic solvents are discussed. Solid phase extraction is attracting increasing attention for the isolation of pesticide residues from aqueous solutions and constitutes an alternative to liquid–liquid extraction. Microwave assisted extraction of stable pesticide residues from soil samples appears to be a viable alternative to conventional Soxhlet extraction. Thermal desorption involving high temperature distillation is an innovative method for efficiently extracting certain pesticides from soil, sediment and plant samples. Supercritical fluid extraction is emerging as a valuable technique for the isolation of pesticide residues from soil, plant and food samples, using supercritical fluids as the extraction media. It represents an excellent alternative to the potentially hazardous solvents currently used in conventional methods. The feasibility of employing water based systems for the extraction of certain pesticides from fruits and vegetables has also been investigated. A brief survey is given of these extraction procedures and a comparison made with the techniques widely used for the extraction of pesticide residues from biological and environmental samples.

1. INTRODUCTION

Analysis of pesticides from environmental and biological samples typically involves a rigorous extraction method, such as Soxhlet or ultrasonic extraction. The extract is then concentrated to a small volume by using a rotary evaporator. Additional clean-up steps are often required to eliminate the interfering co-extractives from the matrix during the extraction procedure. Traditional methods of extraction, because they are considered to be fully developed techniques, are used widely in many residue laboratories. Unfortunately, these methods are time consuming and

---

* Research carried out in association with the IAEA under Research Agreement No. 8076.

involve the use of large volumes of solvents that are often toxic, expensive and may pose problems in their safe disposal. Renewed awareness of the problems associated with classical Soxhlet and sonication extraction has resulted in initiatives being taken by the scientific community to develop viable alternatives which are simple, inexpensive, use less solvent, and are selective and compatible with a wide range of extraction methods. This paper provides an overview of those techniques that have been available for some time but are now on the verge of widespread acceptance in residue laboratories.

2. SOLID PHASE EXTRACTION

Solid phase extraction (SPE) is a relatively new technology that is gaining popularity, particularly for water analysis [1–4], where low concentrations of analyte can be concentrated from a large sample volume. A typical SPE consists of four major steps: (1) conditioning the sorbent beds with solvent to improve the reproducibility of the pesticide retention and to reduce the concentration of any contaminants present; (2) sorbing the pesticide on the bed, together with undesirable matrix constituents; (3) rinsing the column with weak solvent to remove undesirable matrix components; and (4) eluting the pesticide with a sufficiently strong solvent, while leaving the undesirable components on the bed. Figure 1 shows the steps

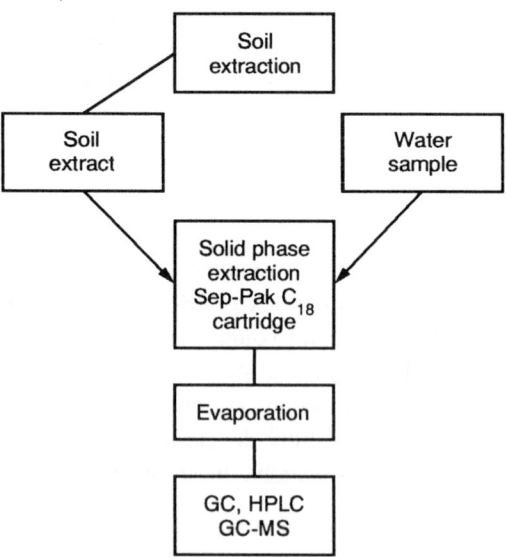

FIG. 1. Flow diagram for the SPE method.

involved in SPE and identifies the pesticide residues in aqueous samples. Useful sorbents in the extraction of pesticides include diatomaceous earth, silica gel and silica supports bonded with ethyl, octyl, octadecyl, cyclohexyl, phenyl and cyanopropyl functionalities. SPE is mostly used off-line, the sorbent being packed in disposable columns or cartridges. In recent years, SPE membrane disks have been preferred to SPE cartridges. In general, the time required to isolate the various pesticides using disks is half that using cartridges (30 min versus 60 min for 1 L of water).

TABLE I. AVERAGE PERCENTAGE PESTICIDES RECOVERED FROM RIVER WATER USING C_{18} EMPORE EXTRACTION DISKS AND DICHLOROMETHANE (LLE) [5][a]

Pesticide	C_{18} Empore disk		LLE	
	Recovery	CV	Recovery	CV
Atrazine	100	5	99	4
Chlorpyrifos	94	4	86	3
Cyanazine	84	7	92	14
Dieldrin	93	6	ND	
Linuron	96	6	94	5
Tetrachlorvinpos	96	6	92	4

[a] Spiking level = 1–5 µg/L for each pesticide; water volume = 1 L; ND = not determined.

TABLE II. MEAN RECOVERIES (%) AND STANDARD DEVIATION (%) (IN PARENTHESES) OF THE PESTICIDES IN SPIKED GROUNDWATER SAMPLES USING XAD-2 RESIN AND SEP-PAK C_{18} CARTRIDGES [6][a]

Pesticide	XAD-2		Sep-Pak C_{18}	
	0.5 µg/L	0.1 µg/L	0.5 µg/L	0.1 µg/L
Atrazine	82 (12)	85 (9)	80 (9)	78 (3)
Simazine	83 (15)	80 (9)	75 (6)	74 (6)
Monocrotophos	82 (5)	68 (7)	93 (8)	90 (7)
Ethoprophos	76 (8)	68 (9)	104 (18)	103 (5)
Isophenphos	80 (5)	69 (8)	103 (15)	98 (13)

[a] Water volume = 1 L; determinations were performed by GC-nitrogen phosphorus detection (NPD).

The Environmental Protection Agency (EPA) has approved various methods based on the use of SPE disks containing C_{18} for determining pesticides in drinking water. Table I shows the recoveries obtained from a 1 L water sample with the Empore extraction disk and from the liquid–liquid extraction (LLE) procedures [5].

Table II gives an example of the determination of some pesticides in ground and drinking water by SPE on XAD-2 columns and Sep-Pak C_{18} cartridges, subsequent elution with an organic solvent, and determination by gas chromatography (GC) and mass spectrometry (MS) in the selected ion monitoring mode (MS–SIM). The water volumes of 1–2.5 L at concentration levels of 0.1–0.5 µg/L were used to apply the method [6]. The limits of detection (LOD) for some pesticides are shown in Table III [6].

TABLE III. LIMITS OF DETECTION FOR PESTICIDES FOLLOWING EXTRACTION WITH C_{18} CARTRIDGES AND DETERMINATION BY GC-NPD OR GC-MS-SIM [6][a]

Pesticide	Limits of detection (µg/L)	
	GC-NPD	GC-MS-SIM
Atrazine	0.30	0.05
Simazine	0.60	0.06
Monocrotophos	0.10	0.08
Ethoprophos	0.20	0.03
Isophenphos	0.20	0.05

[a] Spiking level = 1–0.1 µg/1 L; water volume = 2.5 L; peak measured when the signal/noise ratio was 5.

3. MICROWAVE ASSISTED EXTRACTION

Microwave energy to enhance the extraction of pesticides from solid matrices such as soil and sediments has been attempted by using conventional household microwave ovens to irradiate solvent/sample suspension [7–9]. Environment Canada has patented the process of microwave assisted extraction (MAE) [10]. The technique has been refined through the development of focused microwave instruments specifically designed for a temperature controlled closed system. Extraction units of 12 sample capacity such as the MES-1000 microwave solvent extraction system are commercially available. The sample and solvent are placed in a closed vessel. Microwave radiation heats the solvent to a temperature higher than its boiling point, and

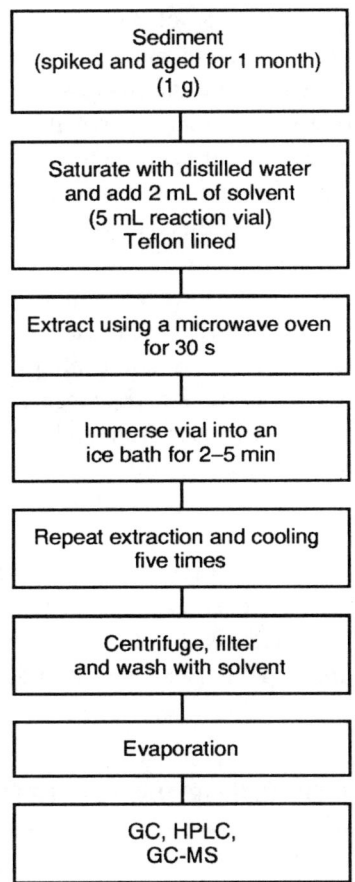

FIG. 2. *Extraction of pesticide residues from sediments using the microwave technique [7].*

TABLE IV. COMPARISON OF EXTRACTION METHODS AND MOISTURE LEVEL ON RECOVERY FROM A REFERENCE SEDIMENT SAMPLE [7]

Extraction method	State	Mean % recovery and relative standard deviation	
		Endrin	Dieldrin
Microwave	Wet	95.7 ± 3.1	95.3 ± 3.8
	Dry	16.7 ± 0.9	15.2 ± 0.6
Soxhlet	Wet	94.8 ± 4.9	90.8 ± 2.7
	Dry	96.2 ± 2.2	96.4 ± 3.1

the hot solvent provides rapid extraction of the pesticide under moderate pressure (usually about 10 kPa). Recently, Onuska and Terry [7] described the extraction of pesticide residues from sediments using the microwave technique (Fig. 2). It was noted that sediment moisture was a significant parameter for good recovery. A minimum water content is necessary to perform microwave extraction. There was no evidence of any breakdown or alteration in the pesticide. The recovery values from sediment extracted by the two methods are compared in Table IV [7]. MAE uses less solvent than conventional Soxhlet or LLE. Extraction can be optimized by varying the experimental parameters, including the heating time, stirring versus non-stirring, closed container versus open container, pulsed heating versus continuous heating, and outside cooling of the vessel versus non-cooling. Multiple samples can be extracted simultaneously. Fish and Revesz [8] recently reported that microwave solvent extraction of chlorinated pesticides from soil can yield recoveries equal to or exceeding the recoveries attained using current EPA methods, with a significant reduction in solvent usage and time. They found that approximately 50–65 mL of solvent per sample were used in this technique, and extractions of as many as 12 samples can be attained in 20 min. Table V [8] shows the result for a 3 g sample using 50 mL of 3:2 (vol./vol.) acetone/hexane heated in the microwave extraction unit at 120°C for 20 min. These authors reported that extraction could be performed with no breakdown in thermally labile pesticides such as endrin and DDT [8].

TABLE V. RECOVERIES OF CHLORINATED PESTICIDES FROM SOIL [8]

Pesticide	Microwave extraction[a] (% recovery)		EPA methods (% recovery)
	Mean	% relative standard deviation[b]	Mean[c]
Aldrin	92.4	2.6	88.4
DDE	92.1	2.0	89.3
Heptachlor	89.6	1.1	85.7
Lindane	88.0	4.6	81.4
Dieldrin	98.8	3.0	80.8
Endrin	95.7	2.0	90.4
DDT	90.0	5.2	78.6

[a] 50 mL of 3:2 (vol./vol.) acetone/hexane at 120°C for 20 min.
[b] n = 6.
[c] Values obtained from ten EPA laboratories.

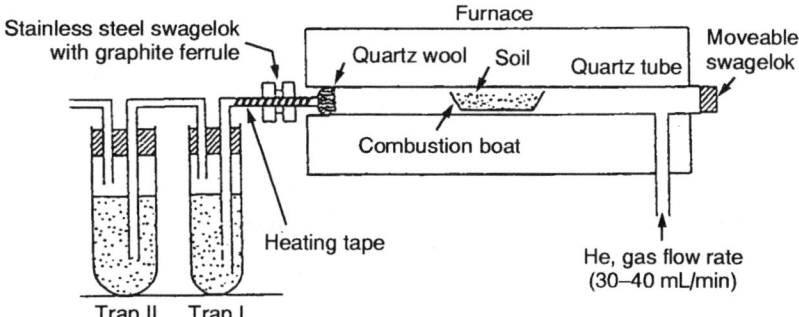

FIG. 3. HTD apparatus for thermal desorption of pesticides [11].

4. HIGH TEMPERATURE DISTILLATION

High temperature distillation (HTD) or thermal desorption may be considered as a viable alternative extraction technique for determining pesticide residues in biological and environmental samples. Effective thermal desorption will depend on the nature of the pesticide residues, the nature of the matrix, and the interaction forces between them. In this technique, the sample is heated, with the carrier gas flushing away the absorbed material. The desorbed residue is collected in a solvent and identified by the chromatographic technique.

A schematic diagram of HTD using a Lindberg tube furnace (Sola basic S/B) is shown in Fig. 3 [11]. The moist soil sample (200–300 mg air dry basis) containing pesticide residues is placed in a tared porcelain boat (0.2 cm width, 0.5 cm depth, 4 cm length), inserted into the middle of the quartz tube (30 cm length, 0.9 cm inner diameter) and the assembly is placed in the groove of the furnace. One end of the tube is closed with a swagelok, while the other end is connected to two traps containing hexane (25 mL) and methanol (25 mL), respectively. Helium is purged through the system and is used as the sweep gas at a flow rate of 30–40 mL/min. The furnace is heated from room temperature to 500°C (ca. 10–15°C/min) and maintained at this temperature for about 15 min. At the end of the experiment, the solution in the traps is processed for identification of the material using chromatographic techniques.

The HTD method has been found to be very useful for the extraction of organochlorine pesticides from field treated soils. Figure 4 shows the extraction of DDT and lindane from a field soil [12]. Most of the residues were desorbed in the temperature range of 250–300°C. Table VI [12] shows that, although the extraction processes involved may be similar to other methods, there are some differences in the efficiency of the methods. HTD appears to remove some of the DDT residues not extracted by other methods (Table VI) [12].

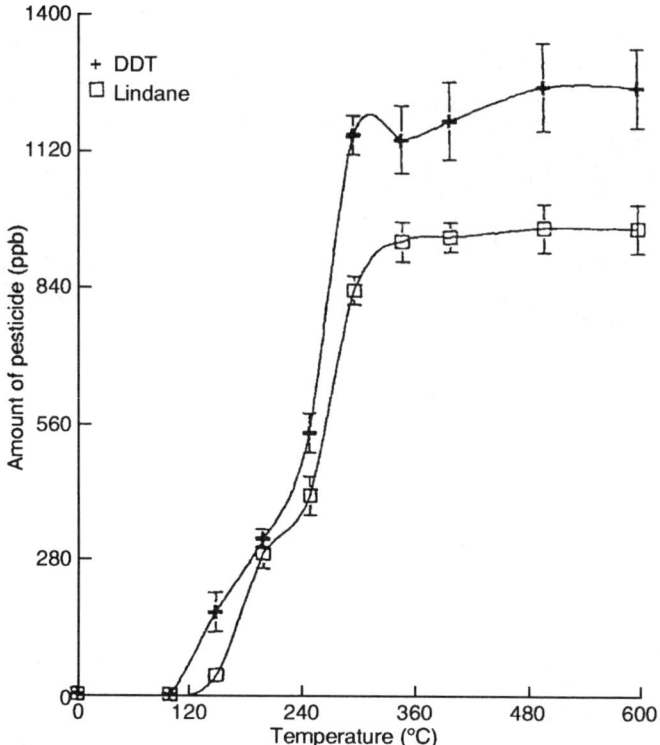

FIG. 4. *Temperature dependent extraction of DDT and lindane from a field soil [12].*

TABLE VI. RE-EXTRACTION OF DDT WITH HTD FROM A FIELD SOIL PRE-EXTRACTED WITH OTHER METHODS [12]

Method	First extraction (ppb)	Re-extraction by HTD (ppb)
SFE	980 ± 90	50 ± 10
Soxhlet	910 ± 90	160 ± 20
Sonication	950 ± 80	110 ± 10
HTD	1140 ± 70	—

Preliminary experiments must be carried out to optimize the method, since each new analyte and matrix combination will require slightly different conditions. Recovery is affected by the moisture content of the sample and the purity of the purge gas, which must be oxygen free. The solutions for trapping the analyte should be selected according to the nature of the compound of interest. HTD requires only small amounts of solvents, thereby reducing the amount of waste generated.

5. SUPERCRITICAL FLUID EXTRACTION

Supercritical fluid extraction (SFE) has recently been used by numerous investigators to extract pesticides from soil, plant, food, and other biological and environmental samples. This technique has shown great potential in offering shorter extraction times, with higher recoveries and low consumption of organic solvents. The technique is based on the enhanced solvating power of supercritical fluids above their critical points. The low viscosity and high diffusivity of supercritical fluids make the mass transfer during extraction rapid. The solvating power of a supercritical fluid can be controlled by manipulating some of the extraction conditions. The most frequently used supercritical fluid is CO_2, which is non-toxic and can be vented harmlessly into the atmosphere. Because of its extreme volatility, CO_2 can be completely separated from any solutes. It has low critical points (74 bar, 31°C), and is non-flammable and inexpensive. Modifiers such as methanol can be added to supercritical CO_2 to alter the solvating power of the fluid. The technique has recently been applied to the extraction of pesticides and/or metabolites from soil [13–15] and from food products [16–18].

Figure 5 shows a schematic diagram of the SFE unit. The solvent is compressed by the pump through the preheated capillary into the sample to be extracted.

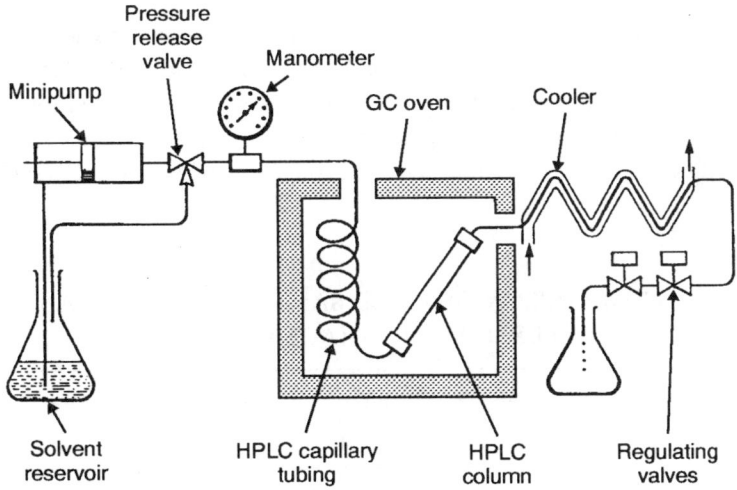

FIG. 5. Schematic diagram of the SFE unit: minipump with a flow rate adjustable to 0.3–2.7 mL/min; stainless steel HPLC capillary tubing (0.5 mm inner diameter); and two regulating valves: Sno-Trik 10 000 psi valve, and Whitey 5000 psi low volume valve (pound force per square inch (= psi): $1 \, lbf/in^2 = 6.895 \times 10^3 \, Pa$).

Preheating of the capillary and the sample is carried out in a gas chromatograph oven maintained at a certain temperature and purged continuously with nitrogen. The latter is necessary in order to avoid the formation of an explosive mixture of air–organic solvent (methanol). The extracts are passed through the cooler and then through the regulating valve, finally collecting in an Erlenmeyer flask with a flow rate of 1 mL/min. The extract is concentrated to a small volume and analysed by chromatographic techniques.

SFE instruments are available from the manufacturer with a high level of automation. These instruments could be used for SFE in various modes. Supercritical CO_2 has been the choice for most SFE studies. Modifiers such as methanol can be added to supercritical CO_2 to alter the solvating power of the fluid.

Table VII [19] shows the solvent extraction and SFE for organochlorine pesticides using an extraction time of 30 min, 20 MPa and 50°C. In Table VIII [19], concentrations are shown after addition of the modifiers acetonitrile and methanol. These modifiers give an increase in concentration, but methanol gives the best overall results. This suggests that in order to overcome the interactions between the analyte and the matrix, introduction of a modifier is essential.

Multiresidue methods have been developed for the analysis of pesticides in food commodities by SFE [17, 18, 20]. Table IX shows the recovery of pesticides from spiked potatoes [20] using supercritical CO_2. The extracts were analysed by gas chromatography–ion trap mass spectrometry (GC–ITMS). Recoveries were >80% for most of the pesticides. The extracts obtained were sufficiently cleaned for subsequent GC–ITMS analysis.

Recently, we used supercritical carbon dioxide with and without methanol as a modifier to extract bound ^{14}C pesticide residues from soil, plant and wheat samples [21]. For extraction, optimal SFE conditions were obtained for each pesticide by varying the temperature, pressure and amount of modifier. Supercritical CO_2 modified with methanol improved the recovery of bound ^{14}C residues from

TABLE VII. COMPARISON OF FIELD SAMPLES WITH LLE AND SFE FOR ORGANOCHLORINE PESTICIDE EXTRACTION [19][a]

Component	Grassland soil (ng/g)		Orchard soil (ng/g)	
	LLE	SFE	LLE	SFE
p,p'DDE	4.1	5.5	63.1	56.2
o,p'DDT	2.3	4.0	44.8	43.7
p,p'DDT	11.4	29.7	260.6	296.4

[a] SFE condition: 20 MPa; 50°C; CO_2; extraction time, 30 min.

TABLE VIII. EFFECT OF DIFFERENT MODIFIERS ON THE SFE OF DDT RESIDUES FROM ORCHARD SOILS [19][a]

Component	CO_2		Control
	Acetonitrile extractable residues (ng/g)	Methanol extractable residues (ng/g)	CO_2 extractable residues (ng/g)
p,p'DDE	68.4	72.0	56.2
o,p'DDT	57.2	61.8	43.7
p,p'DDT	325	335	296

[a] SFE conditions: 20 MPa; 50°C; CO_2; extraction time, 30 min.

TABLE IX. RECOVERY OF PESTICIDES SPIKED AT THE 0.5 µg/g LEVEL IN POTATOES BY SFE AND THE GC-ITMS METHOD [20][a]

Pesticide	Recovery (%)	Pesticide	Recovery (%)
PCB	91 ± 6	Phosphamidon	91 ± 4
Mevinphos	92 ± 2	Dacthal	85 ± 5
HCB	93 ± 2	Carbaryl	91 ± 1
Diphenylamine	87 ± 2	Malathion	87 ± 4
Phorate	82 ± 4	Parathion	91 ± 3
Chlorpropham	91 ± 2	Endosulfan I	93 ± 2
PCNB	90 ± 4	DDE	91 ± 1
Diazinon	86 ± 5	Methidathion	90 ± 4
Lindane	89 ± 2	DDT	93 ± 2
Dicloran	91 ± 4	Ethion	97 ± 16
Carbofuran	90 ± 2	Myclobutanil	83 ± 10
Atrazine	92 ± 2	Methoxychlor	90 ± 1
Dimethoate	83 ± 8	Iprodione	102 ± 28
Chlorothalonil	93 ± 2	Phosmet	88 ± 4
Vinclozolin	91 ± 2	Azinphos-methyl	94 ± 6
Parathion-methyl	85 ± 6	Cis-permethrin	93 ± 3
Chlorpyrifos	72 ± 5	Fenvalerate	93 ± 2

[a] Data are means ± SD of three replicate extractions.

TABLE X. SFE OF BOUND RESIDUES IN THE ^{14}C PESTICIDE TREATED SAMPLE [20][a]

Sample	Pesticide	Recovery of ^{14}C after SC–CO_2 extraction (% of bound ^{14}C)	Relative standard deviation (%)	Recovery of ^{14}C after SC–CO_2/MeOH extraction (% of bound ^{14}C)	Relative standard deviation (%)	Recovery of ^{14}C after SC-MeOH extraction (% of bound ^{14}C)	Relative standard deviation (%)
Organic soil	Atrazine	83.2	2.7	89.2	3.6	72.4	5.8
	Prometryn	91.4	3.8	93.4	1.2	78.8	8.9
Mineral soil	Atrazine	86.9	2.1	90.8	1.6	65.3	5.4
	2,4-D	57.4	10.1	93.3	5.3	80.7	9.2
Wheat	Deltamethrin	86.8	4.0	91.9	4.2	57.4	11.4
Beans	Pirimiphos-methyl	89.9	4.9	95.1	5.6	86.2	5.2
Onion	Fonophos	85.7	2.8	95.4	5.2	73.4	10.1
Radishes	Dieldrin	90.8	4.0	91.8	7.0	94.9	10.3
Canola	Atrazine	10.4	14.5	86.6	8.4	72.3	10.9

[a] SC–CO_2: supercritical carbon dioxide; SC–CO_2/MeOH: supercritical carbon dioxide/methanol (30%); SC-MeOH: supercritical methanol.

soil and plant samples (Table X) [20]. Supercritical methanol was found to be less efficient than supercritical CO_2 or methanol modified supercritical CO_2 for the extraction of bound residues. Thus, SFE could be effectively applied to the extraction of bound (non-extractable) pesticide residues in a variety of matrices.

6. AQUEOUS SOLVENT SYSTEMS FOR THE EXTRACTION OF PESTICIDES

Moye [22] discussed the potential and alternatives to the organic solvents for the extraction of pesticide residues from vegetables and fruits. Fresh fruits and vegetables are composed predominantly of water. If 100 g of fruit or vegetable is

TABLE XI. WATER CONTENT OF FRUIT OR VEGETABLE AND EXTRACTING SOLUTION (100 mL OF ACETONE, 100 g OF CROP) [22]

Food	% water	
	Fresh crop	Extract solution
Apples	84.1	48
Peas	82.7	45
Carrots	28.2	47
Potatoes	77.8	44
Radishes	93.6	48
Celery	93.7	48
Spinach	92.7	48

TABLE XII. EXTRACTION OF DIELDRIN FROM ACETONITRILE/WATER MIXTURE [22]

% water	% recovery
0	67
10	69
20	73
30	71
40	75
50	78

extracted with 100 mL of acetone, the corresponding amount of water in the extract approaches 50% (Table XI) [22]. It was demonstrated that high percentages of water, in conjunction with acetonitrile and methanol, were more effective in extracting a chlorinated hydrocarbon insecticide (dieldrin) from radishes grown in soil treated with the pesticide than were less aqueous solvent pairs. The recoveries of field incurred dieldrin from radishes are shown in Table XII [22] with increasing water content, a maximum occurring at 50:50 water/methanol or water/acetonitrile. Moye [22] suggested that further research with field incurred residues of other pesticide classes is needed to determine the extent to which water is effective.

REFERENCES

[1] BENFENATI, E., et al., Simultaneous analysis of 50 pesticides in water samples by solid phase extraction and GC–MS, Chemosphere **21** (1990) 1411–1421.

[2] SCHUETTE, S.A., SMITH, R.G., HOLDEN, L.R., GRAHAM, J.A., Solid-phase extraction of herbicides from well water for determination by gas chromatography-mass spectrometry, Anal. Chim. Acta **236** (1990) 141–144.

[3] MOLTO, J.C., PICO, Y., FONT, G., MANES, J., Determination of triazines and organophosphorus pesticides in water samples using solid-phase extraction, J. Chromatog. **555** (1991) 145.

[4] TURIN, H.J., BOWMAN, R.S., A solid-phase extraction based soil extraction method for pesticides of varying polarity, J. Environ. Qual. **22** (1993) 332–334.

[5] BARCELO, D., et al., Solid-phase sample preparation and stability of pesticides in water using Empore disks, Trend Anal. Chem. **13** (1994) 352–360.

[6] PSATHAKI, M., MANOUSSARIDOU, E., STEPHANOU, E.G., Determination of organophosphorus and triazine pesticides in ground and drinking water by solid-phase extraction and gas chromatography with nitrogen–phosphorus or mass spectrometric detection, J. Chromatog. **667** (1994) 241–248.

[7] ONUSKA, F.I., TERRY, K.A., Extraction of pesticides from sediments using the microwave technique, Chromatographia **36** (1993) 191–194.

[8] FISH, J.R., REVESZ, R., Microwave solvent extraction of chlorinated pesticides from soil, LC-GC **14** (1996) 230–234.

[9] LOPEZ-AVILA, V., YOUNG, R., Microwave-assisted extraction of organic compounds from standard reference soils and sediments, Anal. Chem. **66** (1994) 1097–1106.

[10] US PATENT 5,002,784, Environment Canada, Ottawa, ON (1991).

[11] KHAN, S.U., Bound pesticide residues in soil and plants, Res. Rev. **84** (1982) 1–25.

[12] GITHIRA, P.N., Extraction of pesticide residues from soils using different methods, PhD Thesis, Carleton University, Ottawa, ON (1995).

[13] SNYDER, J.L., GROB, R.L., McNALLY, M.E., OOSTDYK, T.S., Comparison of supercritical fluid extraction with classical sonication and Soxhlet extraction for selected pesticides, Anal. Chem. **64** (1992) 848–853.

[14] LOCKE, M.E., Supercritical CO_2 fluid extraction of fluometuron herbicide from soil, J. Agric. Food Chem. **41** (1993) 1081–1084.

[15] LOPEZ-AVILA, V., DODHIWALA, N.S., BECKERT, W.F., Development in the supercritical fluid extraction of chlorophenoxy acid herbicides from soil samples, J. Agric. Food Chem. **41** (1993) 2038–2044.

[16] THOMSON, C.A., CHESNEY, D.J., Supercritical carbon dioxide extraction of 2,4-dichlorophenol from food crop tissues, Anal. Chem. **64** (1992) 848–853.

[17] WIGFIELD, Y.Y., LANOUETTE, M., Supercritical fluid extraction of the fortified residues of fluazifop-p-butyl (fusilade II) and its major metabolite, fluazifop-p, in onions, J. Agric. Food Chem. **41** (1993) 84–88.

[18] WIGFIELD, Y.Y., SELWYN, J., KHAN, S.U., McDOWELL, R., Comparison of supercritical fluid extraction and solvent extraction of twenty-two organochlorine pesticides from eggs, Chemosphere **32** (1996) 841–847.

[19] VAN DER VELDE, E.G., DIETVORST, M., SWART, C.P., RAMLAL, M.R., KOOTSTRA, P.R., Optimization of supercritical fluid extraction of organochlorine pesticides from real soil samples, J. Chromatog. **683** (1994) 167–174.

[20] LEHOTAY, S.J., ELLER, K.I., Development of method of analysis for 46 pesticides in fruits and vegetables by supercritical fluid extraction and gas chromatography/ion trap mass spectrometry, J. Assoc. Off. Anal. Chem. Int. **71** (1995) 1–10.

[21] KHAN, S.U., Supercritical fluid extraction of bound pesticide residues from soil and food commodities, J. Agric. Food Chem. **43** (1995) 1718–1723.

[22] MOYE, H.A., "Opportunities for pesticide residues analytical method development: The potential for aqueous extraction of pesticide residues from fruits and vegetables", Pesticide Chemistry (Proc. 8th Int. Congr. Washington, DC, 1995), American Chemical Society, Washington, DC (1995) 193–203.

IAEA-SM-343/34

MOVEMENT OF ^{14}C-CARBOFURAN IN A SILT CLAY SOIL
*A laboratory study**

I. GHANEM, S. BALI, F. MOHAMAD
Department of Agricultural Applications,
Atomic Energy Commission of Syria,
Damascus, Syrian Arab Republic

Abstract

MOVEMENT OF ^{14}C-CARBOFURAN IN A SILT CLAY SOIL: A LABORATORY STUDY.

Carbofuran is used in the Syrian Arab Republic to control the sitona weevil, *Sitona crinitus* Herbst (*S. macularius* Marsh), on legume crops. It is sprayed on soil to kill the immature stages, which attack the leaves close to the soil surface and the nitrogen fixing nodes on the roots in the soil. A laboratory study was conducted to examine the movement of this pesticide into the soil. A known amount of ^{14}C-carbofuran (specific activity of 2 μCi/mg) was applied to the top of the soil columns inside hard polyvinyl chloride (PVC) cylinders inserted into PVC tanks filled with soil of the same origin. The columns were sampled at varying intervals for up to 120 days. Each column was divided vertically from top to bottom into five zones. Each zone was 5 cm high. The radioactive residues were Soxhlet extracted for 6 hours. Extracts were concentrated and the aliquots counted for radioactivity using a liquid scintillation counter (LSC). In addition, samples of Soxhlet extracted soil were combusted in a biological oxidizer and the radioactivity was counted using LSC. The results show that recovery was good. Twenty-four hours after treatment the total radioactivity recovered was 71% of the total amount of insecticide applied. The major radioactive residues were confined to the top 5 cm of the soil column 24 hours after application. The amount of radioactive residues present in the lower layers increased gradually with time. However, this was accompanied by a relatively rapid and gradual loss in the total radioactivity recovered from the whole soil column. At the end of the experiment, most of the recovered radioactive residues (29%) were bound to the soil surface, whereas only 2% of the applied dose was extractable from the whole soil column.

---

* Research carried out in co-operation with the IAEA under Technical Co-operation Project No. SYR/5/015.

1. INTRODUCTION

In view of the growing concern over environmental contamination, it has become extremely important to know about the behaviour and movement of pesticides in soil. Such knowledge has been complicated in the past by the now accepted fact that pesticides can be bound to the soil and are non-extractable by normal extraction techniques [1]. Soil studies involving the use of radiotracer techniques are useful in quantifying extractable and bound pesticide residues. Numerous investigations have used such techniques in studies dealing with pesticide residues in soils [2–5].

Carbofuran is of a class of compounds, the methylcarbamates, that has insecticidal and nematicidal properties. It is used in the Syrian Arab Republic mainly as a soil insecticide to control the sitona weevil, *Sitona crinitus* Herbst (*S. macularius* Marsh), which attacks legume crops [6]. Soil bound residues ranging from 15 to 57% have been reported for some methylcarbamates [7, 8]. Ample information is available in the literature on plant bound and insect bound residues of carbofuran [9, 10], as is some information on its soil bound residues [11] and the potential uptake by microbes and plants.

This paper reports on laboratory work carried out on the movement of carbofuran in a defined type of Syrian soil.

2. MATERIALS AND METHODS

2.1. Chemicals

Carbon-14-carbofuran (2,2-di(^{14}C)methyl-2,3 dihydro-(3-^{14}C)benzofuran-7-yl-N-methylcarbamate), with a specific activity of 0.085 mCi/mg, was purchased through the IAEA from Izotop (Budapest, Hungary).[1] The radiochemical purity was more than 99%, as checked by radiochromatography. Analytical grade carbofuran was obtained from Greyhound Chromatography (Merseyside, United Kingdom). All the other chemicals used were of analytical grade and all the solvents were reagent grade.

2.2. Carbon-14-carbofuran dilution

The radiolabelled carbofuran was diluted with cold carbofuran (technical material) in methylene chloride. The specific activity was 2 µCi/mg. Each millilitre of insecticide preparation contained 100 mg of carbofuran (0.2 µCi).

---

[1] 1 Ci = 3.7 × 10^{10} Bq.

2.3. Experimental set-up

The soil used has the following physiological characteristics: source: Khan Ash-sheeh, near Damascus; pH (saturated paste): 8.03; electrical conductivity: 0.32; organic matter: 1.58%; clay: 32.5%; silt: 50%; sand: 17.5%; and texture: silt clay.

The soil was dried at room temperature and placed in two plastic tanks; no draining holes were fitted to the tanks. Twelve polyvinyl chloride (PVC) cylinders (30 × 5 cm in diameter) were driven into the soil. Each column was filled with 450 g of the same soil. The soil surface was 3 cm lower than the edge inside each column to avoid runoff. Using a microsyringe, 50 μL of insecticide preparation (10 μCi) were applied to each column. The surface of the soil was covered with a very thin layer of soil to avoid evaporation from the surface. The soil moisture was kept at 60% of its water holding capacity by watering every alternate day. Samples were taken at 1, 10, 20, 40, 80 and 120 days. At each sampling time, two soil packed PVC pipes were taken from the tank and frozen at $-20\,°C$ in a horizontal position until analysed. To prepare for analysis, the soil columns were pushed out of the PVC pipes. Each column was sectioned into five zones (5 cm each), keeping the zones in their right order. Soil in each section was weighted and Soxhlet extracted with methanol. For bound residue analysis, 50 mg Soxhlet extracted soil samples were combusted using an OX-600 Harvey biological oxidizer. Each sample was replicated twice.

At the end of the experiment, the soil samples were taken from the tanks to assess the amount of radioactivity that had leached from the soil columns to the tanks. Four samples were taken from each tank and each sample was replicated twice. For Soxhlet extraction, 110 g soil samples were taken. Following extraction, 50 mg soil samples were taken to determine the bound residues.

2.4. Determination of extractable residues

The soil samples were air dried, thoroughly mixed and Soxhlet extracted in 500 mL of methanol for 6 hours, at a rate of three cycles per hour. The extract was concentrated to 60 mL using a rotary evaporator (Rotavapor R-114, Buchi, Switzerland). Aliquots from each concentrate were analysed for ^{14}C radioactivity.

2.5. Determination of radioactivity

Aliquots (1 mL) from the soil extracts were dissolved in 9 mL of Insta-Gel scintillation fluid (Packard Instrument International, Switzerland). For the bound residues, the $^{14}CO_2$ generated from combustion is received in 10 mL of ^{14}C cocktail for the Harvey biological oxidizer. The radioactivity was determined using a Packard Tri-Carb liquid scintillation counter. The external standardization technique was used to correct for the quenching effect.

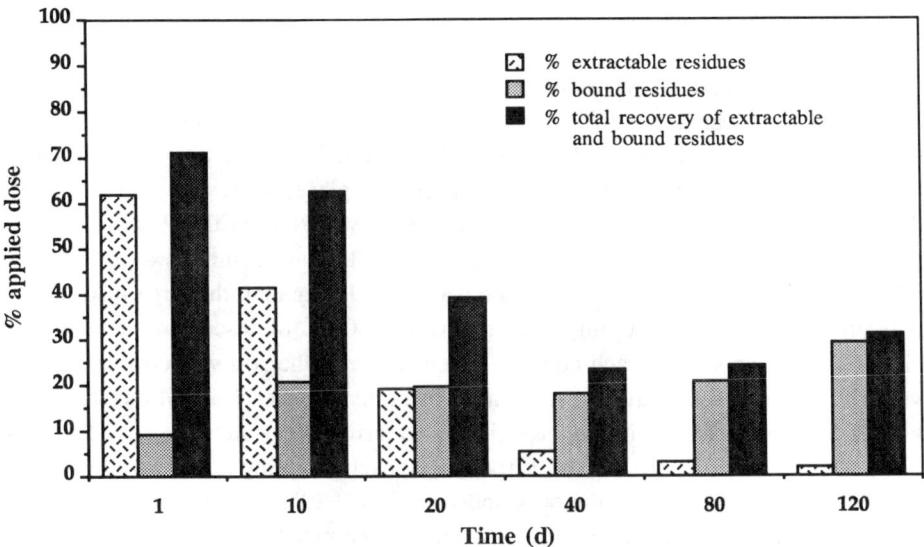

FIG. 1. *Total recovery and distribution of ^{14}C in extractable and bound residues of ^{14}C-carbofuran in soil at various time intervals after application.*

3. RESULTS AND DISCUSSION

The total recovery and distribution of ^{14}C-carbofuran in its extractable and bound forms, expressed as percentage of the applied dose, are shown in Fig. 1. The recovery of radioactivity was quite good. Twenty-four hours after treatment, the total recovered radioactivity represented 71% of the applied dose. However, about 30% of the applied dose was unaccounted for. The results of the biomineralization study of the pesticide by the soil biomass using a biometer flask (not shown here) do not explain the decrease in recovery of the applied insecticide. Nevertheless, this loss could partly have been due to the time of Soxhlet extraction (6 hours), which may be inadequate for complete recovery of the applied insecticide from the soil. One method for verifying this would be to combust the soil samples before extraction in order to quantify the $^{14}CO_2$ corresponding to the applied dose of ^{14}C-carbofuran. Unfortunately, this parameter was not assayed in our first trial. However, it is worth noting that some published literature mentions the extraction of soil in a Soxhlet system for up to 24 hours [3].

Twenty-four hours after application, the average total extractable radioactive residues from the soil column represented about 60% of the applied dose. Afterwards, a gradual and relatively rapid loss of extractable radioactive residues in the whole soil column was observed. By day 40 after extraction, the residues represented, on average, 5% of the applied dose. This level continued to drop, reaching an average of 2% of the applied dose at the end of the experiment, 120 days after application (Fig. 1).

The total bound residues represented only 9% of the applied dose 24 hours after application. Twenty days after application, the bound residues rose to 20% of the applied dose, and reached about 30% by the end of the experiment, 120 days after application (Fig. 1).

The amount of radioactive residue present in the tanks was very small. Only 0.09 and 0.5% (n = 8) of the applied dose were found in the extractable and bound forms, respectively. Under the present experimental conditions, the results indicated that there did not seem to be any significant leaching of the pesticide out of the column into the tanks. Indeed, movement of the pesticide seemed to be confined to the top 25 cm (the length of the column). The data also show that whatever the pesticide percentage leaching into the tank, most of it was bound to the soil surface.

The results, which express the form and movement of the radioactive residues in the soil columns in relation to time, are shown in Table I. Twenty-four hours after application, most of the radioactive residues were confined to the top 5 cm (first zone) of the soil column. Of the total radioactive residues found in the soil column as a whole (i.e. 71% of the applied dose), recovery in the first zone was, on average, 69% of the applied dose. Sixty per cent was in the extractable form, whereas only 9% was bound. Only about 2% of the applied dose was found in the second zone (5–10 cm in depth) of the soil column 24 hours after application. No radioactivity was detected beyond the second zone (Table I). Starting from day 20 after application until the end of the experiment, extractable and bound radioactive residues were found in almost all zones of the soil columns (Table I). However, the level of extractable residues dropped sharply within 10 days. Although the bound residues increased during this period, the increase was not comparable to the drop in extractable residues. There were indications that dissipation was the reason for this loss in pesticide from the soil. At the end of the experiment, 120 days after application, a very small percentage of the applied dose was found in the extractable form. The extractable residues found in all the zones were less than 1%, whereas a substantial percentage of the applied dose was bound to the soil. The percentage bound residues ranged from 17% in the upper zone (top 5 cm) to 4% in the fourth zone.

The present results show that within the experimental set-up there is a substantial loss in the applied dose of radioactivity, as evident from the 24 hour samples (Fig. 1). Some of this was due to the extraction procedures, i.e. the length of the Soxhlet extraction process, and could be overcome by lengthening the process. However, the loss in recovery of radioactive residues observed later, compared with the recovery observed at the beginning of the study, is a clear indication of biodegradation. Further work using a biometer flask should be carried out to verify this process. The identity of the radioactive residues recovered from the soil zones at various time intervals after application is as yet unknown. Thin layer chromatography

TABLE I. PERCENTAGE RECOVERY OF APPLIED DOSE FROM SOIL ZONES AT VARIOUS TIME INTERVALS AFTER APPLICATION OF ^{14}C LABELLED CARBOFURAN

Soil zones	Days after application									
	1		10		20		40		80		120	
	E[a]	B[a]	E	B	E	B	E	B	E	B	E	B
First (0–5 cm)	60.25 (±5.05)[b]	8.85 (±3.03)	15.30 (±4.85)	10.73 (±0.63)	2.43 (±0.39)	7.40 (±1.3)	0.75 (±0.03)	7.15 (±1.78)	0.53 (±0.08)	6.68 (±0.32)	0.76 (±0.12)	17.45 (±1.35)
Second (5–10 cm)	1.60 (±0.06)	0.55 (±0.17)	10.68 (±2.33)	4.93 (±0.41)	1.85 (±0.06)	4.03 (±0.54)	0.50 (±0.00)	3.63 (±0.32)	0.90 (±0.12)	7.55 (±0.18)	0.55 (±0.09)	8.73 (±2.25)
Third (10–15 cm)	0.00 —	0.00 —	9.80 (±0.42)	3.45 (±0.22)	3.65 (±0.12)	3.50 (±0.42)	0.60 (±0.0)	2.25 (±0.21)	1.05 (±0.09)	3.65 (±0.22)	0.45 (±0.03)	5.25 (±0.48)
Fourth (15–20 cm)	0.00 —	0.00 —	5.90 (±0.23)	1.75 (±0.18)	5.03 (±0.48)	2.55 (±0.18)	1.15 (±0.09)	2.10 (±0.1)	0.70 (±0.23)	3.03 (±0.54)	0.30 (±0.00)	3.68 (±0.23)
Fifth (20–25 cm)	0.00 —	0.00 —	0.00 —	0.00 —	6.38 (±0.31)	2.23 (±0.31)	2.20 (±0.23)	2.53 (±0.22)	— —	— —	— —	— —

[a] E = extractable residues; B = bound residues.
[b] Numerals in parentheses represent the SE of the mean (n = 4).

analysis of the soil extracts should be of help in revealing the identity of the extractable residues and the persistance of the parent compound in the soil.

ACKNOWLEDGEMENTS

The authors would like to thank I. Othman, Director General of the Atomic Energy Commission of Syria, and N.E. Sharabi, Head of the Department of Agricultural Applications, for their continuous support. The authors are grateful to the IAEA for technical support.

REFERENCES

[1] KHAN, S.U., Bound pesticide residues in soil and plants, Res. Rev. **84** (1982) 1.
[2] LICHTENSTEIN, E.P., KATAN, J., ANDEREGG, B.N., Binding of "persistent" and "nonpersistent" ^{14}C-labeled insecticides in an agricultural soil, J. Agric. Food Chem. **25** (1977) 43.
[3] HUSSAIN, K.A., "Bound residues of ^{14}C-carbofuran in soil", Quantification, Nature and Bioavailability of Bound ^{14}C-Pesticide Residues in Soil, Plants and Food (Proc. Panel Gainesville, 1985), IAEA, Vienna (1986) 23–29.
[4] SORENSON, B.A., et al., Formation and movement of ^{14}C-atrazine degradation products in sandy loam soil under field conditions, Weed Sci. **41** (1993) 239.
[5] BARRIUSO, E., KOSKINEN, W.C., Incorporating nonextractable atrazine residues into soil size fractions as a function of time, Soil Sci. Soc. Am. J. **60** (1996) 150.
[6] A Guide to Pesticides Used in Syria, Ministry of Agriculture, Damascus (1991) (in Arabic).
[7] KAZANO, H., KEARNEY, P.C., KAUFMAN, D.D., J. Agric. Food Chem. **20** (1972) 975.
[8] MARSHALL, T.C., DOROUGH, H.W., Bioavailability in rats of bound and conjugated plant carbamate insecticide residues, J. Agric. Food Chem. **25** (1977) 1003.
[9] SONOBE, H., CARVER, R.A., KRAUSE, R.T., KAMPS, L.R., Extraction of biologically incorporated ^{14}C-carbofuran residues from root crops, J. Agric. Food Chem. **31** (1983) 96.
[10] KHAN, S.U., STRATTON, G.D., Jr., WHEELER, W.B., Characterization of bound residues of dieldrin, J. Agric. Food Chem. **32** (1984) 1189.
[11] TALEKAR, N.S., LEE, M.E., SUN, L.T., Absorption and translocation of soil and foliar applied ^{14}C-carbofuran and ^{14}C-phorate in soybean and mungbean seeds, J. Econ. Entomol. **70** (1977) 685.

IAEA-SM-343/18

FATE OF ^{14}C-PIRIMICARB IN CHINESE CABBAGE AND SOIL*

Jiarong PAN, Xianfang WEN
Institute for the Application
  of Atomic Energy,
Chinese Academy of Agricultural Science,
Beijing, China

Presented by I.G. Ferris

Abstract

FATE OF ^{14}C-PIRIMICARB IN CHINESE CABBAGE AND SOIL.
  A pot trial was conducted to investigate the persistence and metabolism of ^{14}C-pirimicarb in Chinese cabbage and meadow soil, a common loam soil type of North China. Radioassay and thin layer chromatography techniques were used. During incubation, the temperature was maintained at $25 \pm 1°C$, and sufficient light and moisture were provided to grow the cabbage. The results showed that ^{14}C-pirimicarb was slightly persistent on the surface of the leaves, the majority of which penetrated the shoots rapidly; its half-life on the surface of the leaves was approximately 6 days. Carbon 14-pirimicarb was persistent in meadow soil, with a half-life of about 9 days. For the whole soil–cabbage system, the half-life was approximately 11 days, and about 82% of the applied ^{14}C-pirimicarb disappeared within 15 days, mainly by degradation. Two metabolites of ^{14}C-pirimicarb were determined, both in cabbage and in meadow soil, one being chloroform soluble and the other methanol soluble. The ^{14}C-pirimicarb that penetrated the cabbage converted to metabolites relatively slowly, within a few days after being sprayed, but penetration was slightly faster after this period; in turn, the metabolites degraded to CO_2 very rapidly. It was also found that a very small amount of the ^{14}C-pirimicarb that penetrated the shoots could be assimilated to become bound ^{14}C residues. In soil, the ^{14}C radioactivity in the chloroform extractable residues decreased steadily during the course of incubation, while the methanol extractable residues increased steadily up to the end of incubation. It was also found that ^{14}C-pirimicarb was weakly sorbed by soil.

1. INTRODUCTION

  Pirimicarb-(2-dimethylamino-5,6-dimethylpyrimidine-4-)-N,N-dimethylaminoformate, which is a fast acting selective aphicide that kills by contact, vapour or systemic action, is particularly effective against the organophosphorus resistant

---
  * Research carried out in co-operation with the IAEA under Technical Co-operation Project No. CPR/5/008.

aphids that infest fruit trees, cola crops, beans, lettuce, flowers and certain ornamentals [1]. It has a relatively low mammalian toxicity and is practically non-toxic to predatory beetles and honey bees [2]. It has been found that Chinese cabbage is greatly affected by aphids and that pirimicarb has a strong controlling effect on these insects. A very low concentration of pirimicarb can be very effective against cabbage aphids [3], therefore it is a promising approach for controlling such insects in China. Although the persistence, degradation and metabolism of pirimicarb in lettuce and alfalfa have been studied [4, 5], little information is available on cabbage. This paper reports the results of a study carried out to determine the fate of ^{14}C-pirimicarb in Chinese cabbage and on the soil in which it was grown.

2. MATERIALS AND METHODS

2.1. Chemicals

Carbon-14-pirimicarb (ring labelled) was synthesized at our institute and had a radiochemical purity of 96% and a specific radioactivity of 3.18 μCi/mg.[1] The ^{14}C-pirimicarb solution, with a concentration of 30 μg/mL, was prepared after an aliquot of ^{14}C-pirimicarb was dissolved in distilled water.

2.2. Soil

The light coloured meadow soil, obtained from our experimental farm in the northwest suburbs of Beijing, was a sandy loam on which wheat is produced. It had a pH of 8.6 and organic matter of 1.8%, contained 0.14% nitrogen, and had a bulk density of 1.45 g/cm^3 and a field capacity of 80%.

2.3. Plant

A cultivar of Chinese cabbage, Qingbang No. 1, mainly grown in North China and frequently affected by aphids, was the plant tested in this experiment.

2.4. Incubation experiment

Soil (100 g on an oven dry weight basis) was placed in pots (10 cm in diameter, 15 cm high and with a hole at the bottom of the pot for irrigation) in which the cabbage was sown. The pots were placed inside a temperature–light controlled incubator. Two seedlings were maintained for each pot until the six leaf period, when

---

[1] 1 Ci = 3.70 × 10^{10} Bq.

the ^{14}C-pirimicarb solution was sprayed on to the leaves with a microapplicator; 77 and 23% of the ^{14}C-pirimicarb were applied to the leaves and soil, respectively. During the course of this experiment, the temperature was maintained at 25 ± 1°C, and sufficient light and water were provided to grow the cabbage. Each treatment was replicated three times. The leaves were washed with distilled water several times, then the soil and cabbage samples were collected at 0, 1, 2, 3, 4, 5, 6, 7, 8, 9, 12 and 15 days after treatment. The washed water was collected, concentrated to a small volume and its radioactivity determined using a liquid scintillation counter (LSC). The soil and cabbage samples were combusted to $^{14}CO_2$ and their radioactivity determined by LSC.

2.5. Residue extraction

The remaining soil and cabbage samples were extracted with chloroform and methanol, respectively, using the procedure outlined in Fig. 1. The filtrate was combusted to $^{14}CO_2$ and its radioactivity determined. Extraction recovery for the soil and cabbage samples ranged between 78.90 and 99.8%, indicating that the extraction procedure had removed most of the ^{14}C-pirimicarb residues from the soil and the cabbage.

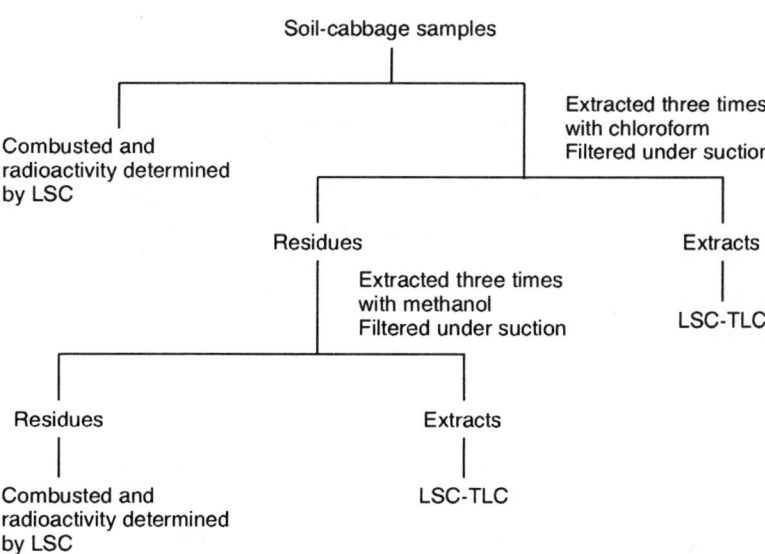

FIG. 1. Extraction procedure for ^{14}C-pirimicarb.

2.6. Thin layer chromatography (TLC) analysis

The R_f values of all the extracts and the standard ^{14}C-pirimicarb were determined using TLC with a solvent mixture of acetone, diethylether and chloroform (5:3:2 by volume). The spots observed under UV light were removed from the TLC plates, extracted with acetone and their radioactivity determined by LSC.

3. RESULTS AND DISCUSSION

3.1. Disappearance of ^{14}C-pirimicarb from cabbage and soil

3.1.1. *Persistence of ^{14}C-pirimicarb on cabbage leaves*

The results showed that only 8.6% of the applied ^{14}C-pirimicarb was removed from the surface of the leaves within the first 3 days, but 35.66% within the next 3 days, and 24.72% between days 6 and 9 after treatment. Only 1.1% of the applied ^{14}C-pirimicarb remained on the surface of the leaves 15 days after treatment (Fig. 2). It was suggested that the amount of ^{14}C-pirimicarb remaining on the surface of the leaves with time could be described by an exponential equation with a half-life of 6.4 days. Nine days after treatment, 60% of the applied ^{14}C-pirimicarb had penetrated the cabbage (Fig. 3). These data indicate that ^{14}C-pirimicarb was only slightly persistent on the leaves; the majority rapidly penetrated the shoots.

3.1.2. *Persistence of ^{14}C-pirimicarb in meadow soil*

Within the first 6 days, only 5.1% of the applied ^{14}C-pirimicarb was lost from meadow soil, but loss occurred more rapidly between days 6 and 9 after treatment. Thereafter, 9.3% of the ^{14}C-pirimicarb residues in the soil remained constant (Fig. 4). A straight line best describes the relationship between the amount of ^{14}C-pirimicarb remaining in meadow soil and the time in days; its half-life was predicted to be 8.9 days.

3.1.3. *Disappearance of ^{14}C-pirimicarb from the soil–cabbage system*

The results showed that initially the ^{14}C-pirimicarb disappeared slowly from the soil–cabbage system, within 9 days of treatment; the amount in the soil–cabbage system only decreased by 20%, while 41% of the applied ^{14}C-pirimicarb was lost

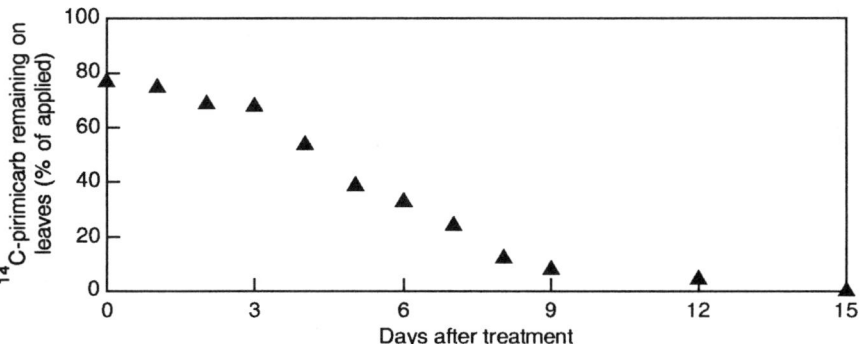

FIG. 2. Persistence of ^{14}C-pirimicarb on Chinese cabbage leaves.

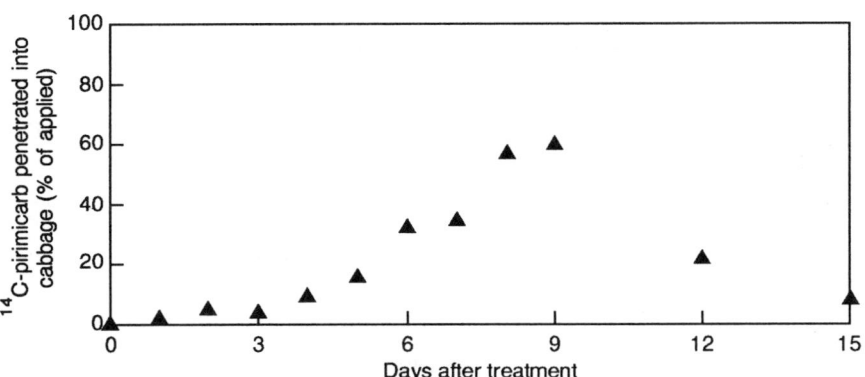

FIG. 3. Percentage ^{14}C radioactivity in the shoots over the total radioactivity of the applied ^{14}C-pirimicarb.

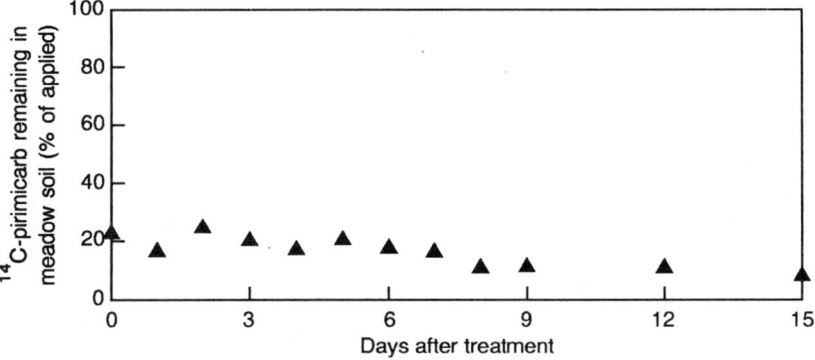

FIG. 4. Persistence of ^{14}C-pirimicarb in meadow soil.

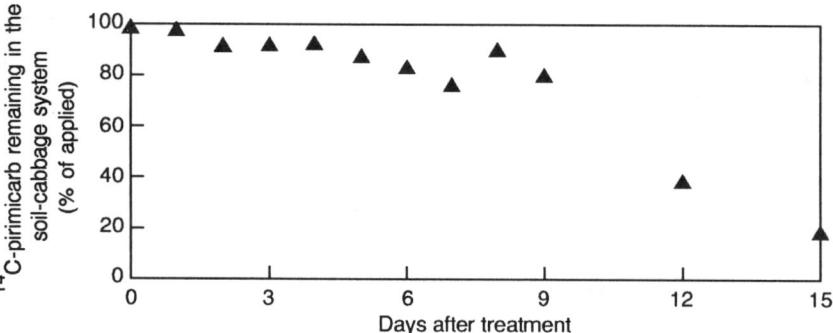

FIG. 5. Persistence of ^{14}C-pirimicarb in the soil–cabbage system.

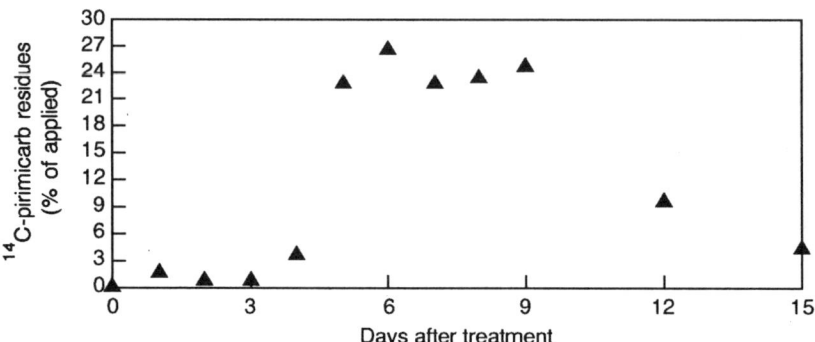

FIG. 6. Chloroform extractable residues of ^{14}C-pirimicarb in Chinese cabbage.

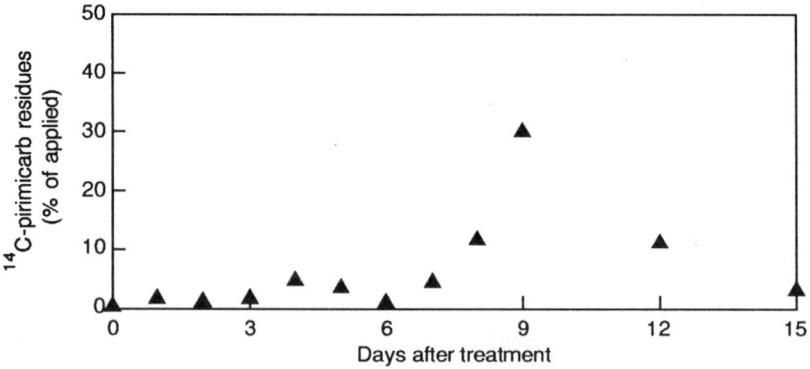

FIG. 7. Methanol extractable residues of ^{14}C-pirimicarb in Chinese cabbage.

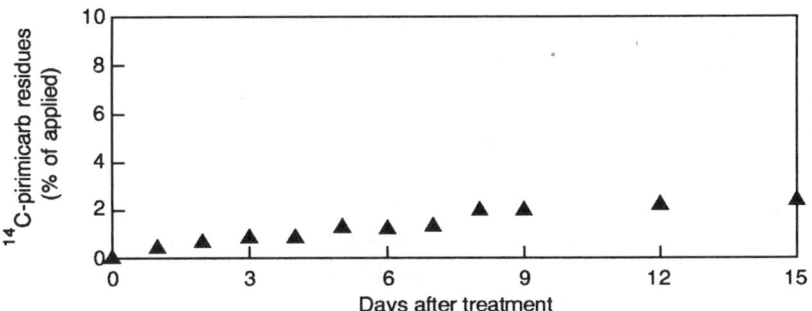

FIG. 8. Bound residues of ^{14}C-pirimicarb in Chinese cabbage.

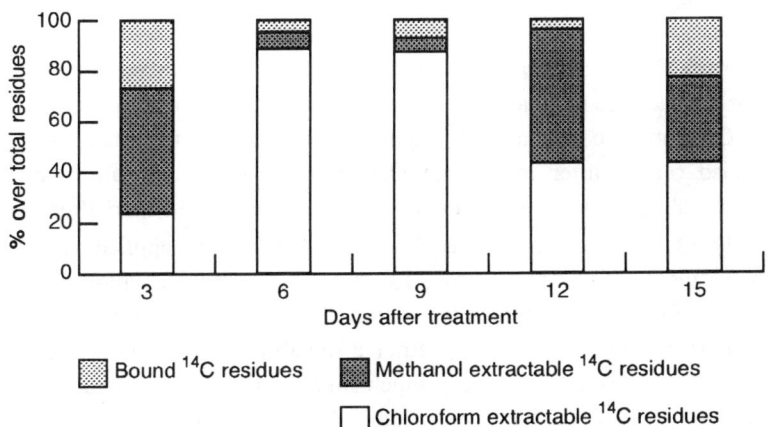

FIG. 9. Residue components of ^{14}C-pirimicarb in Chinese cabbage.

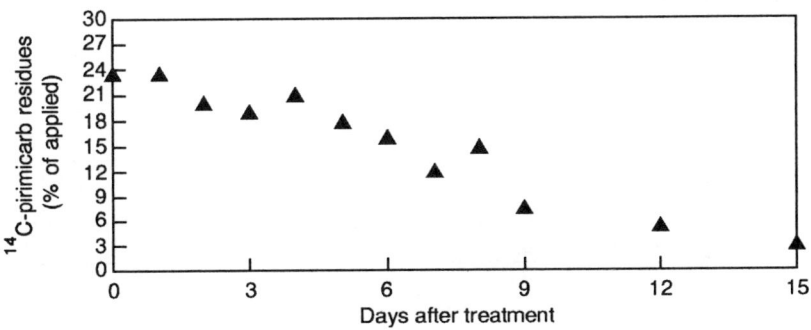

FIG. 10. Chloroform extractable residues of ^{14}C-pirimicarb in meadow soil.

between days 9 and 12 and a further 20% between days 12 and 15 (Fig. 5). Over 15 days, about 82% of the applied ^{14}C-pirimicarb disappeared from the soil–cabbage system, most probably as $^{14}CO_2$. As with the persistence of ^{14}C-pirimicarb on the surface of the leaves, the change in the amount of ^{14}C-pirimicarb remaining in the soil–cabbage system is best described by an exponential curve, but with a longer half-life (10.9 days). Pesticide losses from the soil-cabbage system occur by degradation and volatilization. A previous study [6] suggested that ^{14}C-pirimicarb in meadow soil was mainly lost by degradation; the data in this paper show that degradation was also a prominent pathway for ^{14}C-pirimicarb losses from the soil–cabbage system.

3.2. Metabolism of ^{14}C-pirimicarb in the soil–cabbage system

3.2.1. Cabbage

Figure 6 shows that within 3 days of treatment only a small amount of ^{14}C-chloroform extractable residues was determined in cabbage, equivalent to 1% of the applied ^{14}C-pirimicarb, but a peak, equivalent to 26.8% of the applied pesticide, was obtained 6 days after treatment. The amount of ^{14}C-chloroform extractable residues in cabbage decreased slowly up to 9 days, but thereafter decreased very sharply. Only a small amount, equivalent to 4.7% of that applied, was measured at the end of incubation. Analysis by TLC showed that the ^{14}C-chloroform extractable residues of cabbage contained two compounds. One was the parent compound, ^{14}C-pirimicarb ($R_f = 0.87$), and the other a metabolite ($R_f = 0.22$, metabolite I), the amount of the former being very much higher than that of the latter. Six days after treatment, the ^{14}C-methanol extractable residues of cabbage amounted to only 2.2% of the applied ^{14}C-pirimicarb (Fig. 7), while a peak, equivalent to 30.4% of

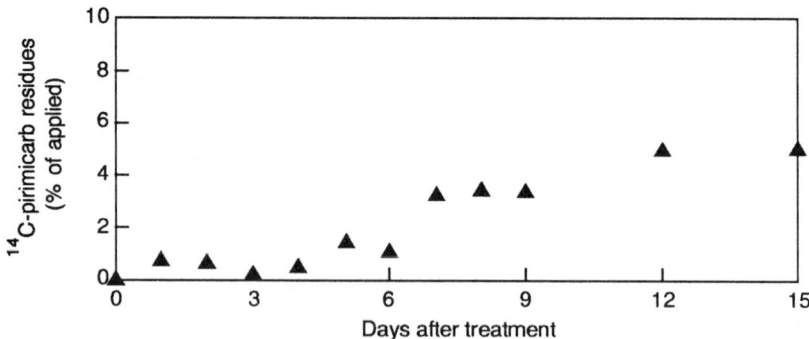

FIG. 11. Methanol extractable residues of ^{14}C-pirimicarb in meadow soil.

that applied, was obtained 9 days after treatment. Thereafter, these residues decreased very rapidly, and only a small amount, equivalent to 3.7% of that applied, was determined at the end of incubation. TLC analysis of the ^{14}C-methanol extractable residues of cabbage showed that there was only one radioactive spot ($R_f = 0.33$), indicating that another metabolite had been formed (metabolite II). Three days after treatment, bound residues of ^{14}C-pirimicarb were detected in cabbage, equivalent to 0.9% of the applied ^{14}C-pirimicarb. These residues then accumulated very slowly up to the end of incubation, when they were equivalent to 2.6% of the total applied ^{14}C-pirimicarb (Fig. 8).

The percentage change in the ^{14}C-chloroform extractable residues, ^{14}C-methanol extractable residues and bound ^{14}C residues over the total residues is shown in Fig. 9.

It was concluded that: (1) ^{14}C-pirimicarb penetrated only slowly into cabbage and was converted to metabolites I (chloroform soluble) and II (methanol soluble) a few days after treatment. These metabolites were in turn rapidly degraded to CO_2, some of which was re-assimilated by cabbage tissues, as evidenced by the formation of unextractable bound residues.

3.2.2. Soil

During the course of this experiment, the ^{14}C radioactivity in the chloroform extractable residues of the soil decreased steadily (Fig. 10), while the methanol extractable residues increased steadily (Fig. 11). At the end of incubation, the ^{14}C-methanol extractable residues were equivalent to 5.1% of the applied pesticide. TLC and LSC revealed that the ^{14}C-chloroform extractable residues of soil contained two compounds: one was the parent compound, ^{14}C-pirimicarb ($R_f = 0.87$), and the other was metabolite I ($R_f = 0.22$). There was only one radioactive spot ($R_f = 0.33$) for the methanol extractable residues of soil, corresponding to metabolite II. During incubation, negligible amounts of bound residue were formed from ^{14}C-pirimicarb.

It can be concluded from the above results that in meadow soil: (1) there was a 3 day lag period before any significant degradation of ^{14}C-pirimicarb occurred; (2) ^{14}C-pirimicarb degradation involved the formation of chloroform soluble and methanol soluble metabolites; and (3) formation of bound residues was not an important pathway for ^{14}C-pirimicarb degradation.

REFERENCES

[1] SHEPPARD, H.H., Pesticide Dictionary, Farm Chemicals Handbook, Minister Publishing, Willoughby, OH (1975) 43–47.

[2] ABDULSALAM, K.S., ABDEL, M.M.I., MOHAMED, M.I., Relative toxicity of certain pesticides against honeybees, Ann. Agric. Sci. (Cairo) **33** 2 (1988) 1309–1320.

[3] GUPTA, P.R., MISHRA, R.C., GUPTA, J.K., Effectiveness of some insecticide sprays against the cabbage aphid *(Brevicoryne brassicae)* and thrips *(Thrips flavus)* on seed crop of radish, Indian J. Agric. Res. **19** 1 (1985) 6–10.

[4] CABRAS, P., et al., Determination and persistence of pirimicarb in lettuce, Riv. Soc. Ital. Sci. Aliment. **17** 1 (1988) 61–64.

[5] CABRAS, P., et al., Pirimicarb and its metabolite residues in lettuce: Influence of cultural environment, J. Agric. Food Chem. **38** 3 (1990) 879–882.

[6] PAN, J., WEN, X., "The persistence of ^{14}C-pirimicarb in three Chinese soils", Development of Nuclear Agricultural Science in China: Nuclear Agricultural Science for Young Scientists (Proc. Conf. Beijing, 1993) (LIANG, Q., HUA, L., Eds), Chinese Agricultural Science–Technology Press, Beijing (1995) 143–145.

IAEA-SM-343/26

DISTRIBUTION AND FATE OF ^{14}C-ACEPHATE IN TOMATO PLANTS AND SOIL

M.S. TUNGGULDIHARDJO, E. ANWAR
Centre for the Application
  of Isotopes and Radiation,
National Atomic Energy Agency,
Jakarta, Indonesia

Abstract

DISTRIBUTION AND FATE OF ^{14}C-ACEPHATE IN TOMATO PLANTS AND SOIL.
 This experiment was carried out using tomato plants (3–8 weeks old) that were planted in soil inside a PVC tube (inner diameter, 8.5 cm; height, 20 cm). Ten millilitres of ^{14}C-acephate (O,S-dimethyl acetyl phosphoramidothioate) solution (0.0214 µCi/mL, 26.5 ppm) were applied to the soil. At the vegetative phase, the ^{14}C-acephate residues were higher in the older leaves than in the younger leaves. At the generative phase, 0.06% ^{14}C-acephate was detected in fruits on day 1, 0.9% on day 4, 1.5% on day 7 and 3% on day 14. The ^{14}C-acephate residues and their metabolites in soil were determined using thin layer chromatography (TLC) and gas–liquid chromatography (GLC) on a packed column of 3% CHDMS on Chromosorb WAW. The ^{14}C-acephate results from TLC and GLC, respectively, were 51.2 and 59.1% on day 1, 24.1 and 24.5% on day 4, 11.8 and 17.1% on day 7 and 3.1 and 3.1% on day 14. The metabolite, methamidophos, showed 2.5% on day 1, 9.2% on day 4, 11.3% on day 7 and 2.2% on day 14, obtained by TLC. TLC analysis showed that the acephate and methamidophos, respectively, in leaves were 3.9 and 0.8% on day 1, 4.5 and 1.5% on day 4, 7.1 and 7.5% on day 7, and 5.5 and 10.3% on day 14.

1. INTRODUCTION

 Acephate (O,S-dimethyl acetyl phosphoramidothioate), or orthene, is an organophosphorus insecticide that is systemic, and is readily absorbed by the root system of plants and distributed to all parts of the plant. It is a broad spectrum insecticide that controls a large number of insects, e.g. *Nothopeus* sp. on the clove tree, *Heliothis* sp. on the tomato fruit, *Agromyza* sp. and *Artona catoxantha*. On the other hand, the insecticide has a low mammalian toxicity, with an acute oral LD_{50} of 945 mg/kg in male rats and an acute dermal LD_{50} of 2000 mg/kg in rabbits. Its primary metabolite, methamidophos (O,S-dimethyl phosphoramidothioate), is also an insecticide, but it is more persistent and more toxic than acephate in plant tissue, with an acute oral LD_{50} of 118 mg/kg in male rats and an acute dermal LD_{50} of 118 mg/kg in rabbits [1]. The objective of this research was to study the distribution and fate of ^{14}C-acephate in tomato plants and soil (the ^{14}C-acephate aqueous solution was applied to the soil on which the tomato plants were cultivated).

2. MATERIALS AND METHODS

2.1. Materials

Tomato plants of the Gondol variety were used for the experiment. The plants were about 3 weeks old at the vegetative phase and 7–8 weeks old at the generative phase. The ^{14}C-acephate was purchased from International Isotopes (Germany), with a specific activity of 26.2 mCi/mmol and a purity of 99.9%.[1]

PPO, dimethyl POPOP and scintillation grade toluene from Merck (Germany) were used for the scintillator mixture. Ethanolamine and scintillation grade methoxyethanol were added to the scintillator when used for the combustion experiment. Other analysis grade chemicals from Merck (Germany) were also used.

A Harvey OX-400 biological oxidizer, a Beckman LS-1801 liquid scintillation counter (LSC) and Shimadzu GC-7A gas–liquid chromatography (GLC) were used.

2.2. Distribution of ^{14}C-acephate at the vegetative phase

The tomato plants were planted in soil inside PVC cylinders (inner diameter, 8.5 cm; height 20 cm). When the plants were 3 weeks old (three leaf stage), 10 mL of an aqueous acephate solution containing 0.214 μCi of ^{14}C-acephate and 265 μg of cold acephate were applied to the soil. Three plants were harvested at days 1 and 4 after application. The plants were divided into three parts: roots, stems and leaves. The roots were rinsed with water and dried with tissue paper. The leaves were divided into third (youngest), second and first (oldest) leaves. Each of these was cut into small pieces, and about 100–200 mg of the samples were combusted with the biological oxidizer. The $^{14}CO_2$ was trapped with the scintillator mixture, ethanolamine and methoxyethanol, and then counted using LSC.

2.3. Distribution of ^{14}C-acephate at the generative phase

The methodology used was the same as that outlined in Section 2.2, except that the three plants were 7–8 weeks old and harvested at 1, 4, 7 and 14 days after application of the insecticide. The plants were divided into five parts: roots, stems, leaves, flowers and fruits. The roots were rinsed with water and dried with tissue paper.

2.4. Carbon-14-acephate residues in leaves

Leaves at the generative phase were extracted with acetone (3×), and the extract was passed through a column containing anhydrous sodium sulphate and

---

[1] 1 Ci = 3.70×10^{10} Bq.

charcoal. The column was eluted with 1 mL of acetone, and the acephate in the sample was separated and identified by thin layer chromatography (TLC) of the silica gel coated plate. A mixture of isopropyl alcohol, benzene and ammonium hydroxide (6:5:2 (vol./vol./vol.)) was used as the developing solvent [2], and the unextractable residues were combusted using the biological oxidizer.

2.5. Carbon-14-acephate residues in soil

Fifty grams of the soil on which the tomato plants were cultivated were extracted by Soxhlet apparatus using a mixture of n-hexane and acetone (1:1 (vol./vol.)) as the solvent for 6 hours. The extracts were concentrated and cleaned up in a Florisil column, then determined using TLC and GLC on a packed column of 3% CHDMS on Chromosorb WAW. The unextractable soil was combusted with the biological oxidizer, and the $^{14}CO_2$ released was counted using LSC.

3. RESULTS AND DISCUSSION

3.1. Distribution of ^{14}C-acephate residues at the vegetative phase

Figure 1 shows that the ^{14}C-acephate residues were distributed throughout the plant, although most had accumulated in the leaves, the level in the older leaves being higher than that in the younger leaves. This is in agreement with other studies, where the plants were transplanted in a ^{14}C-carbofuran aqueous solution [3] or in a ^{14}C-monocrotophos aqueous solution [4]. As the insecticides were systemic, the ^{14}C insecticides were rapidly distributed to all parts of the plant.

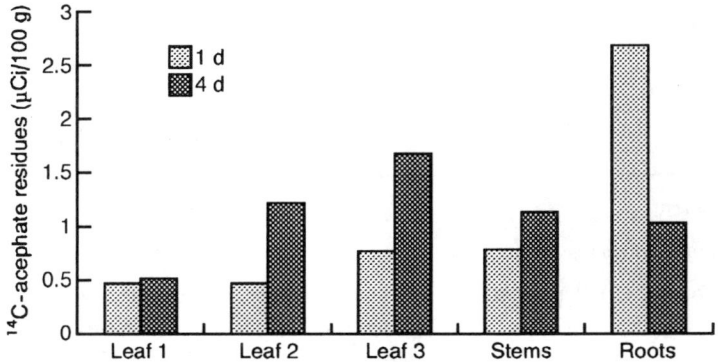

FIG. 1. Carbon-14-acephate residues in tomato plants at the vegetative phase.

The total ^{14}C-acephate residues in the leaves on day 4 were higher than those on day 1. According to previous studies [3–5], the total radioactivity of the insecticides on day 1 was higher than that on day 2.

3.2. Distribution of ^{14}C-acephate residues at the generative phase

Table I shows that the ^{14}C-acephate residues were distributed throughout the plant. The percentage of applied ^{14}C-acephate was 0.06% in fruits on day 1, 0.9% on day 4, 1.5% on day 7 and 3% on day 14.

TABLE I. CARBON-14-ACEPHATE RESIDUES IN TOMATO PLANTS AND SOIL[a]

Day(s) after application	^{14}C-acephate residues (% of applied)						
	Soil	Roots	Stems	Leaves	Flowers	Fruits	Total
1	68.7	6.9	4.5	8.7	0.03	0.06	88.9
4	49.6	14.1	5.0	11.7	0.02	0.90	81.3
7	33.6	14.1	5.9	24.2	0.02	1.50	79.3
14	22.5	22.5	6.1	31.6	0.01	3.00	76.9

[a] Data are the means of triplicate analyses.

TABLE II. CARBON-14-ACEPHATE AND ITS METABOLITES IN LEAVES[a]

Components	^{14}C-acephate residues (% of applied)			
	1 d	4 d	7 d	14 d
Acephate (parent)	3.9	4.5	7.1	5.5
Methamidophos (primary metabolite)	0.8	1.5	7.5	10.3
Unknown (total) ^{14}C metabolites	0.5	0.9	2.1	4.0
Unextractable ^{14}C	2.7	4.9	7.0	11.7

[a] Data are the means of triplicate analyses.

TABLE III. CARBON-14-ACEPHATE AND ITS METABOLITES IN SOIL[a]

Components	^{14}C-acephate residues (% of applied)			
	1 d	4 d	7 d	14 d
Acephate (TLC)	51.2	24.1	11.8	3.1
Acephate (GLC)	59.1	24.5	17.1	3.1
Methamidophos (primary metabolite)	2.5	9.2	11.3	2.2
Unknown (total) ^{14}C metabolites	1.1	2.1	5.3	1.2
Unextractable ^{14}C	14.9	14.1	15.3	16.0

[a] Data are the means of triplicate analyses.

3.3. Fate of ^{14}C-acephate in leaves

The ^{14}C-acephate residues in leaves on days 1, 4, 7 and 14 are given in Table I, while Table II shows that the ^{14}C-acephate degraded to many compounds, methamidophos being the primary metabolite. This is in agreement with Bull [2].

3.4. Fate of ^{14}C-acephate in soil

Table I shows that the ^{14}C-acephate residues in soil were 68.7% on day 1, decreasing to 22.5% by day 14. The table also shows that the total ^{14}C-acephate residues in the soil and plants were less than 100%. Dissipation may have been caused by evaporation of part of the acephate. Table III shows that the primary metabolite of acephate was methamidophos.

4. CONCLUSIONS

(1) Carbon-14-acephate was absorbed by the plant through the roots and rapidly distributed to the stems, leaves, flowers and fruits. At the vegetative phase, the ^{14}C-acephate content in the older leaves was higher than that in the younger leaves.
(2) On day 1, 0.06% ^{14}C-acephate was detected in fruits at the generative phase, increasing to 3% by day 14.

(3) The highest amount of ^{14}C-acephate residues was obtained in the leaves, followed by the stems, roots and fruits.
(4) The ^{14}C-acephate residues in soil ranged from 51.2 to 59.1% on day 1.
(5) Acephate was easily degradated to methamidophos, the primary metabolite.

REFERENCES

[1] MEISTER, R.T., ZILENZIGER, A.W., FITZGERALD, G.F., Farm Chemical Handbook, Meister Publishing, Miami, FL (1982).
[2] BULL, D.L., Fate and efficiency of acephate after application to plants and insects, J. Agric. Food Chem. **27** (1979) 268.
[3] SULISTYATI, M., KUSWADI, A.N., "Absorption and distribution of carbofuran in soybean plants at the vegetative and generative phases", Application of Isotopes and Radiation (Proc. Conf. Jakarta, 1994), Centre for the Application of Isotopes and Radiation, Batan (1994) 593.
[4] SULISTYATI, M., Absorption and distribution of monochrotophos in mungbean plants at the vegetative and generative phases (in preparation).
[5] BOUCHARD, D.S., LAVY, T.L., Fate of acephate in the cotton plant, J. Econ. Entomol. **75** (1982) 921.

IAEA-SM-343/41

PERSISTENCE IN SOIL OF ENDOSULFAN AND LINDANE APPLIED TO SOYABEAN AND MAIZE PESTS IN A FIELD TRIAL AGROSYSTEM IN ZIMBABWE*

M.F. ZARANYIKA, P. MUGARI
Chemistry Department,
University of Zimbabwe,
Mount Pleasant, Harare,
Zimbabwe

Abstract

PERSISTENCE IN SOIL OF ENDOSULFAN AND LINDANE APPLIED TO SOYABEAN AND MAIZE PESTS IN A FIELD TRIAL AGROSYSTEM IN ZIMBABWE.
 The persistence of endosulfan and lindane in soil was determined following their application to control soyabean and maize pests, respectively, under Zimbabwean weather conditions. The results show that degradation of the pesticides is pronounced only during the rainy season, and that when the pesticides are applied at rates of 600 g/ha for endosulfan and 810 g/ha for lindane, carry-over of the pesticide residues in the soil to the next rainy season can be expected.

1. INTRODUCTION

In 1989 it was estimated that Africa accounted for 5% of the world's pesticide use [1]. The greater portion of these pesticides is used to control pests in agriculture. Although use of pesticides in agriculture has been reported to have adverse effects on both man and other non-target organisms in the environment, agricultural development and the need to increase food production demand the use of pesticides. It is, however, necessary to assess the impact of such usage on the environment.

The effect of pesticides on the environment depends on several factors, including the temperature, rainfall, soil type, biotic activity, light intensity, land cultivation and other agricultural practices. These factors determine the persistence of a pesticide in a specific environment, and in this respect, as a group, organochlorine pesticides have been found to be the most persistent. Thus, although the effect of

---

* Research carried out with the support of the IAEA under Research Contract No. 6759.

pesticides on the environment has been studied extensively in developed countries, there is still a need to assess their impact on developing countries, where agricultural practices differ.

The organochlorine pesticides registered for use in agriculture in Zimbabwe include aldrin, endosulfan, chlordane, chlorthalmethyl, dicofol, dieldrin and lindane [2]. The aims of this study were to determine the persistence of organochlorine pesticides in the agrosystem. The pesticides selected for the study were endosulfan and lindane. Endosulfan is used extensively to control the heliothis bollworm, the semi-looper caterpillar and aphids in soyabean and groundnut, aphids in potatoes, and the cutworm and red mite in maize [1], while lindane is used as a seed dresser [2].

Soil is the major sink for the chemicals applied to crops. Persistence patterns vary from one climate to another [3] and also depend on the chemical nature of the pesticide and the type of soil. It has been shown that degradation depends on the climatic conditions. In cold climates, degradation is generally slower than in tropical climates [3, 4].

Lindane undergoes microbial degradation by dechlorination to pentachlorocyclohexane [5]. The bacteria *Clostridium sporogenies* and *Bacillus coli* produce benzene and monochlorobenzene from lindane [6]. Other possible transformations of lindane that occur, especially in wet and submerged conditions, include isomerization into the alpha-HCH and beta-HCH isomers [7]. The isomers rapidly degrade in soil [8]. Endosulfan was found to undergo epoxidation, which is enzyme catalysed [9], and other transformations, including hydrolysis, reduction and hydroxylation [10]. Mechanical processes such as volatilization and runoff also deplete the level of these chemicals in soil.

The persistence of endosulfan in soil environments has attracted the attention of several research workers. Burns [11] studied degradation in soils. Kathpal et al. [12] and Martens [13] studied the kinetics of its degradation in soils, and reported rates of 63% loss in 2–3 months and 5.4% in 15 weeks, respectively. Martens also studied the non-microbial degradation of the pesticide in soils under varying conditions of pH, sample wetness and temperature, and found that degradation was faster in wet samples with a higher pH. El Beit et al. [14] studied the kinetics of endosulfan leaching in sediments and found that the rate of leaching was slow compared with the rate of degradation by microorganisms.

The work reported on in this paper was carried out between 1991 and 1995. The trials conducted during the 1991/1992, 1993/1994 and 1994/1995 growing seasons, November to March in Zimbabwe, were adversely affected by severe droughts. As a result, meaningful data were only collected during the 1992/1993 growing season.

2. MATERIALS AND METHODS

2.1. Experimental field site preparation

The 1992/1993 field experiments were conducted on specially designated land in the experimental section of the University of Zimbabwe farm, approximately 10 km from the university. After mechanical land preparation, the trial field (approximately 1 ha) was divided into four plots (approximately 0.25 ha each), and four subplots within each plot (see Fig. 1), planted with commercially dressed soyabean and maize seed (23 December 1992). The experimental plot was bordered by cattle pastures to the north and east, while plot A was bordered by a maize crop and plot B by a roadway. The crops were separated by a 1.5 m buffer zone. This was done to simulate as closely as possible actual agricultural practice.

2.2. Pesticide application

Soyabean and maize spraying was carried out on 20 February 1993. Thiodan (endosulfan) 50 WP and Multi-benhex (lindane) 75 WP were used at 1200 and

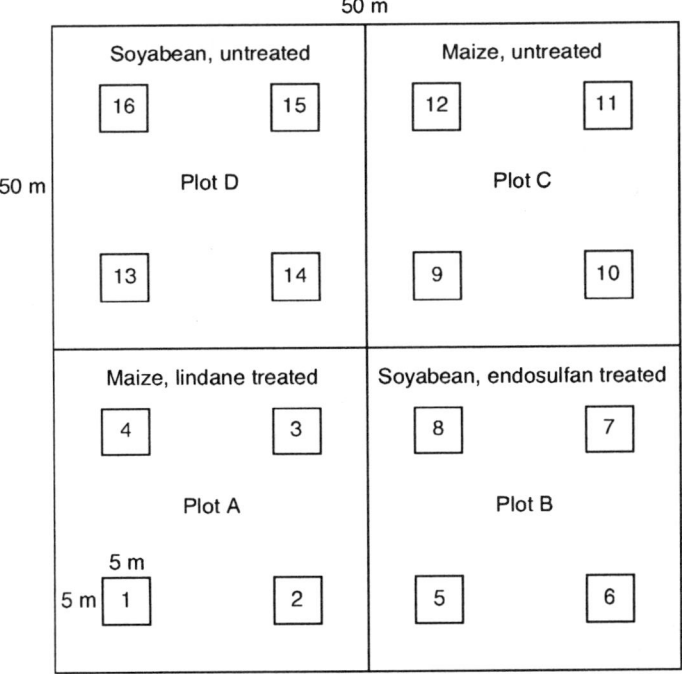

FIG. 1. Experimental field layout (1–16 = subplots).

TABLE I. LINDANE AND ENDOSULFAN RESIDUE LEVELS IN SOIL SAMPLES[a]

Weeks	n	Residue levels (ng/g)		
		Lindane	Endosulfan I	Endosulfan II
Plot A: Maize, lindane treated				
Pretreatment	8	ND	20 ± 7	6 ± 2
2	11	227 ± 71	29 ± 13	17 ± 10
5	14	137 ± 33	4 ± 3	7 ± 6
7	12	53 ± 13	8 ± 6	3 ± 5
10	12	37 ± 8	5 ± 4	0.4 ± 0.8
25	11	24 ± 4	3 ± 2	1 ± 2
Plot B: Soyabean, endosulfan treated				
Pretreatment	8	16 ± 7	ND	ND
2	9	3 ± 2	185 ± 5	119 ± 121
5	15	6 ± 5	83 ± 10	40 ± 7
7	15	7 ± 8	71 ± 1	23 ± 2
10	—	—	—	—
25	14	1 ± 2	40 ± 3	13 ± 2
Plot C: Maize, untreated				
Pretreatment	8	ND	ND	ND
2	13	1 ± 2	1 ± 2	ND
5	10	ND	ND	ND
7	—	—	—	—
10	13	ND	ND	ND
25	—	—	—	—
Plot D: Soyabean, untreated				
Pretreatment	—	—	—	—
2	14	3 ± 3	1 ± 1	—
5	10	4 ± 4	2 ± 1	—
7	9	2 ± 3	4 ± 4	—
10	14	1 ± 1	1 ± 2	—
25	—	—	—	—

[a] ND = not detected; — = analysis not conducted.

1080 g/ha, respectively, as a suspension of 300 g of endosulfan and 270 g of lindane in 120 L of water. Spraying was done using 15 L knapsack sprayers (Taurus Spraying Systems, Harare, Zimbabwe). To ensure that the pesticide was applied evenly, preliminary spraying tests were carried out using water.

2.3. Sample collection

In each plot, four subplots (each 5 × 5 m, and 10 m from the edge of the plot) were marked out, as shown in Fig. 1, for sampling activities. Four soil samples were collected from each of the four subplots within each plot. Each subplot was divided into four quadrants, with one sample collected from each quadrant. Sampling was done by means of a cylindrical soil corer (5 × 15 cm). Each soil core was placed

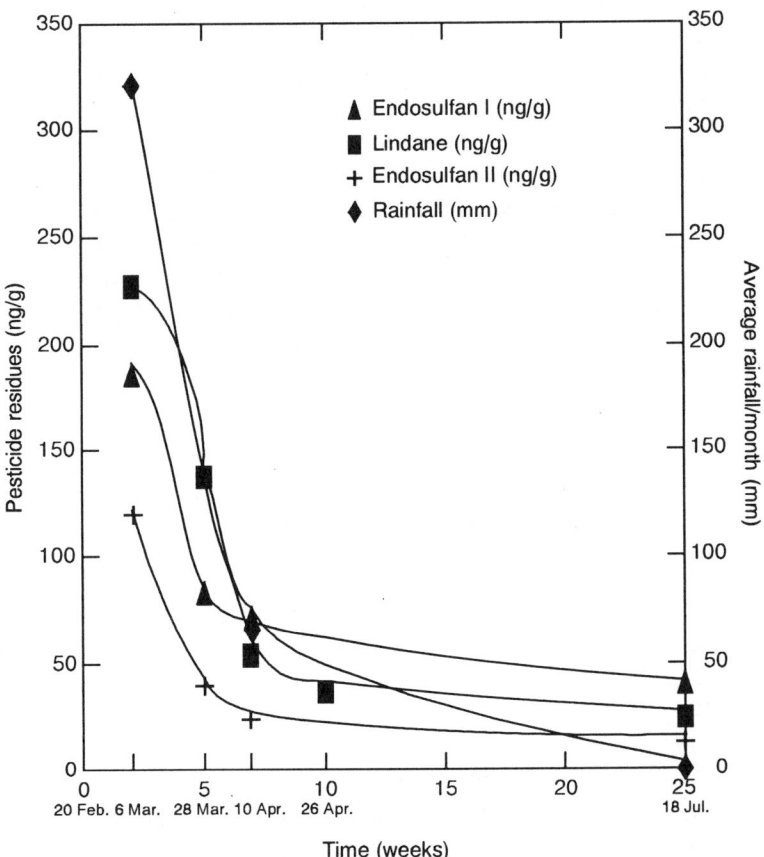

FIG. 2. Variation in the pesticide and rainfall levels with time (monthly average rainfall for February–July 1993).

in a clean polythene bag, which was then sealed and labelled with the subplot number and the sampling point number. Samples from each plot were then placed in a single bag, which was sealed and labelled with the plot letter, for transportation to the laboratory, where they were stored in a deep freeze until analysis.

2.4. Residue analysis

Each soil core was extracted using the method previously described by Japenga et. al. [15], while cleanup was done using a modification of the Environmental Protection Agency method for analysis of soils and house dust [16]. The frozen sample was thawed, ground and sieved through a 0.25 cm sieve. A mass equivalent to 1 g dry weight of soil was taken from the core and weighed in a 10 mL beaker. Then 0.5 mL of acetic acid was pipetted into the beaker and the mixture stirred using a glass rod. The acetic acid was used to hydrolyse the pesticide colloid particle bonds so as to free the soil bound pesticides; 0.5 mL of nonane was then added to the slurry and the mixture again stirred with the glass rod. This was done in order to thin the slurry so that it was suitable for ultrasonic shaking, which is necessary to facilitate free bound pesticide molecules. The resulting slurry was exposed to ultrasonic shaking for 30 min. The mixture was then allowed to stand and 5 g of pesticide grade silica gel (Merck, Darmstadt, Germany) were added and the mixture stirred. The finely divided powder was transferred to a cellulose extraction thimble (Merck, Darmstadt, Germany) containing 5 g of silica gel. The thimble was then placed in a Soxhlet extraction apparatus and extracted for 4 h with a 2:1 mixture of hexane and benzene. The crude extract was concentrated in a Kuderna–Danish apparatus to 1 mL and then subjected to cleanup on a Pyrex glass column (1 × 11 cm) containing

TABLE II. DETERMINATION OF LINDANE AND ENDOSULFAN IN SOIL: EXTRACTION RECOVERY EFFICIENCIES AT THREE FORTIFICATION LEVELS[a]

Lindane			Endosulfan I			Endosulfan II		
F/L	R	%R	F/L	R	%R	F/L	R	%R
3.01	2.50	83	2.48	2.10	85	2.53	2.20	87
6.32	5.00	80	5.00	4.20	84	5.04	4.40	87
9.315	9.04	97	9.00	8.93	99	9.83	9.77	99

[a] F/L = fortification level (ng/g); R = recovery (ng/g); %R = % recovery.

about 5 cm of Florisil and 5 cm of anhydrous sodium sulphate. The column was prewashed with 20 mL of hexane and then 20 mL of 5% methanol in hexane, dried at 180°C, and activated at this temperature overnight. The column was cooled to room temperature and then prewetted with hexane before placing the extract on the column and eluting with 20 mL of hexane (first fraction), followed by 20 mL of 5% methanol in hexane (second fraction). The fractions were separately concentrated to 1 mL in the Kuderna–Danish apparatus and stored in 1.8 mL glass vials with Teflon lined screw caps (Supelco, Bellefonte, PA, United States of America) at 4°C until analysis.

The concentrated fractions were analysed separately using a Varian Model 3300 GC (Varian AB, Solna, Sweden) fitted with a microprocessor, a split/splitless capillary injector, a Varian Model 4400 integrator and a ^{63}Ni electron capture detector (ECD), after separation on a DB-1701 30 m × 0.25 mm refined silica capillary column for pesticides and chlorinated aromatics (J. and W. Scientific, San Jose, CA, USA). Ultrapure nitrogen carrier gas was used at a flow rate of 5 mL/min (make-up gas at 25 mL/min). The injector was maintained at 150°C and operated in the splitless mode for the first 30 s only, while the ECD detector was maintained at 300°C and operated at a range of 10 mV and an attenuation of 8. Single step linear temperature programming was employed from the 100°C initial temperature (held for 2 min) to the 250°C final temperature (held for 3 min). None of the pesticides was detected in fraction 2. The averages for individual samples from each plot are shown in Table I, and plotted as a function of time in Fig. 2. The figure also shows a plot of the average monthly rainfall during the study period. Validation of the results was done by spiking 1 g soil samples at three fortification levels with solutions containing a mixture of lindane, endosulfan I and endosulfan II standards and stirring to ensure even distribution of the pesticides in the sample. They were then stored in the dark for 24 h to equilibrate (before extraction) cleanup, concentration and analysis, as described above. Recoveries of 83–99% were obtained (see Table II). The detection limit was 1 ng/g for both lindane and endosulfan.

3. RESULTS AND DISCUSSION

Figure 2 shows the persistence of lindane and endosulfan I and endosulfan II in soil. Both pesticides showed rapid initial rates of disappearance during the first 5–7 weeks. Thereafter, the rates were much slower. The curves suggest a pseudo first order degradation rate law. Similar results have been obtained by Lichtenstein and Schultz [17] for a single application of 25 lb of dieldrin per acre in Wisconsin (USA) soil.[1] The rapid initial rate of disappearance in the first 5–7 weeks was

---

[1] 1 lb = 0.4536 kg; 1 acre = 4.047 × 10^3 m^2.

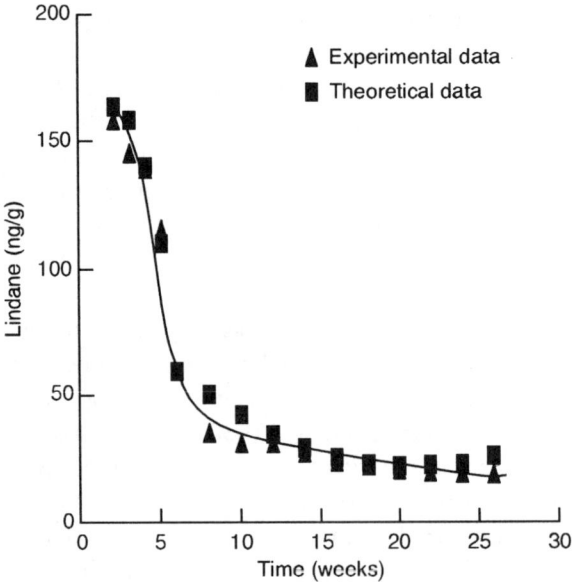

FIG. 3. *Disappearance of lindane from soil in plot A.*

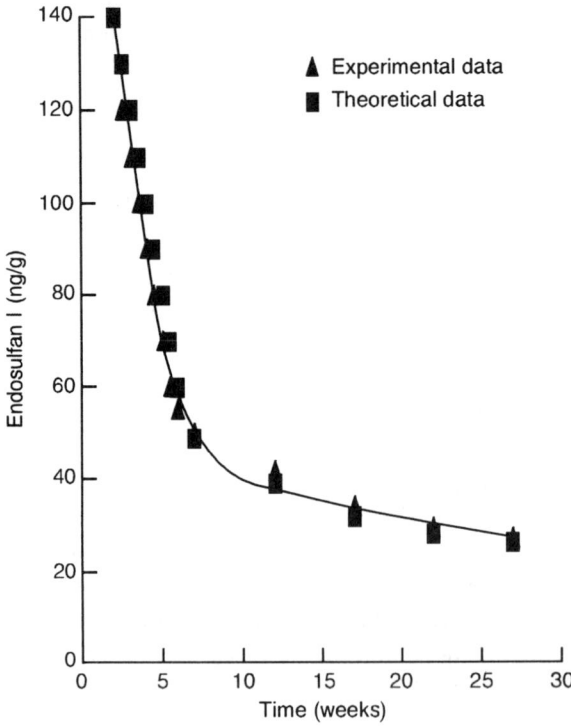

FIG. 4. *Disappearance of endosulfan I from soil in plot B.*

mainly attributed to enzymatic degradation during the weeks of high rainfall when the soil was moist. The rainfall profile during the study period closely resembled the persistence curves of the three pesticides. Loss of pesticide by runoff was minimal, as shown by the extremely low level of pesticides detected in the untreated plots.

Table I shows that lindane was also detected in plot B, while endosulfan was also detected in plot A, i.e. those plots that were not treated with the specific pesticide. This can be attributed to the fact that these plots had been treated with these pesticides in the previous year. The results show the likely extent of carry-over each year, and can be used to access the rate of pesticide buildup after successive annual applications of pesticides.

Regression analysis of the loss of pesticide for the initial rapid rate of disappearance and the subsequent slower rate leads to

$$P_{t_1} = P_0 + \sum_{t=t_1}^{n_1} (a - k_1 t_1) \Big|_{t=t_0}^{n_1} \qquad (1)$$

$$P_{t_2} = P_0 + \sum_{t=t_1}^{n_1} (a - k_1 t) - \sum_{t=t_2}^{n_2} (b - k_2 t_2) \Big|_{n_1}^{n_2} \qquad (2)$$

where P_0 is the initial pesticide concentration, P_t is the pesticide concentration after t weeks, and subscripts 1 and 2 denote the initial period of rapid disappearance and

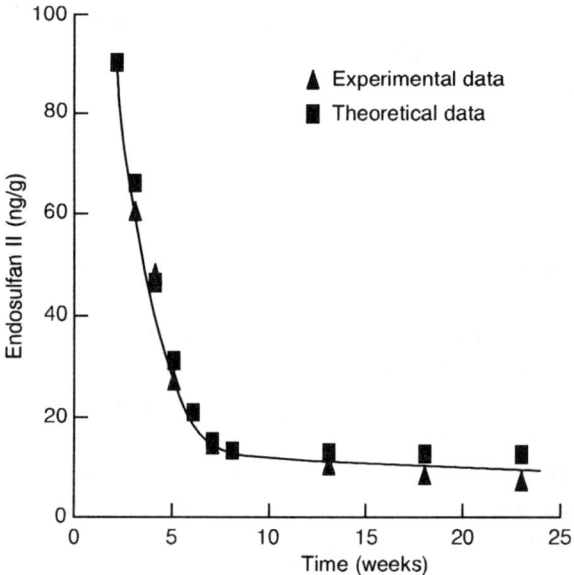

FIG. 5. *Disappearance of endosulfan II from soil in plot B.*

TABLE III. VALUES OF THE CONSTANTS a, b, k_1 AND k_2, AND n_1 AND n_2 FROM REGRESSION ANALYSIS[a]

	a	k_1	b	k_2	n_1	n_2
Lindane	−42	14	15	−0.2	5	25
Endosulfan I	10	0	8.7	−0.24	6	27
Endosulfan II	−39	4.6	0.8	0.03	8	23

[a] a, b, k_1 and k_2 are constants obtained after the regression analysis of loss per interval versus time.

the subsequent period of lower disappearance, respectively; n_1 and n_2 are the upper limits of t_1 and t_2, respectively, and a, b, k_1 and k_2 are constants obtained from regression analysis.

Equations (1) and (2) can be used to predict the rates of disappearance from subtropical soils, provided the mean values of the constants a, b, k_1 and k_2 are used for several seasons. The values of these constants will depend on the type of pesticide, the soil and climatic conditions, physical factors such as land preparation, and chemical factors such as the soil pH, moisture content and microbiological activity.

Figures 3–5 show that the persistence curves of lindane and endosulfan I and endosulfan II, respectively, can be reproduced with very good agreement when the model is used to predict the disappearance of pesticide using the values of a, b, k_1 and k_2 shown in Table III and the n_1 and n_2 values of 6 and 25 weeks, respectively.

4. CONCLUSIONS

It can be concluded that the rate of disappearance of lindane and endosulfan from soil depends on rainfall, and when applied at the rate of 600 g/ha for endosulfan and 410 g/ha for lindane, a buildup of pesticides in the agrosystem can be expected because of carry-over of the two pesticides from previous season application. The results showed that there is a need to monitor the possible buildup of lindane and endosulfan in soil as a result of using these pesticides in agriculture. In addition, it was shown that there is a need to extend these studies to cover other organochlorine pesticides in use, not only in Zimbabwe but also in other subtropical regions of the African continent.

ACKNOWLEDGEMENTS

This work was carried out as part of the FAO/IAEA Co-ordinated Research Programme on the Adverse Side Effects on Flora and Fauna from the Use of Organochlorine Pesticides on the African Continent. It was supported in part by a grant from the IAEA, in part by a grant from the Swedish Agency for Research Co-operation with Developing Countries (SAREC), and in part by a grant from the Research Board of the University of Zimbabwe. The authors wish to thank the University of Zimbabwe Farm Board for use of their farm.

REFERENCES

[1] MHLANGA, A.A., MADZIVA, J.J., Pesticides residues in Lake McIlwaine, Ambio **19** (1990) 368–372.

[2] MATHUTHU, A.S., A Feasibility Study on the Use of Agrochemical Pesticides in Zimbabwe: SADC–ELMS Sub-programme 4.4, Final Report, Southern African Development Co-ordination (Aug. 1993).

[3] EDWARDS, C.A., Environmental impact of pesticides, Parasitis **86** (1987) 309–329.

[4] EDWARDS, C.A, Persistent Pesticides in the Environment, 2nd edn, CRC Press, Cleveland, OH (1976).

[5] CHIBA, M, MORLEY, H.V., Factors influencing extraction of aldrin and dieldrin residues from different soils, J. Agric. Food Chem. **16** (1968) 916–922.

[6] ALLAN, J., Loss of biological efficiency of cattle dipping wash containing benzene hexachloride, Nature (London) **175** (1955) 1130–1132.

[7] NEWLANDS, L.W., CHESTERS, G., LEE, G.B., Degradation of γ-BHC in simulated lake impoundments as affected by aeration, J. Water Pollut. Contr. Fed. **41** (1969) R174–R188.

[8] McRAE, I.C., RAGHU, K., CASTRO, T.F., Persistence and biodegradation of four isomers of benzene hexachloride in submerged soils, J. Agric. Food Chem. **15** (1967) 911.

[9] MENZI, C.M., Metabolism of Pesticides, United States Fish and Wild Life Service, Special Science Report, Washington, DC (1969) 127 pp.

[10] MILES, J.R.W., MOY P., Degradation of endosulphan and its metabolites by a mixed culture of soil microorganisms, Bull. Environ. Contam. Toxicol. **23** (1978) 1.

[11] BURNS, R.G., Degradation of thiodan in soil, Bull. Environ. Contam. Toxicol. **6** 4 (1971) 361–322.

[12] KATHPAL, T.S., et al., Degradation of thiodan in soil under Indian conditions, Indian J. Entomol. **43** 4 (1981) 420–427.

[13] MARTENS, R., Degradation of endosulphan-8,9-^{14}C in soil under different conditions, Bull. Environ. Contam. Toxicol. **17** 4 (1977) 438–446.

[14] El BEIT, I.O.D., WHEELOCK, J.V., COTTON, E.D., Factors involved in the dynamics of pesticides in the soil: The effect of pesticide concentration on leachability, Int. J. Environ. Stud. **16** 3–4 (1981) 187-7.

[15] JAPENGA, J., WAGENAAR, W.J., SMEDES, F., SALOMONS, W., A new rapid clean up procedure for the determination of organic micropollutants: Application in two European estuarine sediment studies, Environ. Technol. Lett. **8** 1 (1987) 9–20.

[16] ENVIRONMENTAL PROTECTION AGENCY, "Sample preparation and analysis of soils and house dust", Manual of Analytical Methods for the Analysis of Pesticide Residues in Human and Environmental Samples, Section 11A, Revised 11 February 1972, Environmental Toxicology Division, EPA, Washington, DC (1978).

[17] LICHTENSTEIN, E.P., SCHULTZ, K.R., Residues of aldrin and heptachlor and their translocation into various crops, J. Agric. Food Chem. **13** 1 (1965) 57.

IAEA-SM-343/23

PERSISTENCE AND FATE OF ^{14}C-LINDANE APPLIED TO SOIL IN A MAIZE ECOSYSTEM*

P.O. YEBOAH**, K.G. MONTFORD***,
F.E. APPOH+, D.K. DODOO+

**Department of Chemistry,
  National Nuclear Research Institute,
  Legon

***Biotechnology and Nuclear Agriculture
   Research Institute,
   Legon

+Department of Chemistry,
 University of Cape Coast,
 Cape Coast

Ghana

Abstract

PERSISTENCE AND FATE OF ^{14}C-LINDANE APPLIED TO SOIL IN A MAIZE ECOSYSTEM.
  Radiolabelled (^{14}C) lindane applied to the soil surface in a maize ecosystem (1 month after planting) was found to be taken up by the plant. Within the first 25 days after application, ^{14}C-lindane was distributed throughout the entire plant, with the greatest concentration in the lower leaves (from ground level) and a sharp buildup towards the tip of each leaf. The radioactivity and, hence, the pesticide concentration were uniformly distributed throughout the plant with time, to the extent that measurable levels of ^{14}C-lindane were detected in the tassel, cob and grain. This indicated that soil applied ^{14}C-lindane was available to the maize plant. The persistence of ^{14}C-lindane in soils of variable organic matter content was also studied. Evidence is presented to show that ^{14}C-lindane dissipated faster in soils with a lower organic matter content. The levels of surface applied ^{14}C-lindane that were bound in the soil increased with time and also with the increasing organic matter content. The radioactivity was mainly associated with the top soil layer (0–3 cm).

---

* Research carried out with the support of the IAEA under Research Contract No. 6754.

1. INTRODUCTION

As a result of the increasing world population, there has been a proportionate increase in the demand for food. Currently, 58% of the world's population is undernourished [1], therefore there is a need to increase food production. This can be achieved using products of modern technology such as high yielding crop varieties, fertilizers, pesticides and irrigation. Pesticides help boost food production by reducing pest populations to tolerable levels. In 1986 it was found that crop losses due to pests were 40%. Cotton yields have also increased threefold through effective use of chemical pesticides for pest control in the Sudan [2].

Despite the benefits derived from pesticides, some tend to persist in the environment and pose a health hazard to living organisms. Organochlorine pesticides constitute one such group of pesticides that are persistent; examples of these include dichloro-diphenyl trichloro ethane (DDT), lindane and endosulfan. They have also been found: (1) to be highly fat soluble and, hence, to accumulate in the fatty tissues of animals; (2) to accumulate in the food chain; and (3) to degrade into other toxic residues that can have long term adverse effects on the environment [3, 4].

In Ghana, the organochlorines in use are lindane, endosulfan, aldrin and dieldrin. Use of lindane, formulated under the trade name Gammalin 20, has been restricted in the cocoa growing industry. However, owing to mixed cropping, other crops are sprayed with lindane during application, e.g. maize, cocoyam and cassava. Under such conditions, lindane serves to control insects that bore into the stems of maize plants, thereby destroying them.

Notwithstanding the hazards posed by organochlorine pesticides to the environment (hence their being banned and/or restricted in temperate regions), many developing countries, including Ghana, continue to use them to boost food production. They are also used in public health to control vector diseases. Continued usage has been attributed to their high efficiency, low cost and easy availability. In addition, owing to the intense heat and high humidity in the tropics, organochlorine pesticides dissipate at a faster rate than in temperate regions. Consequently, it is suggested that there should be no possibility of their residues accumulating in soils, plants and animal matter, leading to the precipitation of environmental hazards.

Whereas considerable data on pesticide residues are available in developed countries to assist in decision making on the continued use of particular pesticides, knowledge on pesticide residues and their effects is rarely available in many developing countries. Thus, there is a need to generate sufficient data on the effects of these pesticides in various agroecosystems in the tropics to justify their worth, or otherwise. These should include extensive knowledge on their fate, persistence, distribution, effect on yield, and the way in which they affect target and non-target organisms in order to enable easy assessment of their potential for human and environmental hazards.

Maize, a staple food of Ghana, is taken in many forms. It may be boiled, roasted or ground into dough. Since lindane is applied to maize, knowledge of its residues in this crop would be of immense importance to Ghanaians. This paper reports on the persistence and fate of radiolabelled lindane in a maize ecosystem.

2. MATERIALS AND METHODS

2.1. Chemicals

Uniformly labelled ^{14}C-lindane (with a specific activity of 647.5 MBq/mmol) was obtained from the Sigma Chemical Co., St. Louis, United States of America, through the IAEA in Vienna. The radioactive concentration of the lindane (supplied as toluene solution) was 37 MBq/mL. The scintillators, 2,5-diphenyloxazole (PPO) and 2.2-phenylenebis (4-methyl-5-phenyoxazole) (DMPOPOP), were purchased from Eastman Kodak Co., Rochester, USA. All the other chemicals were of analar grade.

2.2. Persistence of ^{14}C-lindane

To study the persistence of ^{14}C-lindane in a maize ecosystem, 51 PVC tubes (with an external diameter of 7 cm and a length of 30 cm) were buried in the soil in rows between plants (in another plot), with 3 cm projecting above the ground. The tubes were arranged in groups of 15 according to the organic matter content of the soils they contained. The soil in each of these group tubes was treated with 185 kBq of ^{14}C-lindane at the same time as the plants in the field were sprayed with commercial lindane. Two tubes in each soil group were not treated with ^{14}C-lindane and served as the control. The tubes were dug out carefully at intervals, wrapped in polythene bags and stored in a refrigerator pending extraction and determination of the pesticides.

2.3. Uptake of ^{14}C-lindane in maize plants

To study the uptake of ^{14}C-lindane from treated soils by the maize plants, 20 plastic buckets (open at both ends, with an internal diameter of 24 cm and a length of 22 cm) were filled with soil from one of the three sources. They were buried in the crop row, with 3 cm projecting above the ground. Maize was planted in each bucket on the same day as planting was done in the field. The soil in the buckets was treated with 370 kBq of ^{14}C-lindane on the same day as the plants on the experimental farm were sprayed with commercial lindane. Three plants from three separate buckets were taken periodically and analysed for their uptake of radioactivity. The anatomical and sectional distribution of the ^{14}C-lindane residues in the maize plants were studied.

2.4. Extraction and analysis

2.4.1. Soils

The soil contained in the PVC tubes was scooped out from the top (end exposed to the atmosphere) in successive 1.5 cm layers. Each soil layer was weighed, dried, ground separately (using a pestle and mortar) and sieved through a 2 mm mesh sieve. From each sample of ground soil, 40 g were weighed in cellulose extraction thimbles (43 × 123 mm) and Soxhlet extracted for 2.5 hours using methanol. The extracts were concentrated at 50 ± 5°C under vacuum to a volume of 5 mL, which was then transferred to scintillation vials and analysed for radioactivity. Recovery was 80.2%.

2.4.2. Plants

The anatomical sections of the maize plants above ground (leaves, stem, cob, etc.) were removed at intervals and stored in a refrigerator until analysis. The stem and leaves were carefully removed from the points of attachment, cut into 3 cm segments and analysed for radioactivity using the dry combustion method.

2.5. Radioactivity measurements

The total and bound ^{14}C residues in soils and in the plant parts were determined by combusting the weighed samples in a model 600 biological material oxidizer (Harvey Instrument Corporation, New Jersey, USA). The $^{14}CO_2$ was absorbed in a 2 mL absorber/cocktail (ratio 1:1 (vol./vol.)). The absorber was made up of 12.5% ethanolamine in 87.5% methanol, while the cocktail consisted of 5 g of PPO and 50 g of DMPOPOP dissolved in 1 L of toluene. The radioactivity was measured with a liquid scintillation counter in a Tri-Carb liquid scintillation analyser model 100 using the sample channel ratio method for quench correction.

3. DISCUSSION

3.1. Uptake and translocation of ^{14}C-lindane in maize plants

Figures 1–4 show the fate and distribution of ^{14}C-lindane in maize plants. Within 25 days of pesticide application to the soil it was demonstrated that the plants took up a measurable amount of ^{14}C-lindane, with the greatest amount (0.08%) being in the lowest leaf (first leaf above ground). This was followed by a gradual decrease in the ^{14}C-lindane concentration in the higher leaves and the stem. The upward movement of the radioactivity in the maize plant was further shown by the gradual buildup of ^{14}C-lindane towards the tip of the lower leaves (Fig. 3).

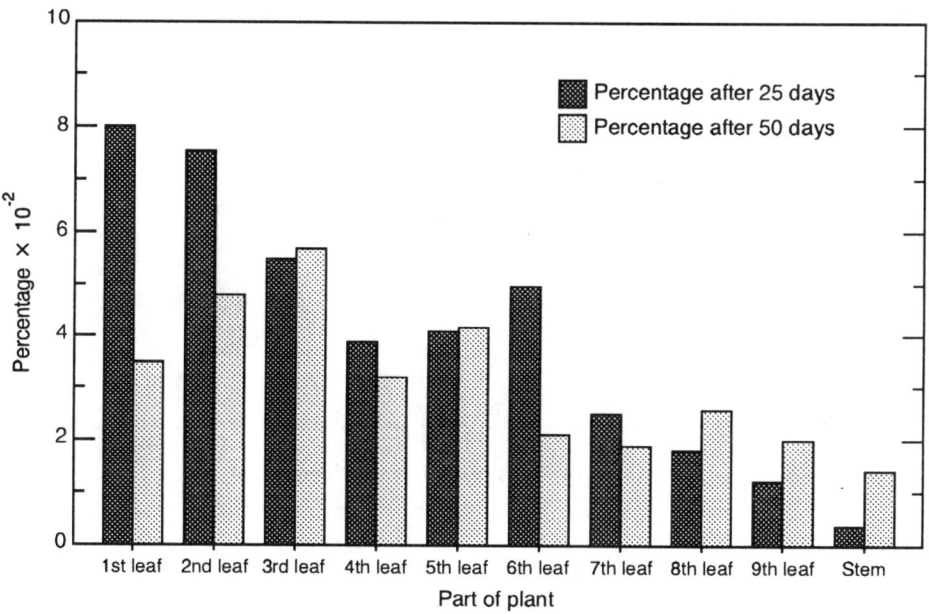

FIG. 1. Percentage initial ^{14}C-lindane recovered from the leaves and stem.

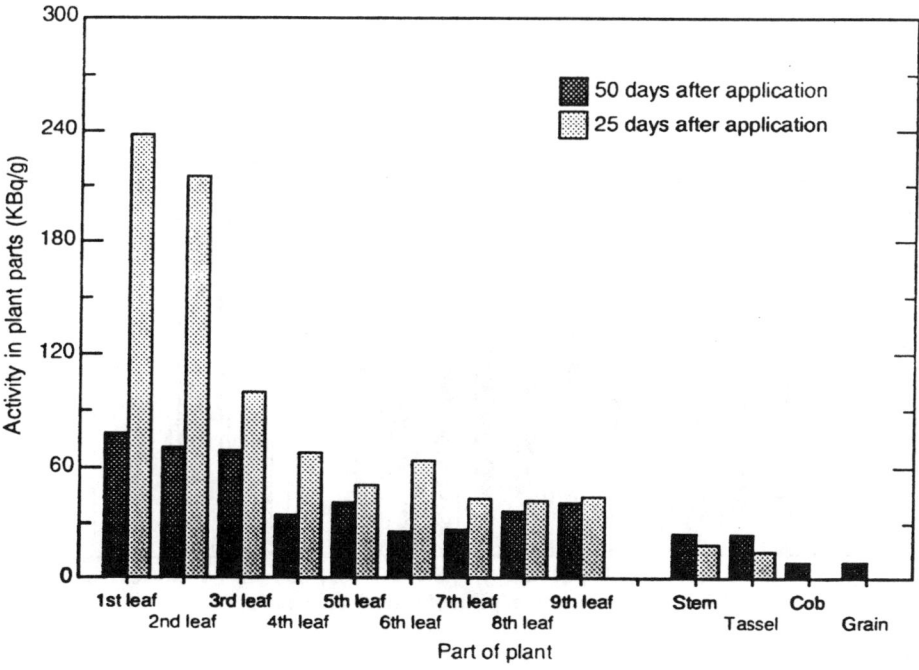

FIG. 2. Distribution of ^{14}C-lindane in maize plants.

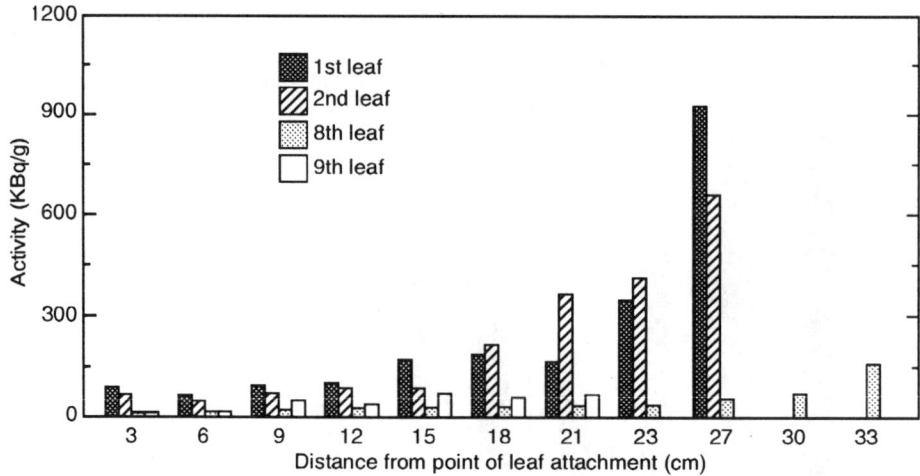

FIG. 3. Distribution of ^{14}C-lindane in maize leaves (25 days after application).

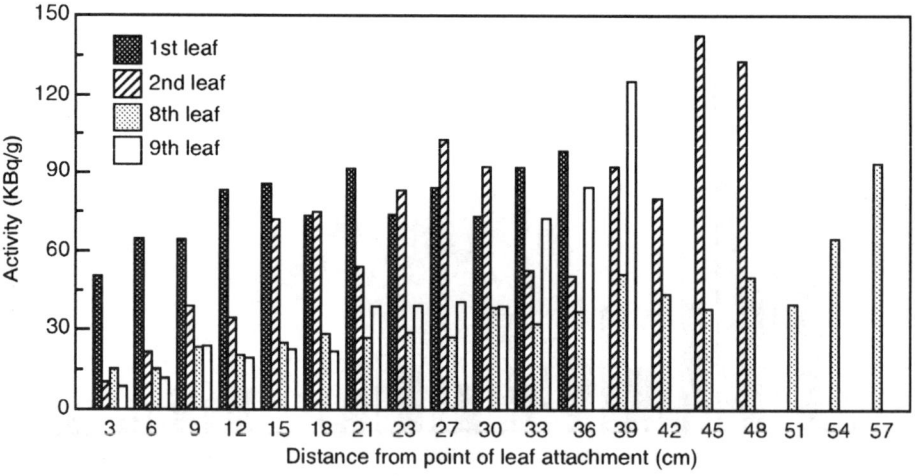

FIG. 4. Distribution of ^{14}C-lindane in maize leaves (50 days after application).

With time, i.e. within 50 days of pesticide application, the radioactivity tended to spread out uniformly throughout the plant, with the greatest concentration of activity in the third leaf and a gradual decrease in the uppermost leaves and the stem. Figure 2 shows that some radioactivity was found in the tassel, cob and grain 50 days after application, indicating that soil applied ^{14}C-lindane is bioavailable to the plant. These results are not surprising. The translocation of soil applied ^{14}C-lindane and its metabolites has been investigated in detail [5–8]. Neither ^{14}C-lindane nor its metabolites were evenly distributed in the plants. Comparatively high ^{14}C-lindane residue levels were always detected in the leaves, whereas small amounts were translocated to the stems, leaves and fruits. Paasivirta et al. [8] have shown that in water plants the lindane concentration is similar in the roots and stem. Differences in residue levels have been shown to be dependent on plant species. Of a series of edible crops grown on soils containing lindane, carrots were found to have higher levels than beans, tomatoes and potatoes, in that order. In soils with little organic matter content, as in this case, the ^{14}C-lindane residues were more mobile and, hence, susceptible to uptake by plants [9].

3.2. Movement of ^{14}C-lindane in soils

Movement of ^{14}C-lindane through the soil profile was also studied. Figure 5 indicates that ^{14}C-lindane dissipated faster in soils with a lower organic matter content. In the three soils tested, there was a sharp rate of dissipation of ^{14}C-lindane within the first 10 days of application, followed by a gradual decrease. These results

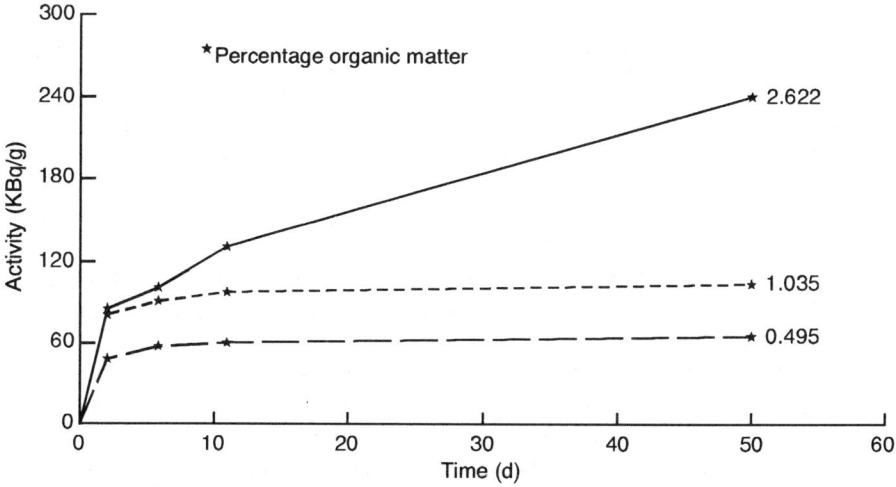

FIG. 5. Bound residues of ^{14}C-lindane in soils of varying organic matter content.

suggest that ^{14}C-lindane dissipated faster from the three soils investigated than has been observed in other tropical regions.

Faster dissipation was attributed to the low organic matter content of the soils. In spite of copious rainfall during the period after pesticide application (104 mm in 2 months), residues of ^{14}C-lindane were found in the upper 0–3 cm layer of the soil surface. This is not surprising, considering the sandy, loamy, clayey nature of the soils examined. Cliath and Spencer [10] observed no mobility of lindane in clay soils.

ACKNOWLEDGEMENTS

This work was supported by the Swedish International Development Agency (SIDA) through the IAEA. The authors acknowledge the contributions of B.Q. Modzinuh and S. Afful of the Chemistry Department of the NNRI for their assistance during sample collection.

REFERENCES

[1] WARE, G.W., Pesticide, Theory and Applications, Freeman, San Francisco, CA (1978) 14 pp.

[2] FOOD AND AGRICULTURE ORGANIZATION OF THE UNITED NATIONS, International Code of Conduct on the Distribution and Use of Pesticides, FAO, Rome (1986) 8 pp.

[3] INTERNATIONAL ATOMIC ENERGY AGENCY, Isotope Techniques for Studying the Fate of Persistent Pesticides in the Tropics, IAEA-TECDOC-476, IAEA, Vienna (1988) 13 pp.

[4] INTERNATIONAL ATOMIC ENERGY AGENCY, Laboratory Training Manual on the Use of Nuclear and Associated Techniques in Pesticide Residues, Technical Reports Series No. 329, IAEA, Vienna (1991) 144 pp.

[5] ITOKAWA, H., SCHALLA, A., WEIGERBER, I., KLEIN, W., KORTE, W., Contributions to ecological chemistry. XXII. Metabolism and residue behaviour of lindane-^{14}C in higher plants, Tetrahedron **26** (1970) 763–773 (in German).

[6] BALBA, M.H., SABA, J., Metabolism of lindane-^{14}C by wheat plants grown from treated seed, Environ. Lett. **7** 3 (1974) 181–194.

[7] BRADBURY, F.R., The systemic action of benzene hexachloride seed dressings, Ann. Appl. Biol. **52** (1963) 361–370.

[8] PAASIVIRTA, J., PALM, H., PAUKKA, R., AKHABUHAYA, J., LODENIUS, M., Chlorinated insecticide residues in Tanzanian environment: Tanzadrin, Chemosphere **17** 10 (1988) 2055–2062.

[9] FÜHRMANN, T.W., LICHTENSTEIN, E.P., A comparative study of the persistence, movement, and metabolism of six carbon-14 insecticides in soils and plants, J. Agric. Food Chem. **28** (1980) 446–452.

[10] CLIATH, M.M., SPENCER, W.F., Dissipation of pesticides from soil by volatilization of degradation products. 1. Lindane and DDT, Environ. Sci. Technol. **6** 10 (1972) 910–914.

DEGRADATION OF CHLORPYRIFOS IN TURKISH SOIL*

Ü. YÜCEL, M. ILİM, K. GÖZEK
Ankara Nuclear Research and Training Center,
Turkish Atomic Energy Authority,
Ankara, Turkey

Abstract

DEGRADATION OF CHLORPYRIFOS IN TURKISH SOIL.
Surface (0–5 cm) and subsurface (40–60 cm) soils were collected from a field not treated with pesticide and 100 g triplicate samples of the soils were weighed in glass biometer incubation flasks. Other than the control samples, each soil was uniformly treated with ^{14}C-chlorpyrifos to yield a soil concentration of 2 µg/g of chlorpyrifos and 7.4×10^4 Bq (2 µCi) of radioactivity and incubated for 97 days at 25°C. The KOH traps were sampled for the evolved $^{14}CO_2$. Soil samples were taken periodically, and the solvent was extracted and subjected to supercritical methanol extraction. The extracts were analysed by thin layer chromatography. The total ^{14}C and the unextractable soil bound ^{14}C residues were determined by combustion in a biological oxidizer. From the surface and subsurface soils, up to 40.6 and 42.6%, respectively, of the applied radiocarbon evolved as $^{14}CO_2$ during the 97 day incubation period. The chlorpyrifos half-lives in the surface and subsurface soils were calculated as 7.2 and 4.3 days, respectively. The major products of degradation, in order of decreasing concentration, were $^{14}CO_2$, chlorpyrifos, soil bound ^{14}C residues and TCP, regardless of soil depth, and the percentages found were 28.9, 13.1, 11.7 and 3.8 of the applied ^{14}C for the surface soils and 22.1, 12.4, 11.4 and 4.8 of the applied ^{14}C for the subsurface soils, respectively.

1. INTRODUCTION

Chlorpyrifos (0,0-diethyl 0-(3,5,6-trichloro-2-pyridinyl)phosphorothioate) is an organophosphorus insecticide that is widely used to control insect pests of maize, potato, tomato and other vegetable crops in Turkey. The pathway of chlorpyrifos degradation in the soil involves both chemical and microbial processes. The major products of degradation were identified as the hydrolysis product, 3,5,6-trichloro-2-pyridinol (TCP), the secondary metabolite, 3,5,6-trichloro-2-methoxy-pyridine (TCMP), and eventually CO_2 resulting from the mineralization of the aromatic ring.

---

* Research carried out in co-operation with the IAEA under Technical Co-operation Project No. TUR/5/015.

The degradation half-life of chlorpyrifos in the soil varies tremendously, depending on the soil type. It can be greatly influenced by environmental factors such as moisture, pH, organic carbon content and pesticide formulation [1].

In other studies, potato, tomato and maize plants were grown in lysimeters under outdoor conditions and ^{14}C-chlorpyrifos was applied according to good agricultural practice to determine the chlorpyrifos ^{14}C residues. The soil in which these plants had been grown was collected from the fields surrounding the research centre. The purpose of this study was to determine the degradation rate and the products of chlorpyrifos, as a function of soil depth, using the same type of soil in which the above mentioned plants were grown.

2. MATERIALS AND METHODS

2.1. Chemicals

The specific activity of the original radiolabelled (2,6-pyridinyl-^{14}C) chlorpyrifos was 1.09 MBq/mg. Radiolabelled and unlabelled chlorpyrifos (Greyhound, chlorpyrifos ethyl, 99% purity) were mixed to yield a specific activity of 370 kBq/mg, with a chlorpyrifos concentration of 200 μg per flask. Rotizsint was used as the scintillation cocktail. All the chemicals used were of analytical reagent grade.

2.2. Soils

The surface (0–15 cm) and subsurface (40–60 cm) soils used for the studies were collected in November 1995 from a field not treated with pesticide at the research centre. They were passed through a 2 mm sieve to remove any stones and debris. The soil properties are given in Table I. All the soil data are expressed on a dry weight basis.

TABLE I. PROPERTIES OF THE SOILS USED FOR DEGRADATION STUDIES

Soil	pH	Texture			Organic content (%)	No. of microorganisms/g	% H$_2$O at 1/3 bar
		Sand	Silt	Clay			
Surface	8.1	18.5	41.8	39.7	0.6	3.8 × 10^7	34.3
Subsurface	7.9	19.6	46.6	33.8	0.5	4.0 × 10^7	31.8

2.3. Soil treatment and incubation

To determine of the rate of chlorpyrifos degradation, 100 g (oven dried equivalent) triplicate samples of the soils were weighed in individual glass biometer incubation flasks. The moisture contents of the surface and subsurface soils were determined as 20.2 and 17.4%, respectively. Other than the control samples, each soil was uniformly treated with ^{14}C-chlorpyrifos in hexane to yield a soil concentration of about 2 μg/g of chlorpyrifos and 7.4×10^4 Bq (2 μCi) of radioactivity. After the soil was mixed thoroughly, air was blown into the biometer flask to remove the hexane, and distilled water was added to raise the soil moisture to 75% of the field capacity. The side arm of each biometer flask was filled with 10 mL of 0.1N KOH to serve as a CO_2 trap, and the flasks were placed in an incubator in the dark at 25°C [2].

2.4. Methanol solvent extraction

The KOH traps were sampled for the evolved $^{14}CO_2$ every day in the first week, semi-weekly until day 28, and then weekly thereafter. After 1–4, 6, 8, 14, 21, 28 and 97 days of incubation, 5 g soil samples were removed and extracted three times by shaking with 10 mL of methanol. Extracts were combined and filtered through Whatman No.1 filter paper, then evaporated to dryness. The residues were redissolved in 4 mL of acetone.

2.5. Supercritical fluid extraction (SFE)

One gram of solvent extracted soils was subjected to SFE with methanol. Methanol was compressed by a high pressure liquid chromatography pump (Waters 600E) to 15.2 Mpa through a preheated capillary (stainless steel, 0.5 mm inner diameter, 2 m length) into a 5 mL extraction vessel that contained the soil sample to be extracted. Preheating of the capillary and the extraction vessel was carried out in an oven (Elektromag 3025) maintained at 250°C and purged continuously with nitrogen. The extracts were passed through a cooler (water cooled stainless steel tube, 3.18 mm), then through a regulating valve (Whitey SS-31RS4-A) and finally collected in a 100 mL measuring cylinder with a flow rate ranging from 1 to 1.5 mL/min in order to maintain optimum pressure. Extraction was carried out until a 100 mL extract was collected, and an aliquot was subjected to liquid scintillation counting (LSC) [3]. The extracts were then evaporated in a Buchi rotary evaporator at 35°C under vacuum and the residues redissolved in 2 mL of acetone.

2.6. Determination of the bound ^{14}C residues

The total ^{14}C and unextractable soil bound ^{14}C residues (after solvent and supercritical fluid extractions) were determined by the combustion of 0.15–0.25 g soil samples to $^{14}CO_2$ in a Harvey biological oxidizer OX600. The radiocarbon in the KOH traps, soil extracts and combustion traps was determined using an LSC Packard Tricarb 1500.

2.7. Thin layer chromatographic (TLC) analysis

The qualitative and quantitative determinations of the ^{14}C residues in the soil extracts were done by TLC using 0.250 mm thick TLC plates (Merck, silica gel F_{254}, 20 × 20 cm). The chlorpyrifos and TCP standards and the 0.050–0.150 mL soil extracts in acetone were applied as spots to the thin layer plates. The plates were developed with the toluene/methanol/hexane (18:1:1, vol./vol./vol.) solvent system. The R_f values for chlorpyrifos and TCP were 0.76 and 0.24, respectively. The plates were then scraped in 1 cm increments and analysed by LSC. After obtaining the radiochromatogram, the percentage of each compound in the soil extracts was calculated.

Distribution of ^{14}C in the incubated soil samples was expressed as the percentage ^{14}C recovered in relation to the initially applied ^{14}C-chlorpyrifos. Calculation of the chlorpyrifos half-lives was based on first order plots of the natural logarithm of chlorpyrifos concentration versus time.

3. RESULTS AND DISCUSSION

Mineralization of the ^{14}C ring carbon, an indicator of microbial catabolism, indicated that, from the surface and subsurface soils, up to 40.6 and 42.6%, respectively, of the applied radiocarbon evolved as $^{14}CO_2$ during the 97 day incubation period (Fig. 1). A twofold increase in the evolved CO_2 was observed within the first 2 weeks for the surface soil, while there was a similar increase in the first 10 days for the subsurface soil. This may be attributed to the difference in the degradation half-lives of chlorpyrifos at the two soil depths. The chlorpyrifos half-lives in the surface and subsurface soils were caculated as 7.2 and 4.3 days, respectively. The degradation rate in the subsurface soil was somewhat higher than that of the surface soil.

The major products of degradation that were detected included, in order of decreasing concentration, $^{14}CO_2$, chlorpyrifos, soil bound ^{14}C residues and TCP after 97 days of incubation. Regardless of soil depth, the percentages found were 28.9, 13.1, 11.7 and 3.8 of the applied ^{14}C for the surface soils and 22.1, 12.4, 11.4 and 4.8 of the applied ^{14}C for the subsurface soils, respectively (Figs 2 and 3). The percentages of chlorpyrifos and TCP are given as the sum of the amounts

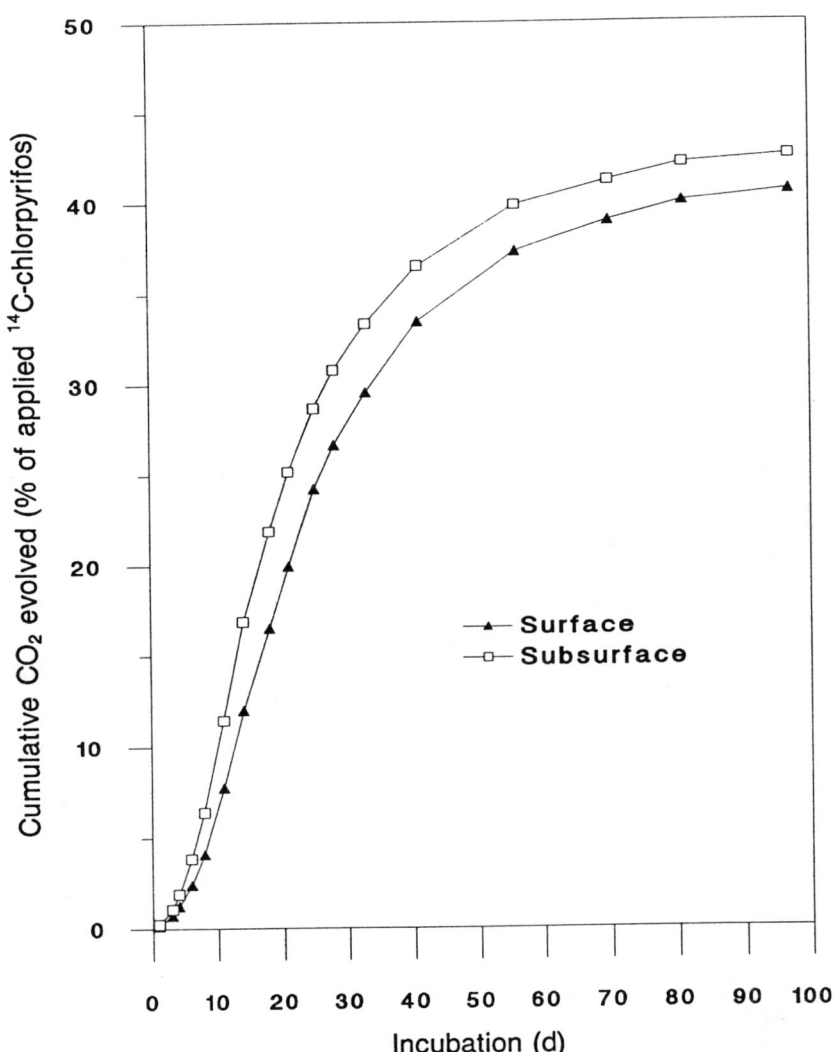

FIG. 1. Mineralization of chlorpyrifos in Turkish soil.

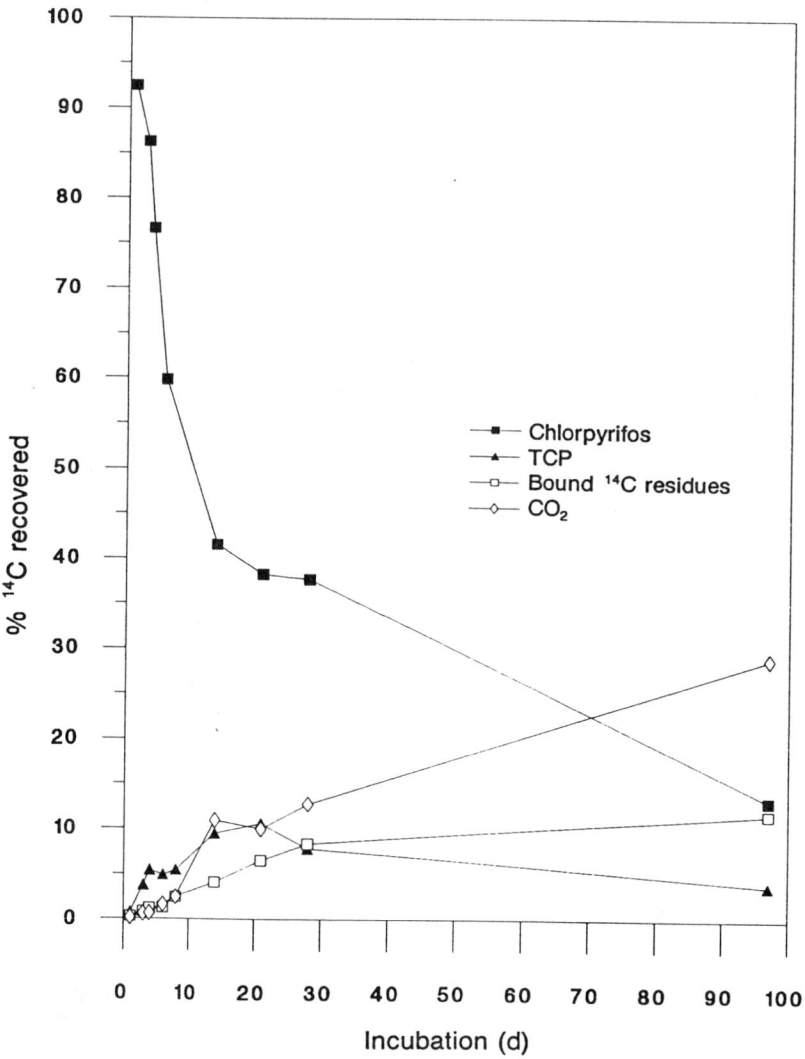

FIG. 2. *Distribution of chlorpyrifos and its degradation products in surface soil (0–15 cm) under laboratory conditions.*

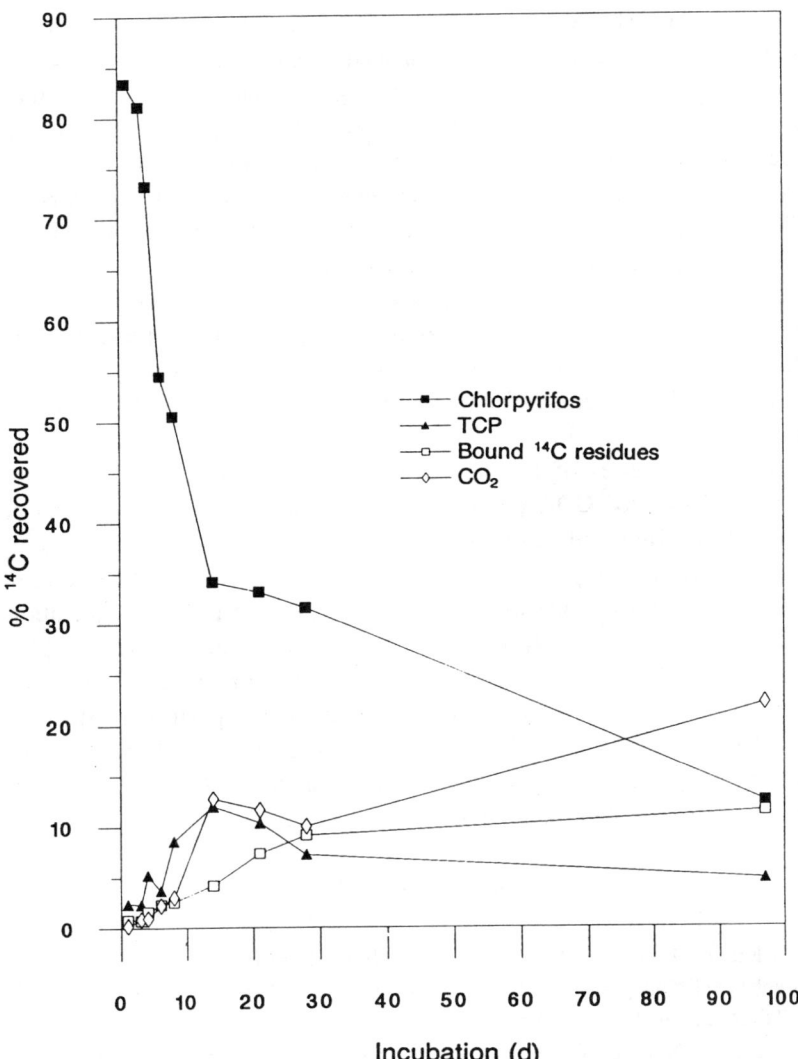

FIG. 3. Distribution of chlorpyrifos and its degradation products in subsurface soil (40–60 cm) under laboratory conditions.

recovered from solvent extraction and SFE. Most of the chlorpyrifos (85–98%) was extracted by solvent extraction. A reverse phenomenon was observed for TCP, most of which (82–99%) was recovered by SFE.

TLC studies showed that for the SFE extracts 40–50% of the applied ^{14}C-chlorpyrifos on the TLC plates remained at the origin. When these amounts were calculated for initial activity, the percentage amount that remained at the origin was in the range of 0.8–17.8%, and increased with incubation time. This may be explained as those amounts of ^{14}C-chlorpyrifos that were bound to the soil structure. Although SFE removed some of the bound ^{14}C residues, it was impossible to identify them with the TLC solvent system used for chlorpyrifos and TCP. Printz et al. [4] investigated the bound residues in fulvic and humic acid fractions by re-extracting the solvent extracted soils with suitable solvents. The portion of radioactivity extracted by SFE but which remained at the origin of the TLC plate is thought to be the residue bound to fulvic and humic acids. If this is the case, the values given in Figs 2 and 3 for the bound ^{14}C residues would increase at least twofold.

Overall recoveries of the radioactivity were calculated as the sum of the percentages of evolved CO_2, chlorpyrifos, TCP, bound ^{14}C residues and activity remaining at the origin and was found to be in the range of 85–99.5%. Activity may have been lost during sampling of the KOH traps and/or removal of the soil samples.

The results showed that the major pathway of chlorpyrifos degradation is by microbial degradation; hydrolysis may partially be attributed to the high pH of the soils. The high pH and low organic content of the surface and subsurface soils increased hydrolysis and decreased the adsorption of chlorpyrifos [1, 5]. Because of the short half-lives and the relatively high mineralization rates of chlorpyrifos in the soils tested, the possibility was reduced of accumulating chlorpyrifos in the soils in which potato, tomato and maize plants were grown.

REFERENCES

[1] RACKE, K.D., LASKOWSKI, D.A., SCHULTZ, M.R., Resistance of chlorpyrifos to enhanced biodegradation in soil, J. Agric. Food Chem. **38** (1990) 1430–1436.

[2] INTERNATIONAL ATOMIC ENERGY AGENCY, Laboratory Training Manual on the Use of Nuclear Techniques in Pesticide Research, Technical Reports Series No. 225, IAEA, Vienna (1983) 215–219.

[3] CAPRIEL, P., HAISCH, A., KHAN, S.U., Supercritical methanol: An efficacious technique for the extraction of bound pesticide residues from soil and plant samples, J. Agric. Food Chem. **34** (1986) 70–73.

[4] PRINTZ, H., BURAUEL, P., FÜHR, F., Effect of organic amendment on degradation and formation of bound residues of methabenzthiazuron in soil under constant climatic conditions, J. Environ. Sci. Health **B30** 4 (1995) 435–456.

[5] FELSOT, A., DAHM, P.A., Sorption of organophosphorus and carbamate insecticides by soil, J. Agric. Food Chem. **27** 3 (1979) 557–563.

USE OF LYSIMETERS FOR DETERMINING PESTICIDE FATE IN AGROECOSYSTEMS*

F. FÜHR, W. MITTELSTAEDT,
T. PÜTZ, A. STORK, M. DUST
Institut für Radioagronomie,
Forschungszentrum Jülich,
Jülich, Germany

Abstract

USE OF LYSIMETERS FOR DETERMINING PESTICIDE FATE IN AGROECOSYSTEMS.

A lysimeter system is presented in which ^{14}C labelled pesticides are applied in accordance with agricultural practice. The long term behaviour of pesticides in the soil–plant system can thus be determined under realistic climatic, cropping and soil conditions. In these experiments, pesticides and their metabolites are measured in soil, soil solutions, plants and drainage water over several growing seasons. In addition, a wind tunnel, combined with sensitive analytical methods, allows the determination of pesticide residues, their metabolites and $^{14}CO_2$ as the mineralization product released into the air. Complementary standardized laboratory experiments yield information on the role of individual processes leading to binding, bioavailability and translocation. Important conclusions are thus obtained for the appropriate practical use of pesticides. As a final step in the validation of the lysimeter, a large scale experiment is in progress to compare lysimeter and field studies, with emphasis placed on the translocation of pesticides in deeper soil layers and groundwater.

---

* Only an abstract appears here, since the full text of the paper is to appear in Integrating Pesticide Environmental Fate and Transport Data from Laboratory and Field Studies (EFFLAND, W., WOLF, J.K., HORNSBY, A.G., Eds), ASA Special Publication, American Society of Agronomy, Madison, WI (in preparation).

A MICROCOSM SYSTEM TO EVALUATE THE FATE OF ^{14}C LABELLED PESTICIDES IN SOILS FROM DIFFERENT CLIMATE ZONES*

I. SCHEUNERT, R. SCHROLL
GSF-Institut für Bodenökologie, Neuherberg,
Oberschleissheim, Germany

G. CAO
Chinese Academy of Agricultural Sciences,
Beijing, China

Abstract

A MICROCOSM SYSTEM TO EVALUATE THE FATE OF ^{14}C LABELLED PESTICIDES IN SOILS FROM DIFFERENT CLIMATE ZONES.
   A method for quantifying and comparing the processes relevant to the fate of pesticides in soils from various climate zones — volatilization, mineralization, transport and leaching — was developed. A microcosm, consisting of an undisturbed soil column in a round metal cylinder (height 30 cm, 15 cm inner diameter), is fitted with devices to collect and analyse the leachate and to investigate the gas phase above the soil surface. To include all the conversion products in the investigations, ^{14}C labelled pesticides are used. Thus, determination of the complete ^{14}C mass balance is possible. As the first model herbicide, ^{14}C-terbuthylazine was used. Ventilation was based on a wind speed of 1 m/s; irrigation, based on an average rainfall of 600 mm/a, was applied discontinuously, interrupted by dry periods. Soils from Brazil, China and Germany were investigated in four replicates each. Soil from Brazil, with the highest organic matter and clay contents, exhibited the lowest volatilization of ^{14}C-terbuthylazine, the highest mineralization to ^{14}CO$_2$, the lowest vertical transport and the lowest leaching of ^{14}C into the percolate water, followed by the soils from China and Germany. Soil from Germany showed the highest volatilization, the highest vertical transport and the highest leaching rates. These findings demonstrate that the properties of soils, as formed under different climatic conditions, influence the fate and persistence of a pesticide in the soil environment.

1. INTRODUCTION

Numerous data on the fate of pesticides in soils, including persistence and degradation, transport and mobility as well as transfer to groundwater, in the atmosphere and in plants, have been published [1–3]. However, nearly all this information has been obtained using the soils and environmental conditions of temperate

---

   * Research carried out in association with the IAEA under Research Agreement No. 8081.

climate zones. Extrapolation of the results obtained in temperate soils to tropical areas has proved to be unsatisfactory. Subtropical and tropical soils differ from temperate soils mainly in the quality of the clay minerals, and in the amount and quality of the iron oxides and organic matter. Moreover, one can expect that the types of microbial community differ from those in temperate soils. All these factors, combined with different soil moisture and temperature regimes, may influence the rates of volatilization, degradation, accumulation and transport of pesticides.

A test system is described that allows a comparative study to be made of various soils under standardized conditions. Thus, the influence of soil properties, as formed under different climatic conditions, can be studied under identical environmental conditions. In a second step, the influence of climatic conditions such as temperature and precipitation regimes can be investigated. For example, studies of the herbicide terbuthylazine in soils from three countries in different climate zones (Brazil, China and Germany) are presented. To include all the conversion and degradation products in the investigation, the herbicide was applied in a ^{14}C labelled form.

2. MATERIALS AND METHODS

2.1. Test system

The test system (Fig. 1) consists of a round metal cylinder with a height of 30 cm and an inner diameter of 15 cm. Filter paper and a metal sieve are placed at the bottom of the cylinder to keep the soil in the cylinder. A defined underpressure maintained by a membrane pump — necessary to counteract the capillary forces — at the bottom of the sieve permits leaching of the percolate out of the soil. The leachate is collected in 1 L glass bottles.

The metal cylinder is closed with a special cover that allows irrigation as well as intensive gas exchange. Fourteen small, specially formed openings at the bottom of the microcosm cover allow drop application of the precipitation water. A ventilator placed in the middle of the microcosm cover gives a wind speed of about 1 m/s over the soil surface. A membrane pump draws the air through the cover, resulting in an air exchange of 23 L/h, which corresponds to an air exchange rate of more than 60 times per hour in the microcosm atmosphere.

Between the microcosm cover and the membrane pump several traps are placed to absorb separately the volatile ^{14}C labelled organic compounds (parent compound and metabolites) and the $^{14}CO_2$ resulting from total degradation of the pesticides. A special piece of equipment on top of the traps guarantees that volatile organic solvents, such as ethyleneglycol-monomethylether, which is added to the first two traps to fix the volatile organic compounds, do not evaporate from the traps. To absorb the $^{14}CO_2$, traps 3 and 4 are filled with a mixture of diethyleneglycol-monobutylether and ethanolamine.

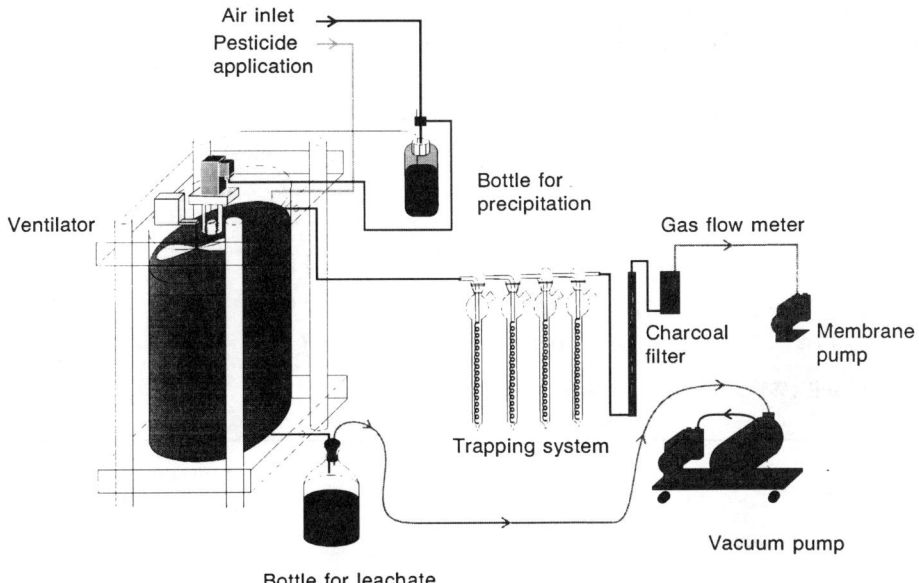

FIG. 1. Microcosm system for studying the fate of ^{14}C labelled pesticides in soils.

An Erlenmeyer flask filled with a saturated salt solution is placed in front of the inlet of the microcosm cover to achieve a defined air humidity for the air drawn into the microcosms. To guarantee a constant temperature, the microcosms can be placed in refrigerators.

2.2. Soils

Soils were sampled at sites in Brazil, China and Germany; their properties are shown in Table I.

TABLE I. PROPERTIES OF THE SOILS USED

	Soil from Germany	Soil from China	Soil from Brazil
Sand (%)	85	63	58
Silt (%)	10	21	9
Clay (%)	5	16	33
pH value	4.9	8.5	4.3
Organic matter (%)	1.8	1.7	2.9

TABLE II. MASS BALANCE OF ^{14}C, 50–60 DAYS AFTER APPLICATION OF ^{14}C-TERBUTHYLAZINE TO THE SOIL SURFACE OF MICROCOSMS (MEAN OF FOUR MICROCOSMS) (% OF ^{14}C APPLIED)

	Soil from Germany (60 d)	Soil from China (50 d)	Soil from Brazil (60 d)
Volatilization (%)	4.90	2.24	1.49
$^{14}CO_2$ (%)	0.09	0.02	0.14
Leaching (%)	0.2	<0.1	<0.1
Rest in soil (%)	≈95	≈98	≈99

2.3. Test procedure

The metal cylinder was pressed carefully into the soil surface to a depth of 28 cm, resulting in a free space of 2 cm above the soil in the column (= atmosphere of the microcosm), and then removed carefully. Before application of the ^{14}C labelled pesticide dissolved in 10 mL of water, precipitation, based on a uniformly distributed average rainfall of 600 mm/a, was simulated for 2 weeks (three times per week) to condition the soil column, followed by a week of no precipitation to allow the soil surface to dry again, as in a real field situation. After application of the ^{14}C labelled terbuthylazine (10 μCi per test)[1] at a dose of 980 g/ha, and another week without precipitation, rainfall was simulated again three times per week.

Aliquots of the cocktail for volatilization and the cocktail for $^{14}CO_2$ were taken about three times per week, mixed with the scintillation cocktails (Ultima Gold XR and Permablend) and measured in a liquid scintillation counter. The main amount of volatilization cocktail was stored each week; the tubes were washed with water and dichloromethane. The organic phase was concentrated to dryness, resolved in a mixture of methanol and water (1:1 (vol./vol.)) and injected into the high performance liquid chromatograph to identify the volatilized ^{14}C labelled organic compounds.

Radioactivity in the leachate was determined each week. At the end of the test period, the soil was pressed out of the metal cylinder with special equipment and cut into several soil layers. Radioactivity in the different soil layers was determined by combustion of the soil aliquots, followed by liquid scintillation counting of the $^{14}CO_2$ formed.

---

[1] 1 Ci = 3.70 × 10^{10} Bq.

3. RESULTS

The mass balance of ^{14}C-terbuthylazine in the microcosms is presented in Table II. Volatilization of ^{14}C-terbuthylazine was highest from the German soil and lowest from the Brazilian soil. This is in line with data published in the literature, which states that volatilization is higher from sandy soils than from loamy soils, and higher from soils with a higher organic carbon content than from soils with a lower organic carbon content [4]. In contrast, evolution of the $^{14}CO_2$ resulting from mineralization (total degradation) of the pesticide was highest in the Brazilian soil, with the highest clay and organic matter contents. However, mineralization is not directly dependent on the physicochemical soil properties but on the quantity and quality of the soil microflora performing the degradation, which is not constant but varies in space and time, depending not only on the soil properties but also on many other factors such as temperature, humidity, aeration status and nutrients. The variability in the mineralization rates of terbuthylazine, depending on the living conditions of the microflora, is reflected by the fact that in outdoor lysimeters the mineralization was more than one order of magnitude higher than that in these microcosms [5].

Leaching of ^{14}C residues originating from ^{14}C-terbuthylazine was negligible in Brazilian and Chinese soils. In the soil from Germany, which had a higher sand content, 0.2% of the ^{14}C was transferred to the percolate. This amount is not important for the mass balance of terbuthylazine in soil; however, it is important for the potential transfer of the residues to groundwater and for the quality of the potable water.

Since the three output processes determined in the microcosms obviously have no quantitative importance for the mass balance of terbuthylazine in these soils,

TABLE III. VERTICAL DISTRIBUTION OF ^{14}C, 50–60 DAYS AFTER APPLICATION OF ^{14}C-TERBUTHYLAZINE TO THE SOIL SURFACE OF MICROCOSMS (MEAN OF FOUR MICROCOSMS) (% OF ^{14}C PRESENT IN THE SOIL AFTER THE EXPERIMENTAL PERIOD)

Soil depth (cm)	Soil from Germany (60 d)	Soil from China (50 d)	Soil from Brazil (60 d)
0–1	62.99	70.8	72.46
1–2	22.58	11.3	17.3
2–3	5.86	6.82	3.71
3–4	2.32	4.42	1.5
4–5	1.66	2.06	1.08
5–28	4.59	4.6	3.95

further efforts were undertaken to investigate the residues remaining in the soil. To study vertical mobility and transport, the soil columns were cut into 1 cm layers, which were analysed for ^{14}C. The results are presented in Table III.

The table reveals that most of the radioactivity remained in the upper 1 cm layer, with a smaller portion found in the 2 cm layer. Transport of residues to deeper layers was low. However, significant differences between the three soils were apparent: the soil from Brazil, with the highest organic matter and clay contents, had the highest ^{14}C residue content in the upper layer, followed by the soil from China and then from Germany. Correspondingly, the ^{14}C residue content in the deepest layers of this soil was lower than that of Germany and China. These findings are in line with the amount of radioactivity leached in the percolate water. Thus, soil from Brazil has the lowest capacity for transport and mobility of the herbicide, whereas the soil from Germany has the highest capacity.

4. CONCLUSIONS

The microcosm system represents a useful tool for comparing the fate of ^{14}C labelled pesticides in different soils under comparable environmental conditions. The results demonstrate that the properties of soils, as formed under different climatic conditions, influence the fate and persistence of a pesticide in the soil environment. By varying the environmental conditions, the influence of climate may also be quantified.

REFERENCES

[1] SCHEUNERT, I., "Physical and physico-chemical processes governing the residue behaviour of pesticides in terrestrial ecosystems", Chemistry of Plant Protection, Vol. 8 (EBING, W., Ed.), Springer-Verlag, Berlin and Heidelberg (1992) 1–22.

[2] SCHEUNERT, I., "Transformation and degradation of pesticides in soil", ibid., pp. 23–75.

[3] SCHEUNERT, I., "Fate of pesticides in plants and in soil fauna", ibid., pp. 77–103.

[4] DÖRFLER, U., et al., A laboratory model system for determining the volatility of pesticides from soil and plant surfaces, Chemosphere **23** (1991) 485–496.

[5] SCHEUNERT, I., et al., "Fate of ^{14}C-terbuthylazine in outdoor lysimeters", Lysimeter Studies of the Fate of Pesticides in the Soil, BCPC Monograph No. 53 (FÜHR, F., HANCE, R.J., Eds), The British Crop Protection Council, Farnham, Surrey (1992) 125–131.

IAEA-SM-343/22

LEACHING AND DEGRADATION OF PESTICIDES IN GROUNDWATER LAYERS

L. VOLLNER*, D. KLOTZ**

*Institut für Ökologische Chemie

**Institut für Hydrologie

GSF-Forschungszentrum für Umwelt und Gesundheit,
Neuherberg, Germany

Abstract

LEACHING AND DEGRADATION OF PESTICIDES IN GROUNDWATER LAYERS.
  To date, most of the published data on pesticides and soil describe the fate of environmental contaminants in the upper soil layers, where low transportation, high adsorption and high microbial degradation occur. In contrast, few data are available for groundwater layers (to a depth of 5 m) with filtration media such as sand or stony layers. Laboratory testing systems were developed that allow examinations to be carried out under conditions that are close to natural. These include original sands and stones from the relevant groundwater filtration areas, original water from the same areas, and measurement of the velocity of water migrations. The temperature and pH were determined and exact analyses of the sand quality and particle size were made prior to the experiments. To produce accurate data on leaching and adsorption, and to identify the degradation products more easily, ^{14}C labelled pesticides were applied. To determine the column characteristics, tritium labelled water was used. The following 12 pesticides were investigated: the carbamate insecticide, carbofuran; chlorinated hydrocarbons, lindane, DDT and DDE; phosphoric acid esters, chlorpyrifos, diazinon, malathion and parathion; phenylurea herbicides, diuron and monolinuron; the triazine herbicide, terbutylazine; and the novel chloronicotinyl insecticide, imidacloprid. Although the recovery data of the individual pesticides were similar in the eluates of the two different sand types, in most cases the elution curves differed significantly. The different degradation rates and procedures for the same pesticide were clearly shown by the significantly different concentration values. Concerning elution and degradation, the behaviour of the chemicals differed, even if they belonged to the same class of bioactive compound. Furthermore, the behaviour of the same chemical differed in different types of sand. Because of the low transportation velocity, and the subsequent long retention times of the pesticides in the sand media, microbial degradation (metabolisms) can also take place.

1. INTRODUCTION

Stable or radioactive isotopic tracers are extensively and commonly employed in many fields of research. They are also widely used to study trace contaminant problems. Suitable isotopic labelling of compounds allows their chemical and physical fate to be followed in the environment, in food and in plant or animal organisms.

The GSF-Forschungszentrum für Umwelt und Gesundheit near Munich has used such materials for more than 30 years, in many experiments in different fields.

This paper demonstrates the co-operation between two institutes at the centre (Institut für Ökologische Chemie and Institut für Hydrologie) to determine the rates of leaching and to identify the types of degradation product (metabolites) of various pesticides in the lower sand layers.

In recent years it has been found that drinking water contains increasing amounts of pesticide residue (to date, about 50 different pesticides, or their degradation products, have been found in groundwater) [1]. Therefore, it was felt that the rates of migration and the types of contamination of groundwater after pesticide application should be investigated.

Most of the published data on pesticides and soil describe the fate of environmental contaminants in the upper soil layers, where low transportation, high adsorption and high microbial degradation occur. In contrast, few data are available for groundwater layers (to a depth of 5 m) with filtration media such as sand or stony layers.

Our interest was in the fate of pesticides after they have passed through the thin humus range (30–50 cm). This occurs after heavy rains and erosion, or in soils with less humus and in the presence of gaps resulting from dryness, plant roots and earthworms.

Therefore, the two institutes developed laboratory testing systems that allow examinations to be carried out under conditions that are close to natural. These include original sands and stones from the relevant groundwater filtration areas, original water from the same areas, and measurement of the velocity of water migrations. The temperature and pH were determined and exact analyses of the sand quality and particle size were made prior to the experiments.

To produce accurate data on leaching and adsorption, and to identify the degradation products more easily, ^{14}C labelled pesticides were applied. To determine the column characteristics, tritium labelled water was used.

2. MATERIALS AND METHODS

Special V2A alloy steel columns, 50 cm long and an internal diameter of 5 cm (with a cross-section of 19.64 cm^2 and a volume of 981.74 cm^3), were used [2].

Monitoring of the conductivity and the pressure during the experiments was maintained by sensors that were installed in holes made at different places in the column walls. A constant flow rate was maintained using peristaltic pumps. The columns were protected by containers placed at the effluent level in case difficulties should arise with the pumps or with lack of water.

Injection of tracers and pesticides was ensured by a non-stop flow, using a special injection system with a septum. A special solvent distributor was installed to ensure uniform application of the injected sample to the top surface of the column [2].

Two different types of sand that are responsible for groundwater filtration in the Munich area were used. These were: (1) Tertiärsand (Tertiary sand, hereafter referred to as tert), with a particle size of 0.19–0.31 mm and collected from 5 m below the ground surface (northern Munich area); and (2) Quartärkies (Quaternary gravel, hereafter referred to as quart), with a particle size of 0.67–4.70 mm and collected from 2 m below the ground surface (eastern Munich area). The mass of this filtration material, the density (dry weight), the total porosity and the volume of the water were determined for each chemical.

The columns were packed with sediment using techniques that were specially developed for these experiments [2]. The columns were packed with about 5 mm of silica sand before the experiment began and the sediment layers were packed in the same way (about 70 g each) to avoid irregular application and elution of the dissolved pesticides.

Groundwater was taken from the same locations and analysed for ions prior to the leaching experiments.

Elution was maintained at rates that were close to the natural movement of water (2×10^{-4} cm/s and 6×10^{-4} cm/s); the total elution volume was 2.8–3.2 L in the tert columns and 0.8–1.2 L in the quart columns.

Radioactivity in the eluate was analysed by liquid scintillation counting, and the quality of the substance mixtures (parent material and degradation products) by thin layer chromatography (TLC), gas–liquid chromatography and high performance liquid chromatography after extraction from the water and the sand.

The following 12 pesticides were investigated: carbofuran, lindane, DDT, DDE, chlorpyrifos, diazinon, malathion, parathion, diuron, monolinuron, terbuthylazine and imidacloprid.

Stock solutions of 50 µCi and a total of about 10 mg of each pesticide were used.[1] Aliquots, dissolved in water, were applied.

Eluates were collected hourly and a 1 mL aliquot was taken for counting. For qualitative investigations, residues were extracted by solid phase extraction using RP-C_{18} columns (e.g. Bond Elute C_{18}).

---

[1] 1 Ci = 3.70×10^{10} Bq.

To investigate the adsorption profile, the sand was removed from the columns layer by layer: 0–5, 5–10, 10–15, 15–20, 20–30, 30–40 and 40–50 cm. After homogenization of these fractions, aliquots (tert 10 g and quart 15 g) were transferred to the liquid scintillation vials. After adding the liquid scintillation fluids, direct measurement of radioactivity was carried out. This procedure was verified by sample oxidization of the sediments. Since the quenching rates were the same for all the samples, the data were used, without further corrections, for preparing the adsorption curves. The quantity of adsorbed material was determined by the difference between the applied and the eluted amounts.

For qualitative analytical work, the sand layers were first extracted with hexane and acetone mixtures, and then by extraction steps with methanol. After evaporation of the solvents, mainly TLC analysis was carried out.

3. HYDRAULIC PARAMETERS

In general, the mobility of pesticides in soils and sediments is related to the adsorption and mass flux of the dissolved fractions. Adsorption influences the mass flux, which consists of diffusion, convection and dispersion [3]. Diffusion is a physical process by which molecules move from sites of higher concentration to those of lower concentration. Convection is the passive movement of solutes within moving water. Dispersion is the distribution of solutes within moving pure water, which results from the different flow velocities of the individual water volumes.

TABLE I. HYDRAULIC PARAMETERS FOR LINDANE AND CARBOFURAN[a]

Pesticide	Column	Q (mL/h)	V_f (cm/s)	V_d (cm/s)	n_{eff}
Lindane	Tert	17.31	2.45×10^{-4}	6.68×10^{-4}	0.367
	Quart	54.63	7.72×10^{-4}	4.94×10^{-3}	0.156
Carbofuran	Tert	15.10	2.24×10^{-4}	5.705×10^{-4}	0.375
	Quart	39.95	5.65×10^{-4}	3.35×10^{-3}	0.169

[a] Q is the velocity of the tracer; V_f is the velocity of the filtration (relation Q to total cross-section = $V_f = QCS^{-1}$); V_d is the distance velocity of tritium (the relation at distance 1 and retention time t, between two points of measurement = $V_d = 1t^{-1}$); n_{eff} is the effective porosity.

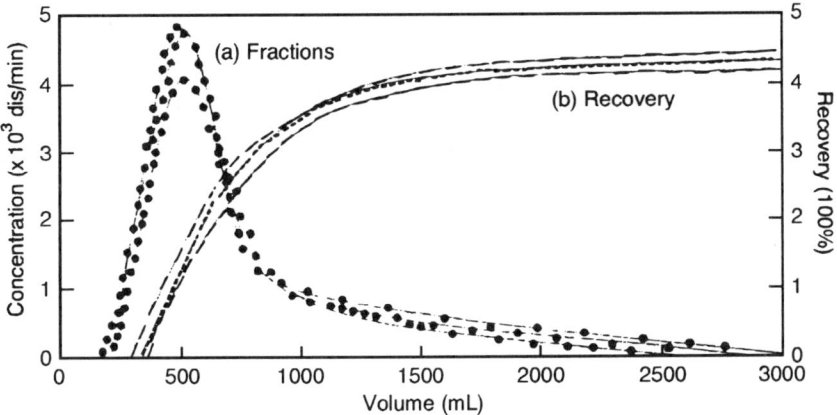

FIG. 1. Variations in replicates for terbuthylazine leaching (results of triplicate analyses).

To identify all these important parameters and to standardize the columns, we used tritiated water (tracer) prior to each leaching experiment. Table I gives the hydraulic parameters for lindane and carbofuran.

A triplicate experiment (with terbuthylazine) showed good correlation of data (Fig. 1), which proves the reproducibility of the experiments.

4. RESULTS

4.1. Leaching results

Table II shows the applied radioactivity, the eluted radioactivity and the recovery of elution. Preparation of the elution curves (curve (a) in Fig. 1 and curve (b) in Fig. 2) was done by relating the elution volume in millilitres and the concentration of pesticides in the fractions in disintegrations per minute. The sum of radioactivity of all the fractions gives the total amount eluted (curve (b) in Fig. 1 and curve (c) in Fig. 2). We call this value 'recovery of leaching' (see also Table II).

Figure 2 compares two significantly different elution curves. In the case of imidacloprid, rapid elution was observed, while for chlorpyrifos one broad peak mainly appeared, indicating slow elution.

Although the recovery data of the other individual pesticides were similar in the eluates of the two different sand types, in most cases the elution curves differed. The different degradation rates and procedures for the same pesticide were clearly shown by the significantly different concentration values (Table III).

TABLE II. LEACHING OF PESTICIDES IN PACKED SOIL COLUMNS[a]

Pesticide	Column	Applied radioactivity ($\times 10^6$ counts/min)	Eluted radioactivity ($\times 10^6$ counts/min)	Recovery of elution (% of applied)
DDE	Tert	4.98	0.29	6.0
	Quart		0.31	6.6
DDT	Tert	19.60	0.88	4.5
	Quart		0.56	2.9
Lindane	Tert	8.90	4.20	47.2
	Quart		2.78	31.3
Chlorpyrifos	Tert	16.31	8.63	52.9
	Quart		7.71	47.3
Diazinon	Tert	5.50	5.22	95.0
	Quart		3.79	69.1
Malathion	Tert	19.44	13.84	71.2
	Quart		15.92	81.9
Parathion	Tert	12.50	9.00	72.6
	Quart		10.87	87.1
Carbofuran	Tert	16.32	13.18	80.8
	Quart		13.92	85.3
Diuron	Tert	25.56	21.57	84.4
	Quart		21.39	83.7
Monolinuron	Tert	20.12	14.16	70.4
	Quart		14.06	69.9
Terbuthylazine	Tert	8.16	6.96	85.0
	Quart		7.26	89.0
Imidacloprid	Tert	20.0	19.78	98.9
	Quart		19.72	98.6

[a] The radioactivity of the pesticides was 2.2×10^6 dis/min, which corresponds to 0.2 mg.

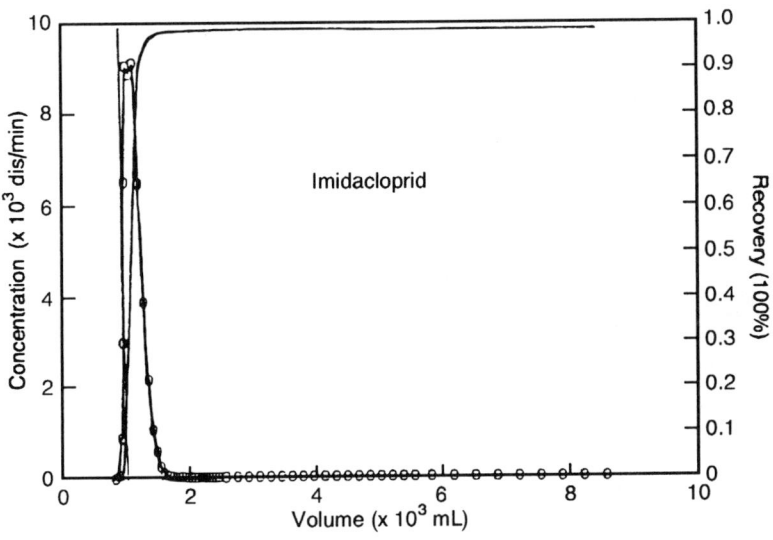

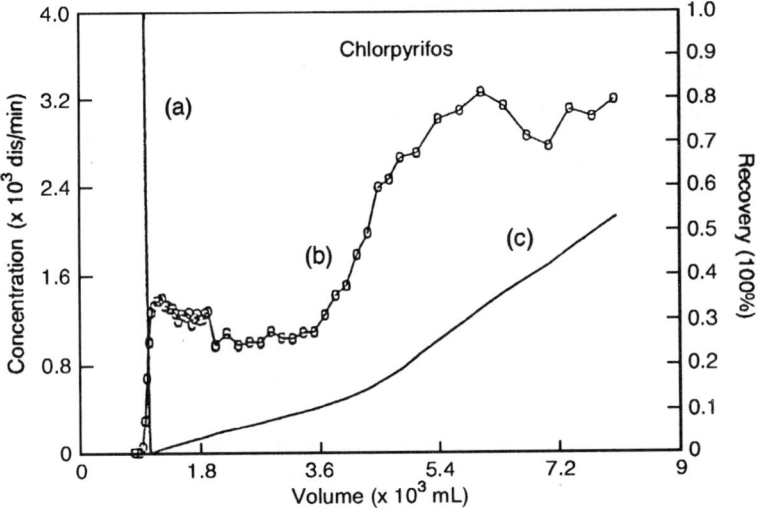

FIG. 2. Leaching of imidacloprid and chlorpyrifos: (a) front of tracer (tritium); (b) concentration in fractions; and (c) recovery (% of applied).

TABLE III. CONCENTRATION OF THE ELUTION CURVES FROM TWO DIFFERENT SOIL TYPES

Pesticide	Eluate (mL)	
	Tert	Quart
DDE	230	270
DDT	266	320
Lindane	320	1590
Chlorpyrifos	1596	3200
Diazinon	530	530
Malathion	320	260
Parathion	800	1330
Carbofuran	290	260
Diuron	345	260
Monolinuron	320	400
Terbuthylazine	320	746
Imidacloprid	290	260

4.2. Adsorption results

The adsorption profiles of the different pesticides in the columns were examined by stepwise removal of the sediments, as described in Section 2. Table IV shows these data. Table V shows the sum of the total adsorbed values and the extractable and non-extractable amounts.

4.3. Interpretation of degradation products

We investigated the extracted fractions using TLC to distinguish between the parent materials and their degradation products. Figure 3 shows the lindane extracts from eluted waters. The two sand types showed significantly different degradation rates.

TABLE IV. ADSORPTION OF PESTICIDES AND THEIR DEGRADATION PRODUCTS IN SEDIMENT LAYERS (RELATIVE % OF APPLIED)[a]

Pesticide	Column	Depth (cm)									
		5	10	15	20	25	30	35	40	45	50
DDE	Tert	58	20	12	3	3	0	0	1	1	1
	Quart	75	10	7	4	1	1	0	0	0	0
DDT	Tert	64	12	7	2	3	3	2	2	2	2
	Quart	57	13	9	4	3	3	4	4	2	1
Lindane	Tert	63	10	5	5	3	3	2	2	3	3
	Quart	12	9	13	13	10	10	6	6	10	10
Chlorpyrifos	Tert	11	9	7	7	11	11	12	12	10	10
	Quart	11	13	13	15	11	11	5	5	8	8
Diazinon	Tert	36	13	13	7	6	6	3	3	7	6
	Quart	34	12	12	10	8	8	4	4	4	4
Malathion	Tert	37	20	18	11	4	4	2	2	1	1
	Quart	14	12	10	8	9	9	10	10	9	10
Carbofuran	Tert	28	21	14	10	6	6	3	3	4	4
	Quart	32	29	12	7	5	5	3	2	2	2
Diuron	Tert	63	9	4	5	2	2	3	3	4	4
	Quart	11	11	11	11	11	11	8	8	9	9
Monolinuron	Tert	75	16	4	1	0	0	0	1	1	1
	Quart	62	12	5	3	4	4	3	3	2	2

[a] Parathion, terbuthylazine and imidacloprid are still under investigation.

TABLE V. ADSORPTION OF PESTICIDES ON ALL LAYERS OF SAND (DIFFERENCE IN THE APPLIED AND ELUTED AMOUNTS) (% OF APPLIED)

Pesticide	Column	Total adsorbed	Extractable	Non-extractable
DDE	Tert	94.0	15.0	79.0
	Quart	93.4	12.1	81.3
DDT	Tert	95.0	15.2	79.8
	Quart	97.1	13.5	83.6
Lindane	Tert	52.8	5.8	47.0
	Quart	68.7	5.5	63.2
Chlorpyrifos	Tert	47.1	15.9	31.2
	Quart	53.7	27.9	25.8
Diazinon	Tert	5.0	1.4	3.6
	Quart	3.5	1.7	1.8
Malathion	Tert	28.8	18.1	10.7
	Quart	18.1	3.9	14.2
Carbofuran	Tert	19.2	10.1	9.1
	Quart	14.7	11.4	3.3
Diuron	Tert	15.6	4.0	11.6
	Quart	16.3	1.8	14.5
Monolinuron	Tert	29.6	3.5	26.1
	Quart	30.1	2.4	27.7

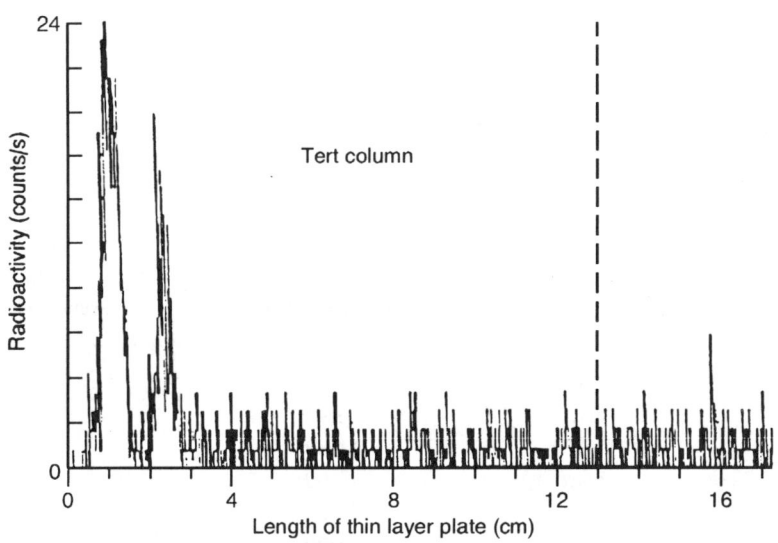
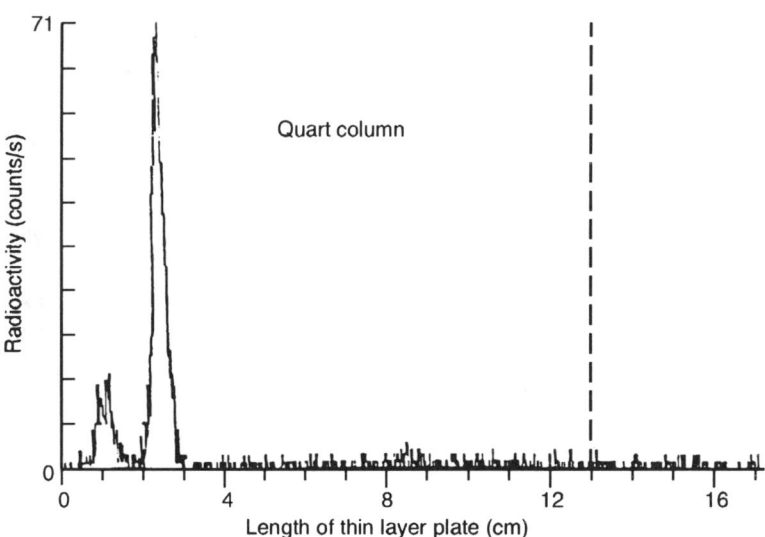

FIG. 3. TLC analysis of leachate extracts of lindane.

TABLE VI. TLC ANALYSIS OF EXTRACTS: RATE OF PARENT MATERIALS AND SUM OF CONVERSION PRODUCTS (RELATIVE % OF APPLIED) (THE EXTRACTABLE AMOUNT AND THE ELUTED AMOUNT EACH AS 100%)

Pesticide	Column	Adsorbed (parent/conversion)	Eluted (parent/conversion)
DDE	Tert	7/93	0/100
	Quart	6/94	0/100
DDT	Tert	11/89	0/100
	Quart	10/90	0/100
Lindane	Tert	47/53	0/100
	Quart	31/69	82/18
Chlorpyrifos	Tert	36/67	45/55
	Quart	58/42	0/100
Diazinon	Tert	20/80	43/57
	Quart	12/88	39/61
Malathion	Tert	100/0	0/100
	Quart	0/100	0/100
Carbofuran	Tert	40/60	90/10
	Quart	19/81	30/70
Diuron	Tert	95/5	90/10
	Quart	98/2	100/0
Monolinuron	Tert	0/100	88/12
	Quart	15/85	83/17

Table VI summarizes the results of these qualitative investigations for all the pesticides, indicating the rate of unchanged parent materials to the sum of the conversion products (which are usually more polar than the original substance). In almost all cases, the non-extractable amount was higher than the extractable amount. As is well known from soil experiments, these products may be chemically bound to the sediment particles and cannot be easily identified.

4.3.1. DDE

Because DDT is still used in some parts of the world, we also investigated the fate of DDE, which is the main degradation product of DDT, has similar properties, and frequently appears in residue analyses of DDT.

The minor amount of radioactivity that was eluted with water was a mixture of further degradation products. Two polar groups of parent material were found: in the quart columns, the polar material was high; on sand, the degradation product was high. In the 10 cm layer 60% decomposed DDE was found, whereas in the 15 cm layer no parent material was detected.

Degradation products originate in different ways, occurring stepwise, by dechlorination and hydroxylation reactions, or simultaneously, by chemical or microbial actions. These materials have been extensively studied worldwide and are still the subject of many investigations [4]. A good review is given in Ref. [5].

4.3.2. DDT

In a manner similar to DDE, the eluted radioactivity (3-5%) was a mixture of polar degradation products. In the columns, 98% of the parent material had already been degraded in the 5 cm layer.

Besides DDE, we found two other (unidentified) polar substances (1.3 and 6%). In the deeper layers, only DDE and further degradation products were detected. Finally, up to 100% degraded DDE was found in the 15-20 cm level.

On the basis of our observations, we conclude that if DDT passes through the humus layer it can be significantly adsorbed on sand layers as a mixture of decomposition products. This could be a significant sink for DDT. However, only a limited mixture of DDT and degradation products (6%) would be leached to the groundwater.

4.3.3. Lindane

In contrast to DDT, a great deal of lindane (tert 47% and quart 31%) was leached. Although the leaching rates were comparable, the pattern of the quality analysis differed significantly. In the case of the tert column eluate, lindane was quantitatively degraded, while in the case of the quart eluate, 82% was unchanged.

In both cases, the adsorbed material was a mixture of lindane and degradation products. Lindane was distributed along the whole column; this has been observed in earlier investigations [6]. Degradation resulted in even less polar material than lindane, which is unusual. The degradation product was more polar in the tert column, which also explains the higher leaching rate of this column.

The possible types of degradation product are known from soil experiments [7]. Dechlorinations and dehydrogenizations occur at the initial phase of decomposition. Introduction of double bonds results from these reactions. Further oxidation leads to chlorophenols. All these materials are more polar than lindane, but still toxic.

In conclusion, lindane may pass unchanged through the 50 cm sand layers at higher rates. This could result in groundwater contamination, together with degradation products, if sandy soils are involved, or if thin humus layers are either passed through or bypassed.

4.3.4. Chlorpyrifos

High rates (47–53%) of chlorpyrifos were leached. In the eluate of the tert sand, 40% unchanged parent material was present, but only degradation products were detected in the eluate of quart gravel.

In a manner similar to these data, 36% chlorpyrifos remained (along the column) on the tert sand, the rest being polar degradation products. On the quart columns, 55% unchanged parent material was found at a depth of 10 cm, but below this level only degradation products were detected.

The main degradation products of chlorpyrifos are 3,5,6-trichloropyridinol and chlorpyrifos oxon [8], but dechlorination can also occur.

Like lindane, chlorpyrifos and its degradation products moved through the sand layers at high rates. This is different to the observations made by other researchers for organophosphates in humus layers.

4.3.5. Diazinon

High elution rates were observed (95% in the tert sand and 69% in the quart gravel). Eluates of both sand types contained about 40% unchanged molecules, but the degradation patterns differed significantly.

Data on the adsorption patterns on sand were similar. The original molecule was distributed along the whole column.

The main degradation products of diazinon are diazoxon, hydroxydiazinon, pyrimidinol and hydroxypyrimidinol [9]. Diazinon moves at a high rate through the sand layers and can result in groundwater contamination if the site characteristics are appropriate.

4.3.6. Malathion

Eluates contain about seven different degradation products, but only traces of the parent material were detected.

Adsorption and degradation on sand differed. The tert column contained almost unchanged malathion along the whole length. Traces of highly polar degradation products were found in the 15–20 cm layer, with higher polar materials in the 20–30 cm layer. In the quart column, malathion was quantitatively degraded. The main degradation product (about 63%) was highly polar and remained in the 10 cm layer of the column, together with the second, less polar product.

The main polar degradation products are the mono and dimalathion carboxylic acids. A lesser polar degradation product is diethyl maleate [9]. In our case, carboxylic acids could represent high polar degradation products.

On the basis of our observations, most of the degraded molecules within the polar carboxylic acid groups were adsorbed on the higher layers. Only small degradation products that are usually biodegradable moved to the groundwater. Thus, although high rates of radioactivity were leached, groundwater contamination by malathion is unlikely.

4.3.7. Carbofuran

The leaching rate of carbofuran was higher than 80%. The tert eluate contained 90% unchanged molecules, while the quart eluate contained only 30%. High polar degradation products were also found.

Most of the adsorbed material was found in the 0–20 cm layer. In the case of tert sand, 40% of the carbofuran remained unchanged. A high polar product (50%) and two lesser polar products (about 5% each) were detected. Only 19% unchanged carbofuran was found on the quart gravel. The high polar product was only 21% and the lesser polar product about 59%. Significantly different degradation behaviour was again observed.

The main degradation products of carbofuran in aquatic systems are hydroxy and phenol compounds such as 3-hydroxycarbofuran and carbofuran phenol [10]. All contained most of the parent molecule. Because of the high leaching rates of all these materials, water contamination by carbofuran could easily occur.

4.3.8. Diuron

About 84% of the mainly unchanged parent material was leached in both columns. Small amounts (less than 10%) of polar products were detected.

The adsorbed material was distributed along the whole column, with few degradation products. The data were similar in both cases.

The main degradation product of diuron is 3,4-dichloroaniline, which could also represent the polar component of the mixtures in our experiment.

Because of the high leaching rate of the unchanged molecule, groundwater contamination could easily take place if sand layers are involved.

4.3.9. Monolinuron

In a manner similar to diuron, about 70% of the total radioactivity was eluted. Most of the eluate (tert 88% and quart 83%) was unchanged parent material.

On tert sand, monolinuron was quantitatively degraded, while on quart gravel, 15% unchanged parent material was detected.

Like diuron, chloroaniline and other polar compounds could occur, but most of the parent material remained unchanged.

As mentioned previously, groundwater contamination is likely if the appropriate conditions and soil characteristics are present.

4.3.10. Imidacloprid

Imidacloprid moves very quickly through the sediment layers (only 1.5% was adsorbed), which indicates that there is a high potential for groundwater contamination. Qualitative investigations of the eluates have not yet been concluded.

5. CONCLUSIONS

By simulating conditions of groundwater filtration that are close to natural, we reached the following conclusions. Concerning elution and degradation, the behaviour of chemicals differed, even if they belonged to the same class of bioactive compound. Furthermore, the behaviour of the same material differed in different types of sand. With the exception of DDT and its corresponding degradation product, DDE, all the pesticides investigated moved at high rates through the 50 cm sand layer. The degree of degradation and the type of breakdown products obviously depend on the stability of the parent material, but in general high degradation rates were observed on the two types of sand. Because of the low transportation velocity, and the subsequent long retention times of pesticides within the sand media, microbial degradation (metabolisms) can also take place [11]. The great variation in findings does not allow prediction of the trends or the groundwater contaminations for other pesticides. However, the results provide the basis for improved application of the pesticides investigated. Concerning groundwater quality, the data emphasize the need for investigating pesticide behaviour, especially the persistence of DDT and some organophosphate insecticides.

ACKNOWLEDGEMENTS

The authors gratefully acknowledge the ^{14}C pesticides supplied by International Isotopes Munich, Unterschleissheim, and the support of the IAEA.

REFERENCES

[1] COHEN, S.Z., CREEGER, S.M., CARSEL, R.F., ENFIELD, C.G., ACS Symposium Series 259, American Chemical Society, Washington, DC (1984) 297–315.
[2] KLOTZ, D., Report 7/91, Gesellschaft für Strahlen- und Umweltforschung, Neuherberg (1992).
[3] WEBER, W.J., Jr., Physicochemical Processes for Water Quality Control, Wiley, New York (1972).
[4] DDT in the Tropics: Pesticides, Food Contaminants, and Agricultural Wastes, Special issue, Part B, J. Environ. Sci. Health (1994) 12701–15176.
[5] METCALF, R.L., A century of DDT, J. Agric. Food Chem. **4** (1973) 511–519.
[6] BERAN, F., GUTH, J.A., Pflanzenschutzberichte **33** (1965) 65–67.
[7] SCHEUNERT, I., "Physical and physico-chemical processes governing the residue behaviour of pesticides in terrestrial ecosystems", Chemistry of Plant Protection, Vol. 8 (EBING, W., Ed.), Springer-Verlag, Berlin and Heidelberg (1992) 1–22.
[8] RACKE, K.D., COATS, J.R., J. Agric. Food Chem. **36** (1988) 193–199.
[9] BARTSCH, E., Res. Rev. **38** (1974) 58–59.
[10] ARCHER, T.E., STOCS, J.D., BRINGHURST, R.S., J. Agric. Food Chem. **25** (1977) 536–541.
[11] SCHMIDT, K., Gewässerschutz-Wasser-Abwasser **109** (1988) 181–185.

IAEA-SM-343/24

MODEL ECOSYSTEMS FOR PREDICTING THE BEHAVIOUR OF PESTICIDES IN THE ENVIRONMENT*

K. RAGHU, N.B.K. MURTHY,
S.P. KALE, M.G. KULKARNI
Nuclear Agriculture Division,
Bhabha Atomic Research Centre,
Mumbai, India

Abstract

MODEL ECOSYSTEMS FOR PREDICTING THE BEHAVIOUR OF PESTICIDES IN THE ENVIRONMENT.

Prediction of the behaviour of pesticides in the environment aids efficient management of agrochemicals, and use of radioisotopes offers a simple and highly reliable tool for such studies. Various experimental set-ups that simulate ecosystems have been used successfully under laboratory and outdoor conditions. Use of a soil biometer flask, a continuous flow-through system and a rice–fish ecosystem to follow the fate of ^{14}C labelled HCH isomers, carbaryl, nitrofen, carbofuran and glyphosate is discussed. By combining the agronomic practices of flooding and/or use of organic matter amendments such as green manure it was shown that the persistence of HCH isomers, carbaryl and carbofuran in soil could be considerably reduced. Use of labelled pesticides in conjunction with a continuous flow-through system enabled determination of the total ^{14}C mass balance and budgeting of the pesticides in the environmental compartments of flood water, organic volatiles, CO_2, extractable and unextractable (bound) residues. Faster degradation in flooded and/or organic matter amended soils was unequivocally shown by demonstrating mineralization to $^{14}CO_2$ and low extractable ^{14}C residues. Use of ^{14}C labelled pesticides helps to determine the extent of formation of bound residues, and to follow their uptake in plants and their bioavailability through microbial release. Bound ^{14}C residues were formed in soils after using all the above pesticides. A soil biometer flask is of help in following the mineralization of bound residues, as seen with lindane (gamma-HCH) and carbaryl. Another use for a soil biometer flask is to follow the fate of ^{14}C pesticides through $^{14}CO_2$ evolved either in soil or by microbial cultures. The extent of glyphosate degradation in different soil types was followed, since this herbicide is known to undergo co-metabolic mineralization to CO_2. Studies using a rice–fish model ecosystem showed that ^{14}C-carbofuran and ^{14}C-nitrofen did not accumulate in rice or fish and indicated that discriminate use of these pesticides is not hazardous under semi-tropical conditions. The above model ecosystem was also useful in following the fate of ^{14}C-DDT in clams and sediments in the marine environment.

---

* Research carried out with the support of the IAEA under Research Contract No. 7936.

Radiotracer techniques offer a powerful tool for understanding the fate and persistence of pesticides in the environment. Use of labelled pesticides in conjunction with model ecosystems also yields valuable information on the behaviour of pesticides in the environment. A wide variety of experimental approaches has been applied using model ecosystems to study this behaviour [1]. Laboratory model systems include soil perfusion systems, soil biometers, continuous flow-through systems and integrated systems. For field experiments, standardized lysimeters have been effectively used [2]. Another model ecosystem that simulates natural conditions is the rice–fish model ecosystem described by Metcalf [3]. In this paper, the results of work conducted at the Nuclear Agriculture Division, Bhabha Atomic Research Centre, Mumbai, are discussed. During these studies we used various ecosystems such as a soil biometer flask, a continuous flow-through system and a rice–fish ecosystem. The pesticides studied were ^{14}C-gamma-HCH (lindane), ^{14}C-beta-HCH, ^{14}C-carbaryl, ^{14}C-carbofuran and ^{14}C-glyphosate.

Persistence of pesticides is influenced by the soil type, the soil moisture conditions and the agricultural amendments, as evident from studies conducted with HCH isomers, carbaryl and carbofuran. Many of these studies were conducted on a black clay soil (Vertisol), which is one of the major Indian soil types. The pesticides studied are extensively used for the cultivation of rice and upland crops. Since most of the rice is grown under flooded conditions, the studies were conducted under moist (unflooded) and flooded conditions. The first report that flooding results in a decrease in the persistence of gamma-HCH in soils [4] was confirmed by similar observations with other pesticides [5, 6]. Detection of $^{14}CO_2$ evolved from ^{14}C-gamma-HCH treated flooded soils indicated degradation through ring cleavage [7]. Organic amendments such as rice straw and green manure decreased the persistence of HCH isomers under flooded conditions [8, 9]. Soil artificially contaminated with HCH isomers at a concentration of 25 mg/kg could be decontaminated to a large extent by puddling green manure in soil under flooded conditions [9]. To establish the role of microbial degradation, the persistence of HCH isomers was compared in sterilized and unsterilized soils with and without organic matter amendment under flooded soil conditions. The HCH isomers were found to be equally persistent in sterilized, unamended and organic matter amended soils, indicating the absence of chemical degradation by organic matter amendment. The decreased persistence observed in unsterilized, organic matter amended and unamended soils compared with sterilized soil treatments was attributed to the activity of soil microorganisms.

To ascertain the decisive role of organic amendments, especially green manure, in reducing the persistence of pesticides, a continuous flow-through system was used that permitted the ^{14}C mass balance to be determined [10, 11]. Gamma-HCH, an insecticidally active and fairly degradable isomer, and beta-HCH, a non-insecticidal contaminant present in HCH formulations and also the most persistent isomer, were used in these studies. It was found that there was considerable formation of radioactive volatiles, both in the ^{14}C-gamma-HCH and in the ^{14}C-beta-HCH

TABLE I. DISTRIBUTION OF ^{14}C IN $^{14}CO_2$, ORGANIC VOLATILES AND AQUATIC AND SOIL PHASES FROM ^{14}C-GAMMA-HCH OR ^{14}C-BETA-HCH TREATED SOILS[a]

					Recovery (% of applied)		
						Soil	
		$^{14}CO_2$	Organic volatiles	Flood water	Extractable ^{14}C residues	Bound ^{14}C residues	Total ^{14}C residues
^{14}C-gamma-HCH	UA	10.7	4.7	4.4	30.0 (28.9)	27.4	77.2
	GA	12.3	22.6	1.6	4.7 (2.2)	16.9	58.1
^{14}C-beta-HCH	UA	4.3	0.8	4.5	80.0 (78.4)	9.7	99.3
	GA	30.2	5.7	3.5	50.6 (47.1)	10.5	100.5

[a] UA = unamended flooded Vertisol soil; GA = green manure amended flooded Vertisol soil. The total amount of ^{14}C-gamma-HCH or ^{14}C-beta-HCH initially added was 1 mg/kg of soil. Numerals in parentheses indicate the percentage recovered.

treated soils; higher levels were found in the green manure amended soils and in the ^{14}C-beta-HCH treated soils (Table I). Mineralization to $^{14}CO_2$ was faster in the green manure amended soils, although the effect was more pronounced with ^{14}C-beta-HCH. The volatile compounds were identified as benzene and CO_2. Like laboratory experiments carried out in a soil metabolic flask under still conditions (without a continuous flow-through of air) and in pot experiments, it was evident in the present continuous flow-through system that both ^{14}C-beta-HCH and ^{14}C-gamma-HCH were degraded. This was evidenced by the low recovery of extractable ^{14}C residues and the formation of $^{14}CO_2$ and organic volatiles. The total ^{14}C mass balance was better in experiments with ^{14}C-beta-HCH than with ^{14}C-gamma-HCH. The unaccounted radioactivity with ^{14}C-gamma-HCH may have stemmed from volatilized benzene derived from parent material during the soil filtration process. Brahmaprakash et al. [12], working with a closed soil system, also observed a higher recovery of ^{14}C-beta-HCH than of ^{14}C-gamma-HCH.

The fate and persistence of carbaryl in soil has been reviewed by Rajagopal et al. [13]. Work on the fate of ^{14}C-carbaryl using a soil metabolic flask (still conditions) under moist and flooded conditions and soil types with a varying pH showed

TABLE II. PERCENTAGE ^{14}C RADIOACTIVITY RECOVERED FROM ^{14}C-CARBARYL AND ^{14}C-1-NAPHTHOL TREATED VERTISOL SOILS[a]

Soil treatment	$^{14}CO_2$ (cumulative)	Extractable ^{14}C residues	Bound ^{14}C residues	Total ^{14}C residues
^{14}C-carbaryl				
Moist	25.6	5.5	45.0	76.1
Flooded	15.1	28.9	47.1	91.1
^{14}C-1-naphthol				
Moist	1.9	18.2	83.0	103.1
Flooded	0.5	24.3	68.6	93.4

[a] The total amount of ^{14}C-carbaryl or ^{14}C-1-naphthol initially added was 10 mg/kg of soil.

that with the increase in soil pH there was a decrease in the extractable ^{14}C-carbaryl residues and an increase in the soil bound ^{14}C residues [14]. The metabolism of ^{14}C-carbaryl and ^{14}C-1-naphthol in moist and flooded soils was studied using a continuous flow-through system [15]. In such a system, the mineralization of ^{14}C-carbaryl was higher in moist than in flooded Vertisol soils. However, with ^{14}C-1-naphthol mineralization was negligible under both soil moisture conditions (Table II). Carbon-14-carbaryl metabolized extensively and the extractable ^{14}C residues were higher in flooded than in moist soils. Carbon-14-carbaryl metabolized mainly to 5-hydroxycarbaryl in moist soil and to 4- and 5-hydroxycarbaryl in flooded soil. The amount of ^{14}C-1-naphthol present in the extractable ^{14}C residues was less than 1%, most of these being located at origin on the thin layer chromatograph plates, indicating the formation of some complexes of ^{14}C-1-naphthol in soil.

The fate of ^{14}C-carbofuran in soils was followed using a continuous flow-through system; it underwent extensive degradation under both moisture conditions in Vertisol soils. The total ^{14}C mass balance of ^{14}C-carbofuran showed that mineralization to $^{14}CO_2$ and formation of bound ^{14}C residues accounted for almost all the ^{14}C residues. Thus, in moist soil the amount of $^{14}CO_2$ was 30.0 and 33.7% of the applied activity in moist and flooded conditions, respectively. The bound ^{14}C residues accounted for 55.3 and 41.3% in moist and flooded soils, respectively [16]. In the soil metabolic flask, where no provision is made for trapping CO_2, only extractable and bound ^{14}C residues were measured.

Radiolabelled pesticides are helpful in determining the formation of bound residues in soils and plants. This can be demonstrated using a soil biometer flask or a continuous flow-through system and ^{14}C labelled pesticide. Experiments using a soil metabolic flask (still conditions) and with ^{14}C-lindane have shown that bound

^{14}C residues were detected in moist and flooded Vertisol soils 1 year after application of the pesticide. Formation of these residues was higher in moist than in flooded soils. Fewer bound ^{14}C residues were found in green manure amended flooded soils [17]. Further studies with seven Indian soil types have shown that the formation of bound ^{14}C residues in moist and flooded soils ranged between 5.5 and 34.6%. Formation of bound ^{14}C residues was higher in neutral and alkaline soils; it was also related to the organic matter content of the soil. All these experiments indicated that a considerable amount of bound ^{14}C residues was formed, even with persistent pesticides such as lindane. In view of this, the general concept that persistent pesticides such as organochlorines (HCH and DDT) form comparatively fewer bound residues than non-persistent pesticides such as organophosphorus and carbamates needs to be reconsidered. Using the soil biometer flask, the evolution of $^{14}CO_2$ from bound ^{14}C residues of lindane was demonstrated [17]. More bound ^{14}C residues of lindane originating from flooded soil were mineralized to $^{14}CO_2$ than those derived from moist soils.

Bound residues of ^{14}C-1-naphthol were formed under still and continuous flow-through systems [14, 15]. The soil bound residues of ^{14}C-carbaryl and ^{14}C-1-naphthol were released when barley was grown on the soils containing these residues [18]. Soil bound residues of ^{14}C-carbaryl were detected in the shoots and the roots, while soil bound residues of ^{14}C-1-naphthol were only shown in the roots. Flooding enhanced the release of soil bound residues of ^{14}C-carbaryl. Mineralization of the soil bound residues of ^{14}C-carbaryl was rapid with rice straw amendments. With ^{14}C-1-naphthol, there was negligible mineralization, and amendments had no influence [18]. All the above observations indicate that the soil biometer flask system appears to be a useful tool for conducting experiments related to the release of bound residues.

The soil biometer flask (with a CO_2 trap) can be used to compare the metabolism of pesticides in different soil types. In general, the half-life of a pesticide in soil is determined by measuring the residual quantity of the pesticide. Glyphosate residues cannot easily be analysed chemically. Owing to the co-metabolic nature of glyphosate degradation, the amount of $^{14}CO_2$ evolved is assumed to be the amount of glyphosate degraded [19], hence it is easy to measure $^{14}CO_2$ from ^{14}C labelled glyphosate. The fate of ^{14}C-glyphosate in four Indian soils differing in their physicochemical characteristics was followed using a biometer flask [20]. There was rapid degradation in ^{14}C-glyphosate in the alkaline soils of Jalgaon (silty clay loam) and Trombay (clay loam), since 58.2 and 49.2%, respectively, of the applied ^{14}C-glyphosate had metabolized to CO_2 after 40 days (Fig. 1). However, in Kerala (loam) and Ratnagiri (loam) soils, evolution of $^{14}CO_2$ was less than 1 and 5%, respectively, indicating negligible degradation. At the end of the incubation period, soils were extracted with water; in all the soils, the water extractable ^{14}C residues were low. However, almost the entire amount of applied ^{14}C-glyphosate was present as soil bound ^{14}C residues in the acid soils of Kerala and Ratnagiri, while

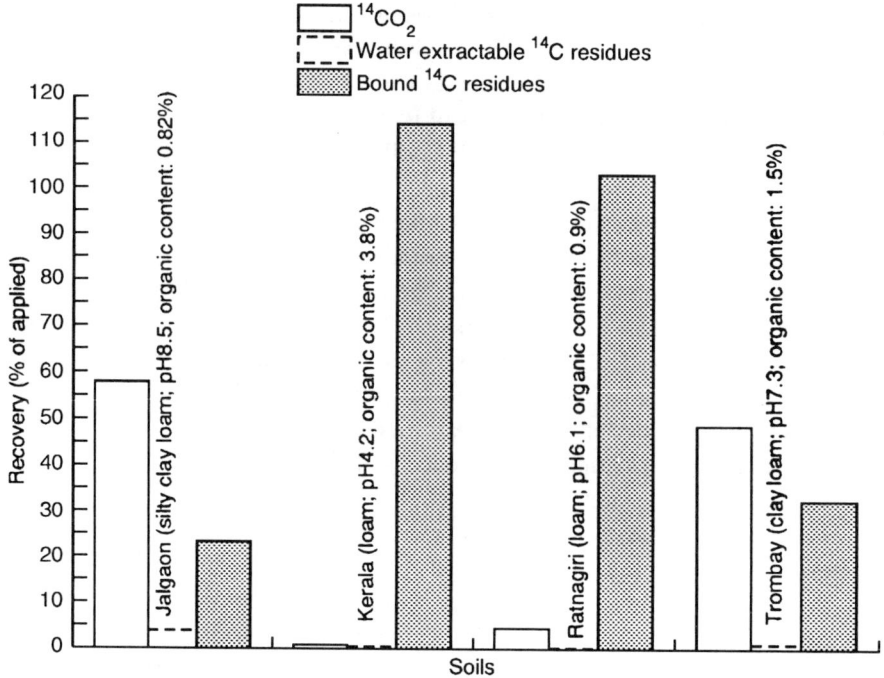

FIG. 1. *Degradation of ^{14}C-glyphosate in four Indian soils.*

TABLE III. PERSISTENCE OF ^{14}C RESIDUES OF ^{14}C-CARBOFURAN AND ^{14}C-NITROFEN IN SOILS IN A RICE–FISH ECOSYSTEM (APPLIED ACTIVITY = 100%)[a]

Days	^{14}C-nitrofen (% of applied)			^{14}C-carbofuran (% of applied)		
	Extractable ^{14}C residues	Bound ^{14}C residues	Total ^{14}C residues	Extractable ^{14}C residues	Bound ^{14}C residues	Total ^{14}C residues
10	26.8	63.3	90.1	10.7	17.5	31.2
40	3.0	22.9	25.9	7.5	63.2	70.7
120	0.8	4.9	5.7	2.5	46.3	48.8
365	—	—	—	1.0	29.0	30.0

[a] Carbon-14 nitrofen was applied at a rate of 2.5 mg/kg of soil, and ^{14}C-carbofuran was applied twice (at 10 and 40 days) at a rate of 0.5 mg/kg of soil.

23.3 and 32.9% of the applied ^{14}C was present as bound ^{14}C residues in Jalgaon and Trombay soils, respectively. All the bound ^{14}C residues were present in the fulvic acid fraction in acid soils, whereas in alkaline soils they were distributed in the fulvic and the humin fractions. Further experiments showed that, while ^{14}C-glyphosate was adsorbed in all four soils, desorption occurred to a greater extent in alkaline soils. Probably, release of these residues enabled microorganisms to mineralize the ^{14}C-glyphosate in the soils.

A laboratory study that provided useful information on the fate of pesticides was the ecological tank, as described by Metcalf [3]; he used an aquarium tank that accommodated terrestrial and aquatic habitats. The fate of ^{14}C-nitrofen and ^{14}C-carbofuran in a rice–fish model ecosystem has been followed using soil and components such as rice plants, *Gambusia* fish, algae and snails [21, 22]. The rice plants, fish and snails showed negligible ^{14}C residues. After 10 days, the soils showed total ^{14}C-nitrofen residues of 90.1%, which decreased to 5.7% after 120 days; bound residues constituted the major share of the ^{14}C residues (Table III). Lee et al. [23] have shown bioaccumulation of nitrofen in the tissues of algae, snails, mosquitoes and fish using a model ecosystem with standard reference water and sand under plant growth chamber conditions. Samples for the present experiments were collected from the field. It is obvious that nitrofen degraded rapidly under semi-tropical conditions. With carbofuran, the extractable ^{14}C residues decreased over a period of 40 days. A considerable amount of bound ^{14}C residues was formed over a period of 40 days (67.8%), but these had decreased to 52.8% after 120 days; most were found in the soil (63.2 and 46.3% after 40 and 120 days, respectively) and to some extent in plants (6.1% after 120 days). It should be noted that 29% of the applied ^{14}C-carbofuran remained as bound ^{14}C residues after 1 year. Regarding bioaccumulation in biological materials, these experiments definitely indicated that ^{14}C-nitrofen and ^{14}C-carbofuran residues were of no consequence under semi-tropical conditions.

The aquarium tank was adopted to develop a marine ecosystem using sea water, sediments (collected from the sea), algae and clams (*Katelysia opima*). The fate of ^{14}C-DDT was followed in water, clams and sediments over a period of 30 days [24]. A negligible amount of ^{14}C residues was present in water after 3 days. In clams, ^{14}C-DDT accumulated during the 3 days of incubation and was found as extractable and bound ^{14}C residues, the major degradation product being DDE. In the case of sediments, DDD was the major degradation product, with no formation of DDT. Also, bound ^{14}C residues were negligible. These experiments show the utility of using an aquarium tank to follow the fate of pesticides in the marine environment.

It is evident that various model ecosystems can be effectively used to study the behaviour of pesticides in terrestrial and aquatic environments.

REFERENCES

[1] GUTH, J.A., "Experimental approaches to studying the fate of pesticides in soil", Progress in Pesticide Chemistry, Vol. 1 (HUDSON, D.H., ROBERTS, T.R., Eds), Wiley, Chichester (1981) 85–114.

[2] FÜHR, F., CHENG, H.H., MITTELSTAEDT, W., Pesticide balance and metabolism studies with standardized lysimeters, Landwirtsch. Forsch., Sonderheft **32** (1976) 272–278.

[3] METCALF, R.L., "A laboratory model ecosystem for evaluating the chemical and biological behaviour of radiolabelled micropollutants", Comparative Studies of Food and Environmental Contamination (Proc. Symp. Otaniemi, 1973), IAEA, Vienna (1974) 49–62.

[4] RAGHU, K., MACRAE, I.C., Biodegradation of the gamma isomer of benzene hexachloride in submerged soils, Science **154** (1966) 263–264.

[5] SETHUNATHAN, N., Microbial degradation of insecticides in flooded soils and anaerobic cultures, Res. Rev. **47** (1973) 143–165.

[6] SETHUNATTHAN, N., ADHYA, T.K., RAGHU, K., "Microbial degradation of pesticides in tropical soils", Biodegradation of Pesticides (MATSUMURA, F., KRISHNAMURTHY, C.R., Eds), Plenum, New York (1982) 91–115.

[7] MACRAE, I.C., RAGHU, K., CASTRO, T.F., Persistence and biodegradation of four common isomers of benzene hexachloride in submerged soils, J. Agric. Food Chem. **15** (1967) 911–914.

[8] FERREIRA, J., RAGHU, K., "Persistence of hexachlorocyclohexane in soils", Nuclear Techniques in Studies of Metabolism: Effects and Degradation of Pesticides, Department of Atomic Energy, Mumbai (1978) 126–133.

[9] FERREIRA, J., RAGHU, K., Decontamination of hexachlorocyclohexane in soil by green manure application, Environ. Toxicol. Lett. **2** (1981) 357–364.

[10] DREGO, J., MURTHY, N.B.K., RAGHU, K., ^{14}C-gamma hexachlorocyclohexane in flooded soil with green manuring, J. Agric. Food Chem. **38** (1990) 266–268.

[11] RAGHU, K., DREGO, J., MURTHY, N.B.K., "Fate of ^{14}C-beta-HCH in green manure amended soils", Nuclear Techniques in the Study of Pesticides in Food, Agriculture and Environment (Proc. Symp. Bangalore, 1989), Department of Atomic Energy, Mumbai (1989) 89–90.

[12] BRAHMAPRAKASH, G.P., REDDY, B.R., SETHUNATHAN, N., Persistence of HCH isomers in soil planted with rice and in rice rhizosphere soil suspensions, Biol. Fertil. Soils **1** (1985) 103–109.

[13] RAJAGOPAL, B.S., BRAHMAPRAKASH, G.P., REDDY, B.R., SINGH, U.D., SETHUNATHAN, N., Effects and persistence of selected carbamate pesticides in soil, Res. Rev. **93** (1984) 1–199.

[14] MURTHY, N.B.K., RAGHU, K., Fate of ^{14}C-carbaryl in soils as a function of pH, Bull. Environ. Contam. Toxicol. **46** (1991) 374–379.

[15] MURTHY, N.B.K., RAGHU, K., Metabolism of ^{14}C-carbaryl and ^{14}C-1-naphthol in moist and flooded soils, J. Environ. Sci. Health **B24** (1989) 479–491.

[16] KALE, S.P., Pesticide–soil microflora interactions with special reference to nitrofen and carbofuran, PhD Thesis, University of Pune, Pune (1991).

[17] RAGHU, K., DREGO, J., "Bound residues of lindane: Magnitude, microbial release, plant uptake and effect on microbial activities", Quantification, Nature and Bioavailability of Bound ^{14}C Pesticide Residues in Soil, Plants and Food (Proc. Panel Gainesville, 1985), IAEA, Vienna (1986) 41–50.

[18] MURTHY, N.B.K., RAGHU, K., Soil bound residues of carbaryl and 1-naphthol: Release and mineralization in soil and uptake by plants, J. Environ. Sci. Health **B23** (1988) 575–585.

[19] TORSTENNSON, L., "Behaviour of glyphosate in soils and its degradation", The Herbicide Glyphosate (GROSSBARD, E., ATKINSON, D., Eds), Butterworths, London (1985) 137–150.

[20] KULKARNI, M.G., Fate and effects of the herbicide glyphosate in tropical soils with special reference to its degradation by actinomycete isolates, PhD Thesis, University of Bombay, Mumbai (1994).

[21] KALE, S.P., RAGHU, K., Fate of ^{14}C-nitrofen in a rice paddy ecosystem, Bull. Environ. Contam. Toxicol. **12** (1989) 544–547.

[22] KALE, S.P., RAGHU, K., "Fate of pesticides in a rice–fish ecosystem", Nuclear Science and Technology in India: Present and Future (Proc. Conf. Mumbai, 1989), Vol. II, Bhabha Atomic Research Centre, Mumbai (1989) 164–169.

[23] LEE, A.H., LU, P.Y., METCALF, R.L., ASU, E.L., The environmental fate of three dichlorophenoxy herbicides in a rice paddy model ecosystem, J. Environ. Qual. **5** (1976) 482–486.

[24] KALE, S.P., RAGHU, K., MOHAN RAO, A., MURTHY, N.B.K., PANDIT, G.G., "Fate of ^{14}C-DDT in the Indian marine environment using microcosm experiments", IAEA Project on Distribution, Fate and Effects of Pesticides on Biota in the Tropical Environment, Paper presented at 2nd Research Co-ordination Meeting, Kuala Lumpur, 1995.

PERSISTENCE OF TERBUFOS AND ITS METABOLITES IN SOIL AND MAIZE*

E. CARAZO**, B.E. VALVERDE***,
O.M. RODRIGUEZ**, M. BARQUERO**

**Centro de Investigación en
  Contaminación Ambiental,
Universidad de Costa Rica,
San José

***Centro Agronómico Tropical de
  Investigación y Enseñanza,
Unidad de Fitoprotección,
Turrialba

Costa Rica

Abstract

PERSISTENCE OF TERBUFOS AND ITS METABOLITES IN SOIL AND MAIZE.

Degradation of ^{14}C terbufos was studied under greenhouse conditions. A mixture of ^{14}C labelled compound (2.48×10^4 Bq of O-ethyl-1-^{14}C) und unlabelled compound (0.09 g of the granule formulation Counter 10 G) was applied to pots containing 750 g of sandy loam clay Ultisol soil with 4.9% organic matter and a cation exchange capacity of 7.6. Two treatments were established, one where maize (cultivar Cristiani) was grown and the other without plants. The soil and plants were extracted at 0, 4, 8, 16, 32 and 64 days and analysed by a liquid scintillation counter, gas chromatography–flame photometric detector (GC–FPD) and autoradiography. The total ^{14}C compounds extracted on day 64 were 31 ± 5.6% of the radioactivity applied in the treatments with plants and 46.1 ± 1.1% without plants. From the autoradiography results it can be concluded that at all times the compounds identified by this technique were terbufos, terbufos sulphoxide and terbufos sulphone. These results will be compared with those to be obtained by GC–FPD and another extraction method. Plant extracts are currently being analysed.

---

* Research carried out with the support of the IAEA under Research Contract No. 7920.

1. INTRODUCTION

Terbufos is an organophosphorus insecticide and nematicide used in maize, bananas, potatoes and other crops. It is applied at planting in bands or directly to the seed furrow. Terbufos controls white grubs, wireworms, maize rootworm and other insect pests and nematodes.

The persistence of terbufos has been studied under different conditions in non-tropical areas. According to Ahmad et al. [1] it is considered moderately persistent in the soil and is rapidly converted to its metabolites, terbufos sulphone and terbufos sulphoxide, which tend to persist in the soil and have been detected at harvest time. This study was performed on silty clay loam soil in South Dakota (United States of America), where the time to 50% disappearance (DT_{50}) of terbufos was about 2 weeks, while the half-life for the metabolite, terbufos sulphone, was two to three times longer. Felsot et al. [2] found that terbufos and its metabolites degraded rapidly during the first 15–30 days after application, and then stabilized gradually. Only 3% of the original application remained in field soils in Illinois after 1 month, with 1.5% of the chemical applied present after 60 days. They also found that dissipation of terbufos was generally faster in soils with very low organic carbon, while binding increased with the increasing organic carbon content. Sandy soils lose more of this chemical than do soils with clay. Over 90% of the applied terbufos were recovered in the top 20 cm of a soil profile, despite heavy rainfall and thorough incorporation. The same researchers stated that soil moisture did not appear to affect the degradation of terbufos. This pesticide breaks down at about the same rate in soils, regardless of the level of wetness.

When applied to a silt loam soil, Szeto et al. [3] calculated the terbufos half-life at 15 days, while the DT_{50} of the total residues (parent and metabolites) was 22 days. After 106 days, the total residues were less then 1 mg/kg. As the temperature increased, terbufos degraded more rapidly. Szeto et al. [3] established that terbufos moved from the soil to the the plants, where it was broken down rapidly. Cobb et al. [4] reported that terbufos degraded slowly in midwestern USA and for about 90 days was present in the environment as its oxidation product. Carazo [5] found high levels of terbufos and its metabolites in the field in silt loam and in loam Maryland soils 56 days after application. Under greenhouse conditions, the same author found the DT_{50} to be between 85 and 95 days. Soil moisture apparently favoured the persistence of terbufos because of the anaerobic conditions created, and neither the disappearance of the compound nor its metabolites changed over the range of temperatures studied.

The purpose of this research was to determine the persistence of terbufos and its metabolites in maize and a local soil under greenhouse conditions in Costa Rica (Fig. 1).

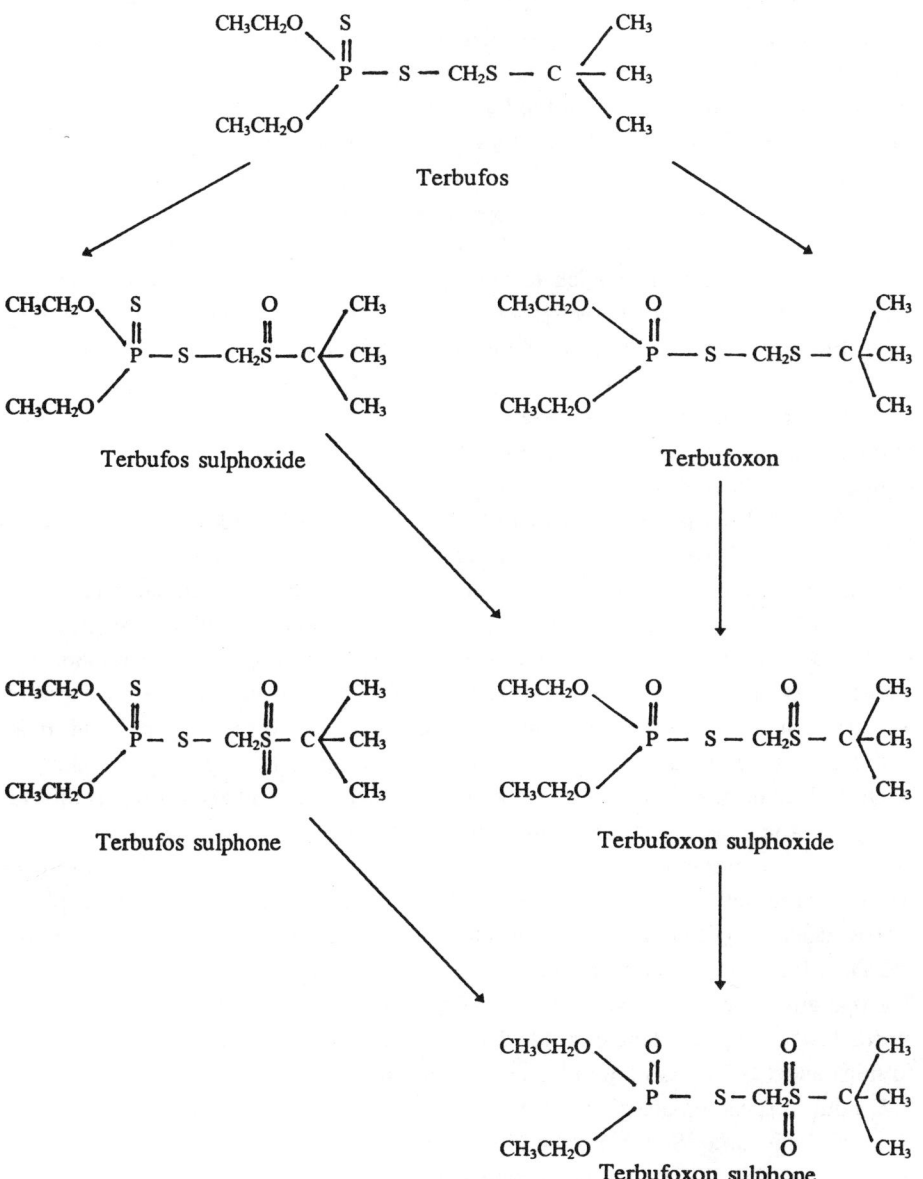

FIG. 1. Terbufos and its metabolites.

2. MATERIALS AND METHODS

The experiment was performed in the greenhouse facilities at the University of Costa Rica. The temperature and relative humidity ranged from 21 to 27°C and from 80 to 100%, respectively. Two treatments were established: soil with plants, and soil without plants. The soil type used was Ultisol, which is a sandy loam clay with 4.9% organic matter and a cation exchange capacity of 7.6. The experimental units were arranged at random, with six replicates per treatment. Plastic pots containing 750 g of disturbed soil were treated with a mixture of 0.09 g of the granule formulation Counter 10 G (American Cyanamid, USA). These granules were treated with 2.48×10^4 Bq of labelled terbufos (O-ethyl-1-^{14}C) in acetone and hexane. The radiochemical purity was better than 95%, and the specific activity 2.1 MBq/mg. After the solvents were evaporated, maize (cultivar Cristiani) was grown at a depth of 2.5 cm.

At sampling, the plants were extracted from the soil and the visible roots removed. Soil from two pots was mixed carefully, quartered several times and a sample of 50 g taken for analysis, with three replicate analytical samples per treatment. Soil analysis is described in Fig. 2. A Beckman LS 600 liquid scintillation counter was used. An aliquot of the extracts was applied to silica gel 60 G254 plates and developed in the following solvent system: acetone/ethylacetate/hexane (25:65:110). The R_fs for the different compounds were: terbufos, 0.80; terbufos sulphoxide, 0.45; terbufos sulphone, 0.69; terbufoxon, 0.62; and terbufoxon sulphone, 0.33. The plates were visualized in UV light (254 nm) and, after exposure to iodine vapour, the R_fs of the soil extract components were compared with those of the analytical standards. Autoradiograms were made by exposing the plates to X-Omat Kodak Xar-5 film. The extracted soil will be oxidized to determine the percentage binding of the ^{14}C compounds to the soil [6].

To analyse terbufos and its metabolites by gas chromatography (GC) following the procedure outlined in Fig. 3, a Shimadzu GC-14A equipped with a flame photometric detector (FPD) and an integrator was used. The column was an SPB-1 (30 m × 0.32 mm inner diameter), with nitrogen as the carrier gas (30 mL/min). The operating parameters were: detector temperature, 250°C; initial column temperature, 120°C; injector temperature, 250°C; final column temperatures, (1) 190°C (0 min) and (2) 250°C (1 min); and temperature rate, 3°C/min (later changed to 5°C/min, to speed up analysis). The detector response was calibrated daily with the analytical standards. Suitable separation of terbufos, terbufoxon and their sulphones is shown in Fig. 3. The sulphoxide metabolites gave no measurable response (as expected) under these chromatographic conditions and need to be oxidized to their respective sulphones (Fig. 3). The retention times were: terbufos, 14.3 min; terbufos sulphone, 22.0 min; terbufoxon, 12.2 min; and terbufoxon sulphone, 19.2 min.

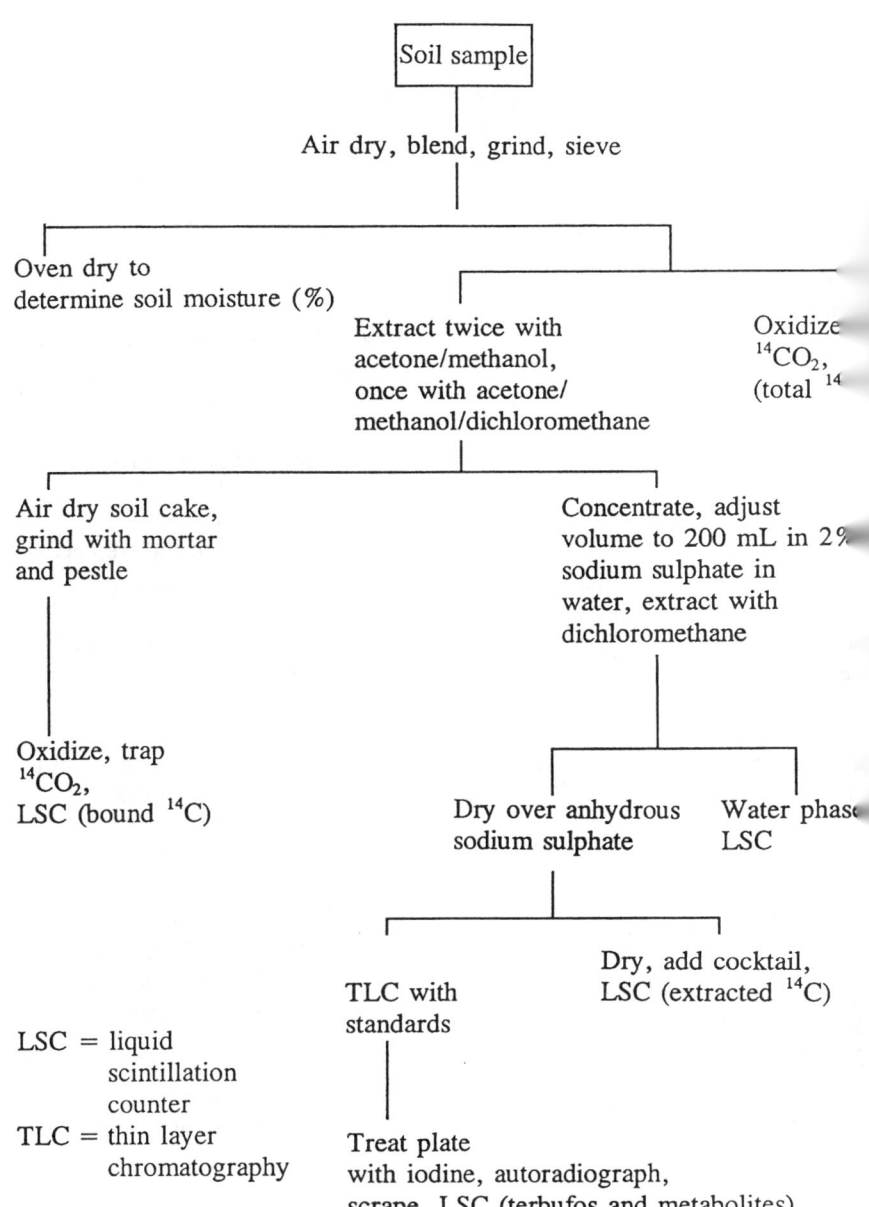

FIG. 2. *Extraction method and determination of ^{14}C-terbufos and its metabolites in using radiometric techniques.*

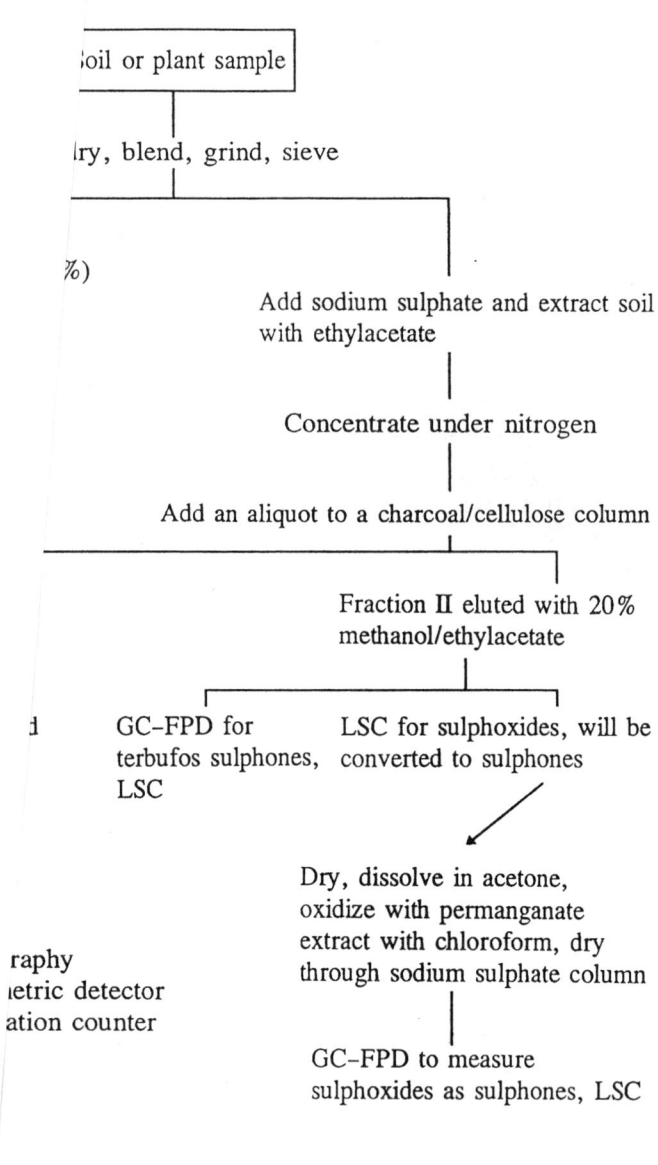

3. RESULTS AND DISCUSSION

All the data from the soil were adjusted to the soil weight after drying at 110°C. The soil moisture varied with the different times of sampling and ranged between 31.8 and 55.9%. Recovery of the total ^{14}C compounds on day 0 was considered to be 100%, with an average of 2779 ± 130 dis·min^{-1}·g^{-1} of soil. The efficiency was between 95 and 100%. On day 4, recovery was 99.9 ± 9.6% (2776 ± 268 dis·min^{-1}·g^{-1}) for the treatments with plants and 122.4 ± 11.2% (3402 ± 311 dis·min^{-1}·g^{-1}) without plants. On day 8, recovery was 79.8 ± 16.7% (2219 ± 463 dis·min^{-1}·g^{-1}) for the treatments with plants and 84.4 ± 21.4% (2346 ± 594 dis·min^{-1}·g^{-1}) without plants. On day 16, recovery was 52.4 ± 14.7% (1609 ± 408 dis·min^{-1}·g^{-1}) for the treatments with plants and 73.3 ± 15.5% (2003 ± 430 dis·min^{-1}·g^{-1}) without plants. On day 32, recovery was 44.5 ± 4.5% (1235 ± 124 dis·min^{-1}·g^{-1}) for the treatments with plants and 59.6 ± 6.0% (1656 ± 167 dis·min^{-1}·g^{-1}) without plants. On day 64, recovery was 31.0 ± 5.6% (862 ± 185 dis·min^{-1}·g^{-1}) for the treatments with plants and 46.1 ± 1.1% (1280 ± 31 dis·min^{-1}·g^{-1}) without plants (Fig. 4). The plates of soil extracts from day 0 showed only a spot with an R_f corresponding to the parent compound, but those of the extracts obtained at 4–16 days showed spots whose R_fs matched the parent compound and terbufos, terbufos sulphoxide and terbufos sulphone. The cleanup of soil and plant extracts is currently being carried out following the procedure outlined in Fig. 3.

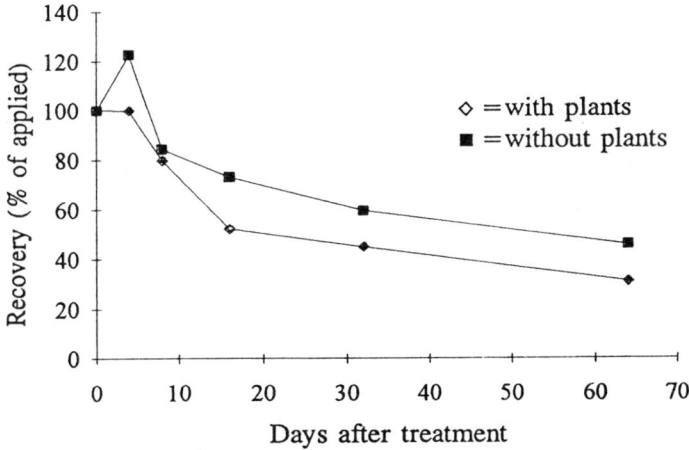

FIG. 4. Recovery of ^{14}C compounds in a sandy loam clay soil after treatment with ^{14}C-terbufos.

REFERENCES

[1] AHMAD, N., et al., Comparative disappearance of fenofos, phonate and terbufos soil residues under similar South Dakota field conditions, Bull. Environ. Contam. Toxicol. **23** (1979) 423–429.

[2] FELSOT, A., et al., Environmental chemodynamic studies with terbufos (Counter) insecticide in soil under laboratory and field conditions, Environ. Sci. Health **B17** 6 (1982) 649–673.

[3] SZETO, S.Y., et al., Degradation of terbufos in soil and its translocation into cole crops, J. Agric. Food Chem. **34** (1986) 876–879.

[4] COBB, G.P., et al., "Degradation of terbufos in soils during drought conditions", Pesticides in Terrestrial and Aquatic Environments (WEIGMANN, D.L., Ed.), Virginia Polytechnic Institute, Blacksburg, VA (1989) 159–170.

[5] CARAZO, E., Behavior, fate and interaction of atrazine and terbufos in soils, PhD Thesis, University of Maryland, College Park, MD (1991).

[6] AMATO, M., Determination of ^{12}C and ^{14}C in plant and soils, Biol. Biochem. **145** (1983) 611–612.

IAEA-SM-343/36

CARBON-14-TRIFLURALIN RESIDUES IN SOIL AND CARROTS*

O. TİRYAKİ, K. GÖZEK
Ankara Nuclear Research and
 Training Center,
Turkish Atomic Energy Authority,
Saray, Ankara, Turkey

Abstract

CARBON-14-TRIFLURALIN RESIDUES IN SOIL AND CARROTS.
 The persistence and movement of ^{14}C-trifluralin and its metabolites were studied in sandy loam soil under outdoor conditions. Uptake by carrots of the herbicide from the treated soil was also investigated. The bulk of the ^{14}C residues was present in the upper 0–7.5 cm layer during the growing period. The non-extractable ^{14}C residues increased with time. Downward movement of the ^{14}C-trifluralin, or its degradation products, was not observed at significant concentrations. After 4 months, 2.7 and 6.1% of the initially applied radioactivity were found in the form of metabolites, designated TR-1 and TR-2, respectively, in the combined soil samples from the three depths. The metabolites TR-4 and TR-1, as well as trace amounts of the metabolite TR-21, were present in the form of soil bound ^{14}C residues. Most of the radioactivity was located in the peel of the carrots. The pulp contained ^{14}C residues of TR-1 and TR-2 (0.010 and 0.004 ppm, respectively).

1. INTRODUCTION

Trifluralin is a pre-emergence, soil incorporated herbicide used to control a wide variety of grass and broad leaf weeds in many agronomic and horticultural crops. A number of workers have investigated persistence, degradation and movement of trifluralin in soil and plants [1–3]. When root crops were grown in soil treated with trifluralin, most of the herbicide residues were found on their surface [4, 5].

In Turkey, about 6570 kg active ingredients of trifluralin are used in the cultivation of carrots [6]. A maximum residue limit of 0.2 ppm has been established for trifluralin in carrots [7], and the tolerance limit for the herbicide has been reported to be 1.0 ppm [8–10].

---

* Research carried out in co-operation with the IAEA under Technical Co-operation Project No. TUR/5/015.

The purpose of this study was to investigate the persistence and degradation of ^{14}C-trifluralin in soil under outdoor conditions. Uptake by carrots of the herbicide from the treated soil was also investigated. A number of metabolites of ^{14}C-trifluralin in soil and carrots was identified and the nature of the non-extractable (bound) ^{14}C residues determined.

2. MATERIALS AND METHODS

2.1. Chemicals

Trifluralin-ring-UL-^{14}C(α,α,α-trifluoro-2,6-dinitro-N,N-dipropyl-p-toluidine) (TR-1) (19.1 mCi/mmol) was supplied by Sigma Chemical Co. (United States of America).[1] The radiochemical purity of trifluralin was 90.15%. The reference standard of trifluralin (purity 99.5%) was a gift from Dow Elanco (USA). The metabolites α,α,α-trifluoro-2,6-dinitro-N-propyl-p-toluidine (TR-2), α,α,α-trifluoro-2,6-dinitro-p-toluidine (TR-3), α,α,α-trifluoro-5-nitro-N^4,N^{4}6-diproyltoluene-3,4-triamine (TR-4), and 4-(dipropylamino)-3,5-dinitrobenzoic acid (TR-21) were supplied by the Biosciences Research Laboratory, United States Department of Agriculture, Fargo, ND, USA.

2.2. Soil

The soil used in this study was a sandy loam with pH7.5 and 3.7% organic matter. Particle size analysis indicated 17.8% clay, 13.4% silt and 68.8% sand. The soil was air dried, screened and passed through a 20 mesh sieve.

2.3. Field experiment

Carrots (Nandor F_1 variety) were grown in boxes (pots) at the Ankara Nuclear Research and Training Center's experimental farm, Ankara, Turkey. The experiments were carried out under outdoor conditions in two galvanized steel boxes measuring 60 × 60 × 60 cm. The base of the boxes contained holes to permit the drainage of excess water [11]. The inside of the boxes (pots) was covered with polyethylene sheets. The bottom 25 mm of the boxes was packed with stone chips of about 25 mm diameter, and the stones were covered with a 25 mm layer of well rotted turf. The boxes were filled with a sand/soil/manure mixture (5:3:3), and placed in pits such that the upper surface of the soil was at the level of the surrounding ground. Carbon-14-trifluralin, applied at the recommended rates of 0.84 kg/ha, was incorporated into the top 7.5 cm of the soil. The specific activities were 3.385

---

[1] 1 Ci = 3.70 × 10^{10} Bq.

and 3.828 µCi/mg for the first and second pot, respectively. The carrots were sown on 28 April 1994. Fertilizer $(NH_4)_2SO_4$ was applied at a rate of 75 g/pot before and after emergence of the carrots. The total rainfall during the growing season was 52.6 mm, with a daily maximum of 18.4 mm and minimum of 0 mm. The mean air temperature from May to August was 20.5°C.

2.4. Soil and crop sampling

Zero time soil samples were removed from the two pots immediately after the application of ^{14}C-trifluralin and thereafter at 1 monthly intervals until harvest. Samples were collected at depths of 0–7.5, 7.5–15 and 15–22.5 cm. The carrots were harvested on 31 August 1994 from the two pots. The tops and roots were separated, the roots washed with cold water to remove any adhering soil particles and the tops discarded.

All the samples were analysed or processed in duplicate; average values are reported.

2.5. Extraction

Soil samples (50–100 g) were shaken with 150–300 mL portions of methanol (3×), followed by an equal amount of 50% aqueous methanol (2×) using an orbital shaker for 30 min. The combined filtrate was then concentrated to a small volume using a Rotary evaporator and partitioned with ethylacetate. The organic phase was passed through a column of dried anhydrous sodium sulphate to remove any water [2, 12].

The carrots were peeled, and the pulp and peel diced into small pieces. A 50–75 g portion of the pulp was extracted with methanol and 50% aqueous methanol, as described above.

2.6. Determination of radioactivity

Combustion of soil and/or carrot samples was done in a Packard sample oxidizer, model 306, to produce $^{14}CO_2$. The latter was absorbed in and admixed with an appropriate volume of Carbosorb E and Permaflour E. Liquids (extracts) were assayed in a Beckman model LS 3801 scintillation spectrometer.

2.7. Chromatography and analysis

2.7.1. Thin layer chromatography (TLC)

TLC was performed with 20 × 20 cm precoated silica gel GF plates, 1.0 mm thick. The metabolite TR-21 was methylated with diazomethane to obtain methylester (TR-21M). The five reference standards were spotted on the plates and then developed in a hexane/benzene (1:1) solvent system. The developed plates were observed under UV light (254 nm). Under these experimental conditions, the following R_f values were obtained: TR-1, R_f = 0.9; TR-2, R_f = 0.79; TR-3, R_f = 0.47; TR-4, R_f = 0.75; and TR-21M, R_f = 0.28. The concentrated carrot extracts were also methylated with diazomethane. The extracts were then applied to the plates and developed in hexane/benzene (1:1). Polaroid photographs of the radioactive areas on the developed plates were taken with a beta camera (Bethold, model LB 292) and the plates also scanned by a TLC scanner for ^{14}C (Bioscan system 2000, autochanger 2000). The plates separated into three radioactive bands with R_f values of 0.-0.14, 0.14-0.52 and 0.52-0.95. The material in these bands was scraped off the plates and extracted with 20 mL of acetone (5×). Each extract was evaporated to a small volume and subjected to further clean-up by column chromatography.

2.7.2. Column chromatography

Column chromatography was carried out with a glass column (20 × 1.5 cm) containing 4 g of Florisil packed in hexane/benzene (2:1). A mixed reference standard (1.6 mL) in hexane/benzene (2:1) was applied to the column and eluted with hexane/ethylether (99:1). Under these conditions, the following recoveries were found for each reference standard: TR-1, 89.4%; TR-2, 84.3%; TR-4, 69.5%; and TR-21M, 87.5%. The carrot extracts from the TLC plates, as described in Section 2.7.1, were brought to the point of dryness and dissolved in a 1.6 mL mixture of hexane/benzene (2:1). The sample was applied to the column and eluted with 4 × 10 mL of hexane/ethylether (99:1). The eluates were combined and evaporated under air to dryness, redissolved in 1.6 mL of hexane/benzene (2:1), and finally analysed by gas chromatography (GC) [12, 13].

2.7.3. Gas chromatography

The gas chromatograph was a Varian model 3400 equipped with a ^{63}Ni detector and a 15 m × 0.545 mm capillary column coated with Carbowax (1.0 μm). To analyse the mixed standard, the column was operated at 160°C, and for TR-21M, at 190°C. The nitrogen carrier flow rate, injector port temperature and detector temperature were 20 mL/min, 180°C and 310°C, respectively. Under these experimental conditions, the retention times for TR-1, TR-2, TR-3, TR-4 and TR-21M were 1.8, 3.1, 5.2, 9.9 and 8.8 min, respectively.

2.8. Confirmation

The identity of the compound was confirmed by comparing the GC retention times with those of the authentic samples and by GC–mass spectrometry. A high resolution mass spectrometer, model VGZAB-ZF, connected to a Varian GC model 3710, was used. The mass spectra were recorded at 70 eV.

2.9. Supercritical fluid extraction (SFE)

The SFE system (Suprex model SFE-50, Suprex Corp., Pittsburg, PA, USA) consisted of a 250 mL syringe pump, a control module for the SFE system, an extraction oven, a 5 mL extraction vessel containing a sample, and a four port valve connected to the outlet restriction (fused silica tubing, 50 μm inner diameter) that was vented into the first of the three glass tubes containing 50 mL of methanol. The three glass tubes containing methanol were connected in series to collect the released material. Extraction was carried out with the modified CO_2 using methanol (30% CH_3OH in CO_2) that was delivered by a high performance liquid chromatograph pump (Varian model 2510). The flow rate of the mobile phase was 2 mL/min, and extraction was carried out at 180°C and 375 atm for 2 h after initial equilibration of the SFE system at 180°C and 150 atm for 5 min.

A schematic diagram for the analysis of ^{14}C trifluralin residues in soil and carrots is given in Fig. 1.

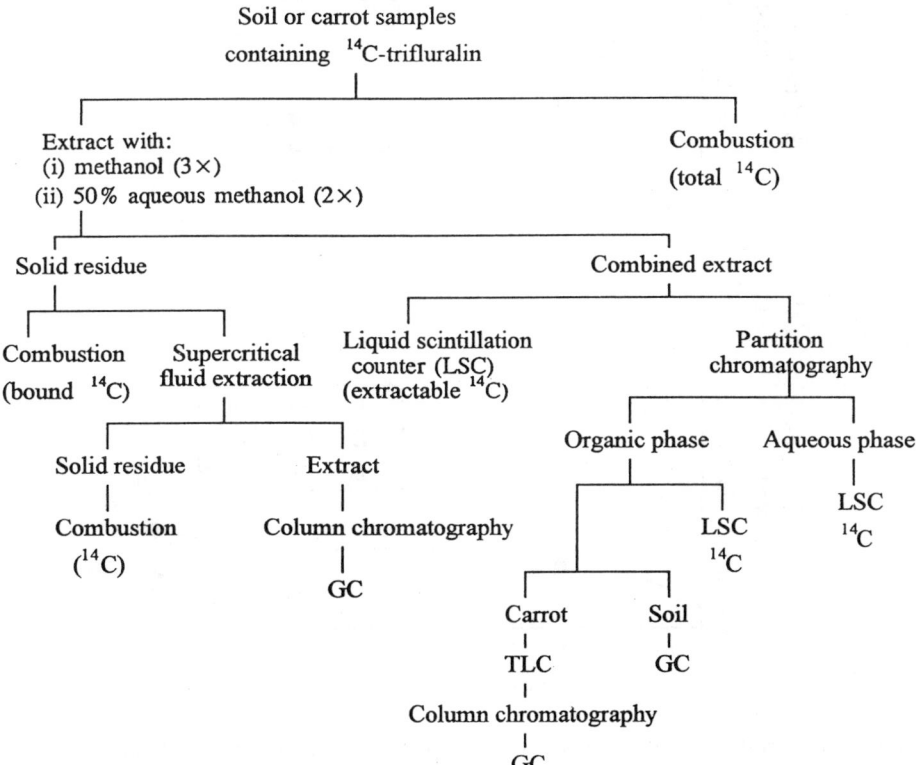

FIG. 1. Schematic diagram for the analysis of ^{14}C-trifluralin residues in soil and carrots.

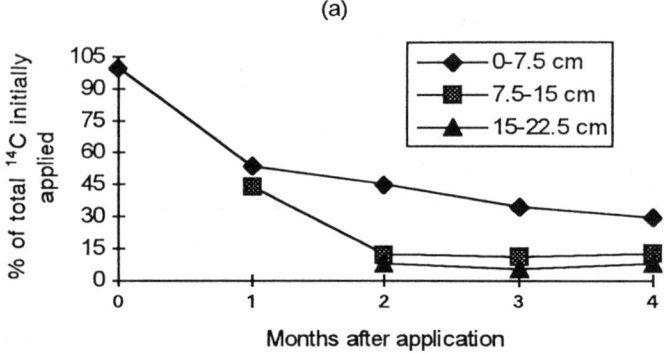

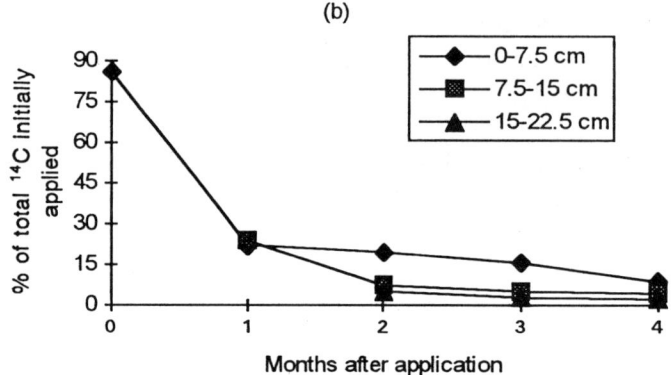

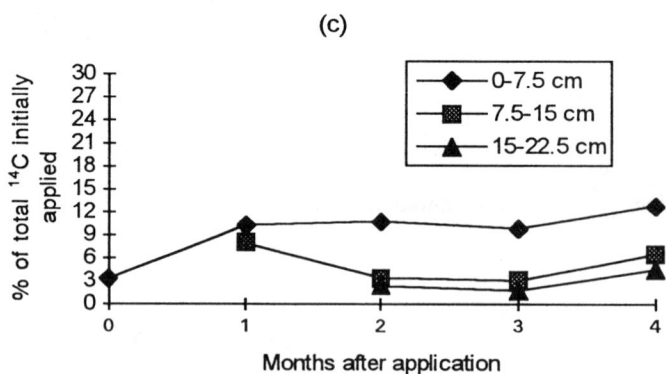

FIG. 2. (a) Total, (b) extractable and (c) soil bound ^{14}C residues (based on the percentage ^{14}C initially applied; 2.687 ppm = 100%).

3. RESULTS AND DISCUSSION

3.1. Carbon-14-trifluralin residues in soil

The total ^{14}C residues recovered from the soil over the growing period of carrots is shown in Fig. 2(a), and the extractable and soil bound ^{14}C residues in Figs 2(b) and (c), respectively. The data show that 4 months after application, the combined three layers of soil (0–22.5 cm) contained 50.4, 15 and 23.8% of the initially applied radioactivity as total, extractable and soil bound ^{14}C residues, respectively. The extractable ^{14}C residues decreased with time; this in turn corresponded with an increase in the soil bound ^{14}C residues (Figs 2(b) and (c)). Golab et al. [2] observed that 3 years after application of ^{14}C-trifluralin, the 0–15 cm soil layer contained 43.5% of the initially applied radioactivity. Smith and Muir [14] found after 45 weeks that the radioactivity recovered from the ^{14}C-trifluralin treated plots was 77% of that applied, while the non-extractable ^{14}C residues accounted for 10%.

TABLE I. EXTRACTABLE ^{14}C RESIDUES IN SOIL (AVERAGE OF TWO POTS)

Time of sampling (months)	Depth of soil (cm)	TR-1 (ppm)	TR-2 (ppm)	TR-3 (ppm)	TR-4 (ppm)	TR-21M (ppm)
Zero time	0–7.5	2.172	0.018	ND [a]	ND	ND
1	0–7.5	0.178	0.027	ND	0.001	ND
	7.5–15	0.163	0.052	ND	ND	0.013
2	0–7.5	0.351	0.007	ND	0.012	0.004
	7.5–15	0.119	0.012	0.029	0.003	ND
	15–22.5	0.071	0.017	0.033	0.014	ND
3	0–7.5	0.222	0.037	0.003	0.001	0.017
	7.5–15	0.061	0.030	0.004	ND	ND
	15–22.5	0.044	0.021	0.003	ND	0.002
4	0–7.5	0.040	0.092	ND	ND	ND
	7.5–15	0.030	0.039	0.009	ND	0.004
	15–22.5	0.004	0.035	0.007	ND	0.002

[a] ND = not detected.

TABLE II. SOIL BOUND ^{14}C RESIDUES EXTRACTED WITH SFE (AVERAGE OF TWO POTS)

Time of sampling (months)	Depth of soil (cm)	Soil bound ^{14}C residues (ppm)	Extractable ^{14}C residues (%)[a]	Non-extractable[b] ^{14}C residues (%)[a]
Zero time	0–7.5	0.091	57.6	41.2
1	0–7.5	0.277	58.6	43.5
	7.5–15	0.214	66.6	37.1
2	0–7.5	0.292	46.2	47.1
	7.5–15	0.093	39.8	54.7
	15–22.5	0.065	26.1	65.4
3	0–7.5	0.266	33.7	57.4
	7.5–15	0.084	19.6	72.0
	15–22.5	0.048	12.5	67.7
4	0–7.5	0.345	51.0	47.4
	7.5–15	0.176	38.3	71.0
	15–22.5	0.121	18.2	71.8

[a] Based on the total soil bound ^{14}C residues.
[b] Determined by combustion.

The persistence and movement of ^{14}C-trifluralin residues in soil changed with time. Thus, after 1 month about 97.5% of the initially applied radioactivity was still present in the combined three depths of the soil, while 4 months later only about half was present. Duseja and Holmes [1] reported that 25 and 120 days after application (at an application rate of 0.75 lb/acre) 48.3 and 25.9% of the initially applied radioactivity, respectively, were present in the combined soil samples from the three depths.[2]

3.2. Movement of ^{14}C-trifluralin residues in the soil

Soil samples collected from the pots at the three depths were analysed for the presence of ^{14}C-trifluralin and its metabolites. The results (Fig. 2(a) and Table I)

---

[2] 1 lb = 0.4536 kg; 1 acre = 4.047 × 10^3 m^2.

indicate a small amount of downward movement of the herbicide and/or its metabolites. Most of the initially applied radioactivity was located in the zone of incorporation. Thus, about 11–44 and 5.5–8.0% remained in the 7.5–15 and 15–22.5 cm soil layers, respectively (Fig. 2(a)). Golab et al. [2] observed that after 12 and 24 months (at an application rate of 1.68 kg active ingredients/ha) 91 and 98.8% of the radioactivity were located in the 0–15 cm zone, and 76 and 95% in the zone of incorporation.

3.3. Identification of extractable ^{14}C residues

Examination of soil extracts by GC indicated the presence of TR-1 (trifluralin) and TR-2 (Table I). After 4 months, the soil extracts showed greater concentrations

TABLE III. EXTRACTABLE ^{14}C RESIDUES RELEASED BY SFE (AVERAGE OF TWO POTS)[a]

Time of sampling (months)	Depth of soil (cm)	^{14}C residues (ppm)	TR-1 (ppm)	TR-2 (ppm)	TR-4 (ppm)	TR-21M (ppm)
Zero time	0–7.5	0.052	ND[b]	ND	ND	ND
1	0–7.5	0.162	0.031	0.001	0.013	ND
	7.5–15	0.142	0.022	ND	0.026	0.010
2	0–7.5	0.134	0.018	ND	0.007	0.001
	7.5–15	0.039	ND	ND	0.003	ND
	15–22.5	0.017	0.010	ND	0.007	ND
3	0–7.5	0.089	0.031	ND	0.014	0.009
	7.5–15	0.016	ND	ND	0.005	ND
	15–22.5	0.006	0.003	ND	0.003	ND
4	0–7.5	0.176	0.021	0.005	0.001	ND
	7.5–15	0.067	0.011	ND	0.008	ND
	15–22.5	0.022	0.009	ND	0.009	0.004

[a] TR-3 could not be eluted through the Florisil column with the hexane/ethylether (99:1) solvent system.
[b] ND = not detected.

in TR-2 than in TR-1. None of the metabolites found had more than 3.5% of the initially applied ^{14}C-trifluralin at any time during the growing period. A similar observation indicating that the metabolites formed were about 3% of the initial herbicide applied has been reported [2]. The data in Table I show the presence of small amounts of metabolites TR-3, TR-4 and TR-21M. Similar findings have also been reported by Golab et al. [4] and Probst et al. [5].

3.4. Identification of soil bound ^{14}C residues

Soil bound ^{14}C residues were considered to be those that were not extracted with methanol and 50% aqueous methanol, as described in Section 2.5. The nature of these ^{14}C residues was determined by subjecting the solvent extracted soil to SFE. It was noted that even after SFE some of the ^{14}C residues still remained nonextractable (Table II). SFE released ^{14}C residues ranged between 12.5 and 66.6% of the total soil bound ^{14}C residues. The SFE extracts obtained were analysed by GC. The extractable ^{14}C residues released by SFE are shown in Table III. The SFE extracts obtained from the samples containing soil bound ^{14}C residues showed the presence of mainly TR-1 and TR-4. None of the metabolites was present in the soil bound form at zero time. Furthermore, some of the SFE extracts also showed trace amounts of TR-2 and TR-21M.

3.5. Carrot

Most of the radioactivity was located in the peel. Distribution of the total ^{14}C residues in the peel and pulp was 73.4% (0.689 ppm) and 24.9% (0.035 ppm), respectively. Similar results have been reported by Golab et al. [4] and Probst et al. [5].

The pulp of carrots is normally consumed as food, therefore further analysis was carried out for ^{14}C residues in this part of the carrot. The total, extractable and bound ^{14}C residues in carrot pulps were 0.035, 0.023 and 0.009 ppm, respectively. It was observed that about 1.3% of the initially applied radioactivity was present in the carrot pulps. The methanol extracts of pulp contained ^{14}C residues of TR-1 and TR-2 (0.010 and 0.004 ppm, respectively). Golab et al. [4] reported the presence of the main compound (TR-1) and a major conversion product (TR-2) in the carrot pulp extracts. Trace amounts of TR-5 and TR-21 were also found in carrots, but identification could not be verified because of the presence of only very small amounts of these compounds.

ACKNOWLEDGEMENTS

The analyses made in this study were carried out at the Centre for Land and Biological Resources Research, Research Branch, Agriculture Canada, Ottawa,

Ontario, Canada, while the senior author was an IAEA fellow. The authors would like to express their deep gratitude to S.U. Khan and his techinician R. McDowell for technical help and valuable suggestions.

REFERENCES

[1] DUSEJA, D.R., HOLMES, E.E., Field persistence and movement of trifluralin in two soil types, Soil Sci. **125** (1978) 41–48.
[2] GOLAB, T., ALTHAUS, W.A., WOOTEN, H.L., Fate of (^{14}C)trifluralin in soil, J. Agric. Food Chem. **27** (1979) 163–179.
[3] HELLING, C.S., Dinitroanilines in soils, J. Environ. Qual. **5** (1976) 1.
[4] GOLAB, T., HERBERG, R.J., PARKA, S.J., TEPE, J.B., Metabolism of carbon-14 trifluralin in carrots, J. Agric. Food Chem. **15** (1967) 638–641.
[5] PROBST, G.W., et al., Fate of trifluralin in soil and plants, J. Agric. Food Chem. **15** (1967) 592–599.
[6] Annual Records, General Directory of Plant Protection, Ministry of Agriculture, Ankara (1993).
[7] Koruma ve kontrol genel müdürlügü zirai mücadele ilaçlari (pestisit) ve bitki gelişimini düzenleyici maddeleri kalıntı limitlerinin kontroluna dair tebliğ, Resmi Gaz. **20624** (1990) 33–36.
[8] SITTING, M., Pesticide Manufacturing and Toxic Materials Control Encyclopedia, Environmental Protection Agency, Washington, DC (1980) 754–755.
[9] ÖZTÜRK, S., ÖZGE, N., Bitki Koruma İlaçları, Eser Press, Ankara (1978) 245–246.
[10] Herbicide Handbook of the Weed Science Society of America, 6th edn, Herbicide Handbook Committee, Geneva and New York (1989).
[11] KOHLI, J., ZARIF, S., WEISBERGER, I., KLEIN, W., KORTE, F.J., Fate of aldrin-^{14}C in sugar beets and soil under outdoor conditions, J. Agric. Food Chem. **21** (1973) 855–857.
[12] LUE, L.P., LEWIS, C.C., MELCHOR, V.E., The effect of aldicarb and its persistence in carrots, soil and hydroponic solution, J. Environ. Sci. Health. **B27** 3 (1984) 343–354.
[13] D'AMATO, A., SEMERARO, I., BICHI, C., Simultaneous determination of liniuron and trifluralin residues in carrots and their pulp by liquid chromatography, J. Assoc. Off. Anal. Chem. **76** (1993) 657–662.
[14] SMITH, A.E., MUIR, D.C.G., Determination of extractable and non-extractable radioactivity from small field plots 45 and 95 weeks after treatment with (^{14}C)dicamba, (2,4-dichloro(^{14}C)phenoxy) acetic acid, (^{14}C)triallate, (^{14}C)trifluralin, J. Agric. Food Chem. **32** (1984) 588–593.

RADIATION INDUCED DEGRADATION OF PARATHION IN AQUEOUS SOLUTION

L.C. LUCHINI
Instituto Biológico,
Centro de Radioisótopos,
São Paulo

M.O. OLIVEIRA DE REZENDE
Instituto de Química de São Carlos,
Universidade de São Paulo,
São Carlos

Brazil

Abstract

RADIATION INDUCED DEGRADATION OF PARATHION IN AQUEOUS SOLUTION.
    Gamma radiation from a ^{60}Co source was used to induce parathion degradation in aqueous solution, and the resulting products were quantified by gas chromatography–nitrogen phosphorus detector and identified by gas chromatography–mass spectrometry. The insecticide was completely degraded in aqueous solution after treatment with a 1.0 kGy dose at a dose rate of 3.12 kGy/h. The metabolites detected after radiolysis were p-nitrophenol, p-aminophenol, paraoxon and aminoparathion, which are the same as those formed by biological degradation. The results showed that the total radiation and the dose rate applied to the aqueous solution had a significant effect on insecticide degradation, and that the formation of metabolites occurred in a selective way, depending on the dose and dose rate applied. The parathion radiation yield (G-parathion) and the gamma radiation dose needed to achieve a 50% reduction in the initial insecticide concentration in aqueous solution were also calculated.

1. INTRODUCTION

    Organic pesticides enter natural waters by direct application, percolation or runoff from agricultural lands (for the control of aquatic weeds and insects), or by runoff from urban and suburban areas, drift from aerial and land applications, discharge of industrial waste waters and discharge of waste waters from the clean-up of equipment used for pesticide formulation and application [1]. However, use of pesticides in agriculture is becoming a problem to the ecosystem because of the residues that remain in the environment.
    Once a pesticide reaches the aquatic environment it can persist for considerable periods of time. Conventional methods of environmental decontamination, e.g.

water treatment by filtration, co-precipitation and absorption on activated coal, only transfer the residues from one location to another, and in many cases do not lead to the desired effectiveness.

One of the most important organophosphorus pesticides is parathion, which is extensively used in agriculture. Because of its very high toxicity and its physicochemical behaviour in the atmosphere, soil and water, parathion is an environmental chemical that produces manifold biological and ecological effects [2].

Attempts have been made to decompose these substances by ionizing radiation [3-10]. However, little information is available on the degradation products resulting from pesticide radiolysis. Using the gamma radiation from a ^{60}Co source, efforts were made to induce parathion degradation in different matrices. The resulting radiolytic products were quantified by gas chromatography-nitrogen phosphorus detector (GC-NPD) and identified by gas chromatography-mass spectrometry (GC-MS).

2. MATERIALS AND METHODS

2.1. Insecticide

Parathion (O,O-diethyl O-4-nitrophenyl phosphorothioate) pure standard was purchased from Greyhound Chromatography and Allied Chemicals, Merseyside, United Kingdom. The corresponding radiolabelled compound was purchased from Amersham International, Amersham, UK, with a specific activity of 451.4 MBq/mmol (12.2 mCi/mmol) and 96.8% radiochemical purity, as determined by thin layer chromatography (TLC).

2.2. Radiation source

Irradiation with gamma rays was performed in a Gammabeam-650 from the Centro de Energia Nuclear na Agricultura, Universidade de São Paulo, with an activity of 6.99×10^{13} Bq at room temperature.

2.3. Radiolysis of parathion

The aqueous solution of technical grade parathion and the corresponding radiolabelled compound prepared at a concentration of 10 mg/L and an activity of 1.73 µCi/L were irradiated at doses of 0, 0.10, 0.20, 0.50 and 1.0 kGy, at different dose rates (1.87, 2.80 and 3.12 kGy/h).[1] After irradiation, the ^{14}C-parathion solution was extracted and analysed according to Fig. 1.

---

[1] 1 Ci = 3.70×10^{10} Bq.

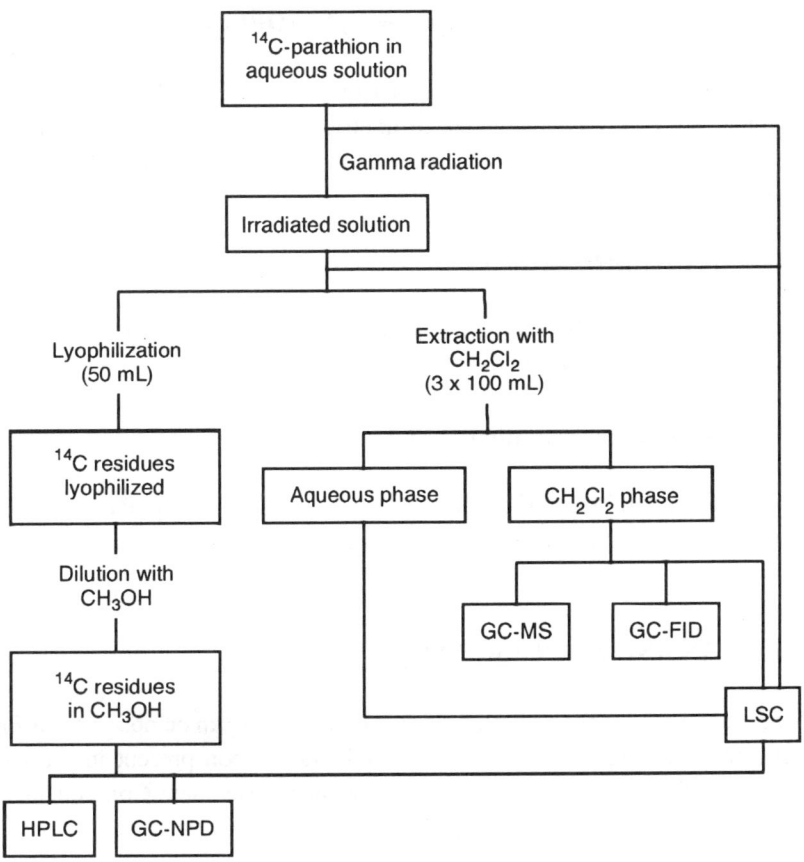

FIG. 1. *Schematic diagram of the irradiation of ^{14}C-parathion in aqueous solution and analysis of the ^{14}C residues.*

2.4. Gas chromatography (GC–NPD, GC–FID)

Determination of parathion and its radiolytic products (paraoxon, p-aminophenol and aminoparathion) was carried out using a gas chromatograph HP 5890 series II equipped with an NPD and a flame ionization detector (FID). It was connected to an HP chemistation analyser, and fitted with an SPB-5 fused silica capillary column (30 m × 0.25 mm).

2.5. High performance liquid chromatography (HPLC)

Determination of p-nitrophenol was performed in an HPLC Shimadzu LC-10A fitted with a variable wavelength spectrometric detector, model SPD-M6A-UV-Visible, and with a Shim-pack C-8 reverse phase column using methanol/water (70:30 (vol./vol.)) as the mobile phase.

2.6. Mass spectrometry (GC–MS)

Capillary column GC–MS was performed in an HP 5890 chromatograph connected to a selective HP 5970 MS detector with an electron energy of 70 eV.

2.7. Liquid scintillation counting (LSC)

Quantification of the ^{14}C compounds was carried out on Packard 1600TR equipment using 10 mL of scintillation cocktail, as proposed by Mesquita and Rüegg [11].

3. RESULTS AND DISCUSSION

The clear effect of the radiation dose and dose rates can be observed in Fig. 2. A significant reduction in the percentage total radiocarbon present in the CH_2Cl_2 was observed, which increased with the dose and dose rate. Consequently, the

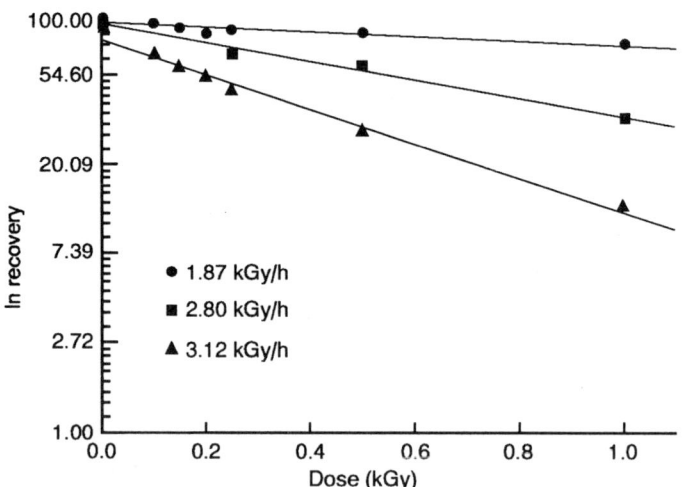

FIG. 2. Recovery of the total radiocarbon from the ^{14}C-parathion in solution after irradiation and liquid–liquid extraction with CH_2Cl_2.

TABLE I. CONCENTRATION OF PARATHION AND ITS RADIOLYTIC PRODUCTS (μg/mL) AFTER IRRADIATION AT DIFFERENT DOSE RATES[a]

Compounds	0			0.10			0.15			0.20			0.25			0.50			1.0		
	I	II	III	I	II	III	I	II	III	I	II	III	I	II	III	I	II	III	I	II	III
Parathion	56.5	42.5	49.0	48.8	*	31.8	49.8	*	29.8	51.5	*	16.1	47.6	21.6	17.9	46.8	15.2	4.5	33.7	5.5	1.8
Aminoparathion	2.0	3.0	2.0	2.1	*	2.2	9.5	*	3.4	4.7	*	0.1	3.0	1.6	1.8	3.6	0.8	—	2.8	0.3	—
Paraoxon	—	—	—	—	*	6.9	—	*	5.4	—	*	12.6	—	2.0	6.5	—	2.7	8.4	—	3.4	3.3
p-aminophenol	—	—	—	—	*	—	—	*	—	—	*	0.9	—	0.4	—	—	—	0.6	—	—	—
p-nitrophenol	1.1	0.6	0.3	b	*	2.7	0.6	*	3.1	b	*	0.5	b	5.7	1.7	3.1	8.6	1.3	4.0	6.3	1.6

[a] I = dose rate of 1.87 kGy/h; II = dose rate of 2.80 kGy/h; III = dose rate of 3.12 kGy/h. * = not irradiated; — = not detected; b = not analysed.

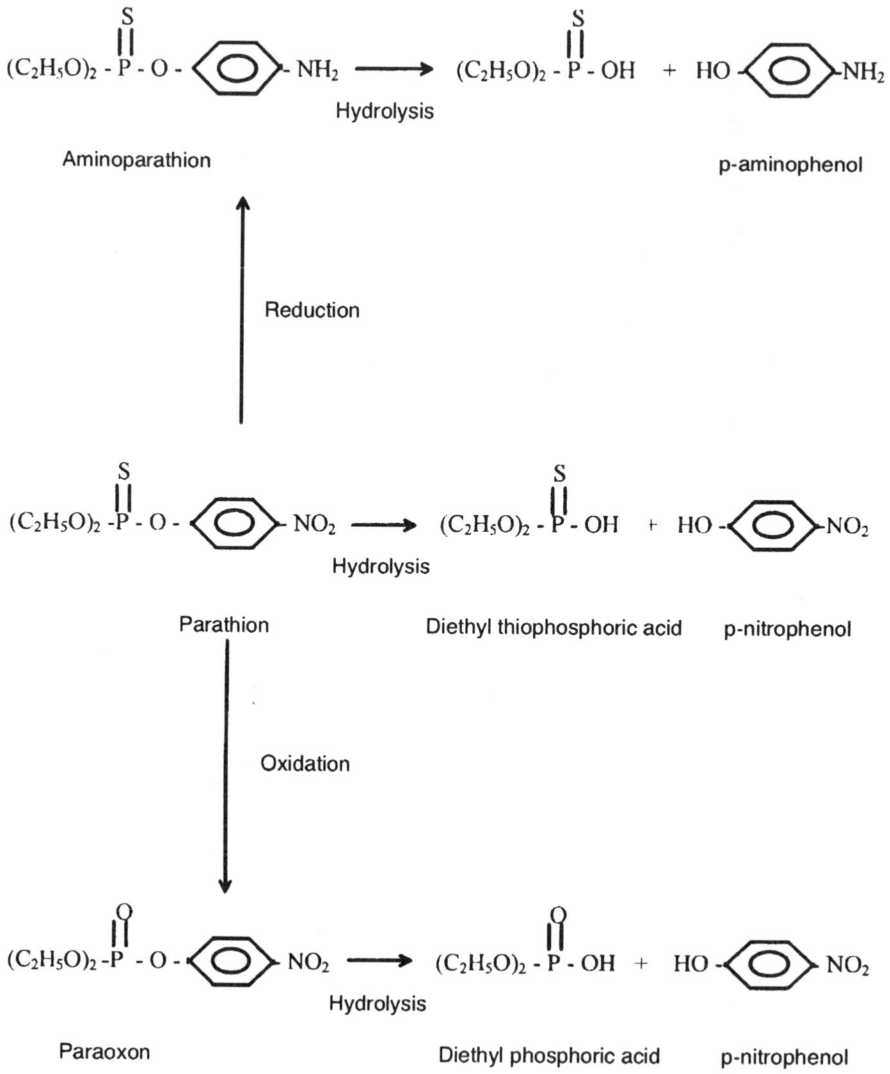

FIG. 3. Schematic representation of the principal path of parathion degradation [12].

remaining radiocarbon in the aqueous phase increased, probably because of the formation of more polar (water soluble) compounds during radiolysis.

Recovery of the total ^{14}C compounds after lyophylization of the aqueous solution and resuspension with methanol (73–97% of the total radiocarbon) proved to be more efficient than extraction with CH_2Cl_2. Therefore, this procedure was used to quantify parathion and its radiolytic products using GC–NPD and HPLC.

Table I presents the concentration of parathion and its radiolytic products as a function of the dose and dose rate. Radiation of 1.0 kGy applied to the parathion aqueous solution degraded approximately 97% of the pesticide when the solution was submitted to a dose rate of 3.12 kGy/h. The first three columns of Table I show the concentration of parathion and its degradation products initially present in solutions not submitted to radiation.

However, there is a dose and dose rate that produce a maximum concentration of each radiolytic product. Aminoparathion had a maximum concentration (9.5 µg/mL) in the solution after being irradiated with 0.15 kGy at a dose rate of 1.87 kGy/h, p-nitrophenol (8.6 µg/mL) irradiated with 0.50 kGy at 2.80 kGy/h, and paraoxon (12.6 µg/mL) irradiated with 0.20 kGy at 3.12 kGy/h. Thus, the combination of dose and dose rate makes possible the selective formulation of various parathion radiolytic products that can be useful for studies of metabolites.

Formation of paraoxon, aminoparathion, p-aminophenol and p-nitrophenol after radiolysis of the parathion aqueous solution denotes that, even though the two systems (biochemical metabolism and gamma irradiation) are totally different, the main products resulting from the two processes are the same, probably because in both parathion degradation involves oxidation, hydrolyosis and reduction mechanisms. The principal path of parathion degradation, either by biochemical metabolism or gamma irradiation, is shown in Fig. 3 [12].

TABLE II. MASS/ELECTRIC CHARGE RELATION AND RETENTION TIME OF THE MAIN IONS OF PARATHION AND ITS RADIOLYTIC PRODUCTS PRODUCED BY 70 eV

Compound	Retention time (min)	Mass/electric charge
p-aminophenol	6.132	109, 80, 53
p-nitrophenol	12.970	139, 123, 109, 81, 65
Paraoxon	19.263	275, 247, 139, 109, 98, 81, 65
Aminoparathion	19.503	261, 233, 205, 125, 109, 80, 65, 53
Parathion	20.359	291, 235, 155, 139, 125, 109, 97, 81, 65

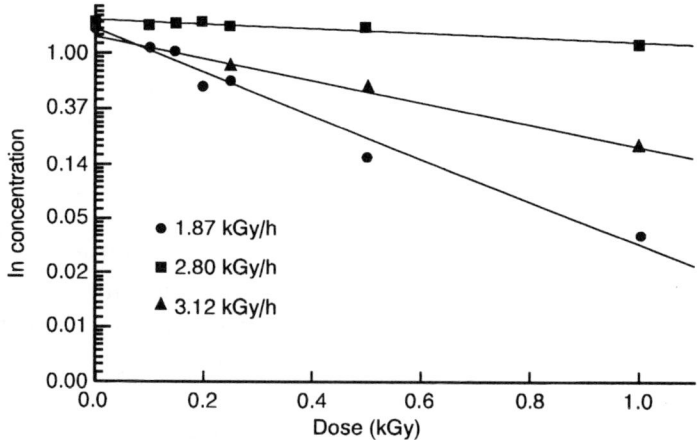

FIG. 4. Radiolytic yield of parathion degradation at dose rates of 1.87, 2.80 and 3.12 kGy/h.

TABLE III. CONCENTRATION OF DEGRADED PARATHION AS A FUNCTION OF DOSE AND DOSE RATE

Dose rate (kGy/h)	ED_{50} (kGy)	Radiolytic yield (G-parathion)	Concentration of parathion (mol/L × 10^{-4})	
			Experimental	Estimated
1.87	1.53	0.64	0.78	0.67
2.80	0.35	1.13	1.27	1.17
3.12	0.18	1.36	1.65	1.41

Identification of parathion and its radiolytic products was made using GC–MS; the mass/electric charge of the main fragments of these compounds are presented in Table II.

The radiolytic yield of parathion degradation by gamma irradiation (G-parathion), i.e. the number of molecules degraded by 100 eV, was calculated from the dose and dose rate, depending on the pesticide degradation curve (Fig. 4), with 1 kGy = 6.24×10^{21} eV/kg. The dose needed to achieve a 50% reduction in the initial pesticide concentration (ED_{50}) in aqueous solution was also calculated (Table III).

According to other authors [13, 14], when the value of the radiolytic yield (G) is equal to 1, then 1.04 mmol of species X is formed or degraded by 1 Mrad (10 kGy). Thus, formation of a determined amount of degraded pesticide by a radiation dose (kGy) can be expressed as

$$Y = 0.140 \times D \times G$$

where Y is the concentration of the degraded pesticide (mmol/L); D is the radiation dose applied to the solution (kGy); and G is the radiolytic yield.

Table III shows the similarity between the estimated values of parathion concentration calculated from the above equation and the values obtained experimentally when 1 kGy of gamma radiation was applied at different dose rates.

Thus, if the radiolytic yield of the degradation is known, the concentration of the pesticide degraded by radiolysis can be estimated for any radiation dose.

4. CONCLUSIONS

Irradiation of parathion could be an important way of environmentally decontaminating this pesticide in aqueous matrix, and it enables parathion metabolites to be produced for ecotoxicological studies.

REFERENCES

[1] GOMAA, H.M., FAUST, S.D., "Removal of organic pesticides from water to improve quality", Pesticides in Soil and Water (GUENZI, W.D., Ed.), Soil Science Society of America, Madison, WI (1974) 413–450.

[2] MANSOUR, M., THALLER, S., KORTE, F., Action of sunlight on parathion, Bull. Environ. Contam. Toxicol. **30** (1983) 358–364.

[3] VOLLNER, L., ROHLEDER, H., KORTE, F., "Degradation of persistent organochlorine pollutants by gamma radiation and its possible use for waste water treatment", Radiation for a Clean Environment (Proc. Symp. Munich, 1975), IAEA, Vienna (1975) 285–296.

[4] FÖLDESOVÁ, M., PIATRIK, M.S., TÖLGYESSY, J., CERVENKA, Z., Contribution to degradation of environmental pollutants. I. Radiation degradation of solid distilled residues and commercial mixtures of polychlorinated biphenyls, Radiochem. Radioanal. Lett. **40** (1979) 73–78.

[5] SINGH, A., KREMERS, W., SMALLEY, P., BENNETT, G.S., Radiolytic dechlorination of polychlorinated biphenyls, Radiat. Phys. Chem. **25** (1985) 11–19.

[6] GETOFF, N., LUTZ, W., Radiation induced decomposition of hydrocarbons in water resources, Radiat. Phys. Chem. **25** (1985) 21–26.

[7] BRUSENTSEVA, S.A., et al., Radiation–adsorption purification of effluents containing pesticides, Radiat. Phys. Chem. **28** (1986) 569–572.
[8] MINCHER, B.J., MEIKRANTZ, D.H., MURPHY, R.J., GRESHAM, G.L., CONNOLLY, M.J., Gamma-ray induced degradation of PCBs and pesticides using spent reactor fuel, Appl. Radiat. Isot. **42** (1991) 1061–1066.
[9] JAVARONI, R. de C., TALAMONI, J., LANDGRAF, M.D., REZENDE, M.O. de O., LUCHINI, L.C., Estudo de degradação de lindano em solução aquosa através de radiação gama, Quím. Nova **14** (1991) 237–239.
[10] YONGKE, H., HONGTAO, C., XIANGRONG, S., JILAN, W., Radiation-induced dechlorination of hexaclorobenzene, Radiat. Phys. Chem. **42** (1993) 715–717.
[11] MESQUITA, T.B., RÜEGG, E.F., Influência de agentes tensoativos na detecção da radiação beta, Ciênc. Cult. **30** (1984) 446–450.
[12] SETHUNATHAN, N., SIDDARAMAPPA, R., RAJARAM, K.P., BARIK, S., WAHID, P.A., Parathion: Residues in soil and water, Res. Rev. **68** (1977) 91–122.
[13] GRONEMAN, A.F., SCHUBERT, J., Mechanisms of action of irradiation on the conditioning of sewage sludge radical scavenging effects, Int. J. Appl. Radiat. Isot. **29** (1978) 301–309.
[14] HOIGNE, J., "Aqueous radiation chemistry in relation to waste treatment: An introductory review", Radiation for a Clean Environment (Proc. Symp. Munich, 1975), IAEA, Vienna (1975) 297–305.

FATE AND BEHAVIOUR OF PESTICIDES IN THE AQUATIC ENVIRONMENT

(Session 4)

Chairpersons

A.M. PECHEN DE D'ANGELO
Argentina

K. RAGHU
India

Rapporteur

F.P. CARVALHO
IAEA

AGROCHEMICAL FATE AND EFFECTS IN TERRESTRIAL, AQUATIC AND ESTUARINE ECOSYSTEMS*

S.J. KLAINE**, P. RICHARDS***,
D. BAKER***, R. NADDY**,
T. BROWN**, B. JOAB**,
R. CASEY**, D. FERNANDEZ**,
J. OVERMEYER**, R. BENJAMIN**

** Department of Environmental Toxicology

and

The Institute of Wildlife and
  Environmental Toxicology

Clemson University,
Pendleton, South Carolina

*** Water Quality Laboratory,
Heidelberg College,
Tiffin, Ohio

United States of America

Abstract

AGROCHEMICAL FATE AND EFFECTS IN TERRESTRIAL, AQUATIC AND ESTUARINE ECOSYSTEMS.
   Characterization and control of non-point source pollution has been less successful than corresponding efforts directed at point source pollution. Sampling of non-point source pollution is complicated by the diffuse and episodic dispersal of chemical residues. Standardized bioassays do not adequately simulate episodic exposures to realistic concentrations of toxic substances. A description is given of a quantitative approach to chemically and physically characterizing non-point source pollutants and an illustration of how this approach could be utilized to design bioassays that characterize the biological impact of non-point source pollutants. The methods described include analysis of pesticide datasets, phytoplankton growth assays and daphnid survival and mobility assays. The atrazine and chlorpyrifos concentrations in the Maumee River during 1983 were used to generate exceedancy plots. These data were used with toxicity data to compute the probable periods of biological stress and recovery. *Selenastrum capricornutum* demonstrated a threshold population growth rate response to 10 μg/L of atrazine. Recovery from exposure to as much as 50 μg/L of atrazine was nearly

---

   * Research carried out in association with the IAEA under Research Agreement No. 8498.

instantaneous once the atrazine had been removed from the overlying water. No cumulative effects were observed during 32 d of exposure to 10 µg/L of atrazine, and the ecological impact of chronic low level exposure appeared negligible. Survival of daphnids to single chlorpyrifos exposures in the 0.125–1.0 µg/L range was dose related, and older organisms were more sensitive. Daphnia subjected to two pulses had a higher mortality if the second pulse was experienced later in life. Organisms with food were more sensitive than organisms without food, so the response to multiple pulses was probably related to the altered feeding rates following initial exposure of the neonates. The mobility data indicated that daphnids could recover from chlorpyrifos exposures below a critical threshold. The overall results suggested that the analytical approaches developed for managing point source pollution can misrepresent the risks associated with non-point source pollution.

1. INTRODUCTION

Manipulation of virgin land to facilitate development results in pressure on both the terrestrial and aquatic ecosystems. This pressure involves destruction of habitat, displacement of species, and mobilization of soil and chemicals into adjacent aquatic systems. Because land development for agriculture, managed forests and urban (industrial and domestic) uses is essential to our accepted human lifestyle, preservation of water resources is essential. The co-existence of land development with water resources requires characterization and management of the pollutants inherent in such development.

1.1. Non-point source pollution

The soil and chemicals moving from lands in a diffuse manner, typically as a result of rain events, are called non-point source pollutants. Point source pollutants are defined as those that enter the environment through a single point such as an effluent pipe. Past efforts to control aquatic pollution have focused on point source pollutants primarily because they were conspicuous and easy to characterize, and because the obvious management strategy was to turn the effluent off. Regulatory efforts such as the Environmental Protection Agency's National Pollutant Discharge Elimination System focused on permits for each discharge. Each permit specified the allowable limits for the contaminants of concern, and each discharge had to be characterized. Initially, this characterization was physical and chemical. In the early 1980s, however, characterization of point source discharges began to include biological responses in standardized bioassays. Ultimately, this characterization process is facilitating the concept of site specific water quality criteria by allowing the toxicity bioassays to incorporate the receiving stream as the dilution water for the effluent being tested.

Non-point sources of pollution are much more difficult to characterize: they are not continuous like point source discharges; they are difficult to sample because of their diffuse nature; and the episodic nature of their occurrence makes them difficult to characterize. Hence, efforts to control non-point source pollutants have been less successful than those directed at point source pollutants.

1.2. Objectives

The paper describes a quantitative approach to chemically and physically characterizing non-point source pollutants and illustrates how this approach could be utilized to design bioassays that characterize the biological impact of non-point source pollutants. From an ecological risk assessment perspective, this paper focuses on assessment of exposure and development of a comparable effects assessment. A general framework for ecological risk assessment has been described elsewhere [1].

2. MATERIALS AND METHODS

2.1. Agricultural pesticide monitoring datasets

The availability of data from monitoring programmes such as those reported by Klaine et al. [2], Richards and Baker [3] and those available through United States Department of Agriculture programs such as MSEA provides a geographically diverse dataset for monitoring the behaviour of agricultural chemicals in watersheds ranging in size from 10 ha to well over 100 000 ha. These data were analysed to provide an estimate of the duration and intensity of contaminant exposure. Roman-Mas et al. [4] reported an optimized non-point source pollutant sampling strategy that is independent of basin size or stream order. They concluded that, for storm runoff sampling, a sampling interval equal to 0.05 of the duration of storm flow as adequate to characterize the concentration distributions in the receiving stream with an error of less than 5%. Hence, only monitoring datasets that meet this sampling frequency were used for this research.

The pesticide concentration was plotted as a function of Julian day for each year and receiving water body. On the basis of the no observable effect levels reported in the literature, the concentration lines of interest were drawn on the chemographs (Fig. 1). These chemographs were analysed to determine the exposure and recovery times. These were defined as follows: exposure occurred when the pesticide concentration was above a critical limit (e.g. 5 μg/L for atrazine in Fig. 1); and recovery occurred when the pesticide concentration fell below the critical limit. The assumption here was that there is a critical threshold above which the organisms were sublethally stressed and below which stress did not occur. The exposure and recovery periods for each critical concentration were catalogued for the entire

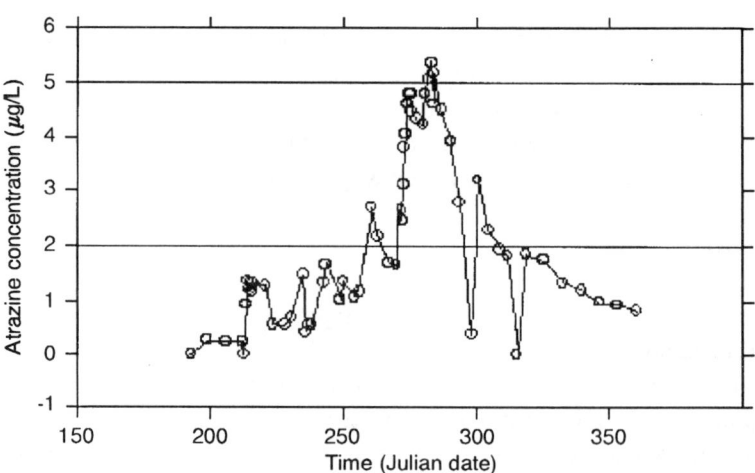

FIG. 1. Chemograph for atrazine in the Maumee River (1983).

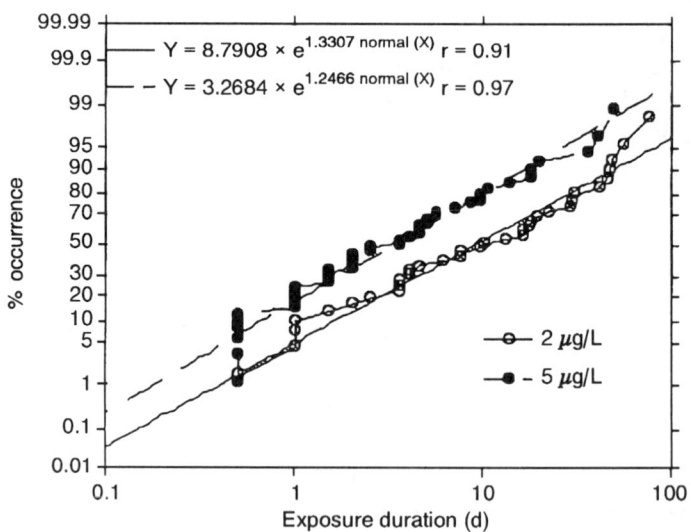

FIG. 2. Exceedancy curves for atrazine exposure in the Maumee River. Curves are composite for the years 1983–1993.

dataset. The exposure and recovery periods were displayed on a log probability plot and exponential regressions calculated for the data (Fig. 2).

This analysis provided a graphical and mathematical way of determining the probability of exposure duration. Toxicity bioassays were then designed on the basis of these analysis. Analysis was performed on atrazine and chlorpyrifos.

2.2. Methods of algal bioassay with atrazine

2.2.1. Atrazine

Atrazine is a herbicide that is commonly used to control grasses and broad leaf weeds in the maize growing regions of the United States of America and Canada. Its primary mode of action is reversible inhibition of photosynthesis, and regulatory limits as low as <5 µg/L have been proposed to protect sensitive aquatic flora [5].

Agricultural use of atrazine, and its subsequent fate in the environment, have both been well documented. The herbicide is applied to crops at label rates of up to 2.8 kg/ha [6] and is relatively persistent in soils [5, 7]. Atrazine is transported in agricultural runoff [5, 7, 8]. Concentrations approaching 250 µg/L have been reported at field edges [2], but the levels in streams and rivers do not usually exceed 20 µg/L for more than a few days. The highest atrazine concentrations are typically observed after storm events, but levels in the range of 10 µg/L sometimes persist for weeks [7, 8].

2.2.2. Organism

Selenastrum capricornutum Printz is a unicellular green alga that has been used extensively as a bioassay organism to measure the fresh water phytotoxicity of pollutants. This organism has been cultured in our laboratory at Clemson University for over 5 years under the conditions described by Hornin and Weber [9].

2.2.3. Bioassay method

While the above static bioassay for toxicity has been standardized [9], it has been criticized because it was originally designed to measure the eutrophication potential of effluents and because it suffers from insensitivity due to the algae growing under nutrient, light and diffusional limitations [10]. It has also been demonstrated that the precision of static algal assays is limited by natural fluctuations in the instantaneous growth rate of algal cultures [11]. These problems, as well as the suspected strain differences, have resulted in extreme differences being reported for the response of *S. capricornutum* to atrazine [7].

A more sensitive stepwise growth balanced dilution (SWGBD) bioassay system was developed for *S. capricornutum* in order to better estimate the threshold of response to atrazine and to describe the responses to episodic atrazine exposures simulating field events [11]. This assay substantially increased the information obtained during 96 h growth tests, and operated in the following manner: (1) cell suspensions were sampled at fixed intervals and counted with a haemocytometer; (2) a fixed volume of cell suspension was removed from each culture vessel; and (3) an identical volume of fresh, sterile medium with the same chemical concentration was then added to each bioassay vessel. Steps (1), (2) and (3) were initiated at time zero and repeated at 12 h intervals. Volumes were chosen so that the stepwise dilution closely balanced the population growth rate of the controls. As a result, the cell counts of the controls remained relatively constant throughout the bioassay, while the cell counts of slower growing treatment flasks tended to decline relative to the controls. All the bioassays used cotton stoppered Erlenmeyer flasks with standard algal medium. They were conducted under cool white fluorescent tubes (continuous, 120 μeinstein$\cdot$m$^{-2}\cdot$s^{-1}) at 25°C on a rotating shaker table.

2.2.4. Acute exposure and recovery experimental designs

Separate atrazine exposure (stress) and atrazine free recovery experiments were designed to simulate a single storm water driven atrazine pulse that might be observed in the surface waters of North America. The objectives were to determine the threshold response and the growth rate recovery once atrazine had been removed from the overlying water.

The stress experiment measured algal growth during 96 h of exposure to atrazine at concentrations ranging from 10 to 50 μg/L. The recovery experiment measured algae growth in atrazine free medium immediately after 96 h of exposure to 10–50 μg/L of atrazine. The experimental end point for the two experiments was the intrinsic rate of population increase ($\mu = N^{-1}dn/dt$) estimated from least squares regression of the cell counts taken at 12 h intervals: N is the number of organisms, and dn/dt is the rate of change in population with time.

The first experiment was initiated with approximately 2×10^6 cells/mL in serially diluted nutrient medium with nominal concentrations of 0, 10, 17, 29 and 50 μg/L. Each treatment level consisted of three replicate culture vessels. Actual concentrations of atrazine were measured with ELISA (Omichron Corporation, USA); correlations between the observed and the nominal concentrations were high (r = 0.99).

The second series of experiments was initiated by growing cells in batch culture for 96 h in serially diluted nutrient medium containing 0, 10, 17, 29 and 50 μg/L of atrazine. The rate data obtained in the first experiment were used to adjust the inoculation densities so that all the treatments contained approximately 4×10^6 cells/mL at 96 h harvest. The atrazine concentrations were confirmed with ELISA, as above. The harvested cells were washed in atrazine free medium and diluted to 2×10^6 cells/mL to begin the 96 h SWGBD recovery assay in atrazine free conditions. No atrazine was detected in the samples of overlying medium from the recovery experiment.

2.2.5. *Chronic exposure experimental design*

A third experiment examined the effects of 32 d of exposure to atrazine. The treatment groups were cultured in 10 µg/L of atrazine for a total of eight successive 4 d incubations. The control groups were cultured for eight successive 4 d incubations in atrazine free conditions. A reference group was split from the controls on day 28 and incubated for an additional 4 d in 10 µg/L of atrazine. An SWGBD assay was initiated on day 33 using the methods described in Section 2.2.4. Growth of the reference and treatment groups (five replicates each) were contrasted in 10 µg/L of nutrient medium. The two groups were compared with the controls (five replicates) grown in atrazine free nutrient medium. Accuracy of dosing was about 10% using ELISA, as described in Section 2.2.4.

2.3. Methods for daphnid bioassay with chlorpyrifos

2.3.1. *Chlorpyrifos*

Chlorpyrifos is a broad spectrum organophosphate insecticide used to treat many agricultural and domestic pests of several grain, nut, fruit, vegetable, turf and ornamental crops [12]. It has been one of the most widely used insecticides in recent years. The 1993 market estimates suggest that approximately 22–33 and 20–25 million kilograms of active ingredients were used in the agricultural and non-agricultural sectors, respectively [13]. Even with this widespread use, chlorpyrifos is not predicted to be of major concern as a pollutant from field runoff because of its hydrophobicity. Literature values support this hypothesis, suggesting that <1% of the applied compound reaches an adjacent water body [14]. While the total percentage chlorpyrifos runoff is low, the concentrations that can result in nearby fields are high enough to cause harm to aquatic organisms. The maximum water concentrations from field runoff can reach values of 3.8 µg/L [3]. Although these values are relatively low, the reported LC_{50} values for both fish and daphnids are within the same range: 1.3 and 0.08 µg/L, respectively [15]. Hence, chlorpyrifos was chosen as a model pesticide in these studies because of its widespread use and its extreme toxicity to aquatic organisms. Chlorpyrifos (98% purity) was obtained by courtesy of Dow Elanco (USA). A stock solution of 100 mg/L of chlorpyrifos was prepared in acetone and used in all the experiments. The maximum amount of acetone added in any experiment was 0.01%. An acetone treatment was included in all the experiments as the control.

2.3.2. Organism

The culturing methods consisted of rearing ten *Daphnia magna* in 1 L of reconstituted hard water [16]. The organisms were cultured following standard procedures in a constant temperature laboratory (25 ± 1°C) with a 16:8 h (light:dark) photoperiod. The average water quality characteristics were as follows: hardness, 165 mg/L of $CaCO_3$; alkalinity, 108 mg/L of $CaCO_3$; and pH8.4. The daphnids were fed a diet of 1.2×10^8 cells of *S. capricornutum* [17] and 10 mg of a yeast-trout (YT) chow mixture [16] daily. The organisms were transferred to fresh media every third or fourth day. The techniques for culturing algae have been reported elsewhere [18].

2.3.3. Bioassay

All the toxicity tests were initiated with <24 h old daphnids and generally followed set guidelines [16], except in the manner of dosing. The organisms were exposed to different concentrations of chlorpyrifos for different durations or at different time intervals, depending on the experiment. The exposure scenarios were determined to model the expected environmental exposure scenarios and also to demonstrate the relationships between pulse concentration, duration and interval. The underlying reason behind this testing procedure was to investigate whether standard toxicity tests accurately represent the manner in which organisms are exposed in situ.

Testing was done by exposing five daphnids to 400 mL of test solution, with either four or five replicates per treatment. The test chambers used in all the experiments were Kerr® pint (0.5 L) canning jars.

2.3.4. Experimental design

The objective of these bioassays was to characterize the response of *D. magna* to pulsed exposures of chlorpyrifos. The chlorpyrifos concentrations ranged from 0.125 to 1.0 µg/L and were measured by ELISA (Omichron Corporation, USA). Single and multiple exposure pulses lasted from 1 to 24 h and were always conducted with both a continuous exposure treatment and a control with no chlorpyrifos present. All the treatments, except as noted, received food daily in the form of 4.8×10^7 cells of *S. capricornutum* and 4 mg of YT chow. The survival and mobility of *D. magna* were monitored every 12 h for the first 4 days, then every 24 h.

3. RESULTS AND DISCUSSION

3.1. Toxicity of atrazine to *S. capricornutum*

3.1.1. Acute bioassay results

The SWGBD bioassay resulted in a 10 µg/L threshold response to atrazine (Fig. 3). The estimated rates of population growth in 10 µg/L of atrazine were statistically less than the controls (multiple first order regression, $P < 0.001$). This confirms some of the more sensitive responses reported for *S. capricornutum* in the literature. These results indicate that SWGBD may be a sensitive bioassay for algal toxicity. Although the growth rates were not uniform during the test, groups with different pre-exposures to atrazine behaved in near unison, and no statistical differences in the growth rates were detected. Algae appeared to fully recover from all levels of atrazine exposure within the first 12 h of resuspension in atrazine free medium (Fig. 4).

3.1.2. Chronic bioassay results

The performance of the reference (acute pre-exposure to atrazine) and the treatment (chronic exposure to atrazine) groups was statistically indistinguishable

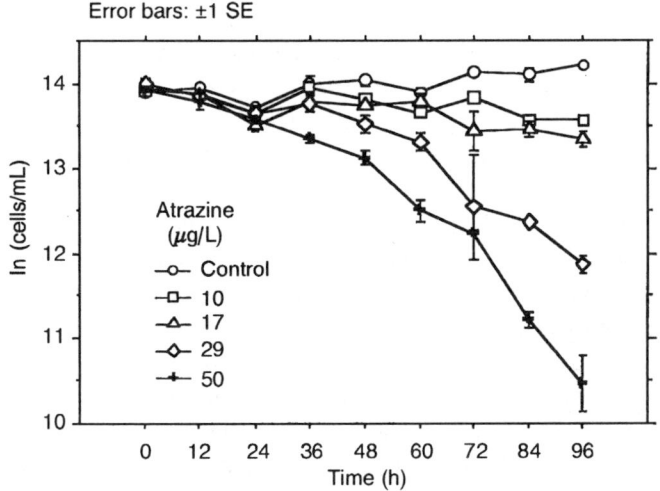

FIG. 3. Dose response of S. capricoruntum *to 96 h of atrazine exposure. Cell densities are natural log transformed to linearize the plots and to stabilize the variance.*

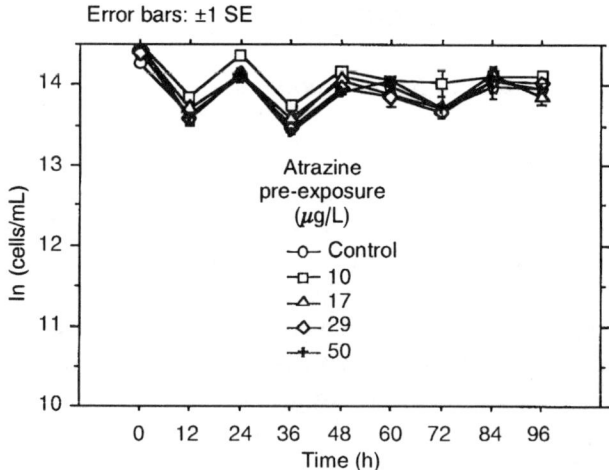

FIG. 4. Natural log transformed cell densities during 96 h of recovery in atrazine free medium immediately after 96 h of atrazine exposure.

during the subsequent 96 h of growth in 10 µg/L of atrazine (Fig. 5). The cell counts of the two groups declined relative to the growth of the controls in atrazine free medium (multiple first order regression, $P < 0.001$). The results suggest no long lasting effects from 32 d of exposure to low levels of atrazine.

3.1.3. Ecological implications

The results of the algal bioassays suggest that low concentrations of atrazine, comparable to those measured in North American watersheds, can inhibit the growth of green algae. Our analysis of atrazine monitoring datasets indicated that this exposure was episodic, followed by recovery periods during which practically no atrazine exposure occurred. The recovery data indicated that the growth rates of *S. capricornutum* return to the pre-stress levels almost immediately after the atrazine is removed. These data are consistent with the fact that atrazine is relatively water soluble, has a reversible site and mode of action, and induces stasis and not senescence in the exposed algae.

Our atrazine exposure analysis also indicated that long periods of low atrazine exposure could occur in some portions of agricultural watersheds. The results of the chronic exposure bioassay suggest that the algal cell physiology does not experience cumulative effects from chronic exposure, and that recovery is rapid. The overall results suggest that the impact on aquatic ecosystems from atrazine residues in agricultural watersheds is minimal. While impacts can be quantified with appropriate test methods, they are not ecologically significant.

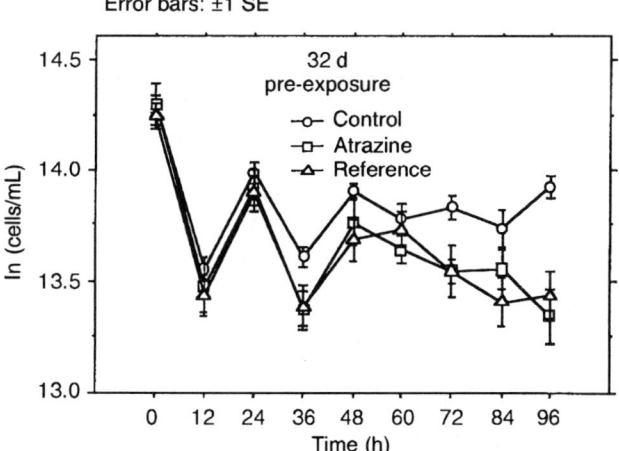

FIG. 5. Natural log transformed cell densities during 96 h of SWGBD bioassay: control groups (atrazine free medium); treatment groups (32 d of pre-exposure to atrazine); and reference groups (4 d of pre-exposure to atrazine) in 10 µg/L of atrazine.

3.2. Toxicity of chlorpyrifos to *D. magna*

Single, short exposures of *D. magna* to chlorpyrifos elicited a dose dependent response (Figs 6–9). As the concentration increased it took less exposure to affect survivability. If the organism survived the initial chemical burden, no observable effects were noticed. In these experiments, all the organisms were less than 24 h old at exposure. The influence of age was opposite to that expected. The *D. magna* were more sensitive when exposed to 0.5 µg/L on day 3 than on day 0 (Fig. 10). Multiple exposures continued this trend, with organisms being more sensitive if the second exposure was after 3 days (Fig. 11). The initial hypothesis was that older organisms have higher feeding rates and take in more food. Because chlorpyrifos is relatively lipophilic it seems reasonable that it would readily partition on to the algae and be more bioavailable to the daphnids. The results of treatments with and without food confirm this hypothesis, suggesting that the *D. magna* were more sensitive to a 12 h pulse of 0.5 µg/L of chlorpyrifos in the presence of food (Fig. 12). Further, experiments are in progress to adequately characterize the bioavailability issue regarding chlorpyrifos.

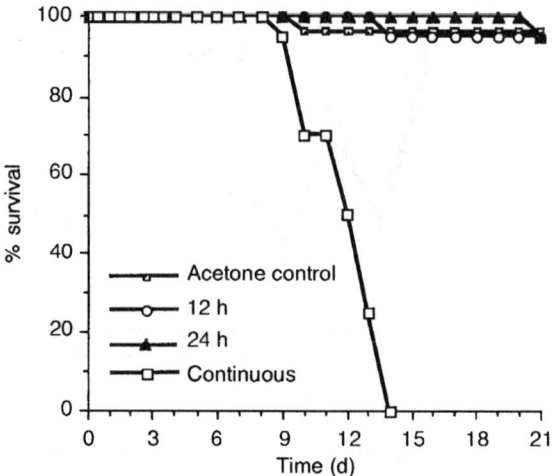

FIG. 6. Survival curves for D. magna *exposed to 0.125 µg/L of chlorpyrifos for different durations.*

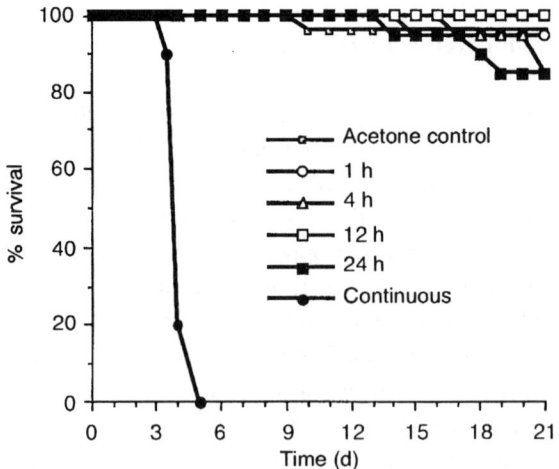

FIG. 7. Survival curves for D. magna *exposed to 0.25 µg/L of chlorpyrifos for different durations.*

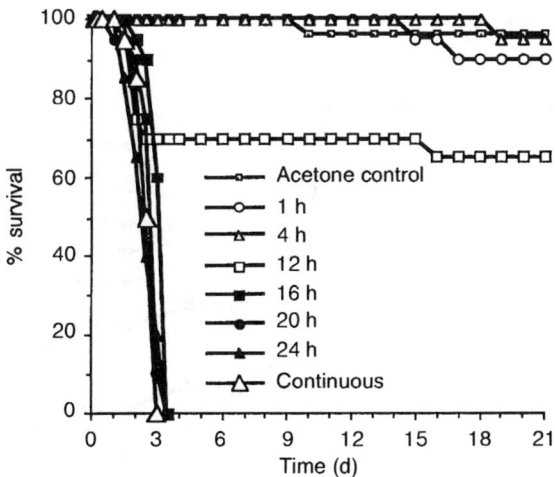

FIG. 8. Survival curves for D. magna exposed to 0.5 µg/L of chlorpyrifos for different durations.

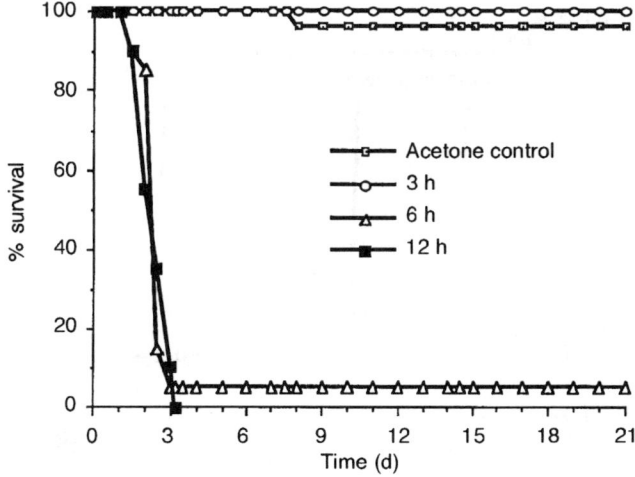

FIG. 9. Survival curves for D. magna exposed to 1.0 µg/L of chlorpyrifos for different durations.

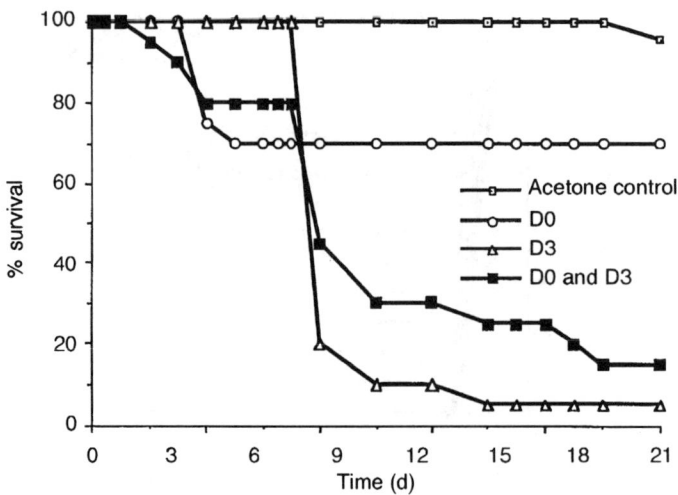

FIG. 10. Survival curves for 0 and 3 d old D. magna *exposed to 12 h pulses of 0.5 µg/L of* chlorpyrifos.

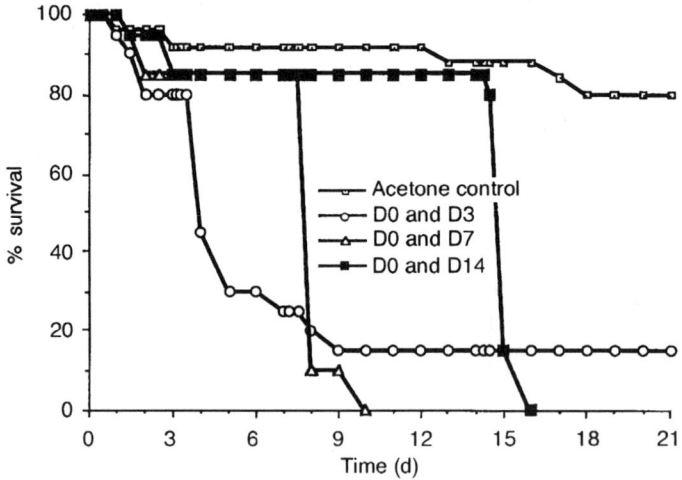

FIG. 11. Survival curves for D. magna *exposed to two 12 h pulses of 0.5 µg/L of chlorpyrifos at different intervals.*

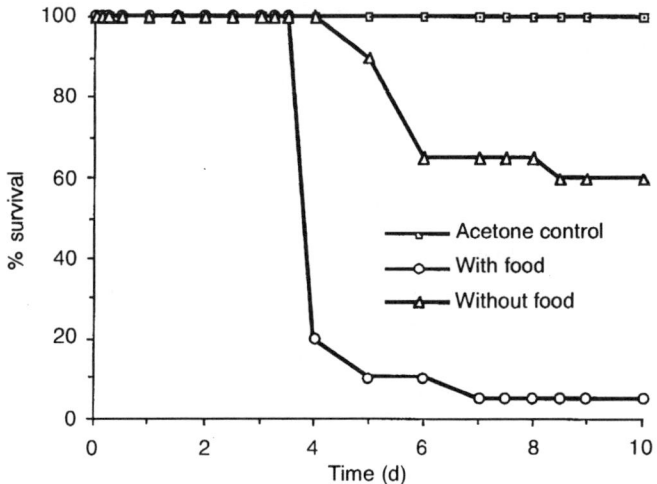

FIG. 12. Survival curves for 3 d old D. magna exposed to a 12 h pulse of 0.5 µg/L of chlorpyrifos with and without food.

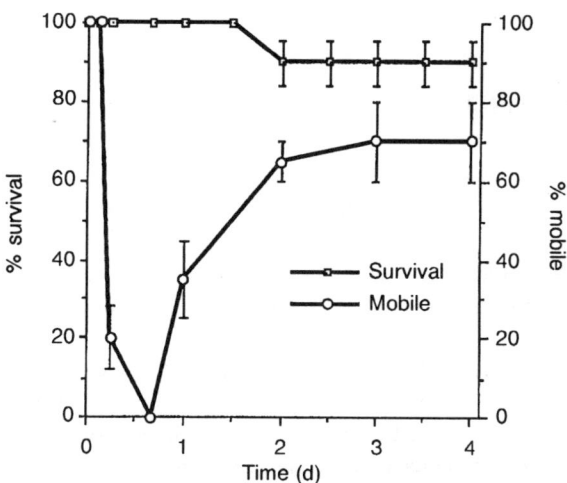

FIG. 13. Recovery of 3 d old D. magna exposed to a 6 h pulse of 1.0 µg/L of chlorpyrifos.

In most of the experiments, if immobility was observed, the daphnids died. In addition, all the surviving daphnids had reproductive rates that were similar to the control organisms during the 21 d test [19]. However, in some instances the *D. magna* were able to recover from initial exposure. Three day old *D. magna* exposed to a 6 h pulse of 1.0 µg/L of chlorpyrifos were all immobilized within 18 h (Fig. 13). Four days after initial exposure, significant recovery had taken place, since 75% of the *D. magna* were mobile and only 10% had actually died. This experiment suggests that organisms do have the ability to recover from some exposures to chlorpyrifos when they do not exceed a potential exposure threshold.

These data suggest that the results from standardized daphnid bioassays do not adequately predict the response of these organisms to the episodic exposures of pesticides that occur in agricultural watersheds. Moreover, several issues regarding age, feeding and recovery must be further characterized if results adequate for definitive risk assessments are to be generated.

4. CONCLUSIONS

The results of the analyses and research presented suggest a method for assessing the exposure and effects of agrochemicals. It is important to note that current approaches, while extremely useful for point source discharges, can misrepresent the risks associated with non-point source discharges.

ACKNOWLEDGEMENT

The authors wish to thank Ciba Crop Protection for partial support of this research.

REFERENCES

[1] ENVIRONMENTAL PROTECTION AGENCY, Framework for Ecological Risk Assessment, Rep. EPA/630/R92/001, Risk Assessment Forum, EPA, Washington, DC (1992).

[2] KLAINE, S.J., et al., Characterization of agricultural nonpoint pollution: Pesticide migration in a west Tennessee watershed, Environ. Toxicol. Chem. **7** (1988) 609–614.

[3] RICHARDS, R.P., BAKER, D., Pesticide concentration patterns in agricultural drainage networks in the Lake Erie Basin, Environ. Toxicol. Chem. **12** (1993) 13–26.

[4] ROMAN-MAS, A., STOGNER, R.W., DOYLE, U.H., KLAINE, S.J., "Assessment of agricultural non-point source pollution and best management practices for the Beaver Creek watershed, west Tennessee", Water Quality (Proc. Symp. Nashville, 1994) (PEDERSON, G.L., Ed.), American Water Resources Association, Chicago, IL (1994) 11–21.

[5] EISLER, R., Atrazine Hazards to Fish, Wildlife and Invertebrates: A Synoptic Review, Biological Report 85 (1.18), Patuxent Wildlife Research Center, United States Fish and Wildlife Service, Laurel, MD (1989).
[6] CIBA–GEIGY CORPORATION, Label for Aatrex Nine-0, EPA Registration No. 100-585, Ciba Crop Protection, Greensboro, NC (1993).
[7] SOLOMON, K.R., et al., Ecological risk assessment of atrazine in North American surface waters, Environ. Toxicol. Chem. **15** (1996) 31–76.
[8] THURMAN, E.M., GOOLSBY, D.A., MEYER, T.T., KOLPIN, D.W., Herbicides in surface waters of the midwestern United States: The effects of spring flush, Environ. Sci. Technol. **25** (1991) 1794–1796.
[9] HORNING, W.B., II, WEBER, C.I., Short Term Methods for Estimating Chronic Toxicity of Effluents and Receiving Waters to Freshwater Organisms, Rep. EPA/600/4-85/014, Environmental Protection Agency, Cincinnati, OH (1985).
[10] KLAINE, S.J., WARD, C.H., Growth optimized algal bioassays, Environ. Toxicol. Chem. **2** (1983) 245–250.
[11] BENJAMIN, R.B., FRAISER, M., JOAB, B.M., CASEY, R.E., KLAINE, S.J., Short term growth rate variation and the sensitivity of static exposure algal assays, Environ. Toxicol. Chem. (in press).
[12] Farm Chem Handbook, Pesticide Dictionary, Chlorpyrifos (MEISTER, R.T., SINE, C., Eds), Meister Publishing Co., Willoughby, OH (1994).
[13] ASPELIN, A.L., Pesticide Industry Sales and Usage: 1992 and 1993 Market Estimates 733-K-94-001, Biological and Economic Analysis Division, Office of Pesticide Programs, Office of Prevention, Pesticides and Toxic Substances, Environmental Protection Agency, Washington, DC (1994).
[14] RACKE, K.D., Environmental fate of chlorpyrifos, Rev. Environ. Contam. Toxicol. **131** (1993) 1–150.
[15] BARRON, M.G., WOODBURN, K.B., Ecotoxicology of chlorpyrifos, Rev. Environ. Contam. Toxicol. **144** (1995) 1–93.
[16] AMERICAN SOCIETY FOR TESTING AND MATERIALS, ASTM Standards on Aquatic Toxicology and Hazard Evaluation, Guide for Conducting Renewal Life-cycle Toxicity Tests with *Daphnia magna*, Rep. E 1193-87, ASTM, Philadelphia, PA (1993).
[17] INGERSOLL, C.G., DWYER, F.J., MAY, T.W., Toxicity of inorganic and organic selenium to *Daphnia magna* (Cladocera) and *Chironomus riparius* (Diptera), Environ. Toxicol. Chem. **9** (1990) 1171–1181.
[18] NADDY, R.B., LA POINT, T.W., KLAINE, S.J., Toxicity of arsenic, molybdenum, and selenium combinations to *Ceriodaphnia dubia*, Environ. Toxicol. Chem. **14** 2 (1995) 329–336.
[19] NADDY, R., et al., in preparation.

IAEA-SM-343/33

FATE OF PESTICIDES IN A MODEL RICE PADDY ECOSYSTEM*

A.W. TEJADA, L.M. VARCA, S.M.F. CALUMPANG,
C.M. BAJET, M.J.B. MEDINA
National Crop Protection Center,
University of the Philippines
  at Los Baños,
College, Laguna,
Philippines

Abstract

FATE OF PESTICIDES IN A MODEL RICE PADDY ECOSYSTEM.
   The fate of ^{14}C-carbosulfan was studied in a model rice paddy ecosystem. Carbosulfan was rapidly converted to carbofuran. Seventy-two hours after application on rice, the radioactivity was distributed as follows: soil > water > plant > fish > air. Rice fodder fortified with ^{14}C-carbofuran was fed to lactating goats. The ^{14}C-carbofuran equivalents were traced in the different organs and tissues of the animals, the highest concentrations being in omental fat and the liver. Carbon-14-carbofuran was metabolized and excreted in urine (77%), faeces (3%) and milk (0.05%). The residues in water were taken up and bioconcentrated by fish, the highest concentration being in the entrails. Even ducks and chickens in the vicinity of the treated rice fields contained pesticide residues. *Ipomoea aquatica* grown in the treated paddy water sorbed ^{14}C-carbofuran residues, with the highest uptake between days 9 and 13, declining thereafter. The pesticide residues used in the rice paddy were not detected in grains, but post-harvest treated rice grains contained high ^{14}C residues after treatment of the jute sacks, as is usual practice in warehouses. Washings, however, greatly reduced (67%) the ^{14}C-chlorpyrifos residues in grains. The bound residues of ^{14}C-isoprocarb on soil were absorbed by rice plants and by the subsequent crop, water melon. Owing to some contamination of well water collected in the vicinity of the treated rice fields, movement of monocrotophos, endosulfan and chlorpyrifos was evaluated in the field. The monocrotophos and endosulfan residues were rapidly lost in the paddy water, but significant amounts leached beyond the surface soil layer, up to a depth of 175 cm. Carbon-14 labelled monocrotophos, however, only reached a depth of 30 cm within 60 days in the soil column experiments. Likewise, ^{14}C-chlorpyrifos was found up to a depth of 20 cm on day 6 after application. Adsorption on soil was 91%, suggesting strong affinity to soil, and hence less leaching. Further experiments are in progress to examine this apparent anomaly.

---

   * Research carried out with the support of the IAEA under Research Contract Nos 8085 and 7977.

1. INTRODUCTION

Use of pesticides is still of major importance for crop protection. In the Philippines, rice production technology can be viewed as one of the contributors to pesticide contamination in the environment because of the usual practice of draining the treated paddy water into irrigation canals 40 days before harvesting the rice.

In rice production, about two to four applications are carried out by low pesticide users and four to eight applications by high pesticide users [1]. The real danger lies in the ability of aquatic organisms to bioaccumulate the residues, especially in a rice–fish culture, which is very popular not only in the Philippines but throughout Asia [2]. Moreover, animals that drink and graze in rice fields may bioaccumulate the pesticide residues. Hence, the fate of these pesticides needs to be thoroughly studied because a large portion of the applied pesticide does not reach the target organism and may contaminate the whole ecosystem.

Radioisotopes are the best method for tracing the metabolic fate of pesticides because one can obtain a total carbon balance.

2. MODEL ECOSYSTEM

The fate of ^{14}C-carbosulfan was determined in a model rice paddy ecosystem. A glass aquarium (74.9 × 40.6 × 48.3 cm) was used that contained 22.1 cm of paddy soil, with water filled to 3.05 cm above the soil (Fig. 1). Twenty rice seeddlings were transplanted in the paddy. A trench beside the rice paddy was

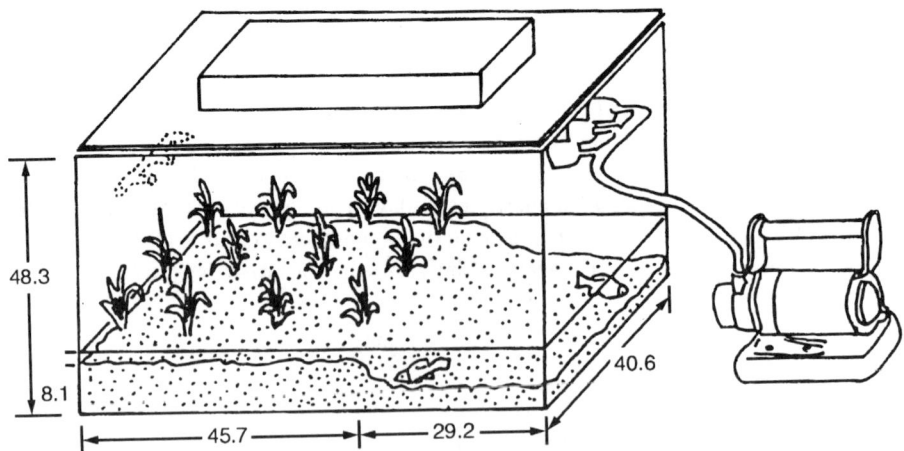

FIG. 1. Model rice paddy ecosystem (in cm).

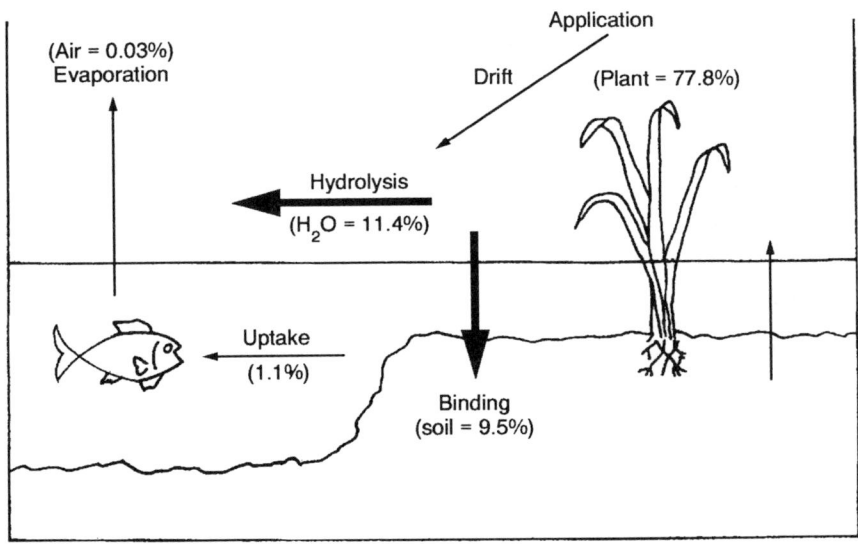

FIG. 2. Fate of ^{14}C-carbosulfan in a model rice paddy ecosystem showing the material balance 3 hours after application of the pesticide.

filled with ten (1 month old) Tilapia fingerlings (*Oreochromis niloticus*). Five millilitres of ^{14}C-carbosulfan (5 μCi) was sprayed on to the rice plants. The whole set-up was covered with a glass top to lessen the loss of radioactive pesticide.

Radioassay showed that carbosulfan was converted rapidly to carbofuran (3 hours after application), except on rice leaves, where it remained for up to 5 days. Of the ^{14}C-carbosulfan applied, 77.8% reached the target plant, 11.4% the paddy water, 9.5% the soil, 1.1% was taken up by fish, and 0.03% evaporated into the atmosphere (Fig. 2). After an equilibrium period of 72 hours, the radioactivity in the plants and fish decreased, and a corresponding increase was found in water (19.1%) and soil (40%). This result agreed with data obtained in the field [3].

3. FATE IN LACTATING GOATS

The residues detected on the rice plants were simulated by fortifying the rice fodder with ^{14}C-carbofuran (ring labelled) at rates of 0.5 and 1 mg·kg^{-1}·d^{-1}. Feeding was carried out for 7 days. Excretions in the faeces, urine and milk were monitored. Accumulation and distribution in various organs and tissues of the goats were determined. The total milk and urine samples were taken twice daily, and the faeces once a day. The goats were sacrificed after 7 days.

TABLE I. RADIOACTIVITY LEVELS IN VARIOUS ORGANS AND TISSUES OF LACTATING GOATS (ppm CARBOFURAN EQUIVALENT)[a]

Organs/tissues	Dose (0.5 mg·kg^{-1}·d^{-1})	Dose (1 mg·kg^{-1}·d^{-1})
Blood	0.08	0.07
Liver	0.55	0.69
Omental fat	0.84	1.42
Subcutaneous fat	0.23	0.41
Kidney	0.25	0.43
Brain	0.27	0.34
Heart	0.17	0.33
Muscle (*biceps femoris*)	0.19	0.34
Muscle (*longissimus dorsi*)	0.02	0.36

[a] Average values from two goats at each dose.

Carbon-14-carbofuran equivalents were detected in various organs and tissues of the goats in decreasing order of concentration: omental fat > liver > kidney > subcutaneous fat > muscles > heart > brain (Table I). Seventy-seven per cent of the ^{14}C-carbofuran applied were excreted in the urine, 3% in the faeces and 0.05% in the milk.

4. BIOCONCENTRATION OF PESTICIDES IN FISH AND DUCKS

Most of the pesticides used in rice production are extremely toxic to fish, *Tilapia* sp. (Table II) [4]. The rice–fish culture that is currently popular in the Philippines could result in the bioaccumulation of pesticide residues in fish and other aquatic organisms.

The bioconcentration of ^{14}C labelled pesticides was tested in a 2 L aquarium containing ten *Oreochromis niloticus* fingerlings. The fish were able to bioconcentrate these residues by as much as 239 times those in water (Table III), with the highest level found in the entrails (Fig. 3). This, however, is still considered low compared with DDT, whose bioconcentration factor (BCF) ranges from 10 000 to 100 000 [5]. Similarly, ducks swimming in the vicinity of the treated paddies were found to contain chlorpyrifos residues (Table IV).

TABLE II. TOXICITY OF SOME PESTICIDES TO *Oreochromis niloticus*

Pesticides	LC_{50} (mg/L) (48 h)	Rank[a]
Azinphos ethyl	1.0×10^{-6}	C
Endosulfan	6.9×10^{-4}	C
Cyfluthrin	1.6×10^{-2}	C
Chlorpyrifos	3.0×10^{-2}	C
Fenvalerate	3.0×10^{-2}	C
Cypermethrin	3.1×10^{-2}	C
Triazophos	3.5×10^{-2}	C
Ethofenprox	9.1×10^{-2}	C
Thiodicarb	0.12	C
Carbosulfan	0.17	C
Alpha cypermethrin + fenobucarb	0.22	C
Monocrotophos + fenvalerate	0.25	C
Fenobucarb + chlorpyrifos	0.28	C
Fenitrothion	0.49	C
Fenobucarb	0.64	C
Malathion	1.48	B
Methamidophos	2.96	B
Methyl parathion	3.50	B
Carbaryl	3.52	B
Monocrotophos	13.8	A

[a] The LC_{50} was determined according to the procedure described by Nishiuchi [4], where A = LC_{50} > 10 ppm: low toxicity; B = LC_{50} 0.5–10 ppm: moderately toxic; and C = LC_{50} < 0.5 ppm: extremely toxic.

TABLE III. BIOCONCENTRATION OF ^{14}C LABELLED PESTICIDES IN FISH AND WATER

Pesticides	Fish (ppb)	Water (ppb)	Bioconcentration factor
Methyl parathion	8.3	1.0	8.3
Carbofuran			
Fillet	10.5	0.09	117
Entrails	17.0	0.09	189
Carbosulfan	1.04	0.004	239
Endosulfan I	0.9	0.011	82
Endosulfan II	1.5	0.007	214
Chlorpyrifos			
Fillet	1.76	0.027	65
Entrails	4.1	0.027	152

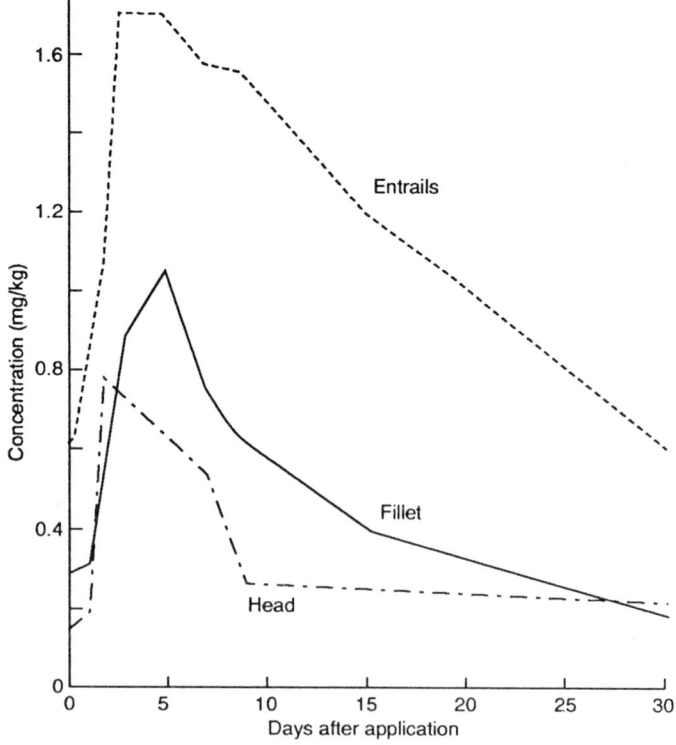

FIG. 3. Residues of ^{14}C-carbofuran in Oreochromis niloticus *fingerlings.*

TABLE IV. DETECTION OF PESTICIDES IN FISH, DUCK AND CHICKEN IN THE VICINITY OF THE EXPERIMENTAL AREA DURING THE PRE-BOOTING STAGE OF RICE[a]

Species	Pesticide(s) used	Residue level (mg/kg)
Tilapia (*Oreochromis niloticus*)	Chlorpyrifos	NDR
	Monocrotophos	NDR
	Decamethrin	NDR
Hito (*Chanca batrachus*)	Fenobucarb	NDR
	Methyl parathion	NDR
	Carbofuran	NDR
Mudfish (*Ophicephalus striatus*)	Chlorpyrifos	NDR
Shrimp (*Marabraechium* sp.)	Isoprocarb	NDR
Duck: heart	Chlorpyrifos	0.020
muscle	Chlorpyrifos	0.027
Chicken: brain	Chlorpyrifos	0.140
heart	Chlorpyrifos	0.030
muscle	Chlorpyrifos	0.025

[a] NDR = no detectable residue(s).

5. UPTAKE IN PLANTS

Aquatic plants such as *Ipomoea aquatica* are usually planted on the sides of paddy canals. These aquatic plants were able to sorb the ^{14}C-carbofuran residues in water for up to 9 days; this levelled off by day 13, and declined thereafter (Fig. 4). This observation is supported by the radiotracer studies of Bajet and Magallona [6], which showed sorption of ^{14}C-isoprocarb through the roots and distribution in the leaves.

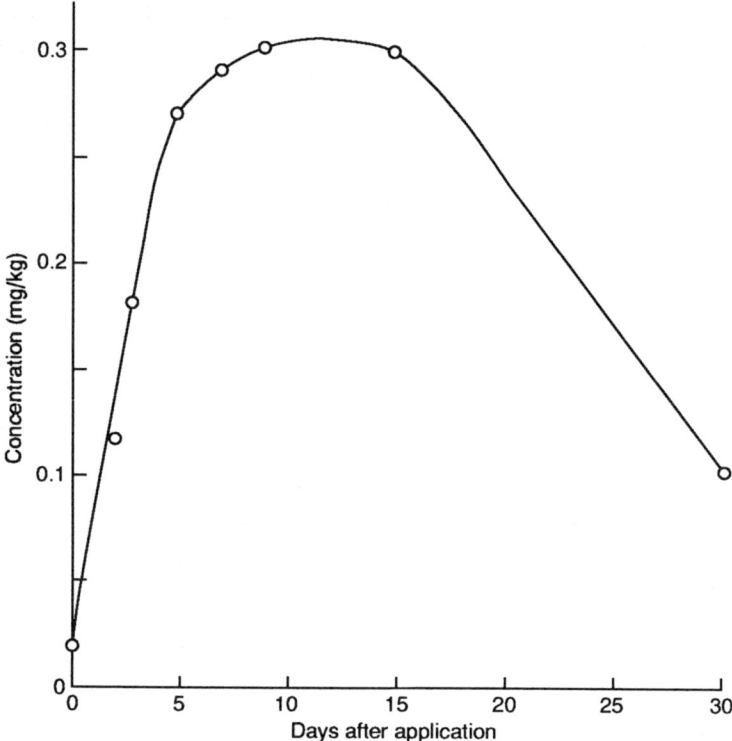

FIG. 4. Update of ^{14}C-carbofuran residues by Ipomoea aquatica.

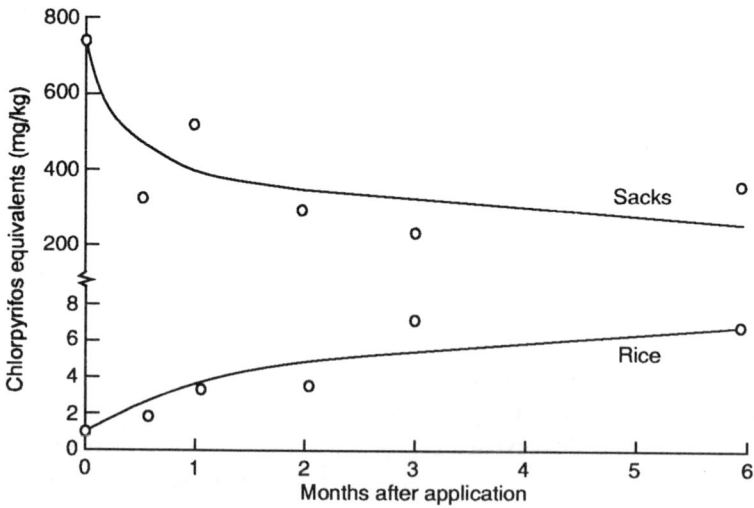

FIG. 5. Residues of ^{14}C-chlopyrifos in rice and sacks.

TABLE V. EFFECT OF WASHINGS ON ^{14}C CHLORPYRIFOS RESIDUES IN RICE

Months after treatment	Chlorpyrifos equivalent (mg/kg)					
	Unwashed	Washings (pH5.2)			Washed rice	% reduction
		1	2	3		
3	6.77	2.01	1.50	0.82	2.23	67
6	5.56	0.79	1.05	0.38	2.81	57

6. RESIDUES IN RICE GRAINS

In the actual rice field experiment, none of the treated grains contained any pesticide residues because there was a long interval between the treatment and the harvesting of the crops. The real danger lies in the post-harvest treatment of rice, which is usually sprayed with a 2% chlorpyrifos solution in jute sacks, or is admixed directly with the grains [7].

Carbon-14-diethyl labelled chlorpyrifos (169 μCi) was diluted with 0.5 mL of chlorpyrifos in water to give a 1.06 mg/mL solution (specific activity of 1.69 μCi/mL). This was sprayed on to the polished rice stored in sacks, as is usual practice in warehouses. The ^{14}C-chlorpyrifos treated sacks containing the rice absorbed the residues for up to 60 days; thereafter, sorption declined (Fig. 5). The residues, however, were greatly reduced with washings. About 0.5 g of rice was washed one to three times. Each washing showed a corresponding decrease in the residues, with a total reduction of 57–67% of the ^{14}C-chlorpyrifos applied (Table V).

7. BIOAVAILABILITY OF SOIL BOUND ^{14}C-ISOPROCARB RESIDUES

As soil is the ultimate repository of pesticides, it contains some bound pesticide residues. The practice of planting water melon after rice was simulated in the laboratory to determine the bioavailability of bound residues to crops.

Carbon-14-isoprocarb was mixed thoroughly with 2 kg of soil in a shallow pan. This was flooded with 2 L of tap water, covered and incubated for 3 months. After incubation, the air dried paddy soil was subjected to Soxhlet extraction for 24 hours, followed by acid reflux for 1 hour. After extraction, the soil was washed with water to remove the excess acid and the residual solvent. The rice plants were planted, followed by water melon.

Autoradiographs made of the rice plants illustrated the distribution of ^{14}C-isoprocarb throughout the whole rice plant, with the highest concentration found in the periphery of the leaf (14 days after transplanting (DAT)) and the leaf tip (28 DAT). After harvesting the rice, the water melons planted on the same soil were able to sorb the ^{14}C-isoprocarb residues, which were assumed to be bound.

8. RESIDUES IN WELL WATER

During the wet season, some pesticide residues were detected in well water in the vicinity of the treated rice fields. Hence, movement of monocrotophos, endosulfan and chlorpyrifos was studied.

Seven selected artesian wells (situated within the experimental site) that had received extensive pesticide applications were monitored for residues of some of the commonly used pesticides for two cropping seasons. Residues of endosulfan,

TABLE VI. RESIDUES (μg/L) OF SOME OF THE COMMONLY USED PESTICIDES IN WELL WATER NEAR RICE PADDIES (WET SEASON, 1989)[a]

Artesian well	Endosulfan	Monocrotophos	Chlorpyrifos
Calauan			
Well 1	0.030 ± 0.001	1.84 ± 0.064	0.032 ± 0.006
Well 2	NDR	NDR	NDR
Well 3	0.011 ± 0.003	NDR	NDR
Well 4	0.011 ± 0.001	NDR	NDR
Well 5	0.002 ± 0.000	NDR	NDR
Calamba			
Well 1	NDR	0.22 ± 0.027	NDR
Well 2	NDR	0.10 ± 0.009	NDR

[a] The minimum detectable levels (μg/L) are: alpha-endosulfan = 0.009; beta-endosulfan = 0.011; monocrotophos = 0.03; and chlorpyrifos = 0.02. The acceptable daily intakes (mg/kg body weight) are: endosulfan = 0.008; monocrotophos = 0.003; and chlorpyrifos = 0.01. NDR = no detectable residue(s).

monocrotophos and chlorpyrifos were detected in wells 1, 3, 4 and 5 (Calauan) using gas chromatography (Table VI) (no diazinon or fenobucarb was detected). Endosulfan ranged from 0.002 to 0.030 µg/L, but did not exceed the acceptable daily intake (ADI) value of 0.008 mg/kg body weight, based on calculations using eight glasses of water per day. Likewise, monocrotophos residues ranging from 0.10 to 1.84 µg/L did not exceed the ADI, which is 0.003 mg/kg body weight; the same applied to chlorpyrifos, with an ADI of 0.01 mg/kg body weight.

9. MOVEMENT OF PESTICIDES ON SOIL

Movement of ^{14}C-monocrotophos, endosulfan and chlorpyrifos was investigated using soil columns to support the findings on contamination of well water within the vicinity of the treated rice paddy fields.

The PVC pipes (20 in) (50.8 cm) were packed with soil and embedded in a simulated rice field with a cement base. The pipes were sampled at various intervals (0–60 days after treatment). The pipes were cut into 10 cm slices and the soil was radioassayed. Carbon-14-monocrotophos reached the 30 cm level 60 days after application (Fig. 6).

A similar experiment was carried out in a farmer's rice field using monocrotophos and endosulfan. Ceramic porous capped samplers fitted to the end of the sampling tubes were used to collect the soil solutions at 25, 50, 125 and 175 cm below the soil surface (Fig. 7). At specific sampling dates, the accumulated solution was collected using small plastic tubes inserted into the porous caps. The water samples were then sucked up with a vacuum pump and drained to the collection jar. Forty soil solution samples and ten surface water samples were collected. The concentration of monocrotophos and endosulfan at different soil depths following pesticide application is shown in Tables VII and VIII, respectively. The monocrotophos in paddy water was 32 µg/L, 6 hours after application; it leached to depths of 25–175 cm during the first week after spraying. This is usually observed for compounds that move by mass flow and diffusion, and are degraded or sorbed by the soil [8].

At a depth of 25 cm, 8.86 µg/L of the monocrotophos were observed on the day of application, but this level declined rapidly, reaching non-detectable levels in the second week. At the 50, 125 and 175 cm soil levels, the dissipation rate of monocrotophos was slower. While it was not detectable at the 25 cm soil level after 2 weeks, it took another 2 weeks to degrade to non-detectable levels at the 50 cm depth, and up to 73 days at the 125–175 cm depths.

Endosulfan was detected at the 125 cm soil layer 1 week after application, and after 1 month at the 175 cm soil layer. Low concentrations persisted for more than 2 months after application.

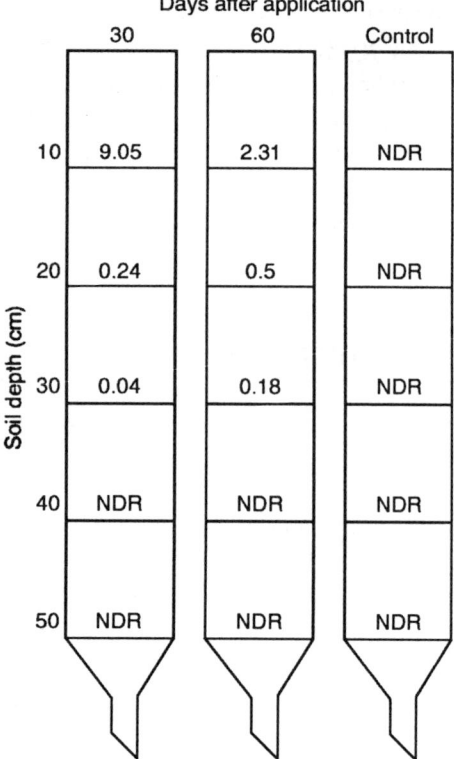

FIG. 6. Movement of ^{14}C-monocrotophos on soil columns (NDR = no detectable residue(s)).

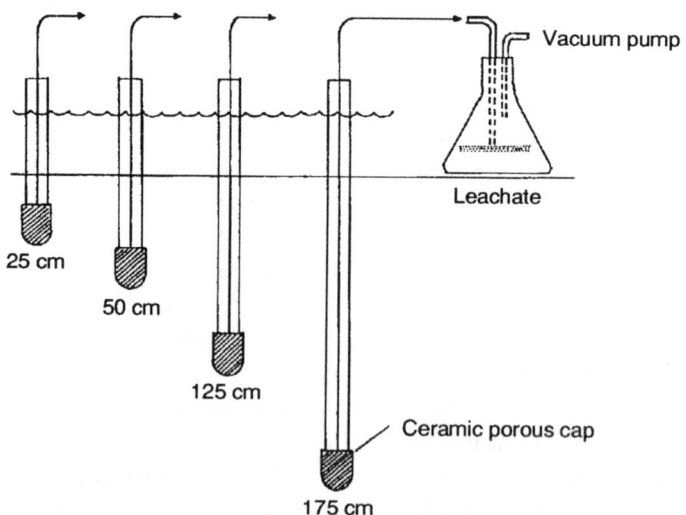

FIG. 7. Method for sampling water at different depths below the soil surface.

TABLE VII. MONOCROTOPHOS RESIDUES (µg/L) AT DIFFERENT SOIL DEPTHS IN A FARMER'S RICE FIELD FOLLOWING THREE APPLICATIONS (DRY SEASON, DEC. 1989–FEB. 1990)[a]

Soil depth (cm)	Days after application					
	13	20	27	34	41	48
0	31.68	1.97	NDR	[b]	NDR	NDR
25	8.86	4.82	NDR	NDR	NDR	NDR
50	3.02	5.70	8.34	0.78	NDR	NDR
125	1.06	2.86	4.99	0.38	1.31	0.29
175	NDR	0.54	0.66	0.28	0.24	0.40

[a] NDR = no detectable residue(s).
[b] No surface water.

TABLE VIII. ENDOSULFAN RESIDUES (µg/L) AT DIFFERENT SOIL DEPTHS IN A FARMER'S RICE FIELD FOLLOWING TWO APPLICATIONS (DRY SEASON, JAN.–APR. 1990)[a]

Soil depth (cm)	Days after application			
	6	31	59	73
0	0.037	Trace	0.038	–
25	2.467	Trace	0.483	–
50	0.390	0.060	0.262	NDR
125	0.104	0.035	0.122	0.153
175	Trace	0.053	0.044	0.009

[a] NDR = no detectable residue(s).

Chlorpyrifos was detected below the 20 cm level of the paddy soil 6 days after application. In the paddy soil, maximum residues of 1.2 mg/kg were found 1 day after application. The coefficients of chlorpyrifos (0.02, 0.10 and 10.0 µg/mL of aqueous chlorpyrifos solution on to the soil) were 44.0, 50.7 and 51.2 mL/g, respectively, with an average of $91 \pm 0.6\%$ sorption. This suggests strong chlorpyrifos affinity to the soil.

REFERENCES

[1] TEJADA, A.W., Pesticide residues in foods and the environment as a consequence of crop protection, Philipp. Agric. **78** 1 (1995) 63–79.

[2] TEJADA, A.W., et al., Toxicity and toxicity indices of pesticides to some fauna of the lowland rice–fish ecosystem, Philipp. Agric. **76** 4 (1993) 373–382.

[3] TEJADA, A.W., MAGALLONA, E.D., Fate of carbosulfan in rice paddy environment, Philipp. Entomol. **6** 3 (1985) 29.

[4] NISHIUCHI, Y., Testing method for the toxicity of agricultural chemicals, Jpn. Pestic. Inf. **19** (1974) 16–19.

[5] METCALF, R.L., SANGHA, G.K., KAPOOR, I.P., Model ecosystem for the evaluation of pesticide biodegradability and ecological magnification, Environ. Sci. Technol. **5** (1971) 709.

[6] BAJET, C.M., MAGALLONA, E.D., Chemodynamics of isoprocarb in the rice paddy environment, Philipp. Entomol. **5** 4 (1982) 355–371.

[7] PRUDENTE, A.D., et al., "Residues of grain protectants on paddy and maize", Pesticide Control Program of the Development Academy of the Philippines (Proc. Workshop Tagaytay, 1995), Fertilizer and Pesticide Authority, Manila (1995).

[8] PANINGBATAN, E.P., MEDINA, J.R., CALUMPANG, S.M.F., MEDINA, M.J.B., Movement of monocrotophos and endosulfan on saturated soil, Philipp. Agric. **76** 3 (1993) 262–269.

DISTRIBUTION AND FATE OF ^{14}C-DDT IN MICROCOSM EXPERIMENTS SIMULATING THE TROPICAL MARINE ENVIRONMENT OF THE BAY OF BENGAL*

M.A. MATIN, E. HOQUE, J. KHATOON
Institute of Food and Radiation Biology,
Atomic Energy Research Establishment,
Dhaka

Y.S.A. KHAN, M.M. HOSSAIN
Institute of Marine Sciences,
Chittagong University,
Chittagong

A.J. MIAN
Department of Chemistry,
Dhaka University,
Dhaka

Bangladesh

Abstract

DISTRIBUTION AND FATE OF ^{14}C-DDT IN MICROCOSM EXPERIMENTS SIMULATING THE TROPICAL MARINE ENVIRONMENT OF THE BAY OF BENGAL.
  Marine microcosm experiments were conducted in aquaria to study the distribution and fate of ^{14}C-DDT in water, sediment, algae and mussels collected from the Bay of Bengal coast at St. Martin Island in the south of Bangladesh. Samples of water, sediment, algae and mussels from the aquarium were taken at 0 h, 2 h, 4 h, 24 h, 6 d, 14 d and 30 d to determine the ^{14}C-DDT residues. A liquid scintillation counter and a biological oxidizer were used to establish the radioactivity in components of the ecosystem. Thin layer chromatographic techniques were used to study the metabolic transformation of DDT. It was found that DDT translocated from water to sediment, algae and mussels at varying rates. No adverse effects of applying ^{14}C-DDT to biota species were observed. At 0 h, 94% activity was present in water, while 3 and 4% were found in sediment and algae, respectively. The ^{14}C-DDT disappeared gradually from water up to 24 h, followed by a sharp reduction; only 1.5% was present on day 14. Sediment in the aquarium gradually accumulated ^{14}C-DDT, reaching 36% on day 6; thereafter, a decline in activity was observed. Sediment retained substantial activity (more than 20%), even on day 30. Algae accumulated ^{14}C-DDT rapidly from the beginning of the experiment, with maximum values of 30–36% during the 3–6 d period, while some

---

  * Research carried out with the support of the IAEA under Research Contract No. 7930.

activity was lost during the subsequent experimental periods. Mussels acquired ^{14}C-DDT very slowly, and a maximum value of 13.28% was found on day 14. During the remaining experimental period, mussels lost some of this accumulated activity. In general, disappearance of ^{14}C-DDT from water was associated with an increase in ^{14}C-DDT in sediment and algae. Similarly, a gradual reduction in ^{14}C-DDT from peak values in one component resulted in an increase in other component(s). Substantial DDT volatilization and water evaporation were observed. The parent ^{14}C-DDT metabolized to DDE and DDD. DDE and DDD were found in algae, while DDD was the only metabolite in the sediment samples.

1. INTRODUCTION

Bangladesh comprises a vast riverine delta situated at the apex of the Bay of Bengal [1], and is a country that is dependent on agriculture. Intensive agricultural systems are practised for food and cash crops involving use of modern agricultural inputs, including pesticides. The widely cultivated, high yielding varieties of crops are very vulnerable to pests and diseases. Moreover, the tropical climate results in high pest pressure, which necessitates the application of pesticides [2].

Generally, vast tracts of productive agricultural land are located on the coastal plains, where rainfall is high. The nearby marine environment includes highly productive shallow coastal lagoons and estuaries. Water in these areas is vital to the nursery grounds of larval fish and other exploitable offshore resources. The lagoons also support important fisheries, and semi-intensive shrimp cultures are gaining importance in some coastal regions.

The nutrients and detritus supplied by runoff from adjacent agricultural lands bring fertility to the marine environment. Such runoff also transports pesticide residues to the marine system. Persistent organochlorine pesticides may accumulate in the marine environment, causing a variety of effects on biotic species. Such a situation could threaten the survival of individual biota and upset the ecological balance.

Organochlorine compounds such as DDT, HCH, toxaphene, aldrin and heptachlor are still used in many developing countries [3–5]. In Bangladesh, use of DDT and organochlorine compounds for agriculture is restricted, but weak regulatory control has led to the misuse of such compounds [4].

Very little information is available on residues and the environmental fate and behaviour of such pesticides in the tropical marine environment. Controlled experiments with radiolabelled pesticides may provide these much needed, valuable data; such an approach has been successfully applied in studies of terrestrial biota [6].

The present study, which is part of an IAEA/IAEA–MEL/SIDA Co-ordinated Research Programme, was designed to develop a marine microecosystem that simulates the local tropical marine environment using radiolabelled pesticides. Accordingly, studies on the distribution and fate of ^{14}C-DDT in the marine environment were undertaken using microcosm experiments. The results of the study are presented here.

2. MATERIALS AND METHODS

2.1. Ecosystems

Duplicate glass aquarium tanks (44 × 30 × 37 cm) were set up, each with 28 L of sea water, 8 kg of sediment (3.5 cm depth), algae (*Helimeda* sp.) and mussels collected from the Bay of Bengal coast at St. Martin Island. The mussels (3/L) weighed 10.5 g/L (1.5 g of soft tissue per litre). The aquaria were aerated using a fish pump and then stabilized for 24 h before introducing the ^{14}C-DDT. Water evaporation from the aquaria was compensated for by adding tap water throughout the experiment. The sediment type (organic matter and grain size), salinity, temperature and pH of the water were determined.

2.2. Radiochemistry

The ring labelled ^{14}C-DDT was obtained from the IAEA Marine Environment Laboratory, Monaco. The insecticide had a specific activity of 24.95 mCi/mmol, with a radiometric purity of 95% (checked by thin layer chromatography).[1] The concentration of the ^{14}C-DDT added to the water of each tank was 5.39 µg/L, with a maximum water solubility of 5.5 µg/L. The total activity introduced to each aquarium was 10.64 µCi (843 dis·min^{-1}·mL of water).

2.3. Sampling

Samples of water, sediment, mussels and algae were collected in triplicate at intervals of 0 h, 2 h, 4 h, 24 h, 3 d, 6 d, 14 d and 30 d for analysis of the ^{14}C-DDT residues in components of the ecosystem. To collect the sediment samples, 27 vials were introduced to the sediment in each aquarium before adding the ^{14}C-DDT. The samples were removed from the aquaria using a remote handling device (tongs).

2.4. Radiometric measurements

The ^{14}C-DDT residues were determined by direct liquid scintillation counting (LSC) (Packard Tricarb 1000). The radioactivity in sediment, mussels and algae was determined by combustion using a Harvey biological oxidizer (Model OX-600) followed by LSC. The soft tissue (edible portion) of mussels was obtained by careful use of forceps. All the solid samples were freeze dried, and the known weights combusted.

---

[1] 1 Ci = 3.70 × 10^{10} Bq.

2.5. Determination of DDT metabolites

Samples were extracted with n-hexane by Soxhlet extraction for ten cycles. Aliquots of the extracts were counted by LSC. The extracts and standards were concentrated and spotted on TLC plates (silica gel G containing a fluorochrome indicator), and identification was done by comparing the R_f values. The respective spots were scraped off and analysed by LSC.

2.6. Bound ^{14}C residues of DDT

The bound ^{14}C residues of DDT in sediment and algae were determined by combustion of the Soxhlet extracted residues in the biological oxidizer.

2.7. Volatilization

A separate experiment was conducted to evaluate the water evaporation and ^{14}C-DDT volatilization under local conditions. Sea water was used in a beaker under the same conditions as those used in the microcosm study. The ^{14}C-DDT in carrier solvent (acetone) was added to the water. A 5 μL solution containing 5.39 μg of ^{14}C-DDT (0.38 μCi) was introduced to the water in the beaker, and water evaporation and DDT loss were determined at various intervals during the 14 d investigation period. Volatilization of ^{14}C-DDT was studied by direct measurement of the radioactivity in water using an LSC.

3. RESULTS AND DISCUSSION

No adverse effects of applying ^{14}C-DDT to biota species were observed during the 30 d experimental period. Each component of the ecosystem was analysed to determine the distribution pattern of ^{14}C-DDT. Figure 1 shows the total ^{14}C-DDT residues in water, sediment, algae and mussels. It was evident that the ^{14}C-DDT applied to water translocated to sediment, algae and mussels at varying rates. Immediately after addition, about 94% activity was found in water, with 3% in sediment and 4% in algae of the microcosm. The ^{14}C activity in water decreased gradually to about one-fourth of its initial value at 24 h. This was followed by a sharp decline on day 3. During the remaining period, there was a gradual reduction in the ^{14}C-DDT levels in water, with only 1.5% remaining on day 30.

A significant amount of ^{14}C-DDT in water translocated to sediment, reaching a maximum value of 36% on day 6. Table I shows the extractable and bound ^{14}C-DDT residues in sediment at various experimental periods. It was evident that some of the activity bound to the soil matrices was not extractable under the Soxhlet extraction process used. At all times, the extractable activity was much higher than the

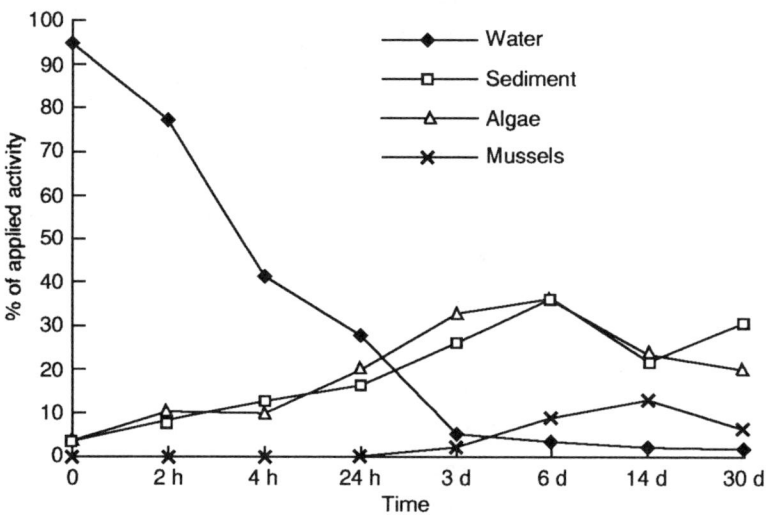

FIG. 1. Distribution of ^{14}C-DDT residues in components of the marine microcosm.

TABLE I. THE ^{14}C-DDT RESIDUES (EXTRACTABLE AND BOUND) IN MICROCOSM SEDIMENT (% OF APPLIED ACTIVITY)

Sampling period	Extractable ^{14}C residues	Bound ^{14}C residues	Total
0 h	2.56	0.86	3.42
2 h	7.72	0.44	8.16
4 h	9.52	2.73	12.23
24 h	10.41	3.38	13.79
3 d	16.89	6.62	23.51
6 d	22.87	7.20	30.07
14 d	15.24	4.65	19.89
30 d	20.37	7.62	27.99

TABLE II. THE ^{14}C-DDT RESIDUES (EXTRACTABLE AND BOUND) IN MICROCOSM ALGAE (% OF APPLIED ACTIVITY)

Sampling period	Extractable ^{14}C residues	Bound ^{14}C residues	Total
0 h	2.83	0.91	3.74
2 h	8.04	1.21	9.25
4 h	8.08	1.27	9.35
24 h	13.59	4.74	18.33
3 d	24.64	6.03	30.67
6 d	25.03	6.77	31.80
14 d	17.00	4.88	21.89
30 d	15.15	3.68	18.83

bound activity, and the total activity was similar to that obtained by direct combustion of the sediment samples (Fig. 1). Brazilian researchers have shown that some of the ^{14}C-DDT applied to soil under field and laboratory conditions bound to the soil, increasing gradually with time [7]. A significant but gradual increase in the formation of soil bound ^{14}C residues has been reported by other workers [8, 9].

Seaweed (algae) accumulated ^{14}C-DDT rapidly from the beginning of the experiment, reaching a maximum value of 30–36% during the 3–6 d period; thereafter, some of the accumulated activity in algae was lost. Some radiocarbon could not be extracted from algae and remained as bound ^{14}C residues, as evidenced by combustion of the extracted ^{14}C residues (Table II). The mussels used in the microcosm experiment acquired ^{14}C-DDT very slowly, accumulating a maximum value of 13.28% on day 14; thereafter, activity declined (Fig. 1). Marine microecosystem studies conducted in India have shown a similar trend for ^{14}C-DDT accumulation in clams [10].

The high rate of volatilization, influenced by solar radiation and the temperature conditions, led to a rapid loss in DDT, as well as to significant evaporation of the aquarium water. A separate experiment revealed that there was a gradual loss in water and disappearance of ^{14}C activity (Fig. 2). The data indicate a substantial loss in ^{14}C-DDT during the experimental period. Studies carried out in India on the persistence and fate of ^{14}C-DDT in soils have shown that volatilization of ^{14}C-DDT increased with time and a rise in temperature [11]. Vollner and Klotz in Germany have shown that the rate of volatilization and degradation of DDT was higher in an open system [12].

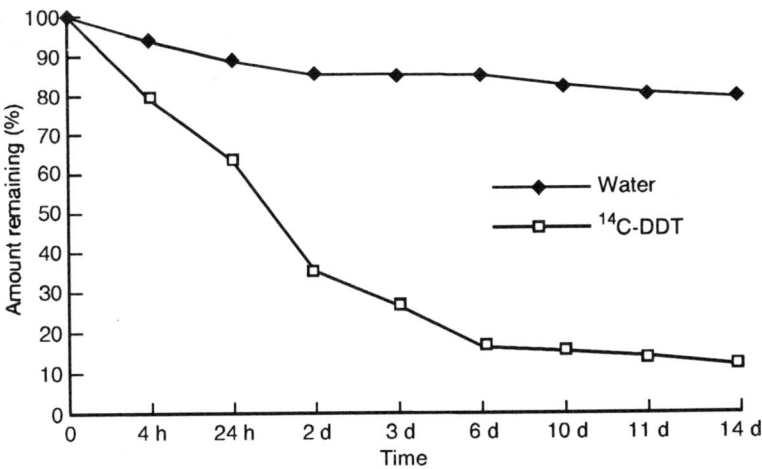

FIG. 2. Evaporation of water and volatilization of the ^{14}C-DDT applied.

TABLE III. DISTRIBUTION OF ^{14}C-DDT AND ITS METABOLITE IN MICROCOSM SEDIMENT (% OF APPLIED ACTIVITY ON TLC PLATES)[a]

Sampling period	DDT	DDD
0 h	63.50	—
2 h	59.30	7.93
4 h	72.22	9.02
24 h	65.20	7.40
3 d	73.78	6.50
6 d	69.87	8.73
14 d	73.77	6.10
30 d	68.16	7.61

[a] No DDE was found.

TABLE IV. DISTRIBUTION OF ^{14}C-DDT AND ITS METABOLITES IN MICROCOSM ALGAE (% OF APPLIED ACTIVITY ON TLC PLATES)

Sampling period	DDT	DDE	DDD
0 h	15.40	—	—
2 h	10.64	8.45	4.25
4 h	13.50	6.50	2.75
24 h	16.25	6.75	4.20
3 d	12.55	7.20	3.75
6 d	16.50	9.40	4.50
14 d	12.20	5.45	5.20
30 d	17.45	7.52	6.65

Metabolic transformation of the parent ^{14}C-DDT in sediment and algae was studied. Table III shows the distribution of the parent ^{14}C-DDT and its metabolite (DDD) in sediment samples. The climatic conditions may well have influenced the metabolic transformation of DDT. DDE and DDD have been detected as the degradation products of DDT in soils under different field conditions. It has been shown that DDD was the predominant metabolite of DDT in Chinese soils, while DDE and DDA were relatively minor products [13]. In Egyptian soils, DDE (30%) was the main degradation product of DDT in sunlight, whereas DDD was the main product in a dark experiment (9% of the applied dose) [14]. In the present study (laboratory conditions), DDD was the only metabolite found in microcosm sediment. This may have been due to the different conditions under which the experiments were carried out (aerobic versus anaerobic).

The hexane extracts of the algal samples used in the microcosm experiments were further analysed by TLC to determine the distribution of metabolites. The R_f values for the unknown spots were compared with those of the standards, and the activity was measured by LSC. DDE was found to be the major metabolite of the parent ^{14}C-DDT in algae of the ecosystem (Table IV).

The results of this study suggest that DDT is translocated to marine water. The sources include runoff from agricultural lands, rainfall, tidal surges, etc. Ultimately, DDT reaches the marine environment, where it is accumulated by biotic and abiotic components.

ACKNOWLEDGEMENT

The authors deeply appreciate the support provided by the IAEA.

REFERENCES

[1] MATIN, M.A., Environmental pollution and its control in Bangladesh, Trends Anal. Chem. **14** 1 (1995) 468–473.

[2] RAHMAN, M.S., MALEK, M.A., MATIN, M.A., Trend of pesticide usage in Bangladesh, Sci. Total Environ. **159** (1995) 33–39.

[3] KANNAN, K., TANABE, S., RAMESH, A.A., SUBRAMANIAN, A., TATSUKAWA, R., Persistent organochlorine residues in food stuffs from India and their implications on human dietary exposure, J. Agric. Food Chem. **40** (1992) 518–524.

[4] MATIN, M.A., et al., DDT residues in dried fish of Bangladesh, Nucl. Sci. Appl. **4** 1 (1995) 61–66.

[5] DIKSHITH, T.S.S., KUMAR, S.N., SRIVATSAVA, M.K., RAIZADA, R.B., RAY, P.K., Pesticide residues in edible oils and oilseeds, Bull. Environ. Contam. Toxicol. **42** (1989) 50–56.

[6] INTERNATIONAL ATOMIC ENERGY AGENCY, Use of Isotopes in Studies of Pesticides in Rice–Fish Ecosystems, Report of the Final Research Co-ordination Meeting, Bangkok, 1991, IAEA-TECDOC-695, IAEA, Vienna (1993).

[7] MERCEDES DE ANDREA, M., FLORES-RÜEGG, E., "Behaviour of ^{14}C-DDT in soil under field and laboratory conditions in Brazil", Isotope Techniques for Studying the Fate of Persistent Pesticides in the Tropics, Report of the Final Research Coordination Meeting, Bangkok, 1988, IAEA-TECDOC-476, IAEA, Vienna (1988) 53–59.

[8] HUSSAIN, A., MAQBOOL, U., ASI, M., Studies on dissipation and degradation of ^{14}C-DDT and ^{14}C-DDE in Pakistani soils under field conditions, J. Environ. Sci. Health **B29** 1 (1994) 1–15.

[9] AGARWAL, H.C., SINGH, D.K., SHARMA, V.B., Persistence, metabolism and binding of P,P'-DDT in soil in Delhi, India, J. Environ. Sci. Health **B29** 1 (1994) 73–86.

[10] KALE, S.P., RAGHU, K., MOHAN RAO, A., MURTHY, N.B.K., PANDIT, G.G., "Fate of ^{14}C-DDT in the Indian marine environment using microcosm experiments", The Distribution, Fate and Effect of Pesticides on Biota in the Tropical Marine Environment: Use of Radiotracers, Paper presented at 2nd IAEA/IAEA–MEL/SIDA Research Co-ordination Meeting, Kuala Lumpur, 1995.

[11] SAMUEL, T., PILLAI, M.K.K., "Persistence and fate of ^{14}C-P,P'-DDT in an Indian sandy loam soil under field and laboratory conditions", Isotope Techniques for Studying the Fate of Persistent Pesticides in the Tropics, Report of the Final Research Co-ordination Meeting, Bangkok, 1988, IAEA-TECDOC-476, IAEA, Vienna (1988) 27–39.

[12] VOLLNER, L., KLOTZ, D., Behaviour of DDT under laboratory and outdoor conditions in Germany, J. Environ. Sci. Health **B29** 1 (1994) 161–167.

[13] XU, B.J., GANG, J.Y., ZHANG, Y.X, LING, H.B., Behaviour of DDT in Chinese tropical soils, J. Environ. Sci. Health. **B29** 1 (1994) 37–46.

[14] ZAYED, S.M.A.D., MOSTAFA, I.Y., EL-ARAB, A.E., Degradation and fate of ^{14}C-DDT and ^{14}C-DDE in Egyptian soil, J. Environ. Sci. Health **B29** 1 (1994) 47–56.

IAEA-SM-343/31

A TROPICAL COASTAL LAGOON AFFECTED BY AGRICULTURAL ACTIVITIES
The importance of radiolabelled pesticide studies*

F. GONZALEZ-FARIAS**
Mazatlán Marine Station,
Institute of Marine Sciences and Limnology,
National Autonomous University of Mexico,
Mazatlán, Sinaloa, Mexico

F.P. CARVALHO, S.W. FOWLER, L.D. MEE***
Marine Environmental Laboratory,
International Atomic Energy Agency,
Monaco

Abstract

A TROPICAL COASTAL LAGOON AFFECTED BY AGRICULTURAL ACTIVITIES: THE IMPORTANCE OF RADIOLABELLED PESTICIDE STUDIES.
  The objective of this work was to integrate the results obtained from laboratory and field radiolabelled pesticide studies on an appropriate management model for a tropical coastal lagoon in which diverse human activities (e.g. agriculture, aquaculture, fisheries and tourism) take place. The tropical coastal lagoon studied is surrounded by agricultural fields on which large quantities of pesticides are used, and pesticide residues commonly enter the lagoon as runoff. Information on the distribution and dynamics of these contaminants is necessary for establishing coastal lagoon management. The distribution of pesticide residues in sediments of the lagoon was evaluated and the dynamics of the pesticides (water:sediment partitioning and bioaccumulation) experimentally assessed using ^{14}C labelled pesticides (chlorpyrifos, DDT and parathion) in model ecosystems. The results of these experiments indicate that partitioning between phases (water:sediment) is very rapid, with the half-life varying from a few hours for chlorpyrifos and DDT to up to 8 days for parathion. In the same way, bioaccumulation of the different pesticides is an active process that varied for the different organisms exposed to sublethal pesticide concentrations for 30 days. The results demonstrate that the persistence and the degree of bioaccumulation of pesticides are a threat to the ecosystem, both from the ecological and the economic point of view. Thus, traditional fisheries in the lagoon, shrimp farms and tourism could be seriously affected by their excessive use. Improved environmental management is urgently needed to reduce the risk of these ecological hazards.

---

* Research carried out with the support of the IAEA under Research Contract No. 7939.
** Present address: Aquaculture and Coastal Management Unit, Center for Research in Nutrition and Development, Mazatlán, Sinaloa, Mexico.
*** Present address: Black Sea Environmental Programme, Programme Co-ordination Unit, Dolmabahce Saray II, Beşiktaş, Istanbul, Turkey.

1. INTRODUCTION

Coastal lagoons are very important, both in ecological and in economic terms. These systems support economically important, detrital based marine food webs. Furthermore, they are considered to be the most productive areas of the world's oceans, since they function as nurseries for a wide variety of organisms. In fact, 90% of the world's fisheries are found in coastal zones; of these, 70% of the catch is composed of organisms that live all, or part of, their lives in coastal lagoons [1].

Tropical coastal lagoons are fringed by mangroves that constitute an important energy source because they provide detritus for lagoon ecosystems and near-shore coastal waters [2, 3]. The importance of mangroves is enhanced in semi-arid coastal regions such as northwest Mexico, because they represent one of the few perennial primary producers that supply energy all year round, both to coastal lagoons and to the surrounding terrestrial ecosystems [4]. The supply of mangrove detritus basically depends on the litterfall, degradation rate, river flow, rainfall, tides, winds and morphology of the lagoon [5].

Owing to their location, coastal lagoons generally receive substantial amounts of contaminants (e.g. pesticides, heavy metals and polychlorinated biphenyls) through discharges from rivers flowing through agricultural lands as well as through waste water drainage from agricultural, aquacultural, industrial and urban sources. Although there is global concern on the fate of certain contaminants, particularly pesticides, in tropical coastal lagoons [6], most of our knowledge of the environmental fate of pesticide residues is based on studies carried out in temperate regions. Thus, the impact of those contaminants that enter tropical coastal lagoons is relatively unknown [7].

Application of substantial amounts of pesticides in tropical regions is common, mainly in areas where intensive agriculture takes place. This practice often occurs on coastal plains, where runoff from agricultural land is rapidly discharged into tropical marine waters [8]. For example, this is the case for the Altata–Ensenada del Pabellón coastal lagoon, which is located in the Sinaloa State of northwest Mexico. The objectives of this work were to examine the ecological and economic importance of this coastal lagoon–mangrove ecosystem, and to integrate the results obtained from a biogeochemical study using radiolabelled pesticides into an appropriate management model.

2. STUDY AREA

2.1. Coastal lagoon

The Altata–Ensenada del Pabellón lagoon system is located between 24°40′ and 24°19′N and 107°28′ and 107°58′W on the coastal plain of Sinaloa State (Fig. 1). The system is situated in a semi-arid, subhumid region according to Köp-

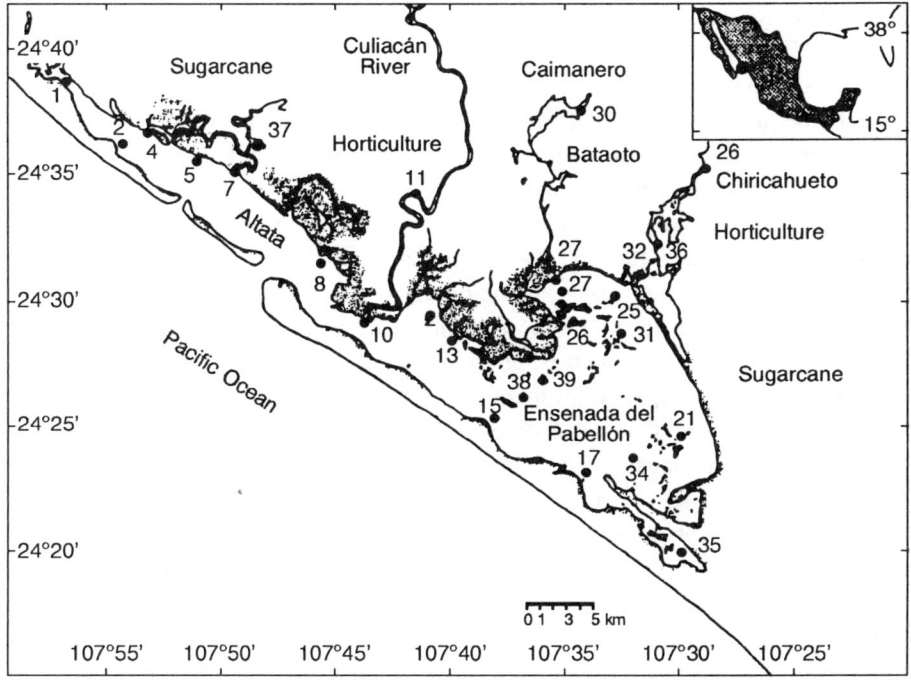

FIG. 1. Sampling sites in the Altata-Ensenada del Pabellón coastal lagoon ecosystem, including the Caimanero, Bataoto and Chiricahueto inland lagoons.

pen's classification (Bw(h')w(e')), with summer rains from July to October (average annual rainfall of 775 mm) and an annual mean temperature of 24°C [9]. The rainy period is commonly associated with the tropical storms and hurricanes that develop in the east Pacific Ocean [4].

In the adjacent coastal area, the tides are semi-diurnal, with a mean of 1.5 m. The surface sea temperature varies from 17°C in winter to 29.8°C in summer, and the average salinity varies from 33.9‰ in summer to 35.0‰ in winter [10].

This lagoon system is 360 km² wide and includes the Bahía de Altata and the Ensenada del Pabellón lagoon (Fig. 1). The former covers an area of about 80 km², and is a long, narrow lagoon running parallel to the coast with sandy sediments and a mean water depth of 5 m. Two deep inlets connect it to the sea, thus marine conditions prevail (32‰). The Ensenada del Pabellón is wider than the Bahía de Altata and covers 274 km² with silt and clay sediments and a mean depth of 1 m. Owing to fresh water inputs, mainly from the Culiacán River, the Ensenada del Pabellón lagoon displays estuarine conditions (15–28‰) throughout the year. The basin of the Culiacán River covers about 17 000 km^2, accounting for an annual discharge into the Altata–Ensenada del Pabellón lagoon of 3.3 × 10^9 m^3 [11].

Three small inland lagoons, the Bataoto, Chiricahueto and Caimanero (Fig. 1), receive discharge from the majority of the waterways, drains and ponds in the agricultural fields adjacent to the system. They also receive organic rich effluents from two of the three sugarcane factories located near the system. The three factories produce approximately 400 000 t of sugar per year [11].

The borders of the lagoon, including those of the lagoon islands, are covered by mangroves (*Rhizophora mangle*, *Laguncularia racemosa*, *Avicennia germinans* and *Conocarpus erectus*), comprising 10 631 ha. Although the climatic characteristics of the area are not optimal for mangrove growth, their energy input (via litterfall) to the system is similar to those observed in other subhumid and humid areas of Mexico. The mean annual litterfall for the mangrove forest of the Altata–Ensenada del Pabellón is 1.44 kg/m^2, an input which is as high as that recorded in well developed mangrove forests in humid regions [12]. As a result, mangroves are one of the most important features that have to be considered in establishing appropriate management of the area because they make a significant contribution to the productivity of the lagoon and also serve as feeding or reproductive grounds for different organisms, including some endangered species such as marine turtles, crocodiles and jaguars. In addition, a significant amount of shrimp larvae and post-larvae are harvested from this lagoon, as well as from other lagoons in the region, and serve to supply the shrimp farms. These mangroves also represent an attractive scenario for tourism [13].

2.2. Agriculture

Sinaloa represents approximately 14% of the total irrigated area of Mexico in which intensive mechanized agriculture takes place. In fact, the state has approximatively 1 429 251 ha of agricultural land, of which 757 380 ha are irrigated. The irrigated fields adjacent to the Altata–Ensenada del Pabellón coastal lagoon system comprise 42% of the state's total area (about 320 000 ha). In this particular horticultural area, large amounts of agrochemicals are used [14].

The total crop production from this region is more than 8 × 10^6 t/a. The principal crops include beans, chick pea, chilli, maize, cotton, cucumber, egg plant, melon, peanuts, rice, sesame seed, squash, soyabeans, sugarcane, tomato, water melon and wheat. For some of these crops, Sinaloa occupies first place in production at the national level. Agriculture contributes some US $730 × 10^6/a, which represents roughly 39% of the state's gross internal product, including all the commercial activities derived from agriculture [15]. Although horticulture production accounts for 18.7% of the total crop production of the state, the value of these crops is about 34% of the state's total income for agriculture [14].

2.3. Fisheries

Local fisheries are also extremely important to the state's economy. Total production is around 140 000 t/a, and includes molluscs, crustaceans and fish, products that represent around US $420 $\times$ 10^6. Fisheries production of the Altata-Ensenada del Pabellón lagoon is more than 3000 t/a, the main products being shrimp (*Penaeus* spp.), oysters (*Crassostrea* sp.), clams (*Anadara* spp.) and fish (*Mugil* spp., *Gerres* sp., *Lutjanus* sp., etc.), representing more than US $10 $\times$ 10^6 [15].

In recent years, active development of shrimp culture farms has taken place in Sinaloa. For the first half of 1995 there were 8284 ha of shrimp (*Penaeus vannamei*) ponds, with a production of 8557 t valued at US $30 $\times$ 10^6. Development of the shrimp farms around the Altata–Ensenada del Pabellón lagoon has to be considered in the management plan for the area because of use of the land (in some cases mangrove areas) and the water drainage (from shrimp ponds) that flows into the lagoon. Also, shrimp larvae and post-larvae are actively removed from the lagoon, and consequently about 30% of the shrimp farms of the state are located close to the study area [16].

3. PESTICIDES

3.1. Pesticide usage

As mentioned in Section 2.2, horticulture is practised in fields adjacent to the Altata–Ensenada del Pabellón lagoon [17]. As a result, relatively large quantities of pesticides are used in the area: roughly 40% of the total pesticide imports of the country [18]. A pesticide usage survey was carried out among farmers, aerial spray pilots, scientists, agricultural associations, local authorities and local pesticide distributors to determine which agrochemicals were being used in the area. More than 100 different compounds were identified, including a wide range of herbicides, insecticides, fungicides, acaricides and nematicides.

Owing to the prohibition of organochlorine pesticides by Mexican law, there have been major changes in the usage patterns of agrochemicals. At present, the compounds applied to crops are mainly organophosphorus, carbamates and pyrethroids rather than the persistent organochlorine pesticides (DDT and 'drins') used in the past. The amount of agrochemicals applied per year in this zone was estimated to be roughly 3.3 kg/ha [7]. The principal agrochemicals used in the fields adjacent to the coastal lagoon are listed in Table I. Carbamates and organophosphorus compounds constitute approximately 27 and 57% of the total, respectively.

TABLE I. PRINCIPAL AGROCHEMICALS USED IN THE CULIACAN VALLEY ADJACENT TO THE ALTATA–ENSENADA DEL PABELLON COASTAL LAGOON

Name	Group	Volume
Malathion	Organophosphorus insecticide	541 387 L
Methamidophos	Organophosphorus insecticide	42 300 kg
Parathion	Organophosphorus insecticide	61 000 kg
Bensulide	Organophosphorus herbicide	36 400 L
Monocrotophos	Organophosphorus insecticide	32 000 kg
Chlorpyrifos	Organophosphorus insecticide	30 000 kg
Dimethoate	Organophosphorus insecticide	6 400 kg
Mancozeb	Dithiocarbamate fungicide	262 900 kg
Zineb	Dithiocarbamate fungicide	2 000 kg
Aldicarb	Carbamate insecticide	66 160 L
Methomyl	Carbamate insecticide	22 290 kg
Carbaryl	Carbamate insecticide	5 000 kg
Chlorothalonil	Phtalamide fungicide	105 712 kg
Copper oxychloride	Inorganic fungicide	80 000 kg
Paraquat	Bipyridylium herbicide	18 200 L
Permethrin	Pyrethroid insecticide	5 000 kg
Fenvalerate	Pyrethroid insecticide	500 kg
Metribuzin	Triazine herbicide	4 000 kg
Camphechlor	Chlorinated camphene insecticide	1 500 kg

3.2. Distribution of pesticide residues in lagoon sediments

Organophosphorus and organochlorine pesticides have been detected in lagoon sediments. The former were found in higher concentrations than the persistent organochlorine compounds, an observation which reflects the reduction in use of the latter. Although DDT was banned for agricultural purposes in Mexico in the 1970s, it is still employed in sanitary campaigns against mosquitoes, and both DDT and its metabolites have been detected in the coastal lagoon. In general, there were higher concentrations of pesticide residues near the riverine and drainage inputs carrying

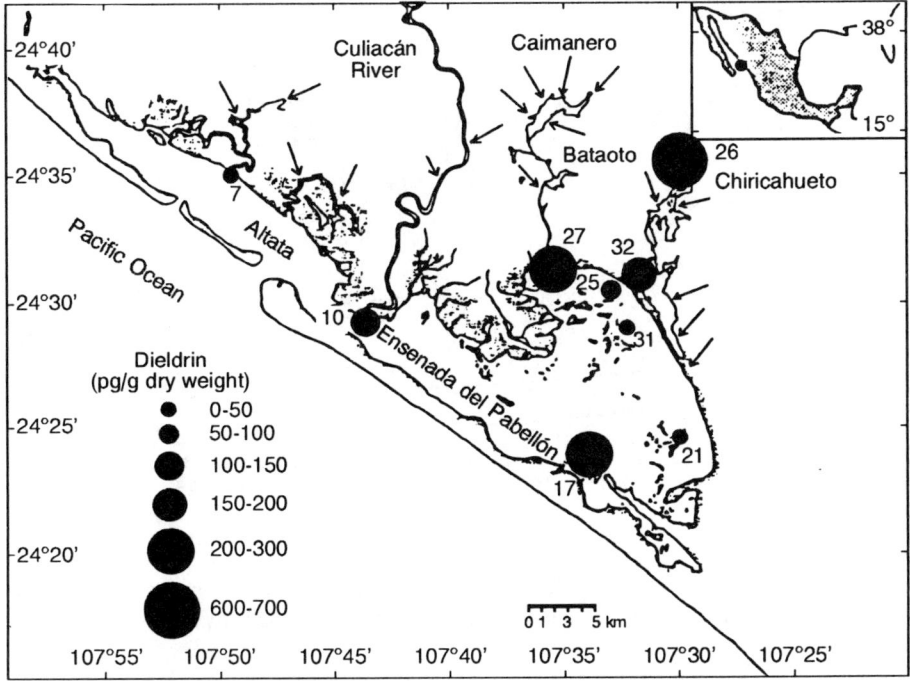

FIG. 2. *Distribution of dieldrin in sediment samples from the Altata–Ensenada del Pabellón coastal lagoon ecosystem.*

agricultural runoff than at the seaward sampling sites. The distribution of dieldrin is shown in Fig. 2; the observed gradient is due to dilution and/or degradation of the pesticides that enter the lagoon [17]. The Chiricahueto, Bataoto and Caimanero inland lagoons (Fig. 1) are very important for the system because they function as a trap for contaminants; therefore, the effluents that leave these lagoons and enter the Ensenada del Pabellón are less contaminated. This has been corroborated for pesticides [8] and also for phosphorus (from fertilizers) and organic matter [19], the concentrations of which are significantly higher in sediments of the inland lagoons than in those of the Altata–Ensenada del Pabellón lagoon.

3.3. Pesticide dynamics

The dynamics of those pesticides that reach the aquatic ecosystems are regulated to a great extent by two processes: partitioning between phases, and degradation. Partitioning of pesticides between water and sediment is influenced by the water

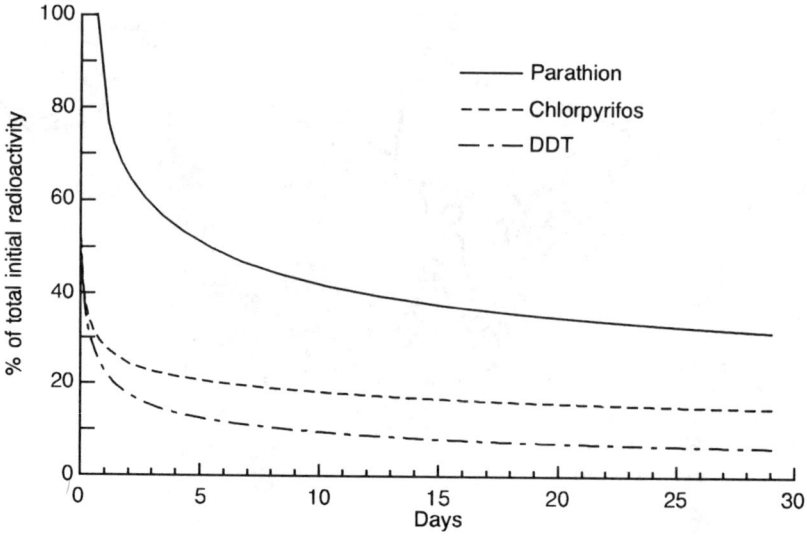

FIG. 3. Temporal variation of the total initial radioactivity in water samples collected from aquaria (water–sediment systems) spiked with ^{14}C labelled pesticides (chlorpyrifos, DDT and parathion).

solubility of the pesticides, the suspended matter and the organic content of the sediments [20, 21]. In terms of organisms, the bioavailability of the pesticides is regulated by these processes, which in turn influence their bioaccumulation potential. In addition, the bioaccumulation of pesticides in organisms is affected by their feeding habits and lipid content [22].

To understand the dynamics of pesticides in a tropical coastal lagoon, ^{14}C labelled pesticides (chlorpyrifos, DDT and parathion) were used to determine the partitioning between water and sediment and also to evaluate the bioaccumulation of these compounds in tropical lagoon organisms. It was observed that partitioning of pesticides between the aqueous phase and the sediments took place rapidly. Chlorpyrifos and DDT were removed from the aqueous phase faster than parathion because of the lower solubility in water of the former. Twelve hours after the introduction of pesticides into the experimental aquaria (water and sediment), the concentrations of chlorpyrifos, DDT and parathion in water were, on average, 22, 50 and 94% of the initial concentrations, respectively (Fig. 3). The results also indicate that the half-life (the time in which 50% of the initial concentration of pesticides is removed from water) varied from a few hours for chlorpyrifos and DDT to up to 8 days for parathion [23]. It appears that absorption of chlorpyrifos and DDT on to sediments reaches a steady state within a few hours, as reported for the same compounds in other water–sediment systems [24].

Several organisms were exposed in aquaria to sublethal concentrations of ^{14}C labelled pesticides (chlorpyrifos, DDT and parathion) in order to calculate the concentration factors (CFs) for each compound: oyster (*Crassostrea corteziensis*), mussel (*Mytella strigata*), clam (*Protothaca asperrima*), snails (*Nassarius bailyi* and *Cerithium stercusmuscarum*), shrimp (*Penaeus vannamei*) and fish (*Diapterus peruvianus* and *Lutjanus argentiventris*). All the organisms accumulated the different pesticides tested. The results indicate that the CF for chlorpyrifos varied from 18 in a snail (*N. bailyi*) to 748 in a clam (*P. asperrima*); for DDT, the CF ranged from 69 in a snail (*C. stercusmuscarum*) to 15 267 in fish brains (*D. peruvianus*); and for parathion it varied from 20 in a snail (*C. stercusmuscarum*) to 858 in a clam (*P. asperrima*). The order of CFs (maximum to minimum values for the different organisms tested) was as follows: chlorpyrifos: clam > fish > oyster > mussel > shrimp > snails; DDT: fish (*D. peruvianus*) > mussel > oyster > clam > fish (*L. argentiventris*) > shrimp > snails; parathion: clam > oyster > mussel > shrimp > snails.

4. RECOMMENDATIONS

The results of our study clearly demonstrated that there is an input of organochlorine and organophosphorus residues from the agricultural fields into the lagoon water. Because of the well known problems arising from the use of organochlorine pesticides, mainly in bird populations, these compounds should be totally prohibited. Likewise, because of the high toxicity of the organophosphorus pesticides and the persistence demonstrated in our studies, careful management of these compounds should be implemented in the future. Further studies on the sublethal effects of organophosphorus pesticides on the larval and juvenile stages of representative lagoon organisms are necessary if their full impact is to be understood.

Owing to the bioaccumulation of organophosphorus and organochlorine pesticide residues in commercial species, a permanent survey should be carried out to avoid health problems to the consumers of such products. The presence of pesticide residues in water and their persistence in sediments could affect both the traditional fisheries and the more recently developed aquaculture farms, resulting in negative economic and social consequences. To maintain sustainable development of the Altata–Ensenada del Pabellón coastal lagoon, there should be a conciliation between the different activities that take place in the region. Furthermore, it is proposed that inland lagoons continue to be used as contaminant 'traps' and that other trap areas around the lagoon be developed to reduce direct discharges from agricultural drainage into the ecosystem.

ACKNOWLEDGEMENTS

The authors wish to thank the European Union, which supported this project through Contracts Nos Cl1-0387-ME(JR) and Cl1-0326-M(A). F. González-Farias expresses his gratitude to the IAEA for financial assistance.

REFERENCES

[1] YAÑES-ARANCIBIA, A., Patrones ecológicos y variación cíclica de la estructura trófica de las comunidades nectónicas en lagunas costeras del Pacífico de México, An. Centro Cien. Mar Limnol. **5** (1978) 287–306.

[2] BENNER, R., HODSON, R.E., Microbial degradation of the leachable and lignocellulosic components of leaves and wood from *Rhizophora mangle* in a tropical mangrove swamp, Mar. Ecol. Prog. Ser. **23** (1985) 221–230.

[3] WOODROFFE, C.D., Studies of a mangrove basin, Tuff Crater, New Zealand. I. Mangrove biomass and production of detritus, Estuar. Coast. Shelf Sci. **20** (1985) 265–280.

[4] FLORES-VERDUGO, F., GONZALEZ-FARIAS, F., ZARAGOZA, U., "Ecological parameters of the mangroves of semi-arid regions of Mexico: Important for ecosystem management", Towards the Rational Use of High Salinity Tolerant Plants, Vol. 1 (LIETH, H., AL MASSOOM, A., Eds), Kluwer, Dordrecht (1993) 123–132.

[5] GONZALEZ-FARIA, F., HERNANDEZ-GARZA, M., "Aspectos ecológicos de la materia orgánica en lagunas costeras de México", Temas de Oceanografía Biológica en México, Vol. 1 (DE LA ROSA-VELEZ, J., GONZALEZ-FARIAS, F., Eds), Universidad Autónoma de Baja California, Ensenada (1990) 79–105.

[6] NEWELL, R.C., CLEGG, D.R., MAUGHAN, D.W., Environmental impact of liquid wastes discharge in coastal waters, Ocean Shorel. Manage. **16** (1991) 327–347.

[7] CARVALHO, F.P., GONZALEZ-FARIAS, F., MEE, L.D., FOWLER, S.W., "The impact of pesticides in coastal lagoons and implications for management of the coastal zone in Mexico", Proc. Int. Water Seminar, Mazatlán, 1994, Comisión Nacional del Agua, European Union and Instituto Mexicano de Tecnología del Agua, Mexico City (1994).

[8] MEE, L.D., READMAN, J.W., CARRANZA, J., GONZALEZ-FARIAS, F., "Fate of agrochemicals in tropical coastal lagoon ecosystems", Consolidated Report of Activities 1986–1990, EEC–Mexico (KESSLER, C., Ed.), Rep. EUR-13970, Directorate General for Science, Research and Development, Commission of the European Communities, Brussels (1991) 113–117.

[9] GARCIA, E., Modificaciones al Sistema de Clasificación climática de Köppen (Adaptación a las Condiciones de la República Mexicana), Instituto de Geografía, Universidad Nacional Autónoma de México, Mexico City (1973) 243 pp.

[10] GONZALEZ-FARIAS, F., HERNANDEZ-GARZA, M., FLORES-VERDUGO, F., CALVARIO, O., "Distribución de la materia orgánica particulada y bacterias heterótrofas en aguas superficiales del Mar de Cortés", Temas de Oceanografía

Biológica en México, Vol. 2 (GONZALEZ-FARIAS, F., DE LA ROSA-VELEZ, J., Eds), Universidad Autónoma de Baja California, Ensenada (1995) 87-110.

[11] INSTITUTO NACIONAL DE ESTADISTICA, GEOGRAFIA E INFORMATICA, Anuario Estadístico del Estado de Sinaloa 1985, INEGI, Mexico City (1986) 909 pp.

[12] FLORES-VERDUGO, F., "Distribución y productividad de los manglares", Ecología de los Manglares, Productividad Acuática y Perfil de Comunidades en Ecosistemas Lagunares-Estuarinos de la Costa Noroccidental de México. Parte 1. Ensenada del Pabellón-Altata y Bahía de Mazatlán (ARENAS, V., FLORES-VERDUGO, F., Eds), Informe Técnico, DGAPA Rep. IN-202389, Universidad Nacional Autónoma de México, Mexico City (1991) Chapter 1.

[13] FLORES-VERDUGO, F., "Algunos aspectos sobre ecología, uso e importancia de los ecosistemas de manglar", Temas de Oceanografía Biológica en México, Vol. 1 (DE LA ROSA-VELEZ, J., GONZALEZ-FARIAS, F., Eds), Universidad Autónoma de Baja California, Ensenada (1990) 21-56.

[14] GOBIERNO DEL ESTADO DE SINALOA, Estado de Sinaloa, Monografía 1990, Gobierno Estatal, Culiacán, Sinaloa (1990) 244 pp.

[15] GOBIERNO DEL ESTADO DE SINALOA, Tercer Informe de Gobierno 1995, Poder Ejecutivo Estatal, Culiacán, Sinaloa (1995) 141 pp.

[16] SECRETARIA DE MEDIO AMBIENTE, RECURSOS NATURALES Y PESCA, Informe Técnico, Delegación Sinaloa, Mazatlán, Sinaloa (1995) 26 pp.

[17] READMAN, J.W., et al., Persistent organophosphorus pesticides in tropical marine environments, Mar. Pollut. Bull. **24** (1992) 398-402.

[18] INSTITUTO NACIONAL DE ESTADISTICA, GEOGRAFIA E INFORMATICA, Anuario Estadístico de los Estados Unidos Mexicanos, 1988-1989, INEGI, Mexico City (1990) 838 pp.

[19] PAEZ, F., BOJORQUEZ, H., IZAGUIRRE, G., OSUNA, I., GONZALEZ-FARIAS, F., Carbono y fósforo en sedimentos de un sistema lagunar asociado a una cuenca de drenaje agrícola, An. Inst. Cienc. Mar. Limnol. **19** (1992) 1-11.

[20] WANG, T.C., HOFFMAN, M.E., Degradation of organophosphorus pesticides in coastal water, J. Assoc. Off. Anal. Chem. **74** (1991) 883-886.

[21] ISNARD, P., LAMBERT, S., Estimating bioconcentration factors from octanol-water partition coefficient and aqueous solubility, Chemosphere **17** (1988) 21-34.

[22] HERMSEN, W., SIMS, I., CRANE, M., The bioavailability and toxicity to *Mytilus edulis* L. of two organochlorine pesticides adsorbed to suspended solids, Mar. Environ. Res. **38** (1994) 61-69.

[23] GONZALEZ-FARIAS, F., et al., in preparation.

[24] CARVALHO, F.P., FOWLER, S.W., READMAN, J.W., MEE, L.D., "Pesticide residues in tropical coastal lagoons: The use of ^{14}C-labelled compounds to study cycling and fate of agrochemicals", Applications of Isotopes and Radiation in Conservation of the Environment (Proc. Symp. Karlsruhe, 1992), IAEA, Vienna (1992) 637-653.

IAEA-SM-343/28

FATE OF ENDOSULFAN IN SOIL AND IN RIVER AND COASTAL WATERS OF JAMAICA*

D.E. ROBINSON, A. MANSINGH, T.P. DASGUPTA
Pesticide and Pest Research Group,
The University of the West Indies,
Mona, Kingston, Jamaica

Abstract

FATE OF ENDOSULFAN IN SOIL AND IN RIVER AND COASTAL WATERS OF JAMAICA.

The runoff, leaching, volatilization and degradation of endosulfan were studied in a coffee plantation in sterilized and unsterilized soils, and in river and coastal waters from closed (harbour) and open seas. After 9 weeks and 27.5 mm of precipitation, the runoff of endosulfan was significantly ($P = 0.05$) higher in weeded than in unweeded plots, and at the of 23° and 38° slopes than at the 5° slope. The percentage losses of residues at the origin of the slopes in weeded and unweeded (data in parentheses) plots were 33.5 (27.8) at the 5° slope, 60.5 (53.9) at the 23° slope, and 59.9 (56.6) at the 38° slope. The percentage residues recovered at various points along the 6 m slopes were 0.42 (0.28) at the 5° slope, 2.4 (3.17) at the 23° slope, and 1.83 (2.98) at the 38° slope. Leaching was significantly ($P = 0.032$) higher at the 5° slope than at the 23° and 38° slopes, and in the unweeded plots than in the weeded plots. At the origin, the percentage leaching at depths of 10–15 cm at the 5°, 23° and 38° slopes in weeded and unweeded plots were 3.5 (5.9), 1.02 (2.0) and 0.86 (0.92), respectively. Volatilization in 12 hours was 17.4 ± 0.42%, 19.8 ± 0.79% and 22.4 ± 0.88% at the 5, 10 and 20% moisture levels, respectively. The hydrolytic rates (half-life in days) in distilled, river, open sea and closed sea waters were 27.5, 260.3, 303.2 and 104.9, respectively, for alpha-endosulfan, and 23.5, 547.5, 151.5 and 86.9, respectively, for beta-endosulfan. Photolysis in the soil was significantly ($P < 0.001$) lower in complete darkness, tree shade and direct sunlight; the half-life values were 215.8, 28.5 and 25.2 days, respectively, for alpha-endosulfan, and 181.3, 24.5 and 8.7 days, respectively, for beta-endosulfan. Microbial degradation in the unsterilized soil was not significantly ($P > 0.05$) different from that in the sterilized soil.

---

* Research carried out with the support of the IAEA under Research Contract No. 8053.

1. INTRODUCTION

Endosulfan has been used extensively in coffee plantations throughout Jamaica to control the coffee berry borer, *Hypothenemus hampei* Ferrari. In the Portland and Hope River watersheds of the Blue Mountains, a total of 2530 L of endosulfan active ingredients is applied annually on about 1260 ha of coffee [1, 2]. The residues eventually run off to rivers and to the sea, contaminating the sediment, water and aquatic fauna; in fact, the residues found in fish and shrimp often exceed their tolerance levels [1-5].

Current investigations were carried out to provide basic data on the fate of endosulfan with a view to assessing environmental risk and to developing strategies for residue management on an island ecosystem.

2. MATERIALS AND METHODS

Vertical leaching and runoff of endosulfan were studied in a coffee plantation in the Blue Mountains on Cuffy Gully gravelly sandy loam soil (64% sand, 22% silt, 14% clay, 4.45% organic matter and pH6.96) at slopes of about 5°, 23° and 38° in 2 × 6 m plots that were weeded or left with their natural vegetation (unweeded). A band of 0.3 × 2 m at the top of each of the three replicates of the experimental plots at the three slopes was sprayed with about 1000 ppm of endosulfan active ingredients (Thiodan EC35) with a Swissmex knapsack sprayer. After 1 hour and at 2, 4, 6 and 9 weeks after treatment, the soil samples taken from the top 5 cm of each treated band and from 1.5, 3, 4.5 and 6 m down the slopes were collected for residue analyses. Soil samples for the vertical leaching were collected at the origin and at each distance down the slopes from a depth of >5 to 15 cm.

Volatilization from the soil was studied under laboratory conditions in a closed system of flasks filled with the treated soil over which air was passed, and the residues were trapped in hexane. The soil was completely dried in an oven and nine 10 g samples were weighed. Each sample was mixed with 10 mL of 1% active ingredient Thiodan EC35 and the required volumes of water (where needed) to obtain 1, 10 and 20% moisture levels in the soil. After thorough mixing, each soil sample was placed at the bottom of a horizontal glass column and air passed over it at a rate of 9.26 mL/s from one end, bubbling it through the hexane in two Erlenmeyer flasks at the other end to trap the residues [6]. The solvent flasks were removed and replaced by new flasks every 2 hours for 12 hours for residue analyses.

Photolysis of the residues in soil was studied in a coffee plantation. Three glass dishes, each containing 100 g of sterilized soil at the 10% moisture level and 20 ppm of endosulfan were placed under the open sky, under a tree that provided constant shade, and inside a light proof box. On days 0.05, 8, 16, 32, 64 and 82 each dish was removed, the soil thoroughly mixed and a 10 g subsample withdrawn for residue determination. Similarly, microbial degradation in sterilized soils (autoclaved at 120°C for 8 hours) and in unsterilized plantation soils was studied in sterilized brown bottles plugged with cotton and held in a sterile chamber.

Hydrolysis of the endosulfan was studied in 500 mL of distilled, river (Rio Cobre), closed sea (Kingston Harbour) and open sea (Caribbean Sea at Buff Bay) waters. The water in the three replicates was mixed with Thiodan EC35 to provide a 10 ppm (active ingredient) concentration in each replicate, and held in open 1 L narrow mouthed brown bottles under laboratory conditions at 30°C. At regular intervals for 130 days, 25 mL aliquots were removed for residue analyses.

The endosulfan residues were extracted from the water samples in hexane and dried in an anhydrous sodium sulphate column. The soil samples were extracted by Soxhlet for 8 hours, filtered and cleaned up by passing them through a Florisil column [5]. A Hewlett-Packard HP 5890 Series II gas chromatograph equipped with a 30 m HP-1 capillary column and a ^{63}Ni electron capture detector was used to separate and detect the residues. The analytical conditions were: carrier gas, nitrogen at a flow rate of 10–15 mL/min; temperature settings, column: at an initial temperature of 100°C, with a rate increase of 20°C/min until 250°C; injector, 280°C; and detector, 300°C. Endosulfan recovery from the soil and the water was 88.9 and 91.7%, respectively. The limit of detection of 0.001 ng was twice the standard deviation of the mean value of the blanks.

3. RESULTS AND DISCUSSION

The weekly and total 6 week losses of endosulfan residue from the treated plots in the weeded slopes were steady, and always significantly ($P = 0.01$) lower at the 5° slope than at the 38° slope (Fig. 1(a)). The losses at the 23° slope did not follow a regular pattern, since they were not significantly ($P > 0.05$) different from those at the 5° slope for the first 2 weeks, and from those at the 38° slope for weeks 3–9. A similar pattern of weekly losses was recorded in the unweeded plots (Fig. 1(b)). The weekly losses of residue from the weeded and unweeded plots (Figs 1(a) and (b)) at each slope were always significantly ($P = 0.05$) different from each other. The overall 6 week losses of 33.5 and 60.5% from bands in the weeded slopes at the 5° and 23° slopes, respectively, were significantly ($P = 0.05$) higher than the 27.8 and 53.8% losses from the unweeded plots at the two respective slopes. Losses from the treated bands of the weeded and unweeded plots at the 38° slope were 58.9 and 56.6%, respectively.

Runoff of the residues down the 6 m slopes was gradual, depending on the slope and the rainfall. At the 5° slope, 0.15% of the residues moved only up to the 1 m level in 9 weeks, regardless of the rainfall. At the other slopes, residues in the weeded and unweeded plots had reached 3 m down from the treatment band within 1 week and 4.5 to 6 m within 4 weeks. Transport of the residues down the slopes

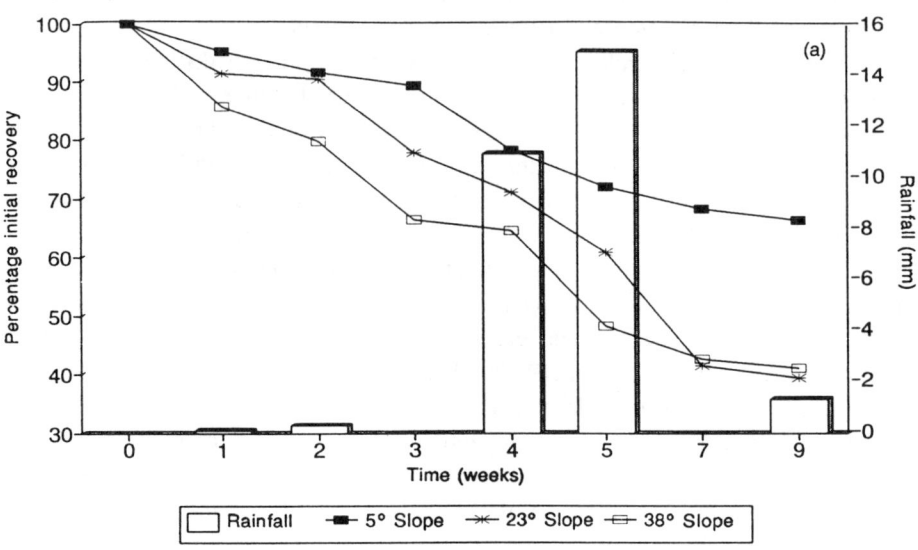
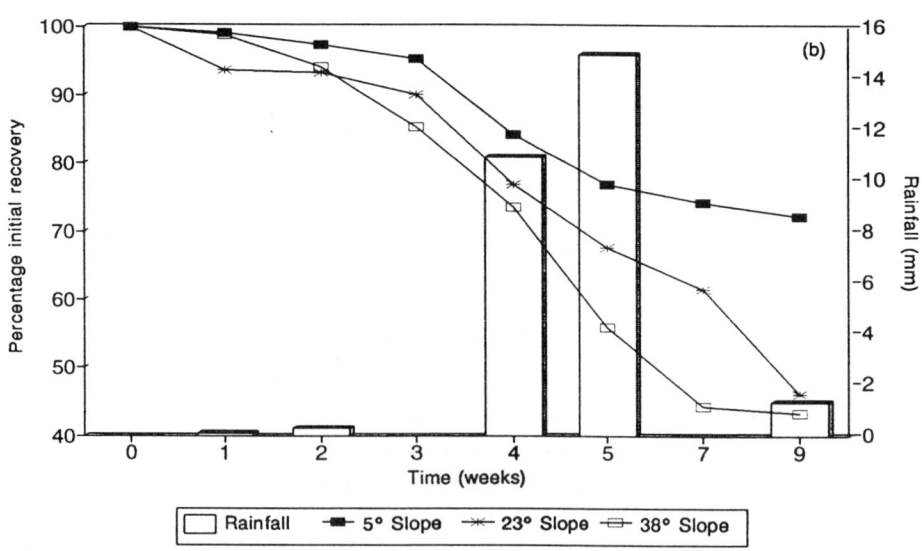

FIG. 1. Loss of endosulfan residues from treated bands in (a) the weeded plots, and (b) the unweeded plots at the 5°, 23° and 38° slopes.

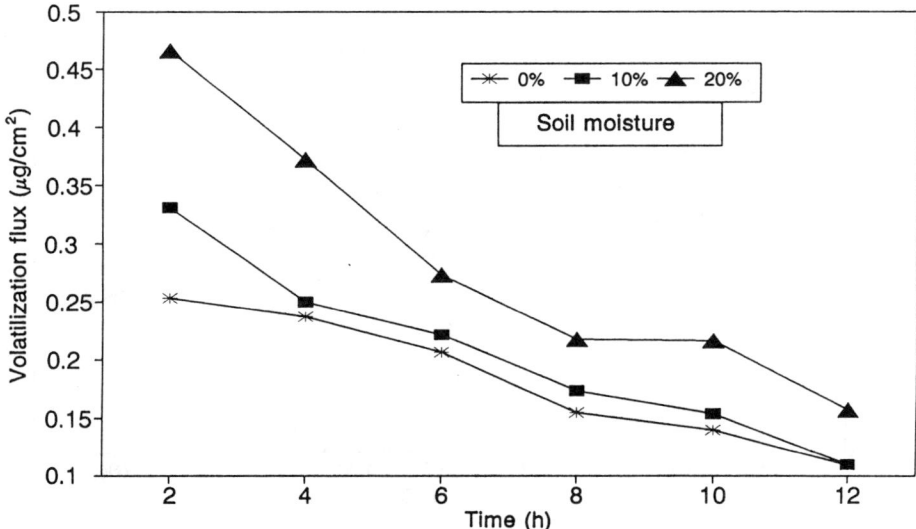

FIG. 2. *Volatilization of endosulfan from soil of a Blue Mountain plantation, Jamaica, at different moisture levels.*

was aided by light rainfall during the first 2 weeks, and 10.9 mm rainfall in the fourth week. The total residues recovered after 9 weeks from the 6 m stretch in the weeded plots were 2.40% at the 23° slope and 1.93% at the 38° slope; these values were significantly ($P = 0.05$) lower than the 3.15 and 2.98% recovered from the unweeded plots at the two respective slopes (Figs 1(a) and (b)).

Runoff of the residues beyond the 6 m mark was not determined, but must have been significant. The current data are in agreement with those of endosulfan (0.35%) in water from a loamy field at the 7° slope [7], and of aldrin (5.2%), DDT (1.6%), endrin (0.9%) and dieldrin (0.7%) from fields at the 2°, 7°, 7° and 12° slopes, respectively [7].

Since endosulfan has a low solubility in water, its runoff is usually through the erosion of residue adsorbed soil particles, which occurs more often in weeded slopes than in unweeded slopes, particularly during rainfall. Vegetative cover could reduce the runoff by checking the soil erosion [8], redistributing the residues within the top soil [9], and through uptake by plants [10].

Leaching was significantly ($P = 0.032$) higher at the 5° slope than at the 23° and 38° slopes, and in the unweeded than in the weeded plots (except at 38°); recovery of the residues at a depth of >5–15 cm was 3.5, 1.02 and 0.86% of the applied insecticide in the weeded plots, and 5.9, 2.0 and 0.92% in the unweeded plots at the

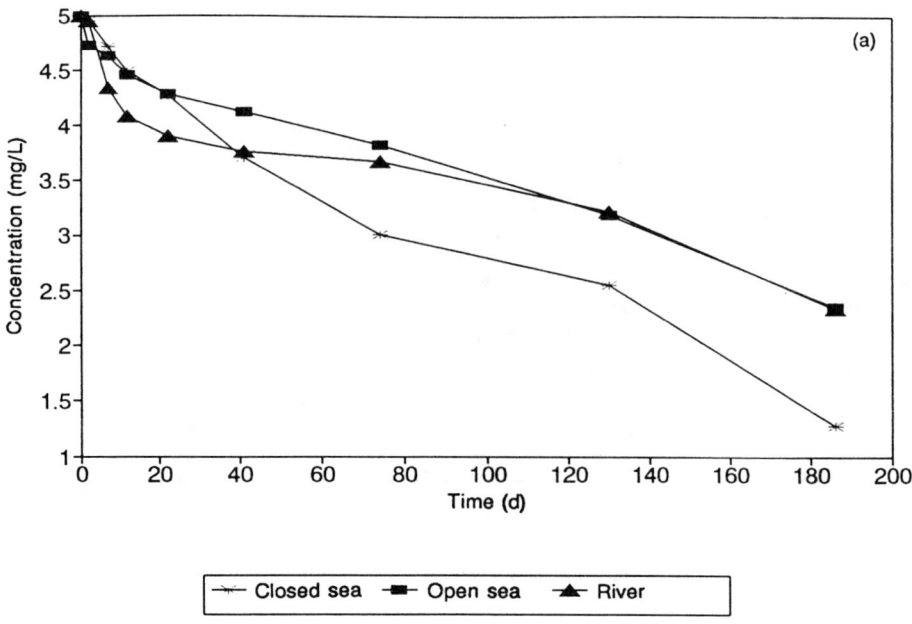
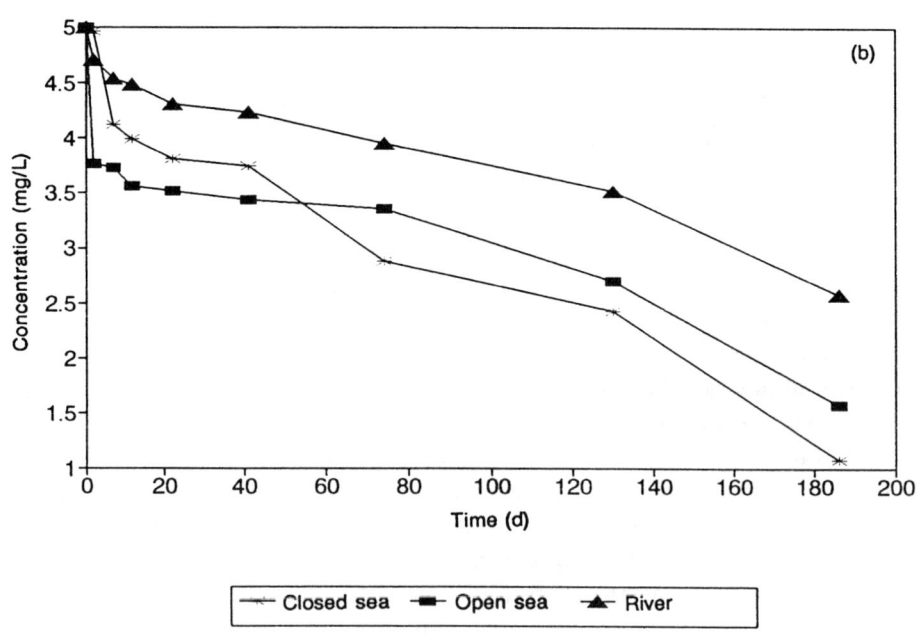

FIG. 3. Persistence of (a) alpha-endosulfan, and (b) beta-endosulfan in water from closed sea, open sea and river waters in Jamaica.

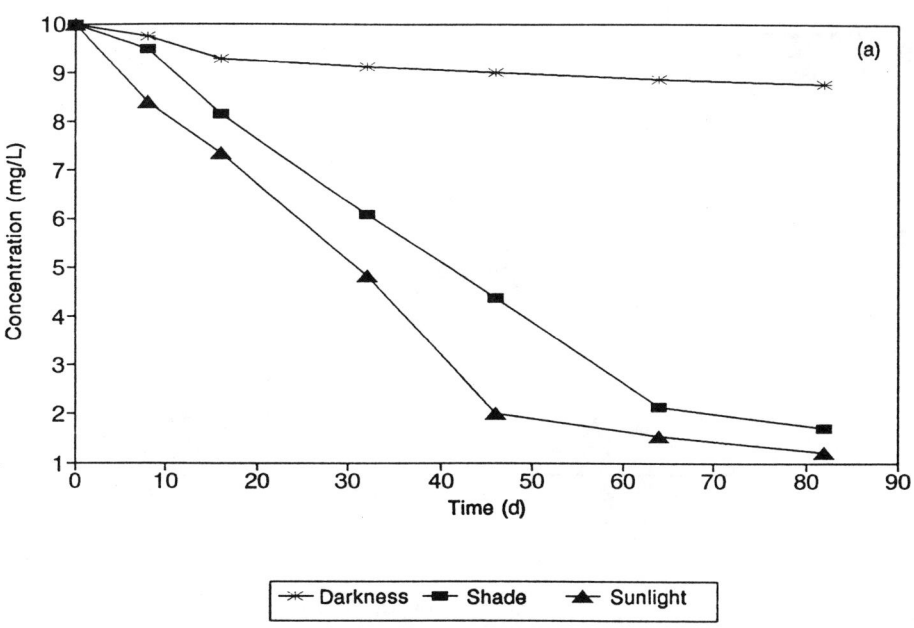

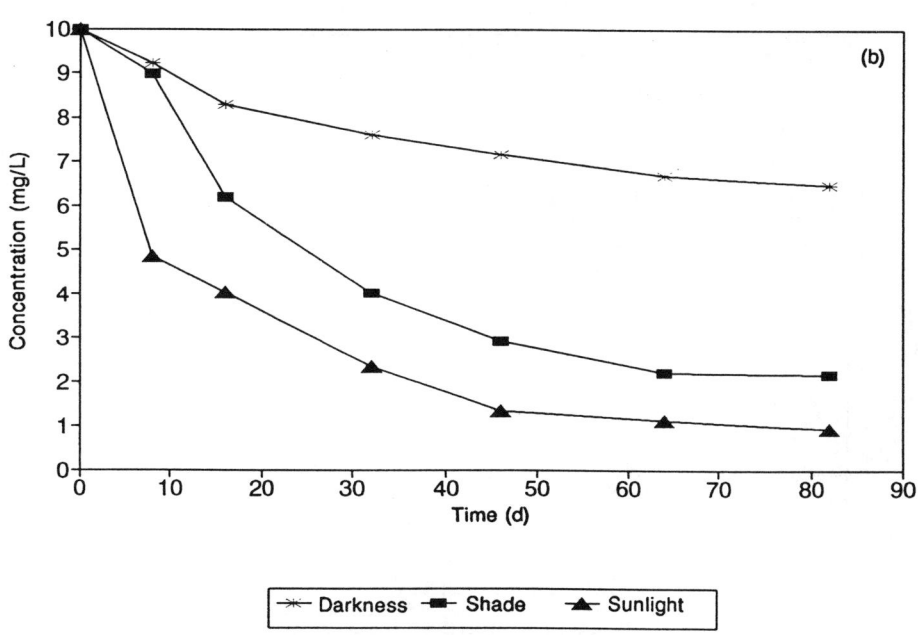

FIG. 4. Persistence of (a) alpha-endosulfan, and (b) beta-endosulfan in soil in total darkness, tree shade and direct sunlight.

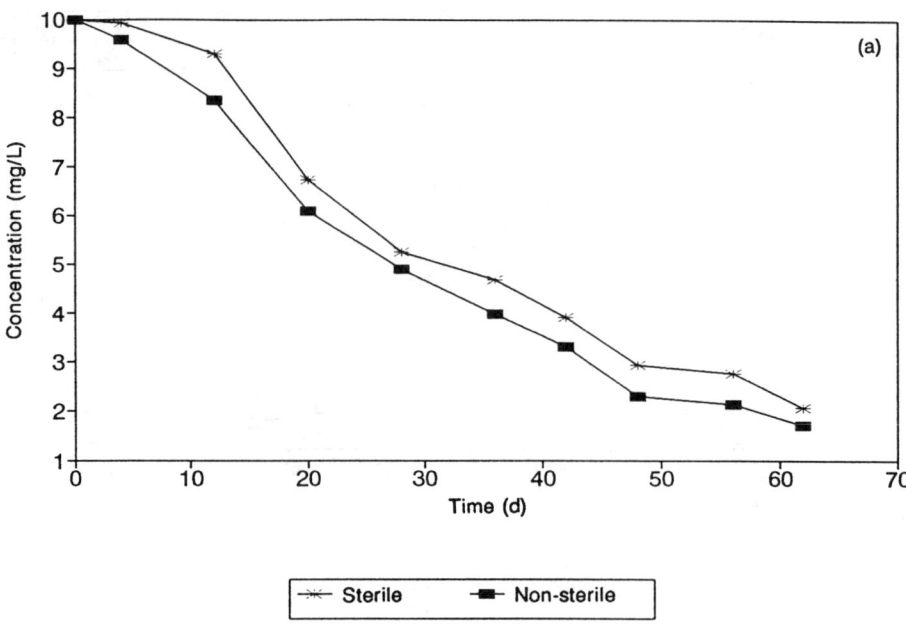

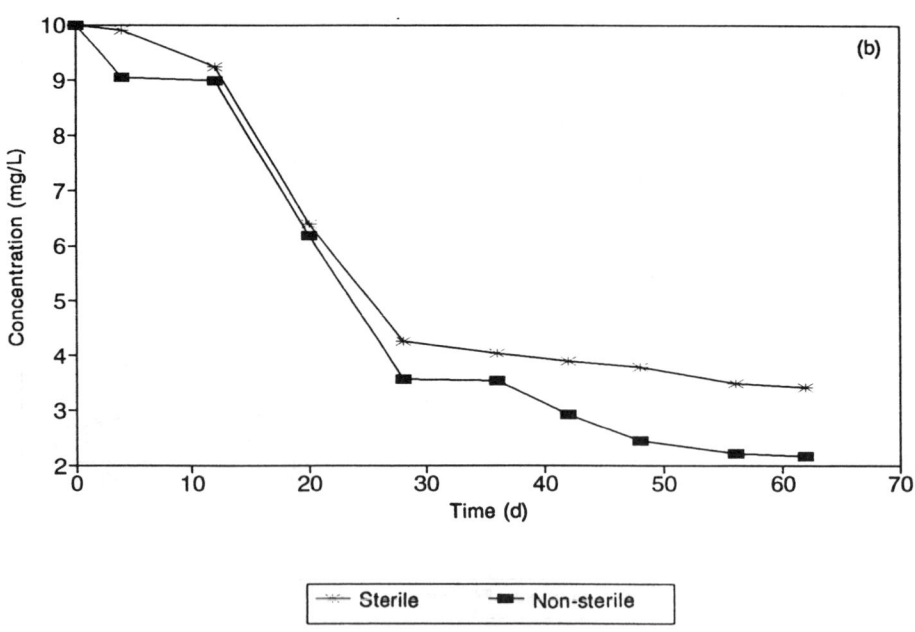

FIG. 5. Persistence of (a) alpha-endosulfan, and (b) beta-endosulfan in non-sterile and sterile soils.

5°, 23° and 38° slopes, respectively. Vegetation also increases percolation of water into the soil [10], although intermittent wet and dry periods determine the downward and upward movement of residues in the soil, particularly in fields without vegetative cover [11].

The volatilization flux of endosulfan was 0.26, 0.34 and 0.47 $\mu g/cm^2$ during the first 2 hours, but declined after 12 hours to 0.12, 0.12 and 0.17 $\mu g/cm^2$ at the 0, 10 and 20% moisture levels, respectively (Fig. 2). The total 12 hour dissipation of the residues from the soil was significantly ($P = 0.025$) different at the 0, 10 and 20% moisture levels, the data being 17.4 ± 0.42%, 19.8 ± 0.79% and 22.4 ± 0.88%, respectively. Our data followed Henry's law constant [12] and were in agreement with Singh et al. [6], whose half-life values were 3.92 days for alpha-endosulfan and 2.06 days for beta-endosulfan on a glass surface.

Persistence of alpha-endosulfan and beta-endosulfan was generally lower in closed sea water, followed by river and open sea waters (Figs 3(a) and (b)). The hydrolytic rates for alpha-endosulfan were significantly ($P = 0.01$) faster in distilled water (half-life of 27.5 days) than in closed sea (half-life of 104.9 days), river (half-life of 260.3 days) and open sea (half-life of 303.2 days) waters. The half-life values for beta-endosulfan (23.5, 86.9 and 151.5 days in distilled, closed sea and open sea waters, respectively) were significantly ($P = 0.01$) lower than the alpha isomer, while the half-life in river water (547.5 days) was significantly higher ($P = 0.001$). Degradation of endosulfan in aqueous medium was faster in alkaline medium than at pH7, or less [6]. The fast hydrolytic rate in distilled water can be attributed to the lack of any organic matter to adsorb the residues [6]. In aquatic ecosystems, the hydrolysis rate is affected by complex adsorption and degradation processes because of organic impurities and salts [13]. The closed sea water of Kingston Harbour may have endosulfan degrading microorganisms that could be used in residue management.

The photolysis of endosulfan in soil was significantly ($P < 0.001$) affected by the quantity and perhaps the quality of light; the half-life values were 215.8, 28.5 and 25.2 days for alpha-endosulfan and 181.3, 24.5 and 8.7 days for beta-endosulfan in total darkness, tree shade and direct sunlight, respectively (Fig. 4). These values reflect soil degradation and photolysis. Photolysis of alpha-endosulfan and beta-endosulfan in hexane under direct sunlight (half-life values of 40.3 and 32.6 days, respectively) [6] was higher than in the soil (Fig. 4).

Microbial degradation of endosulfan in the non-sterile soil with only 4.5% organic matter was not significantly ($P > 0.05$) different from that of the sterile soil (Fig. 5); the half-life values of alpha-endosulfan and beta-endosulfan in the sterile soil were 32.2 and 33.0 days, and in the non-sterile soil, 28.5 and 24.5 days, respectively.

4. CONCLUSIONS

The 33–60% losses in endosulfan from Jamaican soils at different slopes can be attributed primarily to volatilization, photolysis and hydrolysis, although in steep slopes without vegetative cover runoff was the major route for residue transport, even when the rainfall was moderate. Soil cover could be an essential tool in residue management in Jamaica, where slopes of up to 75° are under cultivation.

ACKNOWLEDGEMENT

The authors wish to thank the IAEA for financial support.

REFERENCES

[1] MANSINGH, A., ROBINSON, D.E., Baseline Studies on Pesticide Contamination of Portland Rivers and Formulation of Control Measures — 1991–1992, United Nations Educational, Scientific and Cultural Organization, Paris (1994).

[2] MANSINGH, A., ROBINSON, D.E., WILSON, A., Pesticide contamination of Jamaican environment. II. Patterns of fluctuations in residue levels in the rivers of Hope Watershed in 1989–1991, Jamaica J. Sci. Technol. (in press).

[3] ROBINSON, D.E., MANSINGH, A., Insecticide contamination of Jamaican environment. IV. Transport of residues from coffee plantations in the Blue Mountains to coastal waters in eastern Jamaica, Environ. Toxicol. Chem. (in press).

[4] MANSINGH, A., ROBERTS, E.V., HENRY, C., LAWRENCE, V., ROBINSON, D.E., Pesticide contamination of Jamaican environment. I. Organochlorine residues in the rivers and shrimps of Rio Cobre Basin in 1982–1983, Jamaica J. Sci. Technol. (in press).

[5] MANSINGH, A., WILSON, A., Insecticide contamination of Jamaican environment. III. Baseline studies on the status of insecticide pollution of Kingston Harbour, Marine Pollut. Bull. **30** 10 (1995) 640–645.

[6] SINGH, N.C., DASGUPTA, T.P., ROBERTS, E.V., MANSINGH, A., Dynamics of pesticides in tropical conditions. I. Kinetic studies of volatilisation, hydrolysis, and photolysis of dieldrin and α- and β-endosulfan, J. Agric. Food Chem. **39** 3 (1991) 575–579.

[7] PIONKE, H.B., CHESTERS, G., Pesticide–sediment–water interactions, J. Environ. Qual. **2** (1973) 29–45.

[8] MORGAN, R.P.C., Topics in Applied Geography: Soil Erosion, Longmans, New York (1979) 113 pp.

[9] Pesticide Microbiology (HILL, I.R., WRIGHT, S.J.L., Eds), Academic Press, London (1978) 844 pp.

[10] BEALL, M.L., NASH, R.G., Insecticide depth in soil — Effect on soyabean uptake in greenhouse, J. Environ. Qual. **1** (1972) 283–288.

[11] NICHOLLS, P.H., Factors influencing entry of pesticides into soil water, Pestic. Sci. **22** (1988) 123–137.

[12] SPENCER, W.F., CLIATH, M.M., JURY, W.A., ZHANG, L.Z., Volatilisation of organic chemicals from soil as related to their Henry's Law Constants, J. Environ. Qual. **17** (1988) 504–509.

[13] HAQUE, R., KEARNEY, P.C., FREED, V.H., "Dynamics of pesticides in aquatic environments", Pesticides in Aquatic Environments, Plenum, New York (1977) 39–52.

IAEA-SM-343/40

DISTRIBUTION AND FATE OF ^{14}C-DDT IN THE ESTUARINE ENVIRONMENT OF THE NORTH OF VIET NAM*

Duc Nhan DANG, Van Thuan VO
Institute for Nuclear Science and Technique,
Viet Nam Atomic Energy Commission

Manh Am NGUYEN
The Directorate for Quality and Standardization
  of Viet Nam

Hanoi, Viet Nam

Abstract

DISTRIBUTION AND FATE OF ^{14}C-DDT IN THE ESTUARINE ENVIRONMENT OF THE NORTH OF VIET NAM.

The distribution and fate of DDT in various environmental compartments have been studied using ^{14}C techniques in a microcosm simulating the marine environment conditions of the north of Viet Nam. It was shown that up to 95–96% of the initial amount of ^{14}C-DDT was deposited on sediment 72 hours after application. After 1 month, the bound form of the insecticide in sediment did not exceed 0.1–0.2%. All the ^{14}C-DDT was deposited at the 0–5 cm layer beneath the sediment surface. Accumulation of ^{14}C-DDT in algae raised in the microcosm was about 2% of the total amount 1–2 days after application. Accumulation of ^{14}C-DDT in shrimp muscle was 1.2–1.5% of the initial activity 2–3 days after application. One week after application, 80 and 12% of the ^{14}C-DDT recovered in sediment were found in DDT and its DDD metabolite, respectively. At the same time, not only DDT, but also its DDD and DDE metabolites were observed in shrimp muscle. However, the DDD and DDE contents were less than 20% compared with the total activity.

1. INTRODUCTION

DDT was extensively used in the north of Viet Nam during the 1960s to control vectors of malaria, but since the early 1970s it has been banned in the country. DDT is known to be a persistent, fat soluble substance that accumulates in body fat and in the food chain. High concentrations have been found in carnivores, including humans. One reported adverse effect of DDT is a reduction in bird reproduction.

---

* Research carried out with the support of the IAEA under Research Contract No. 7940.

According to the results obtained by Talekar et al. [1], DDT appears to accumulate slightly in the subtropical soils of Taiwan (China). A study conducted by these authors using a dose of 5 kg active ingredients/ha of DDT twice yearly showed that after 4 years of continuous application only 6.6% of the total amount of chemical applied (80 kg) was recovered. Of the metabolites of DDT, only DDE was detected. In another study [2], the average residual levels of DDT and DDE were determined at soil depths of 12.5, 25 and 37.5 cm at different geographical locations in Taiwan (China). The results were as follows: DDT, 47.9, 10.9 and 0.6 ppb; and DDE, 20.3, 7.4 and 2.6 ppb, respectively. On the other hand, Chen and Zhong from China [3] found that almost all the DDT applied to an experimental microcosm was detected in the 0–5 cm top surface layer of sediment. This indicates wide distribution of DDT and its metabolites in the environment through drainage or as a result of erosion. Therefore, agrochemical residue pollution has become a serious problem in coastal zones, especially in tropical regions, where the rainfall is usually high.

The Red River is the largest in the north of Viet Nam. Annually, it discharges sediment amounting to about 114×10^6 t into the sea. At the same time, along the coastal areas of its estuary farmers use the lagoons to raise marine foods such as mussels and shrimps for domestic use and for export. These organisms are known to be sensitive to agrochemical residues, especially organochlorine (OC) pesticides. Unfortunately, to date no studies on the distribution and fate of DDT in the marine environment have yet been conducted in Viet Nam.

This investigation aimed at studying the distribution of DDT in various environmental compartments of the Balat estuary (106°30'E, 20°22'N), namely, in the water, sediment, fauna and flora. It also aimed at evaluating the possibility of DDT accumulating in the fauna and flora. Work was carried out on a microcosm simulating the conditions of the estuary using radioisotope (^{14}C) labelled DDT as the tracer to follow the chemical distribution in various environmental compartments.

2. MATERIALS AND METHODS

Carbon-14 DDT (1,1,1-trichloro-2,2-bis(p-chloro(ring-U-^{14}C)phenyl)ethane, with a specific activity of 4.144 GBq/μmol (112 mCi/μmol), was obtained from the IAEA Agricultural and Biotechnology Laboratory, Seibersdorf, and the non-labelled DDT, DDD and DDE standards from the IAEA Marine Environment Laboratory, Monaco. The purity of all the chemicals was more than 95%, as checked by thin layer chromoatography (TLC).

Pesticide grade n-hexane (Prolabo, France) was used to extract the insecticide residues from the water and sediment samples. To determine the bound forms of the insecticide in the sediment and biological samples, a tissue solubilizer (soluene 350, Packard) was used.

The experiment was conducted in a glass microcosm with a wooden frame (60 × 60 × 30 cm). The microcosm consisted of 70 L of brackish water (8‰ salinity and a pH of 7.0–7.2), 20 kg of wet sediment and 0.5 kg of green algae. All the water, sediment and flora were taken directly from the estuary. The composition of the sediment was as follows: silt, 80%; clay, 3–5%; sand, 15–17%; the total organic matter (TOM) was determined as 10–12%. To determine the accumulation and fate of the insecticide, 20 shrimps, *Metapenaus ensis*, with a total weight of about 100 g, were first acclimatized in the laboratory for 1 week before being raised in the microcosm. During this experiment, the microcosm was aerated with a fish pump and the shrimps were fed with a formulation used by farmers in the lagoons from which the shrimps were collected for the experiment.

A treating solution of 24.6 GBq of ^{14}C-DDT was prepared in 200 mL of brackish water, and then poured as evenly as possible into the microcosm, giving a nominal concentration of 30 ppb.

The radioanalysis method used to determine the distribution of DDT in various compartments in the microcosm was the same as that followed by Carvalho et al. [4]. The metabolite products of DDT in the fauna and flora samples were identified and quantified using TLC (the silica gel plates were a gift from the IAEA-MEL). The developing solvent was a mixture of hexane/acetone/methanol/diethylamine (100:1:1:1). The analytical procedure followed was the same as that described by Kale et al. [5].

3. RESULTS AND DISCUSSION

Table I gives the dynamics of ^{14}C-DDT distribution in water and at the 0–5 cm layer beneath the sediment surface. All the results presented in the tables are mean arithmetic values, with the standard deviations obtained from three replicates; these were converted to the percentage initial activity applied to the microcosm. Table II shows the distribution of ^{14}C-DDT at various depths of sediment in the microcosm 7 days after application. Table III presents the distribution of the ^{14}C-DDT metabolites in sediment. Table IV shows the accumulation of ^{14}C-DDT residues in green algae and shrimp muscle taken from the microcosm treated with the insecticide. Table V presents the distribution of ^{14}C-DDT metabolites in shrimp muscle.

Because of its low solubility in water (0.002 mg/L), ^{14}C-DDT rapidly deposited on to sediment, as seen in Table I. The ^{14}C-DDT deposited on sediment was mainly concentrated at the 0–5 cm layer beneath the sediment surface (Table II). This shows that DDT and its metabolites can be widely distributed in the environment as a result of erosion, especially in the tropics, where the rainfall is usually high. In fact, at present DDT, DDD and DDE can be found everywhere in Viet Nam, but fortunately their concentrations are rather low [6]. The bound form of ^{14}C-DDT in sediment was not so high: 0.1–0.2% of the initial activity 2 days after application (Table I). The magnitude of the bound form is thought to correlate either with the TOM or the mechanical composition (MC) of the sediment. Using ^{14}C techniques, a study is currently being carried out in our laboratory on the behaviour of DDT in soils with various TOM contents and MCs taken from different locations in Viet Nam.

As can be seen in Table III, in this short term experiment only DDD was detected in sediment. However, in previous work [6], no DDD was found in sediment collected from the same estuary. This contamination may be evidence of DDT application a long time ago.

Tables IV and V demonstrate the slow excretion rate of ^{14}C-DDT from flora and fauna. After 1 month, more than half of the ^{14}C-DDT applied still remained in shrimp muscle. This proves the high sensitivity of shrimps towards OC pesticides. In fauna muscle, the metabolite forms of DDT (DDD and DDE) were detected.

TABLE I. RESIDUES OF ^{14}C-DDT RECOVERED IN WATER AND AT THE 0–5 cm LAYER BENEATH THE SURFACE SEDIMENT[a]

Time after application	^{14}C-DDT activity in water (% of applied)	^{14}C-DDT activity in sediment (% of applied)		
		Extractable	Bound	Total
1 h	61.0 ± 3.0	6.0 ± 2.0	NF	6.0 ± 2.0
2 h	12.0 ± 2.0	22.0 ± 3.0	NF	22.0 ± 3.0
4 h	4.5 ± 0.5	44.0 ± 3.0	NF	44.0 ± 3.0
7 h	3.5 ± 0.7	60.0 ± 3.0	NF	60.0 ± 3.0
1 d	0.2 ± 0.5	82.0 ± 5.0	NF	82.0 ± 5.0
2 d	1.5 ± 0.3	92.0 ± 3.0	0.10 ± 0.10	92.0 ± 3.0
3 d	1.1 ± 0.3	96.0 ± 1.0	0.15 ± 0.10	96.0 ± 1.0
7 d	NA	98.0 ± 1.0	0.12 ± 0.10	98.0 ± 1.0

[a] NA = not analysed; NF = not found.

TABLE II. VERTICAL DISTRIBUTION OF ^{14}C-DDT AT VARIOUS DEPTHS OF SEDIMENT LAYER (7 DAYS AFTER APPLICATION)

Depth from surface (cm)	0–2	2–5	5–7
^{14}C-DDT activity (% of applied)	85.0 ± 5.0	12.5 ± 3.0	1.7 ± 0.5

TABLE III. DISTRIBUTION OF ^{14}C-DDT METABOLITES IN SEDIMENT[a]

Time after application	^{14}C-DDT residues on TLC plates (% of activity recovered)	
	DDT	DDD
2 h	83 ± 6	9 ± 3
4 h	80 ± 2	12 ± 3
7 h	85 ± 5	7 ± 3
24 h	83 ± 2	8 ± 4
2 d	80 ± 3	12 ± 3
3 d	81 ± 3	12 ± 3
7 d	80 ± 6	12 ± 5

[a] No DDE was found.

TABLE IV. ACCUMULATION OF ^{14}C-DDT RESIDUES IN GREEN ALGAE AND SHRIMP MUSCLE

Time after application	^{14}C-DDT residues in biological samples (% of activity applied)			
	Algae (total)	Shrimp muscle		
		Extractable	Bound	Total
2 h	0.26 ± 0.05	0.38 ± 0.05	0.07 ± 0.03	0.45
4 h	0.24 ± 0.02	0.30 ± 0.05	0.12 ± 0.02	0.42
7 h	0.73 ± 0.06	0.50 ± 0.07	0.10 ± 0.01	0.60
16 h	1.23 ± 0.11	0.68 ± 0.03	0.14 ± 0.04	0.80
24 h	1.70 ± 0.20	0.80 ± 0.05	0.15 ± 0.03	0.95
2 d	1.30 ± 0.05	0.83 ± 0.02	0.32 ± 0.02	1.25
3 d	0.95 ± 0.01	0.76 ± 0.03	0.67 ± 0.02	1.43
7 d	0.73 ± 0.04	0.57 ± 0.05	0.55 ± 0.03	1.12
1 month	0.32 ± 0.07	0.50 ± 0.03	0.33 ± 0.01	0.83

TABLE V. DISTRIBUTION OF ^{14}C-DDT METABOLITES IN SHRIMP MUSCLE[a]

Time after application	^{14}C-DDT residues on TLC plates (% of activity recovered)		
	DDT	DDD	DDE
2 h	93 ± 2	7 ± 2	NF
4 h	88 ± 3	12 ± 3	NF
7 h	84 ± 1	11 ± 2	5 ± 1
24 h	81 ± 2	12 ± 1	7 ± 1
2 d	82 ± 3	10 ± 1	8 ± 2
3 d	81 ± 2	6 ± 2	13 ± 2
7 d	80 ± 3	8 ± 1	12 ± 2
1 month	82 ± 1	6 ± 2	12 ± 3

[a] NF = not found.

4. CONCLUSIONS

Some conclusions can be drawn:

(1) In the environment, ^{14}C-DDT is rapidly deposited on to sediment and its residues are concentrated at the 0–5 cm layer beneath the sediment surface;
(2) Like benthic fauna, shrimps appear to accumulate up to 2% of the initial ^{14}C-DDT activity (up to 10 ppb) 2–3 days after application, followed by slow excretion of the insecticide from fauna muscle;
(3) Of the metabolite products of DDT, only DDD was detected in sediment, whereas both DDD and DDE were found in shrimp muscle.

It appears that the ^{14}C technique using liquid scintillation counting is the best way of studying the distribution of pesticide residues in different environmental compartments. The technique is easier to use than chromatography because only the radioactivity has to be measured, so avoiding many clean-up procedures that could necessitate analysis of loss of contaminants. Because this technique is highly sensitive, the results obtained are of high precision. The ^{14}C technique is the only way of determining the total carbon balance and enables measurement of the bound ^{14}C residues.

One of the limitations of this study was that no data were available on the volatilization factor of the insecticide under the experimental conditions. In the tropics and subtropics, volatilization plays an important role in the loss of pesticides [1], therefore it is important that such data be collected in the future.

ACKNOWLEDGEMENTS

This study was carried out with the financial support of the IAEA. The authors wish to thank M. Hussain of the Agricultural and Biotechnology Laboratory, Seibersdorf, and F.P. Carvalho of the IAEA Marine Environment Laboratory, Monaco, for kindly providing the ^{14}C-DDT and TLC plates, as well as the DDT, DDD and DDE standards.

REFERENCES

[1] TALEKAR, N.S., CHEN, J.S., KAO, H.T., Long term persistence of selected insecticides in subtropical soil: Their absorption by crop plants, J. Econ. Entomol. **76** (1983) 207–214.

[2] LI, G.C., Residues and toxicity problems associated with pesticide use in Taiwan, J. Trop. Agric. Res. Ser. **16** (1983) 165–177.

[3] CHEN, Shunhua, ZHONG, Chuangguang, "A preliminary study on the accumulation and excretion of ^{14}C-DDT and ^{14}C-fenvalerate in marine biota", The Distribution, Fate and Effect of Pesticides on Biota in the Tropical Marine Environment: Use of Radiotracers, Paper presented at 2nd IAEA/IAEA–MEL/SIDA Research Co-ordination Meeting, Kuala Lumpur, 1995.

[4] CARVALHO, F.P., FOWLER, S.W., READMAN, J.W., MEE, L.D., "Pesticide residues in tropical lagoons: Use of ^{14}C labelled compounds to study the cycling and fate of agrochemicals", Applications of Isotopes and Radiation in Conservation of the Environment (Proc. Symp. Karlsruhe, 1992), IAEA, Vienna (1992) 637–653.

[5] KALE, S.P., RAGHU, K., MURTHY, N.B.K., "Organochlorine pesticides in the Indian marine environment", The Distribution, Fate and Effect of Pesticides on Biota in the Tropical Marine Environment: Use of Radiotracers, Paper presented at 2nd IAEA/IAEA–MEL/SIDA Research Co-ordination Meeting, Kuala Lumpur, 1995.

[6] DANG, Duc Nhan, et al., "Evaluation of the level of organochlorinated pesticide contamination in the environment of the Red River and its Balat estuary", ibid.

EVOLUTION OF DDT RESIDUES
IN ALL SAINTS BAY, BRAZIL, 1985-1994*

T.M. TAVARES, M. BERETTA, M.A. COSTA
Instituto de Química,
Campus Universitário da Federação,
Universidade Federal da Bahia,
Salvador, Bahia, Brazil

Abstract

EVOLUTION OF DDT RESIDUES IN ALL SAINTS BAY, BRAZIL, 1985-1994.
  Use of organochlorine insecticides has been prohibited in Brazil since 1976. However, later use of existing stock and smuggled parcels are partially responsible for their incomplete elimination. Extended, official use of DDT took place over the entire country at the end of 1984 to eradicate malaria. In 1985, a cross-sectional survey was conducted for DDE at nine intertidal sites of All Saints Bay using local mussels. In 1994, the study was repeated in 19 sites utilizing four selected species of mussel and the same analytical methodology as previously applied. The DDT residues (as ng/g of DDE, wet weight) varied between 0.24 and 5.1 for *Anomalocardia brasiliana*, 2.1 and 44.0 for *Brachidontes exustus*, <0.33 and 15.8 for *Macoma constricta*, and <0.31 and 1.36 for *Littorina* spp. No samples were found that exceeded the various action limits. The highest concentrations were detected on the coastline of Salvador City, in the Aratu Industrial Centre area, in the Subaé estuary and in the Itaparica channel. From 1985 to 1994, increases were observed at all the sites, except for Mataripe, where an oil refinery operates on a large restricted area. Increases of up to 15 fold were observed in the Salvador and Aratu areas. Prohibited use of DDT, together with the carry over of the DDT used in the malaria campaign, seem to be the reasons for these increases.

1. INTRODUCTION

Use of organochlorine insecticides has been prohibited in Brazil since 1976. However, later use of existing stock and smuggled parcels are partially responsible for their incomplete elimination. Extended, official use of DDT took place over the entire country at the end of 1984 to eradicate malaria. Depending on government permission, restricted use includes specific cultures in certain areas, stocking of food for internal consumption and as a domestic insecticide under some circumstances.

---

* Research carried out with the support of the IAEA under Research Contract No. 7931.

All Saints Bay, the largest along the Brazilian coast (1100 km^2), is situated at latitude 13°S and longitude 38°W; it has a large diversity of habitats, and along its coastline different, conflicting activities take place, such as urban settlements (Salvador, the capital city, with 2 million inhabitants), industries (over 300) and farming (land and sea).

In 1985, a cross-sectional survey of DDT (as DDE) was conducted using ten species of local mussel as sentinel organisms at nine intertidal sites around the bay [1]. The levels at that time were low, not exceeding 5.1 ng/g wet weight.

To verify the evolution of DDT residues in the bay, a new cross-sectional survey was undertaken in 1994 at 19 sites, utilizing four selected species of mussel (with different eating habits) and the same analytical protocol as previously applied. To identify recent inputs of DDT into the bay, selected mussel samples, as well as surface intertidal sediments, were analysed.

2. METHODOLOGY

Samples were collected during low tide between January and February 1994. Individuals of medium size were selected for each bivalve species in the field. The samples were obtained from portions of intertidal sediments randomly obtained from a grating that lies within 25 m of the station centre [2]. Detergent washed (alkaline Extran Merck) and water, acetone and dichloromethane rinsed sampling metal tools were used. The samples were wrapped in clean solvent washed aluminium foil, placed in styrofoam boxes with ice, taken to the laboratory on the same day and kept at −18°C until pretreatment. All collected individuals of the same bivalve species were pooled and homogenized to obtain the mean values of the pollutant concentrations [3]. Subsamples of the tissue homogenates (9–10 g wet weight) were treated as described by Albaigés et al. [4]: saponification of the homogenates (6N aqueous NaOH), extraction (ethylether) of the total organics, and further clean-up on 5% water deactivated silica–alumina [5]; DDT in the DDE form was eluted in the second fraction. Separate subsamples, without saponification, were extracted with three portions of dichloromethane for 5 min using ultrasound, followed by the same clean-up to determine the DDT and DDE in the second fraction.

Subsamples of about 8 g of freeze dried sediments in thimbles (previously cleaned up by 24 hours of extraction) were Soxhlet extracted for 24 hours with 180 mL of dichloromethane and methanol (2:1) containing activated Cu wire, followed by clean-up on an activated Cu powder (1 cm) silica–alumina column [6, 7]. The fractions containing DDT residues in all the samples were diluted to 1.0 mL with iso-octane.

The pp'DDE and pp'DDT determinations were accomplished by injecting 1 μL into a GC/^{63}Ni electron capture detector (Varian Star 3400) equipped with an autosampler (Varian Star 8200 cx) using He as the carrier gas and the following tem-

TABLE I. QUALITY CONTROL OF pp'DDE AND pp'DDT DETERMINATION IN A BIOLOGICAL MATRIX AT TWO CONCENTRATION LEVELS (ng/g DRY WEIGHT)

Organochlorine insecticide	Value	Tuna homogenate (IAEA 351)	Mussel homogenate (IAEA 142)
pp'DDE	Certified value	140	9.00 ± 2.9
	Range	77–203	2.43–13.6
	Obtained in this study	152	7.6
pp'DDT	Recommended concentration	53	$1.95 \pm 1.20^a / 2.40 \pm 1.80^b$
	Range	22–84	$0.440-4.45^a / 0.44-6.87^b$
	Obtained in this study^a	22.5	1.32

[a] After the Chauvenet test [8].
[b] After the Dixon test [8].

perature programme: an isothermal injector at 300°; column: an initial temperature of 100°C for 2 min, a temperature gradient of 15°C/min, and a final temperature of 290°C for 6 min; and the detector at 300°C.

The detection limit was 2 pg/μL, and the limit of quantification for an average 8 g sample diluted to 1 mL was 0.25 ng/g. The accuracy of the IAEA reference samples was determined at two concentration levels; the results are given in Table I [8]. The pp'DDE and pp'DDT values lay within the range of accepted values. The highest variation coefficient found for DDE was 15.6%, while the values obtained for DDT tended to be at the lower end of the accepted range.

3. RESULTS AND DISCUSSION

Selection of the four species of mussel was based on their different feeding habits. (1) *Anomalocardia brasiliana*, the dominant and most frequent species of the Brazilian northeast coast, is a suspension feeder that lives in the upper 5 cm of sandy substrate, therefore it filters that part of the water column enriched by sediment. (2) *Brachidontes exustus* is a suspension feeder that lives attached to intertidal rocks, being only partly submersed and distant from the sediment, therefore it filters the average water column (2.5–3.0 m at high tide). (3) *Macoma constricta* is a deposit feeder that obtains its nourishment from the material adsorbed on sediment particles preselected by size, particularly bacteria, therefore it only filters from suspended particles. (4) *Littorina* spp. live in mangroves and scrape algae for food from stems and roots, therefore they do not filter suspended matter. No mussel of the same species was found at all stations, and no station had all the selected mussels.

TABLE II. DDT RESIDUES (ng/g OF DDE, WET WEIGHT) IN *A. brasiliana* AND *B. exustus* IN SOME SITES OF ALL SAINTS BAY (1994)

Site	A. brasiliana	B. exustus	B. exustus/ A. brasiliana
Mapele	2.9	44.0	15.2
Ponta Passé	0.8	22.0	28.2
Coqueiro Grande	2.5	3.7	1.5
São Brás	0.3	28.7	98.0
Acupe	0.5	2.1	4.4
Cabuçu	0.3	4.7	14.0
Cações	0.5	16.8	32.0

Correlation: r = 0.369; P = 0.415.

TABLE III. DDT RESIDUES (ng/g OF DDE, WET WEIGHT) IN *A. brasiliana* AND *M. constricta* AT SOME SITES OF ALL SAINTS BAY (1994)

Site	A. brasiliana	M. constricta	M. constricta/ A. brasiliana
Paripe	0.7	15.8	21.0
Mataripe	0.3	0.6	1.9
Pati Island	0.2	1.6	6.8
Acupe	0.5	<0.3	—
Cabuçu	0.3	0.7	2.1
Jiribatuba	0.9	10.2	11.0
Baiacu	1.1	13.4	12.2

Correlation: r = 0.848; P = 0.0159.

In 1994, the DDT residues (as ng/g of DDE, wet weight) in the studied mussels varied between 0.24 and 5.1 for *A. brasiliana*, 2.1 and 44.0 for *B. exustus*, <0.33 and 15.8 for *M. constricta* and <0.31 and 1.36 for *Littorina* spp. At the time of sampling, the latter mussel was only found at five stations, corresponding to the sites with lower DDT residues. *B. exustus* produced the highest results of all the species studied, followed by *M. constricta*. Between these two species and *A. brasiliana*, the factors can vary between 2 (value accepted by the International Mussel Watch Project [9]) and 99 for *B. exustus,* and between 2 and 21 for *M. constricta* (Tables II and III). These results were unexpected, since DDT and DDE are retained in sedi-

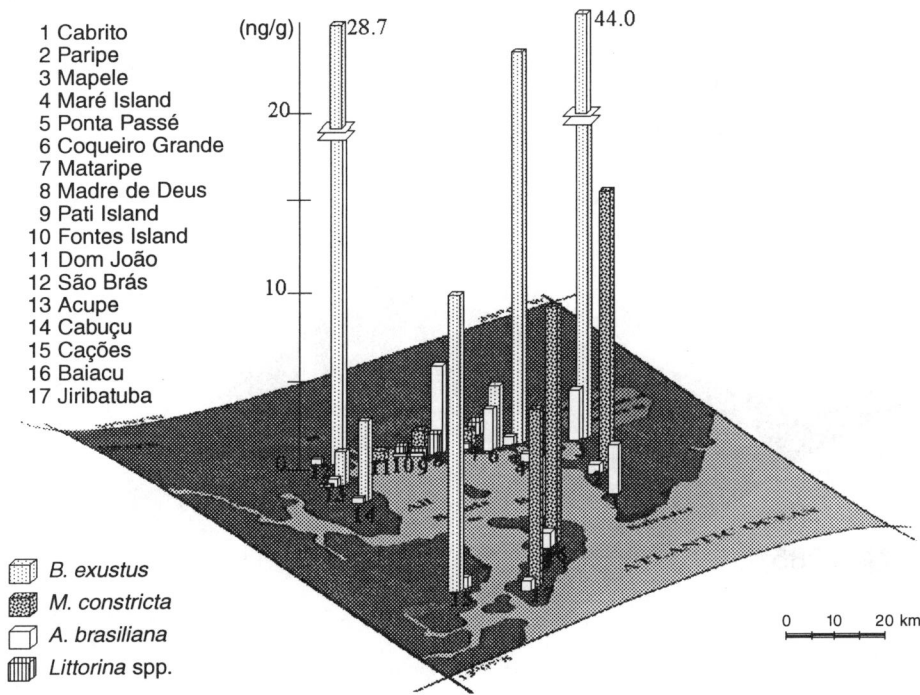

FIG. 1. DDT residues (as ng/g of DDE, wet weight) in mussels of All Saints Bay (1994).

ments, therefore both *M. constricta* and *A. brasiliana* should have been more exposed to them. Although the number of observations was small, a positive correlation of high significance was found between the *A. brasiliana* and *M. constricta* levels. However, no significant correlation was found between *A. brasiliana* and *B. exustus*. More paired data should be obtained for these two species at different sites.

The spatial distribution of DDT residues on the intertidal zone of the All Saints Bay coastline is presented in Fig. 1. The highest concentrations were found near the Aratu Industrial Centre, to the east of the bay, in the Subaé estuary to the north, where sugarcane plantations dominate, and in the Itaparica channel, which has extensive mangrove areas and where palmoil farms and many small villages are located. The International Mussel Watch Project found a range of total DDT components varying from below the limit of quantification to about 215 ng/g dry weight [9]. (considering an average dry weight of 20% of the total). No samples were found that exceeded the various action limits. However, taking into consideration the National Oceanic and Atmospheric Administration (NOAA) evaluation [10, 11], values of 120 ng/g dry weight for molluscs and 37 ng/g dry weight for sediments are defined

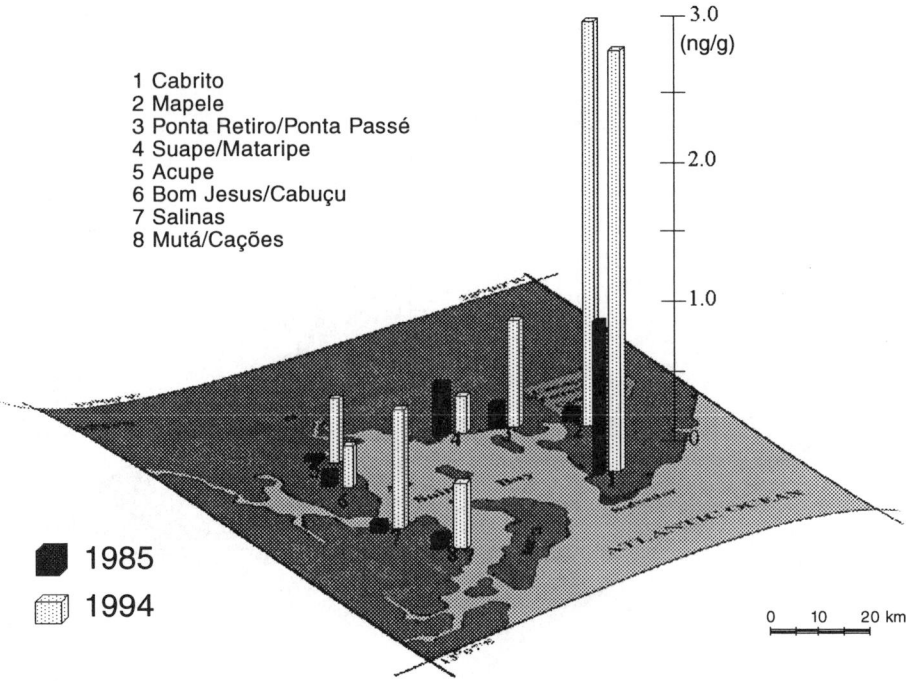

FIG. 2. DDT residues (ng/g of DDE, wet weight) in A. brasiliana of All Saints Bay (1985 and 1994).

as high. On the basis of this evaluation, the DDT concentrations at São Brás and Mapele were high.

The levels found in 1994 were compared with those obtained in 1985 [9], and Fig. 2 gives the data on *A. brasiliana* at the sites where measurements were made in both years. Increases were observed at all the sites around the bay, except for Mataripe, where an oil refinery operates on a large restricted area. From the sites where comparisons were possible, greater increases (as high as 15 fold) were observed along the coastline of Salvador and to the north of the city, where the Aratu Industrial Centre is located. These are the most populated areas of the bay.

The range of analytical intercalibration results for DDT residues is always wide. Comparison exercises conducted by IMW NOAA-NIST showed between-laboratory differences of as high as a factor of four for 4,4′ DDE [9]. One might argue that the differences found for the cross-sectional survey on the two different periods in All Saints Bay could be a result of analytical differences. However, in both years the analytical protocols used were the same, and quality was controlled by using reference samples. In the present study, two reference samples, at two different concentration levels, served as the control. All the reported DDE data above 0.75 ng/g wet weight lay within the range covered by these reference samples. A maximum deviation of 15.6% from the certified DDE reference value was obtained in this study, with accepted deviations ranging from −45 to 26% for IAEA 351 and ±32% for IAEA 142. The chance of obtaining laboratory differences of a factor of 15 seems unlikely.

DDT is very persistent in the environment [12]. Determinations of its half-life were valid for the tested conditions. The half-life values for tropical conditions were determined at 32°C [13]: sediments, about 5×10^3 d; sea water, about 105 d; brackish water, $19°/_{oo}$; and fresh water, about 110 d. The ratios of DDT/DDE may provide a rough age estimation of DDT inputs. Both pp'DDE and pp'DDT were analysed for several mussel and sediment samples.

Figure 3 shows their levels along the coastline of the bay. It was observed that DDT was present in some samples of mussel and sediment, particularly in São Brás at the Subaé estuary, where the concentration of total DDT in sediment reached about 42 ng/g, with a DDT/DDE ratio of about 0.91. If the half-life found by Carvalho et al. [13] holds true for the All Saints Bay conditions, it could be stated that DDT must have been used in the Subaé region over the past 5 years. DDT has also been

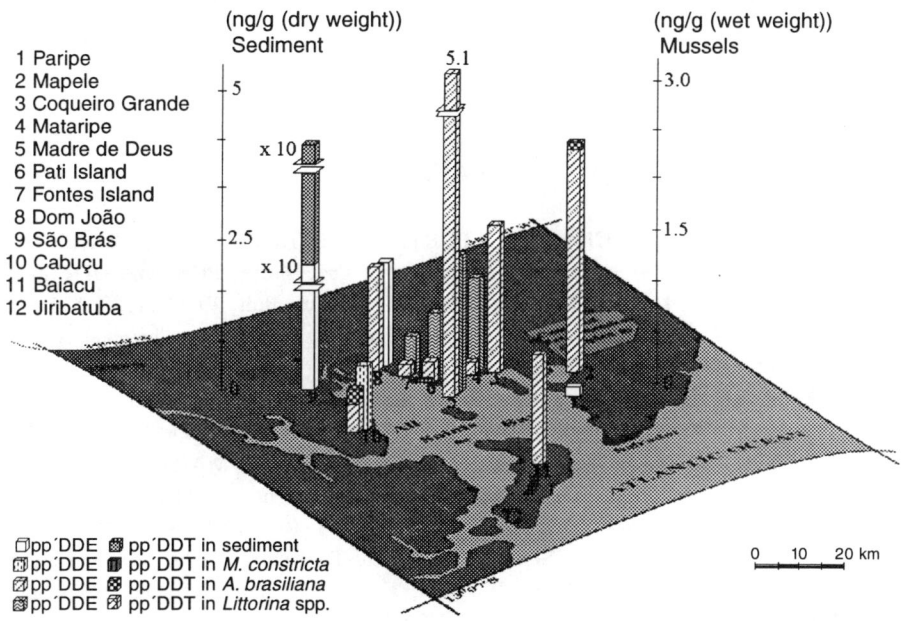

FIG. 3. Levels of pp'DDE and pp'DDT (n/g) in some samples of bivalves (wet weight) and sediments (dry weight) in All Saints Bay (1994).

found on the eastern and western coastlines of the bay. At these sites, the DDT/DDE ratios were much lower, indicating possible DDT application prior to the previously mentioned 5 years. As the last malaria eradication campaign took place in 1984, the observed levels seem to have resulted from the prohibited use of DDT and from carry over of the DDT used against malaria.

4. CONCLUSIONS

The levels of DDT residue (as DDE) in mussels of All Saints Bay, Bahia, have increased since 1985, although general use of DDT has been prohibited in Brazil since 1976. The DDT/DDE ratios in sediments showed recent DDT inputs that seem to originate from the 'prohibited' use of DDT rather than its use for malaria eradication. The highest concentrations were found in the Subaé estuary, along the coastline of Salvador City and in the Aratu area. No values exceeding the action limits were found.

ACKNOWLEDGEMENTS

The authors would like to thank M. Peso-Aguiar, from the Biology Institute of the Unversidade Federal da Bahia, and F.P. Carvalho, from the IAEA Marine Environment Laboratory, for helpful discussions. Financial support from the IAEA and Petrobrás made this study possible.

REFERENCES

[1] TAVARES, T.M., ROCHA, V.C., PORTE, C., BARCELO, D., ALBAIGES, J., Application of the Mussel Watch Concept of hydrocarbons, PCBs and DDT in the Brazilian Bay of Todos os Santos (Bahia), Mar. Pollut. Bull. **19** 11 (1988) 575–578.

[2] KEITH, L.H., Principles of Environment Sampling, American Chemical Society, Washington, DC (1988).

[3] BERNHARD, M., Manual of Methods in Aquatic Environment Research. Part 3. Sampling and Analysis of Biological Material, FAO Fish Technical Report No. 158 (1976).

[4] ALBAIGES, J., FARRAN, A., SOLER, M., GALLIFA, A., MARTIN, P., Accumulation and distribution of biogenic and pollutant hydrocarbons, PCBs and DDT in tissues of western Mediterranean fishes, Mar. Environ. Res. **22** (1987) 1–18.

[5] MACLEOD, W.D., et al., Standard Analytical Procedure of the NOAA National Analytical Facility, 1985–1986 (Revised), Extractable Toxic Organic Compounds (CANTILLO, A., SLOAN, C., LAUENSTEIN, G., Eds), NOAA Technical Memorandum NOS ORCA 71, National Oceanic and Atmospheric Administration, Silver Spring, MD (1993).

[6] ALBAIGES, J., ALGABA, J., BAYONA, J.M., GRINALT, J., "New perspectives in the evolution of anthropogenic inputs of hydrocarbons in the western Mediterranean coast", VIes Journées étude pollution, Cannes, 1982, Conférence internationale pour l'exploitation scientifique de la Méditerranée, Monaco (1982).

[7] ACEVES, M., et al., Analysis of hydrocarbons in aquatic sediments, J. Chromatog. **463** (1988) 503–509.

[8] VILLENEUVE, J.P., HORVAT, M., CATTINI, C., World-wide and Regional Intercomparison for the Determination of Organochlorine Compounds and Petroleum Hydrocarbons in Mussel Sample IAEA-142, Report No. 59, IAEA Marine Environment Laboratory, Monaco (1966).

[9] NATIONAL OCEANIC AND ATMOSPHERIC ADMINISTRATION, International Mussel Watch Project: Final Report (FARRINGTON, J.W., TRIPP, B.W., Eds), NOAA Technical Memorandum NOS ORCA 95, NOAA, Silver Spring, MD (1995).

[10] NATIONAL OCEANIC AND ATMOSPHERIC ADMINISTRATION, Second Summary of Data on Chemical Concentrations in Sediments from the National Status and Trends Program, NOAA Technical Memorandum NOS OMA 59, NOAA, Silver Spring, MD (1991).

[11] O'CONNOR, T.P., BELIAEFF, B., Recent Trends in Coastal Environmental Quality: Results from the Mussel Watch Project, National Oceanic and Atmospheric Administration, Silver Spring, MD (1995).

[12] WORLD HEALTH ORGANIZATION, Environmental Health Criteria 83: DDT and its Derivatives — Environmental Aspects, WHO, Geneva (1989).

[13] CARVALHO, F.P., FOWLER, S.W., READMAN, J.W., MEE, L.D., "Pesticide residues in tropical coastal lagoons", Applications of Isotopes and Radiation in Conservation of the Environment (Proc. Symp. Karlsruhe, 1992), IAEA, Vienna (1992) 637–653.

IAEA-SM-343/1

PERSISTENCE OF TEMEPHOS AND FENITROTHION AND THEIR TRANSFORMATION ON PRODUCTS IN RICE FIELD WATERS*

D. BARCELO
Department of Environmental Chemistry,
Centro de Investigación y Desarrollo,
Consejo Superior de Investigaciones Científicas,
Barcelona, Spain

Abstract

PERSISTENCE OF TEMEPHOS AND FENITROTHION AND THEIR TRANSFORMATION PRODUCTS IN RICE FIELD WATERS.
  A study was made of the persistence of temephos and fenitrothion and their degradation products, temephos sulfoxide, temephos isomer, temephos sulfoxide isomer, fenitro-oxon, 3-methyl-4-nitrophenol and S-methyl isomer, under field conditions. The other breakdown products identified were temephos oxon and temephos oxon isomer. Temephos and fenitrothion were applied to rice fields in the Ebre Delta area (Tarragona, Spain) during July and August 1994, and in 1995 by aircraft, at a spraying rate of 0.25 L/ha of Abate 50 E (temephos) and 0.148 L/ha of Tionfos 50 LE (50% fenitrothion). The concentration in rice field waters ranged between 41 and 125 µg/L. The evolution of the two pesticides and their transformation products was recorded for 72 hours, at sampling intervals of 5 and 10 hours. The concentration of the compounds formed was measured in the rice fields and at the outlet of the two rice field waters, and varied between 0.01 and 2.27 µg/L. A maximum of only 5.5% of the applied temephos was detected immediately after application, whereas for fenitrothion the expected amount (±90%) reached the rice field waters. Analytical determinations were performed using solid phase extraction with C_{18} Empore extraction discs, followed by liquid chromatography–thermospray–mass spectrometry (LC–TSP–MS) and LC–atmospheric pressure chemical ionization–MS. ELISA was also used for fenitrothion.

1. INTRODUCTION

  The fate of organophosphorus pesticides in the aqueous environment has led to numerous investigations in recent years. In this regard, it has been reported that degradation is influenced by hydrolysis, particularly at a pH of > 7, by photolysis and by microbial degradation [1–3]. Under laboratory conditions, many different

---

  * Research carried out in association with the IAEA under Research Agreement No. 7978.

experiments have been performed using capped bottles to avoid the volatilization of pesticides, or waters at various pH (6-8) and at various water temperatures (11-30°C). Many of the reported results cannot be compared and extrapolation to field conditions is difficult; consequently, there is a lack of field studies. We have reported on a field study of fenitrothion following manual spraying [1]. Studies on the aerial spraying of pesticides have also been reported for fenitrothion, applied in New Brunswick, Canada [4], and for temephos [3].

In this paper, we report on the disappearance of fenitrothion and temephos, which are used to control several insect pests and mosquitoes in the Ebre Delta area (Tarragona, Spain). Generally, the analytical techniques used to monitor pesticide degradation in the aquatic environment involve either gas or liquid chromatography coupled to different detectors. Immunoassays, although implemented for several years to monitor pesticides such as atrazine [5] and chlorpyrifos [6] in the aquatic environment, have hardly been applied to degradation studies. One problem is that there is no ELISA immunoassay kit that is sensitive enough to monitor fenitrothion or temephos. The only kit commercially available for fenitrothion has very poor sensitivity (0.15 mg/L) and is used for crops [6]. To solve this problem, we decided to use the EnviroGard parathion plate kit to detect fenitrothion as it offers adequate sensitivity. This is not surprising, since the chemical structure of fenitrothion and parathion-ethyl exhibits many similarities. The data of this ELISA test were compared with automated on-line solid phase extraction (SPE) (Prospekt) liquid chromatography–diode array detection (LC-DAD), which is a robust technique used in different studies of various organophosphorus pesticides [3] and gives results that are similar to gas chromatography–nitrogen phosphorus detection (GC-NPD).

This paper reports the results obtained when the fenitrothion and temephos samples from rice fields after aerial spraying were analysed. The main objectives of the research were to evaluate the dissipation and persistence of fenitrothion and temephos in natural (rice field) waters: (1) by measuring the amount of pesticides in water after helicopter spraying using two independent analytical techniques; (2) by monitoring the possible transformation products (TPs) formed during the sampling period; and (3) by investigating the behaviour of the pesticides after aerial spraying.

2. EXPERIMENTS

2.1. Pesticide application

Abate 50 E (temephos) was applied by aircraft to rice fields in the Ebre Delta area at a rate of 0.25 L/ha. The amount of active ingredients reaching the water surface was estimated on the assumption that the flood depth in the rice plots was maintained at 25 ± 10 cm for the days immediately following application. The temephos residue concentrations were 125 µg/L in water at the 10 cm depth and 41 µg/L in water at the 30 cm depth. Two treatments were carried out over the same area. The first, performed on 6 July 1994, was done in order to eliminate the mosquito larvae that abound in this area, especially during the summer months when the rice fields are flooded. The second was done on 4 August 1994 to eradicate those larvae that might have appeared in July. The amount sprayed over the rice fields was the same for both treatments.

Tionfos 50 LE (50% fenitrothion) was applied by helicopter to rice fields in the Ebre Delta area at a concentration of 4% active ingredients by mixing 40 L of Tionfos with 500 L of water to obtain 3.7% diluted fenitrothion. In this way, 2 L/ha of the formulated diluted product were applied, corresponding to 0.148 L/ha of Tionfos 50 LE. The irrigation ditch monitored is close to the rice fields and was treated in a manner similar to the rice fields, although the water column was somewhat deeper (50 cm) and the amount of active ingredients reaching the water surface was between 125 and 100 µg/L in water at the 5–10 cm depth and between 40 and 30 µg/L in water at the 30 cm depth.

The Ebre Delta area, where rice is the main crop (18 000 ha), underwent aerial applications of fenitrothion for 2 weeks, from early morning until afternoon. The samples were collected between 26 and 28 July 1995.

2.2. Sampling

For Abate 50 E, two different rice fields, Poble Nou and Eucaliptus, were chosen. In the former, there was an irrigation ditch that provided fresh water to the rice field. At Eucaliptus, the fresh water entered from another rice field and exited 50 m from where the samples were collected. The sampling strategy adopted for each site differed. Samples from Poble Nou were taken at 1, 5, 11, 24, 34, 48, 60 and 72 hours after spraying, plus a blank, collected just before the treatment started. During the treatment at Eucaliptus (August 1994), a different sampling approach was undertaken, entailing a study of the mass transfer of temephos through the rice fields. This was performed to evaluate not only the initial amount that had reached the water surface but also the quantity eliminated by runoff, and to achieve better estimation of the TPs formed. Accordingly, samples were taken from the rice field and outlet waters at 0 (immediately after spraying), 1, 2, 5 and 7 hours after spraying.

Three samplings were performed: one at Poble Nou and two at Eucaliptus (rice fields and outlet water). In each case, the water samples consisted of composites of 20 grab samples (1 m from each other); the samples were pooled to obtain a final volume of 2 L. The water samples were collected just below the water surface and at water depths varying between 15 and 30 cm, without spillage. The pH varied between 7.8 and 8.8 and the rice water temperature between 26°C (early morning) and 33°C (during the day). The pH of the paddy water followed a diurnal pattern because of photosynthesis and the respiration of algae.

For Tionfos 50 LE, sampling was carried out in the same irrigation ditch that provided fresh water from another rice field. Collection of water samples consisted of three to five grab samples that were pooled to a final volume of 1 L. The amber bottles used to transport the collected water had previously been cleaned up (two to three times) with the same irrigation ditch water. The samples were collected on the upper layers of the water surface, between 5 and 20 cm, without spillage. Samples were collected at 0, 1, 2, 3, 4, 5, 8, 10, 24, 30 and 48 hours after application, plus a blank, collected just before the treatment started. Also, a sample of the treating solution was taken to determine the concentration of active ingredients.

2.3. Sample preparation

The water samples were passed through filter paper (0.7–1 μm) to remove the coarse material, and then through 0.45 μm glass fibre filters (Millipore, Bedford, MA, USA) to remove the suspended particles. Using SPE, the water samples were extracted immediately after collection; 1 L of of each sample was extracted with C_{18} Empore extraction discs, which were conditioned with 10 mL of methanol and 10 mL of high performance liquid chromatographic (HPLC) grade distilled water prior to percolation of 1 L of water. The discs were then stored at −20°C until analysis. Elution of the discs was carried out with 2 × 10 mL of acetonitrile. The eluate was rota evaporated to dryness and the extract finally dried under a gentle stream of nitrogen. The final extract was diluted in 100 μL of acetonitrile. The extracts corresponding to the samples collected during the first treatment were injected by LC-DAD, LC-TSP-MS and LC-APCI-MS. For more details on use of the LC-MS technique, see Refs [2, 3].

2.4. Sample analyses

2.4.1. ELISA procedures

The procedure for microwell immunoassay is as follows. (1) Pipette 100 μL of standard/sample into antibody microwells; (2) add 100 μL of a parathion–enzyme conjugate, mix, cover the wells with tape or parafilm to prevent evaporation, and incubate for 60 min at room temperature; (3) after incubation, drain the contents well, and then wash five times with distilled water; (4) add 100 μL of substrate-chromogen to each well, mix, cover the wells with tape or parafilm to prevent evaporation, and incubate for 30 min at room temperature; (5) add 100 μL of stopping solution to each well and mix thoroughly; this will turn the solution yellow; and (6) read the absorbances, with the detector set at 450 nm. The standards were prepared by dilution from a 100 mg/L solution of fenitrothion in methanol stored at −20°C. A standard series was prepared by making several dilutions of the stock solutions to yield the following fenitrothion concentrations: 0, 0.1, 1, 10, 100, 1000 and

10 000 µg/L. The standard series was made up with distilled water. All the standards were analysed in triplicate. The test kits were stored at 4°C. Temperature and time are important parameters that have to be controlled if this ELISA procedure is to work properly. In all cases, solutions of immunoassay and samples were allowed to equilibrate to room temperature before use; the reaction times were consistent throughout the experiment.

2.4.2. Prospekt LC–DAD

Ten millilitres of water sample were preconcentrated at a flow rate of 2 mL/min and introduced into the chromatographic system. The C_{18} cartridges were conditioned with 10 mL of acetonitrile and 10 mL of HPLC grade distilled water. Gradient elution was employed, starting from a mobile phase containing 30% acetonitrile and 70% water, to a phase containing 60% acetonitrile and 40% water within 8 min, followed by an isocratic state for 12 min and, from these conditions, to 100% acetonitrile within 10 min at a flow rate of 1 mL/min. Quantification was done using the external standard calibration method, with the detector set at 270 nm. Calibration graphs were constructed for fenitrothion by analysing the spiked aqueous samples prepared with MilliQ water ranging between 0.1 and 160 µg/L.

3. RESULTS AND DISCUSSION

3.1. Determination of temephos and its TPs

The temephos concentrations were 0.49 and 0.07 µg/L at 1 and 5 hours after application, respectively. The low concentration of temephos at 1 hour after application could have been because the site where the samples were collected was perturbed by the entry of fresh water from a channel, and because there was a high dilution of temephos within the first hours after application. Temephos was not detected 11 hours after application. Moreover, temephos sulfoxide was formed at concentrations varying from 0.06 to 0.1 µg/L over the 72 hours it was monitored. Transfer of temephos through the rice fields was evaluated and it was found that the amount reaching the water surface was very low. Similarly, when temephos was determined in stagnant ponds there was a sharp decrease in its concentration under field conditions, with a half-life that varied between 6 and 8 hours. In this experiment, the water was kept in a closed glass reservoir to avoid any losses by vaporization and circulation of the water, in contrast to what happens in rice fields under natural conditions, where volatilization, dilution and degradation occur. Since dissipation takes place faster at higher temperatures, and especially in the upper water layers of the rice field, it was of interest to carry out a short term study on the persistence of temephos under such conditions. Determination of the temephos concentrations in the rice field and outlet waters of a specific rice field, and the monitoring and disappearance of their TPs, were undertaken.

3.2. Dissipation process

It was observed that the amount of temephos reaching the water was 2.27 and 2.23 µg/L in rice field and outlet waters, respectively. This represents only 2–5.5% of that estimated to reach the water, taking into consideration the different depths. The values found are related to the fact that: (1) the formulated compound is an emulsionable liquid that is dissolved in water before it is applied by aircraft spraying; an important problem that experts have encountered is deposition of the compound on rice leaves after application; (2) the adopted water sampling strategy collected a water volume from a depth of 15–30 cm; and (3) the water temperature varied between 26 and 33°C, thus enhancing the dissipation of temephos. In selecting a thinner layer (less than 5 cm) on the water surface, given that temephos is a compound that can form such a thin layer on the surface because of its poor water solubility, the temephos values were higher, so lower losses were expected. Higher temperatures favour fast dissipation of temephos, since it is a compound that can be partly degraded by such water temperatures; in general, higher temperatures accelerate other dissipation processes, e.g. volatilization.

An additional factor to take into account in the dissipation of pesticides in natural waters is the K_{OC} value. The K_{OC} of temephos, which varies between 200 and 500 L/kg, depending on the soil, implies partitioning of temephos in the organic matter of the water sample. However, it has been proved that chlorpyrifos, with a similar K_{OC} value of 498 L/kg, does not absorb on to organic matter because the contact time of 72 hours and the concentration of the pesticide applied are too low. In addition, other factors should be considered: the temephos concentration was low, as was the particulate matter in the rice field water. The situation will differ if high concentrations of temephos are present in the sediment and are released to the water, or if there is continuous spillage. In such cases, binding of temephos to soil particles is of importance. In this respect it has been reported [7] that at least 5 days are needed for equilibration to be reached between the highly lipophilic compound, e.g. mirex, and the dissolved organic matter (DOM). The time of equilibration is important for the associated kinetics between the compounds of interest and the DOM; in the case of temephos, this time was too short. However, we did not directly measure the temephos residues in sediment, so arguments regarding the role of sediment in dissipation have to be regarded as hypothetical.

3.3. Photolysis

The fast disappearance of temephos (1.9 hours) can be attributed to its chemical structure. Owing to the presence of two aromatic rings, this compound has a tendency to exhibit high quantum yields [3]. However, this is not the only aspect that has to be taken into consideration when high rate photolysis constants are exhibited. From the UV spectrum it has been shown that temephos contains a suitable chromophore for absorption at wavelengths longer than 295 nm. Consequently, the observed fate of this compound under natural conditions can be explained by a combination of these two effects [3]. In addition, because running water continuously enters the rice fields, the pH of the water under environmental conditions can also vary slightly (usually between 7.8 and 9). It has been reported that basic pHs promote degradation of organophosphorus pesticides in water samples by stabilizing the more phenolic structures, which in addition show a tendency to absorb more sunlight, thus increasing the overall degradation rate constants [8].

3.4. Factors affecting the formation of temephos TPs

Blank analyses were performed before the treatment started. These blanks did not contain traces of either temephos or its TPs. However, temephos TPs were found immediately after spraying in both rice field and outlet waters. The TPs were identified by LC–TSP–MS [3]. Oxidation is one of the major photolytic routes of pesticide degradation. In this study, it was found to be the first degradative mechanism of temephos. Through oxidation, temephos is converted to oxon and sulfoxide, which are the activated forms of the pesticide, since they have a much higher inhibitory activity towards acetylcholine esterase. Therefore, it is important that not only the parental compound but also the oxon derivatives of the pesticides are measured during treatments against pests because the environmental damage they can cause is much more severe than that of the parental pesticide. Thus, oxidation of the parental compound occurs rapidly, e.g. oxidation can take place during dissolution of the formulation in tanks, since temephos sulfoxide and temephos isomer form immediately after spraying. Oxidation can also occur while temephos is being sprayed on fields, before reaching the water surface. Temephos sulfoxide is mainly formed by photooxidation, which is attributed to the quantum yields of temephos. Temephos oxon also forms immediately after spraying. This compound could not be accurately quantified because a standard was not available, but the amount encountered in outlet waters was twice as high as that in rice field waters, probably as a result of the TP concentration (more compounds were formed in the outlet waters than in the rice field waters). S-methyl isomers are produced by thermally induced isomerization, generally during synthesis and/or storage, but this can also take place during exposure to sunlight. The temephos isomer found immediately after spraying was similar to that of fenitrothion [1]. Its concentration was five times higher in outlet waters than in rice field waters, i.e. 1 hour after treatment, 38 and 7.2% of the initial concentration, respectively; these levels remained constant at 0.05 μg/L until 5 hours after spraying. Detection of S-methyl isomers is important in environmental studies, since their biological activity differs greatly to O-methyl isomers (parent compounds). Temephos oxon isomers were also found immediately after spraying, at levels that were similar in rice field and outlet waters, and decreased with time. The drop in level of these two compounds indicates that the formation of temephos isomers and temephos oxon isomers is not an important route of temephos dissipation.

3.5. Rice fields

The initial concentration of temephos sulfoxide was higher in rice field waters than in outlet waters. However, the concentration decreased with time in rice field waters but increased with time in outlet waters. The large number of temephos degradation products formed in rice field waters can be attributed to: (1) the large surface area of the rice fields, with temephos distributed along the upper water layer; (2) the high temperatures of the water during the days of application (30°C), which enhance thermal degradation and hydrolysis, and increase microbial activity; and (3) the fairly high organic matter content of the flooded soils (3.3%) (caused by rice straw being mixed with the soil samples), which enhances the biodegradation of pesticides in rice field waters. It has been reported [1] that oxidation is typical in organic rich soils under flooded conditions because of the rapid formation of TPs. It is evident that sulfoxide forms in the water matrix at increased concentration levels; consequently, it can persist in water samples for a longer period than the parent compound.

Temephos sulfoxide isomers were only found 1 hour after application at levels of 0.01 and 0.16 µg/L in rice field and outlet waters, respectively. The higher level found in outlet waters matches that of the temephos sulfoxide encountered in this area. The temephos sulfoxide isomer levels in rice field waters showed a slight increase during the 7 hours of monitoring. The temephos sulfoxide levels decreased, which indicated that an isomerization process had taken place in water, so enhancing dissipation.

3.6. Determination of fenitrothion by immunoassay

A new concept for using immunoassay is described. An immunoassay kit should be capable of detecting a target compound with high sensitivity, but it is known that many immunoassays give high cross-reactivity values for structurally related compounds. This has to be taken into account for fenitrothion, whose cross-reactivity is 1.83% when parathion-ethyl and fenitrothion are present (Table I), so this study would not have been possible if rice field waters had contained both these pesticides. Table I summarizes the specificity data of parathion-ethyl, fenitrothion and a metabolite of fenitrothion, fenitro-oxon, in distilled and estuarine water.

TABLE I. SPECIFICITY OF THE PARATHION-ETHYL ENVIROGARD ELISA AND RELATED COMPOUNDS IN DISTILLED AND ESTUARINE WATER (n = 6)

Compound	Distilled water			Estuarine water		
	LDD[a]	IC_{50}[b]	% CR[c]	LDD	IC_{50}	% CR
Parathion-ethyl	0.011 (8)[d]	0.45 (5)	100	0.017 (6)	0.54 (4)	100
Fenitrothion	0.17 (3)	24.6 (7)	1.83	0.2 (7)	25.3 (5)	2.08
Fenitro-oxon	>10 000	>10 000	<0.01	>10 000	>10 000	<0.01

[a] LDD = least detectable dose, calculated at 90% B/B_0 (in µg/L).
[b] IC_{50} = 50% inhibition concentration (50% B/B_0) (in µg/L).
[c] % CR = percentage cross-reactivity determined by estimating the amount of compound required to displace 50% of the enzyme conjugate to the amount of parathion-ethyl.
[d] Numerals in parentheses = coefficient of variation expressed in percentage.

The coefficient of variation within and between assays was less than 8 and 9%, respectively, and the linearity of the assay in distilled water varied between 1.3 and 100 µg/L. A 10% inhibition of signal was considered to be significantly different from the zero analyte concentration, and was used to estimate the sensitivity to the assay, the least detectable dose (LDD) at 90% B/B_0. The assay sensitivity was estimated to be 0.17 µg/L using the 90% B/B_0 concentration.

3.7. Factors affecting the formation of fenitrothion metabolites

The results show that metabolites were formed by various mechanisms. As a result of isomerization, the S-methyl isomer of fenitrothion had already formed in the tank when fenitrothion was mixed with water. Its concentration was higher at the beginning of analysis, but because the fenitrothion concentration decreased, that of the S-methyl isomer also decreased. A similar phenomenon was observed for the temephos isomer. Fenitro-oxon is formed through oxidation, while 3-methyl-4-nitrophenol results from chemical hydrolysis. The two processes took place because of the high temperatures experienced during the summer (air temperatures of 35°C) and because of the high pH of the water, both of which favour fast hydrolysis of fenitrothion.

3.8. Environmental levels

ELISA was used to check the fenitrothion concentration in the commercial product, Tionfos (50% fenitrothion), and the diluted product (3.7% fenitrothion). The results for these products were 48 and 3.5%, respectively. Twelve samples were also measured with ELISA and Prospekt LC–DAD. The concentrations varied from 119.5 to 3.8 µg/L for ELISA and from 178.8 to 1.5 µg/L for LC–DAD. High correlation was obtained with both techniques (r = 0.999) compared with the data recorded before extraction with the C_{18} discs.

Chemical degradation can be described using a first order degradation curve

$$C_t = C_0 e^{-kt}$$

where C_t is the concentration of the pollutant at time t; C_0 is the initial concentration; and k is the rate constant (slope). The half-life is then ln (1/k). By plotting the natural logarithm versus time (hours after application), a straight line can be obtained, and the rate constant k may be derived from it.

3.9. Behaviour of aerially applied fenitrothion

The persistence of fenitrothion is much shorter than that reported under laboratory conditions. Lartigues and Garrigues [9] obtained a half-life of 3 days at a pH of water of 8.1 and exposure to sunlight (22–25°C). Other field experiments in a lake reported a half-life of 0.9–7 days at a water temperature of 11°C and a pH of 6.7 [4]. At the same Ebre Delta ditch area during the winter period, the half-life was 13 hours at a pH of water of 7.8 and a temperature of 11°C. The differences shown in Table II can be explained by the fact that organophosphorus pesticides are particularly sensitive to pH and temperature. At higher pHs and temperatures, higher degradation is expected. This is the explanation for most of the values reported in the table. The lower half-life values obtained are related to the volatilization of fenitrothion from water. Fenitrothion is a compound with a structure that is sensitive to photodecomposition. This is because it has a relatively high quantum yield and a phenolic structure that is very sensitive to changes in pH. The relatively high natural pH of Ebre Delta waters affects fenitrothion, consequently influencing the stability of the molecule.

It has been reported that volatilization was a major degradation pathway in the experiments carried out in Canada [4] and that it was the major source of disappearance from the Canadian lake. This can be explained by Henry's law constant (HLC), which has a value of 0.0036 Pa·m^{-3}·mol^{-1} calculated at 20°C. The HLC can be calculated from the vapour pressure and water solubility. Despite the high temperatures in the area where the experiments were performed, the water solubility of fenitrothion was not greatly affected. However, vapour pressure increases rapidly

TABLE II. COMPARATIVE DATA OF THE HALF-LIFE AND ENVIRONMENTAL CONDITIONS OF FENITROTHION IN VARIOUS WATER SAMPLES

Half-life (h)	Water		Refs
	pH	Temperature (°C)	
72	8.1	22–25	[7][a]
15–168	6.7	11	[4][b]
13	7.8	11	[1][c]
11–19.3	7.8–8.2	25–30	Present paper[d]
70–74	8.2	15–18	Present paper[e]

[a] Samples in closed bottles in the darkness of natural sea water from Arcachon Bay (France).
[b] Samples from Palfrey Lake (Canada) taken in summer.
[c] Samples from an irrigation ditch in the Ebre Delta area (Spain) taken in winter.
[d] Samples from rice field waters in the Ebre Delta area (Spain) taken in summer.
[e] Samples in closed bottles from the degradation study performed in the laboratory.

at higher temperatures, consequently the HLC would have been somewhat higher than the reported value at 20°C. Thus, volatilization from water is considered to be the major reason for loss of fenitrothion after aerial application in Ebre Delta ditch waters during the summer period.

ACKNOWLEDGEMENTS

This work was supported by the Environment R&D Program 1991–1994 (Commission of the European Communities, Contract No. EV5V-CT94-0524) and by the Comisión Interministerial de Ciencia y Tecnología, Madrid (Contract No. AMB95-1230-CE). The Research Agreement from the IAEA is also acknowledged.

REFERENCES

[1] LACORTE, S., BARCELO, D., Rapid degradation of fenitrothion in estuarine waters, Environ. Sci. Technol. **28** (1994) 1159–1163.
[2] LACORTE, S., LARTIGUES, S.B., GARRIGUES, P., BARCELO, D., Degradation of organophosphorus pesticides and their transformation products in estuarine waters, Environ. Sci. Technol. **29** (1995) 431–438.

[3] LACORTE, S., EHRESMANN, N., BARCELO, D., Persistence of temephos and its transformation products in rice crop field waters, Environ. Sci. Technol. **30** (1996) 917–923.

[4] METCALFE, C.D., McLEESE, D.W., ZITKO, V., Rate of volatilization of fenitrothion from fresh water, Chemosphere **9** (1980) 151–155.

[5] GASCON, J., DURAND, G., BARCELO, D., Pilot survey for atrazine and total chlorotriazines in estuarine waters using magnetic particle-based immunoassay and gas chromatography–nitrogen phosphorus detection, Environ. Sci. Technol. **29** (1995) 1551–1556.

[6] SKERRITT, J.H., HILL, A.S., BEASLEY, H.L., EDWARD, S.L., McADAM, D.P., Enzyme-linked immunosorbent assay for quantitation of organophosphosphate pesticides: Fenitrothion, chlorpyrifos-methyl and primriphos-methyl in wheat grain and flour-milliing fractions, J. Assoc. Off. Anal. Chem. **75** (1992) 519–528.

[7] OUBIÑA, A., GASCON, J., BARCELO, D., Determination of cross-reactivities of immunoassays: Effect of common cross-reactants for chlorpyrifos-ethyl in water matrices using magnetic particle-based ELISA, Environ. Sci. Technol. **30** (1996) 513–516.

[8] WAN, H.B., WONG, M.K., MOK, C.Y., Comparative study on the quantum yields of direct photolysis of organophosphorus pesticides in aqueous solution, J. Agric. Food Chem. **42** (1994) 2625–2630.

[9] LARTIGUES, S.B., GARRIGUES, P., Degradation kinetics of organophosphorus pesticides in different waters under various environmental conditions, Environ. Sci. Technol. **29** (1995) 1246–1245.

ORGANOCHLORINE PESTICIDES IN SEDIMENT AND BIOTA IN THE COASTAL REGION TO THE SOUTH OF THE PINAR DEL RIO PROVINCE, CUBA*

G. DIERKSMEIER, R. HERNANDEZ, P. MORENO,
K. MARTINEZ, C. RICARDO
Instituto de Investigaciones de
 Sanidad Vegetal,
Havana, Cuba

Abstract

ORGANOCHLORINE PESTICIDES IN SEDIMENT AND BIOTA IN THE COASTAL REGION TO THE SOUTH OF THE PINAR DEL RIO PROVINCE, CUBA.
 A study was conducted to assess the levels of persistent organochlorine pesticides in the coastal region near to a large rice field to the south of the Pinar del Río Province in Cuba. Samples of sediment and biota (*Isognomon alatus*) were taken periodically over a period of 14 months. Analysis of organochlorines, polychlorinated biphenyls and other related compounds was done using gas chromatography–electron capture detection with a mean recovery of 87% and a detection limit of 0.25 µg/kg. Only DDT, DDE and DDD were found in the sediment and biota samples, with small differences in concentration between sampling sites during the dry season. After the rainy season began, the residues in sediment were very low, while the concentrations in biota remained comparatively unchanged, ranging from 11.2 to 23.8 µg/kg total DDT.

1. INTRODUCTION

Use of pesticides in agriculture is necessary if the quality and quantity of some crops is to be improved. This is the case for citrus, banana, tobacco, sugarcane, legumes and rice crops in Cuba. In particular, rice production is of environmental concern because of the diversity and quantity of the pesticides used, the spraying system (almost exclusively low volume spraying by aircraft) and the proximity of all the rice fields to waterways and coastal regions. All these factors accounted for some ecological damage in recent years, especially fish kill in the drainage channels and rivers near to the rice fields.

---

* Research carried out with the support of the IAEA under Research Contract No. 7934.

For these reasons, a study was undertaken to evaluate the levels of the most persistent pesticides in sediment and biota in the coastal region very near to a large rice field to the south of the Pinar del Río Province. It forms part of a global project on the Distribution, Fate and Effects of Pesticides on Biota in the Tropical Marine Environment in which several countries are taking part and which is financially and technically sponsored by the IAEA and its Marine Environment Laboratory in Monaco.

2. MATERIALS AND METHODS

2.1. Sampling location and sites

To monitor the organochlorine pesticides, a coastal region was selected that is located between two rivers, Carraguao and San Diego, which enter the sea 7 km from each other at Dayanigua beach. Both rivers flow through a flat sandy region near to the rice fields. This region is located to the south of the most western province of Cuba, and was divided into the following sites:

(1) *Site I: mouth of the San Diego River*, which is a very shallow estuarine, with a deep sediment layer.
(2) *Site II: mangrove coast.*
(3) *Site III: mangrove coast*, with a very deep sediment layer.
(4) *Site IV: mouth of the Carraguao River*, which has deep, clean water.

2.2. Sample processing and analyses

All the sediment and biota (*Isognomon alatus*) samples were packed in ice and taken to the laboratory within 24 hours of sampling [1]. The samples were ground with anhydrous sodium sulphate and maintained in the deep freeze. Analyses were done using the IAEA–MEL procedure, which is recommended for organochlorines and polychlorinated biphenyls, except that in the final chromatographic step a wide bore column (DB-5 (diphenyldimethyl-polysiloxane), 30 m length, 0.53 mm internal diameter and 1.5 μm film thickness) was used [2].

3. RESULTS AND DISCUSSION

Table I gives the analytical results of the field samples. Only DDT residues were found in the sediment and biota samples. No organochlorines, pyrethroids or polychlorinated biphenyls were detected. Each value in the table represents the average of two sediment and two biota samples taken at each sampling site.

TABLE I. TOTAL DDT RESIDUES FOUND IN SEDIMENT AND BIOTA (μg/kg)

Date	Sampling sites					
	I	II		III		IV
	Sediment	Sediment	Biota	Sediment	Biota	Sediment
9 Dec. 1994	2.39					
18 Feb. 1995	11.63	14.46	12.96	13.51	19.45	12.44
25 Apr. 1995	4.62	23.15	11.20	17.65	23.80	21.00
25 Jul. 1995	BDL[a]	BDL[a]	12.90	BDL[a]	14.74	BDL[a]
17 Nov. 1995	BDL[a]	BDL[a]	NT[b]	BDL[a]	NT[b]	BDL[a]

[a] BDL = below detection limits of 0.25 μg/kg for organochlorines and 0.5 μg/kg for pyrethroids.
[b] NT = samples not taken.

During the dry season, the DDT concentrations remained comparatively unchanged for each sediment sample at the sampling points. After the rainy season began, the residues in sediment were very low, probably because of intense runoff and flooding of the sampling sites caused by strong tropical storms during the months of October and November. However, the concentrations in biota remained almost unchanged, with only small differences between sampling sites. It was impossible to take biota samples during the last sampling because of an insufficient amount of organisms. The absence of other pesticide residues used in the rice fields near to the sampling sites, e.g. endosulfan and some synthetic pyrethroids, was surprising.

To explain these findings, an experiment was carried out in a micropond (0.3 m^3 capacity) in which sediment from the rice field drainage channels was exposed to the environmental conditions that exist outside the laboratory. The water/sediment ratio in the micropond was about 15:1. The pesticide residues in water were extracted with dichloromethane and the extracts cleaned up using the same procedure as that used for the sediment and biota samples. The results of this study are shown in Table II. They indicate that these pesticides showed a strong tendency to accumulate in the sediment phase. In water, the dissipation rate was high. With the exception of endosulfan, all the pesticides were relatively stable in sediment.

TABLE II. DISTRIBUTION AND DEGRADATION OF SOME INSECTICIDES IN THE WATER/SEDIMENT SYSTEM (CONCENTRATION IN WATER (mg/L) OR IN SEDIMENT (mg/kg))[a]

Pesticide	System	Days after treatment							
		0	5	8	16	21	63	78	140
Endosulfan	Water	1.9	0.23	0.16	0.03	0.02	0.01	ND	ND
	Sediment	5.2	206	—	35	25	—	—	—
		Days after treatment							
		0	12	25	38	61	96	120	150
DDT	Water	0.96	0.08	0.05	0.003	0.0008	0.006	—	0.003
	Sediment	20.8	44.8	49.08	—	—	44.7	42.4	22.3
		Days after treatment							
		0	5	11	19	45	50	136	170
Permethrine	Water	13.6	7.9	5.8	0.27	0.05	ND	—	—
	Sediment	169	177	—	434	395	—	270	367
		Days after treatment							
		0	5	11	19	45	50	136	170
Cypermethrine	Water	16.8	6.95	4.76	0.25	0.06	ND	—	—
	Sediment	161	175	—	424	263	—	187	165

[a] ND = not determined.

This strong tendency to accumulate in sediment could have taken place in the drainage channels and rivers, thus reducing the amount of endosulfan and pyrethroids in coastal waters, which explains their absence in the region. The DDT levels were similar to the values found in the region 2 years earlier as part of a study carried out by the Internaional Mussel Watch Project [3, 4].

4. CONCLUSIONS

(1) Only low concentrations of DDT residues were found in sediment and biota in the coastal region.
(2) Accumulation of endosulfan and some pyrethroids in sediment could have taken place in the drainage channels and rivers, thus preventing these pesticides from reaching the coastal region.

REFERENCES

[1] FOOD AND AGRICULTURE ORGANIZATION OF THE UNITED NATIONS, Sampling of Selected Marine Organisms and Sample Preparation for the Analysis of Chlorinated Hydrocarbons: Reference Methods for Marine Pollution Studies, No. 12, Revision 2, FAO, Rome (1991).
[2] UNITED NATIONS ENVIRONMENT PROGRAMME, Determination of DDT and PCBs by Capillary Gas Chromatography and Electron Capture Detection: Methods for Marine Pollution Studies, No. 40, UNEP, Nairobi (1988).
[3] NATIONAL OCEANIC AND ATMOSPHERIC ADMINISTRATION, International Mussel Watch Project: Initial Implementation Phase, Final Report, NOAA, Silver Spring, MD (1994).
[4] NATIONAL OCEANIC AND ATMOSPHERIC ADMINISTRATION, International Mussel Watch Project: Global Assessment of Environmental Levels of Chemical Contaminants, NOAA, Silver Spring, MD (1992).

BEHAVIOUR OF BHC AND DDT AT THE MOUTH OF THE ZHUJIANG RIVER, CHINA*

Fulong CAI, Zhifeng LIN, Ying CHEN,
Shumei CHEN, Jiadong YIANG,
Feng CAI, Lumin QIAN
Laboratory of Marine Radioactivity,
Third Institute of Oceanography,
State Oceanic Administration,
Xiamen, Fujian, China

Abstract

BEHAVIOUR OF BHC AND DDT AT THE MOUTH OF THE ZHUJIANG RIVER, CHINA.
 The BHC and DDT concentrations in sea water, sediment, aerosol and marine organisms sampled at the mouth of the Zhujiang River, China, were determined by gas chromatography. The results showed that the BHC and DDT concentrations were 11.11 and 25.67 µg/kg dry weight, respectively, in surface sediment, and 0.075 and 0.166 µg/dm^3, respectively, in sea water. In the dry season, the BHC and DDT concentrations in sea water were higher than those found in the rainy season. The BHC and DDT concentrations in surface sea water were lower than those in bottom sea water. However, in the rainy season the difference in pesticide concentration between surface and bottom sea water was not significant. The influence of salinity, dissolved oxygen and chemical oxygen dispersion on BHC and DDT distribution was not correlated. The BHC and DDT concentrations in aerosol were 0.136 and 0.078 ng/m^3, respectively. The BHC and DDT residues in sea water were not from runoff of the Zhujiang River, but resulted from sediment suspended in sea water. The results also showed that the relationship between the BHC and DDT concentrations was negatively correlated to the microbial population (bacteria) found in sediment. Transfer of BHC and DDT was typical in the marine food chain.

1. INTRODUCTION

Numerous surveys of organochlorine pesticides in the marine environment have been undertaken in China [1–7]. In 1994, we participated in an IAEA Co-ordinated Research Programme on the Distribution, Fate and Effects of Pesticides on Biota in the Tropical Marine Environment. We surveyed BHC and DDT residues at the mouth of the Zhujiang River and conducted radiotracer experiments in artificial marine microhabitats.

---

* Research carried out with the support of the IAEA under Research Contract No. 7932.

2. METHODS AND MATERIALS

Sea water and surface sediment at the mouth of the Zhujiang River were collected in November 1994 and July 1995, respectively (Fig. 1). Samples of aerosol and marine organisms in the above area were collected in July 1995, as described in Ref. [8].

The BHC and DDT concentrations were determined in samples of sea water, sediment, aerosol and marine organisms. In addition, the salinity, temperature, dissolved oxygen (DO) and chemical oxygen dispersion (COD) of sea water were measured in the field. Finally, the species and population of microorganisms were determined in sediment, together with the sediment particle size.

Gas chromatography (HP-5890A, with a ^{63}Ni electron capture detector) was used to detect BHC and DDT. Analyses were made according to the relevant Chinese standard procedure [8, 9].

3. RESULTS

3.1. BHC and DDT concentrations in sea water

The BHC concentration averaged 0.066 $\mu g/dm^3$ in surface sea water and 0.083 $\mu g/dm^2$ in bottom sea water. The DDT concentration averaged 0.061 $\mu g/dm^3$ in surface sea water and 0.271 $\mu g/dm^3$ in bottom sea water.

Table I shows that the BHC concentration in sea water was higher in the dry season (1.93 times (surface) and 2.44 times (bottom)) than in the rainy season, as was the DDT concentration (1.45 times (bottom) and 1.95 times (surface)). In the dry season, the BHC and DDT concentrations in bottom sea water were 1.35 and 6.33 times higher, respectively, than in surface sea water. However, these concentrations were not significant in the rainy season.

3.2. BHC and DDT concentrations in surface sediment

The BHC and DDT concentrations were 11.11 and 25.67 $\mu g/kg$ dry weight, respectively, in surface sediment. The concentration factors for BHC and DDT in surface sediment were 148 and 155, respectively. Table I shows that the BHC concentration in the rainy season was similar to that in the dry season, while the DDT concentration in the rainy season was lower than that in the dry season.

3.3. BHC and DDT concentrations in aerosol

The BHC and DDT concentrations were 0.136 and 0.078 ng/m^3, respectively, in aerosol (Table I). In a section of filter membrane, their concentrations appeared to be negligible (Table II).

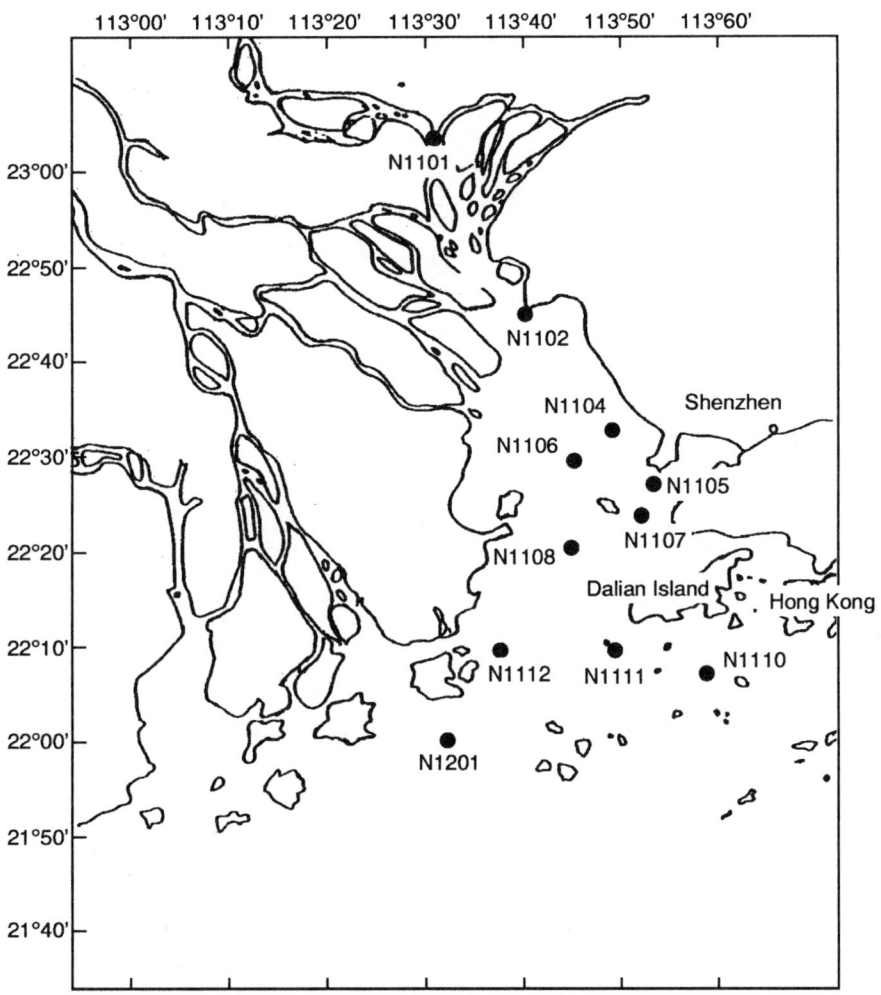

FIG. 1. Sampling sites for BHC and DDT at the mouth of the Zhujiang River.

TABLE I. BHC AND DDT CONCENTRATIONS IN SEA WATER, SEDIMENT AND AEROSOL[a]

Pesticide medium	Dry season		Rainy season	
	BHC	DDT	BHC	DDT
Surface sea water ($\mu g/dm^3$)	0.087 (0.057–0.156)	0.080 (ND–0.263)	0.045 (0.021–0.084)	0.041 (ND–0.086)
Bottom sea water ($\mu g/dm^3$)[b]	0.117 (0.036–0.305)	0.506 (ND–1.221)	0.048 (0.028–0.085)	0.035 (0.010–0.072)
Surface sediment ($\mu g/kg$ dry weight)	11.150 (4.980–20.27)	33.460 (17.79–51.71)	11.070 (2.130–24.65)	17.880 (4.13–83.84)
Aerosol (ng/m^3)			0.136 (0.038–0.262)	0.078 (0.051–0.130)

[a] ND = not detected ($< 0.004\ \mu g/L$).
[b] A depth of more than 10 m.

3.4. BHC and DDT concentrations in marine organisms

The BHC concentration in marine organisms ranged from 0.403 (*Laminaria*) to 8.507 µg/kg fresh weight (sea bird), and that of DDT from 1.770 (phytoplankton) to 207.60 µg/kg fresh weight (sea bird) (Table III).

3.5. BHC and DDT transfer in the marine food chain

Mytilus, a filter feeder, and the sea bird, which is at the top of the food chain, accumulated the most BHC (Fig. 2). Accumulation of DDT by marine organisms was higher than that of BHC (Fig. 2).

TABLE II. BHC AND DDT CONCENTRATIONS IN A SECTION OF FILTER MEMBRANE ($ng \cdot cm^{-2} \cdot min^{-1}$)

Sampling (h)	Section (m^3)	BHC	DDT
10	620.50	2.6×10^{-4}	3.1×10^{-4}
18	1116.95	9.0×10^{-5}	1.3×10^{-4}
24	1490.61	6.2×10^{-4}	1.2×10^{-4}

TABLE III. BHC AND DDT CONCENTRATIONS IN MARINE ORGANISMS (µg/kg FRESH WEIGHT)

Marine organism	BHC	DDT	Marine organism	BHC	DDT
Phytoplankton	1.890	1.770	Prawn	3.161	8.54
Zooplankton	1.505	4.370	Crab	2.396	70.13
Sargassum	1.396	7.437	Collichthys	1.244	72.63
Laminaria	0.403	9.895	Solicidae	1.156	96.05
Mytilus	5.570	49.80	Sea bird	8.507	207.60
Mactra	1.577	3.090			

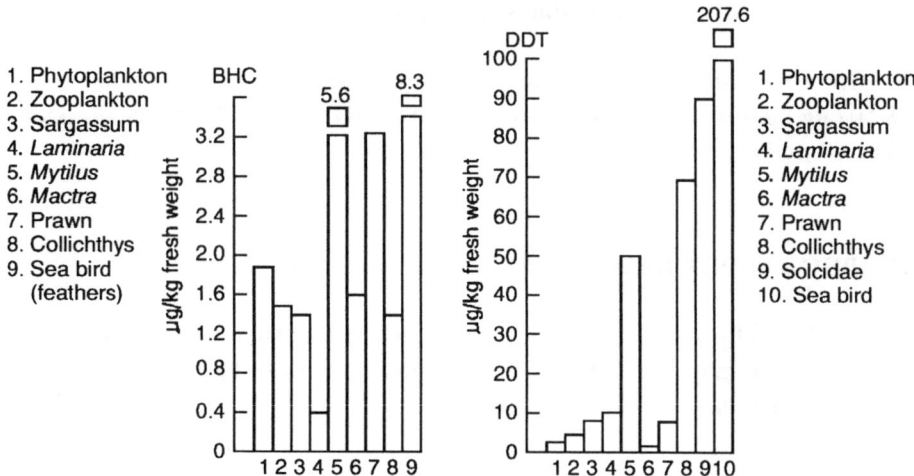

FIG. 2. BHC and DDT accumulation in the marine food chain (numerals above columns are the highest accumulation values).

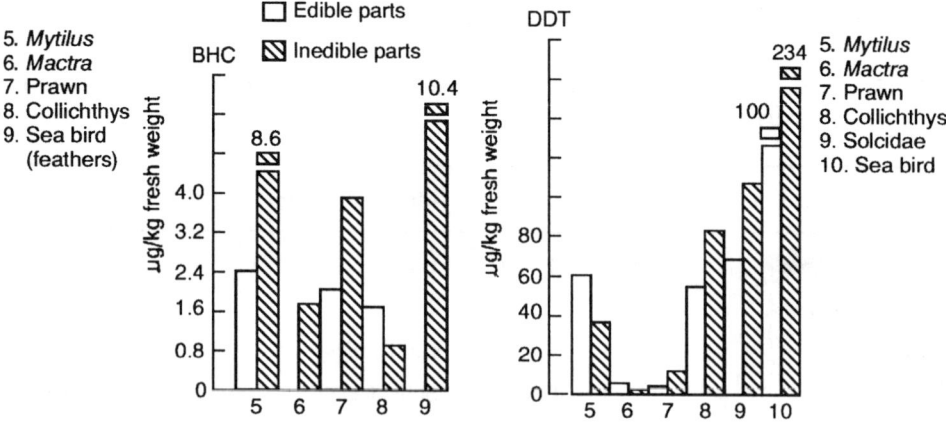

FIG. 3. BHC and DDT distribution in the body of marine organisms, sampled during the rainy season (numerals above columns are the highest distribution values).

TABLE IV. BHC AND DDT CONCENTRATION FACTORS IN THE EDIBLE PARTS OF MARINE ORGANISMS[a]

Marine organism	BHC	DDT	Marine organism	BHC	DDT
Phytoplankton	50	38	*Mactra*	34	116
Zooplankton	40	94	Prawn	60	80
Sargassum	37	161	Collichthys	46	1256
Laminaria	11	213	Solcidae	24	1585
Mytilus	66	1338	Sea bird	ND	2166

[a] ND = not detected (< 0.3 µg/kg fresh weight).

3.6. BHC and DDT distribution in the body of organisms

In most marine organisms, organochlorine pesticides were concentrated in the inedible parts, although *Mactra* did not accumulate (Fig. 3) significant amounts of DDT easily. Table IV shows the concentration factors of BHC and DDT in the edible parts of marine organisms. This phenomenon has been identified in earlier research [10].

The BHC and DDT concentrations were highest in the feathers and digestive tract of the sea bird. The values were 41.33, 0.306, 287.5 and 308.1 µg/kg fresh weight, respectively. In the viscera, bone and muscle of the sea bird, BHC was not detected.

3.7. Influence of some factors on the BHC and DDT concentrations

3.7.1. Microbial population

Despite some variability, the relationship between the BHC and DDT concentrations was negatively correlated to the microbial population (bacteria) found in surface sediment (Fig. 4). Microbial degradation of BHC and DDT may have been the mechanism responsible [1].

3.7.2. Other factors

The influence of salinity, DO and COD on the distribution of BHC and DDT in sea water was not correlated in two seasons, nor was there any significant correlation between the BHC concentration, the DDT concentration and the sediment particle size.

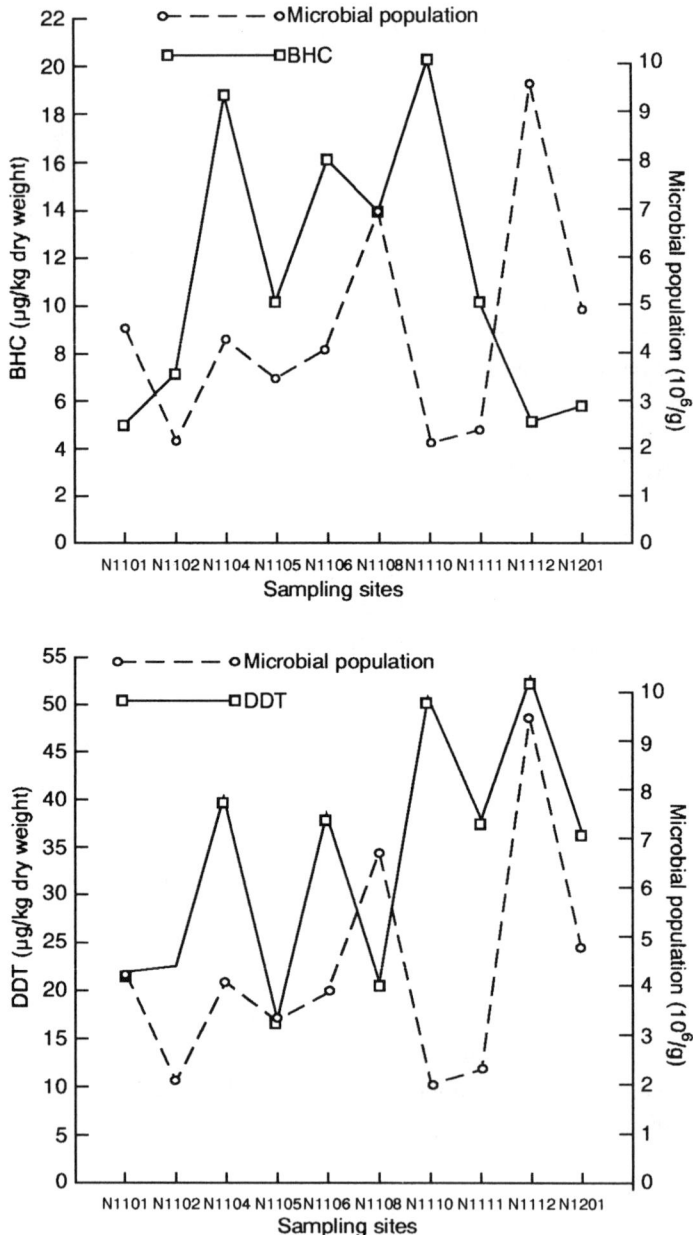

FIG. 4. Relationship between the BHC and DDT concentrations and the microbial population in surface sediment at the mouth of the Zhujiang River.

4. DISCUSSION

4.1. BHC and DDT distribution at the mouth of the Zhujiang River

The results suggest that the BHC and DDT concentrations were diluted by runoff from the Zhujiang River in the rainy season. Similar BHC and DDT concentrations in surface and bottom sea water indicate uniform mixing of the layers in this season. The influence of salinity, DO and COD on the BHC and DDT distribution in sea water was not correlated. The BHC and DDT concentrations in aerosol were negligible. Thus, the BHC and DDT residues in sea water were not from the Zhujiang River but may have resulted from the sediment suspended in sea water. From 1955 to 1982, 31 862 t of organochlorine pesticide were used annually in the Guangdong coastal zone [2]. The distribution isopleth of BHC and DDT in sea water and sediment in the two seasons reflected the hydrodynamic conditions that exist at the mouth of the Zhujiang River [11, 12].

4.2. BHC and DDT concentrations in marine organisms

The marine organisms concentrated more DDT than BHC. The concentration factors for BHC ranged from 11 to 60, and for DDT from 38 to 216.

BHC and DDT accumulated in the high tropic levels of the marine food chain, which is in agreement with previous reports [1, 3, 4]. Regarding the distribution of BHC and DDT in marine organisms, accumulation in inedible parts was higher than that in edible parts. Faster metabolism in edible parts or lower concentrations of fat may have been responsible.

REFERENCES

[1] LU, Yongqi, DING, Meili, Marine Pollution Biology, China Ocean Press, Beijing (1991).

[2] CHINA COASTAL ZONE OFFICE, Survey Report on the Environmental Quality of the Chinese Coastal Zone, China Ocean Press, Beijing (1987) 506–513.

[3] MIAO, Qiang, The pollution of organochlorine pesticide in seawater, sediment and organisms in the sea area of western Guang Dong, Dalian, Mar. Environ. Sci. 2 4 (1983) 37–40.

[4] MIAO, Qiang, Initial survey on the distribution character of residual organochlorine pesticide in benthon by gas chromatography, Dalian, Mar. Environ. Sci. 2 1 (1983) 95–101.

[5] CHEN, Xulong, The accumulation and distribution of BHC and DDT in fish and their migration in the ecological system of the Buohai Sea, Dalian, Mar. Environ. Sci. 2 3 (1983) 14–19.

[6] WU, Shengshan, The residues of BHC pesticide in the Jioulang River mouth and Xiamen harbor, Xiamen, Taiwan Strait, Mar. Environ. Sci. **2** 1 (1983) 29–34.

[7] YE, Xinrong, The PCBs, BHC and DDT in the Chang Jiang River mouth and the near waters, Dalian, Mar. Environ. Sci. **10** 4 (1991) 52–56.

[8] STATE OCEANIC ADMINISTRATION, Specifications of Oceanography Survey, Trade Standards HY003, 1-91-HY/T003, 10-91, SOA, Beijing (1991).

[9] STATE OCEANIC ADMINISTRATION, Specifications of Oceanography Survey, National Standards GB 12763, 6-91, SOA, Beijing (1991) 10–17.

[10] CAI, Fulong, et al., Transfer of radioactive nuclides from seawater to sediment and marine organisms. 2. Function of biological procedure, Acta Sci. Circumst. **5** 3 (1985) 335–340.

[11] ZHOU, Ying, The hydro force condition and suspension motion in the coastal region of the Zhujiang River, Guangzhou, South Chinese Sea **1** (1985) 59–64.

[12] HUANG, Fang, The balance model of water circulation and salinity in the Zhujiang River mouth area, South Chinese Sea **2** (1986) 24–28.

COMPARISON OF FATE AND BEHAVIOUR OF PESTICIDES IN DIFFERENT ENVIRONMENTS

(Session 5)

Chairperson

E. CARAZO

Costa Rica

Rapporteur

C.A. EDWARDS

United States of America

IAEA-SM-343/8

A DECISION-SUPPORT SYSTEM FOR PESTICIDE ENVIRONMENTAL PREREGISTRATION ASSESSMENT

P.H. NICHOLLS
Integrated Approach to Crop Research (IACR)-Rothamsted,
Harpenden, Hertfordshire,
United Kingdom

Abstract

A DECISION-SUPPORT SYSTEM FOR PESTICIDE ENVIRONMENTAL PREREGISTRATION ASSESSMENT.
  Many aspects of the fate and behaviour of organic compounds in the environment can be predicted from a few physicochemical properties. Such predictions give a useful scientific perspective that aids in evaluating a data package. A decision-support system called Physicochemial Evaluation — The Environment was developed for the pesticide registration authorities so that they can obtain a broad theoretical prediction of the behaviour of pesticides against which data provided by applicants for registration can be interpreted. The Windows based system provides data on sorption to soils, the translocation of compounds in plants, and their bioconcentration, volatilization, biodegradation and leaching to surface and groundwaters. A detailed description of sorption to soils is given. Innovative features include predictive routines for ionizable compounds and an empirical method for estimating the trace concentrations of compounds in lysimeter leachates. Operation of the system is fast, and can be learned quickly. A database assembled from the literature and containing most of the properties of over 800 compounds speeds up the preparation of input data. Additional information, such as soil texture diagrams, is stored in the context sensitive Windows Help system.

1. INTRODUCTION

Many factors that quantify the potential for a compound to redistribute in the environment can be predicted from a few physical properties. A program [1] that covers a range of such calculations is described. The theoretical output from the program can be used to provide a scientific perspective against which experimental data can be critically evaluated. In addition, the understanding that develops from using the program will help an evaluator make a greater constructive contribution to interpreting a number of topics, including residues in food and their effects on wildlife.

2. INPUT DATA

One of the strengths of the system is that it requires as input only six main items of physicochemical data (namely, the log K_{ow}, melting point, vapour pressure, molecular weight, pK_a and half-life in soil). Five of these items are usually provided at the beginning of a data package. The most difficult item is the half-life in soil measured at a well defined temperature and soil–water content. Input data for almost all commercial compounds are displayed automatically from a database.

3. METHODS

3.1. Soils

The values of the soil sorption coefficients of non-ionized compounds, including K_d, K_{oc} and the R_f value, on soil thin layer chromatographic plates, are calculated from the 1-octanol/water partition coefficient (K_{ow}). The sorption relationship for non-ionized compounds given below is taken from the well known equation of Briggs [2]

$$\log K_{OM} = 0.52 \log K_{ow} + 0.62$$

where K_{ow} is the octanol/water partition coefficient and K_{OM} is the coefficient of sorption to soil organic matter (OM).

The calculated values of K_d are compared with the literature values for soils S and O in Table I [3] for non-ionized atrazine and its desalkylated degradation products. Unusually in the desalkylated compounds, the measured sorption did not decrease with the decreasing values of $\log K_{ow}$.

Many herbicides are monobasic organic acids, so their sorption is additionally a function of their pK_a and of the pH of the soil. Nicholls and Evans [4] gave the relationships for calculating the soil sorption coefficient (K_d) for acids, shown below, that require the pH and organic carbon (OC) content for soil, and the pK_a and lipophilicity of the undissociated molecule

$$K_d = (K_d^u (1 - P^-)) + (K_{d^-} P^-)$$

$$P^- = (10^{pH - pK_a - 1.8})/(1 + (10^{pH - pK_a - 1.8}))$$

Here, P^- is the proportional of the anionic form. The pH is the measured value of the pH of soil, and pK_a is the acid dissociation constant. The factor 1.8 allows for the pH close to the surface of soil particles to be more acidic than that in the bulk solution. K_d^u is the sorption of the non-ionized form and is calculated from the K_{oc} of the non-ionized form. K_{d^-}, the sorption of the anionic form, is calculated from the $\log K_{ow}$ value of the anion, which is taken as 3.4 units less than the $\log K_{ow}$ value of the non-ionized form using the equation

$$\log K_{OM} = 0.58 (\log K_{ow} - 3.4) + 1.3$$

TABLE I. SORPTION AND PHYSICAL PROPERTIES OF ATRAZINE AND ITS METABOLITES

Soil	OM[a] (%)	pH	Data	K_d (L/kg)			
				Atrazine	Desethyl-atrazine	Desisopropyl-atrazine	Hydroxy-atrazine
S	14	5.0	Literature[b]	14	6.5	8.6	82
S	14	5.0	Calculated	13	4.0	2.2	24
O	1	7.3	Literature[b]	0.4	0.2	0.4	2
O	1	7.3	Calculated	0.9	0.3	0.2	9
			Log K_{ow}	2.6	1.6	1.1	1.2
			pK_a	—	—	—	5.3

[a] Soil organic matter content.
[b] Data of Brouwer et al. [3].

TABLE II. SORPTION OF 2-FLUOROBENZOIC ACID BY A SILTY CLAY LOAM

Soil pH	3.6	4.4	6.6	7.2	9.0
Measured K_d	0.26	0.12	0.06	0.02	−0.03
Predicted K_d	0.81	0.69	0.10	0.08	0.07

The predicted and measured values of sorption for 2-fluorobenzoic acid are given in Table II and illustrate the decrease in sorption with pH as the proportion of the anionic form increases. These and other results indicate that the predicted values for acids can be as accurate as those predictions usually obtained for non-ionized compounds.

Many fungicides, including morpholines, pyrimidines and benzimidazoles, are bases, and ionization influences their sorption to soil. Organic cations and the protonated forms of bases can be strongly sorbed by the mechanism of cation exchange. Experiments [4] using a range of cations and bases show that the intensity of sorption increases with the degree of delocalization of the positive charge on the cation. Sorption of such compounds to soil organic carbon (K_{oc}) is calculated as a function of their pK_a and a measure of the delocalization of electrostatic charge on the cation, together with the pH value of the soil

$$\text{Pseudo } K_{oc} = K_d \, 100/OC$$

$$K_d = (K_d^u \, P^u) + ((1 - P) \, K_d^+)$$

where K_d^u is the sorption coefficient of the non-ionized form and is obtained from

$$P^u = (10^{pH - pK_a - 1.8})/(1 + 10^{pH - pK_a - 1.8})$$

which is the proportion of the non-ionized form in soil solution. OC is the soil organic carbon content (%) and K_d^+ is the sorption coefficient of the protonated form. Sorption of the protonated form is calculated using $K_d^+ = (80 \text{ Catdeloc})$, where Catdeloc is a factor that allows for delocalization of the cationic charge.

Not delocalized: Catdeloc = 0.1, as in aliphatic amines;
Part delocalized: Catdeloc = 0.3, as in anilines;
Delocalized: Catdeloc = 1, as in pyridines.

All relationships are empirical rules of thumb, based on the experience gained from obtaining data for Ref. [4]. The pseudo K_{oc} term is used because sorption of ionizable compounds is a function of pH as well as the OC content. Hydroxyatrazine is a base and its sorption is predicted to be stronger than that of the parent atrazine, which agrees with the observed values (Table I). In the future, a relationship that additionally includes the lipophilicity of the cation will be developed.

3.2. Translocation within plants

Several factors that indicate the systemicity of compounds in plants are calculated using the methods of Briggs et al. [5]. The root concentration factor (RCF) is a measure of the potential for a compound to be taken up from water by roots. RCF is the concentration in roots/concentration in solution, i.e. it is a roots/water partition coefficient. The minimum value of 0.82 represents the volume in living roots not occupied by solids, termed free space. Lipophilic compounds can reach high concentrations in root tissue

$$\log (RCF - 0.82) = 0.77 \log K_{ow} - 1.52$$

Translocation from roots to shoots depends on permeation of the plasmalemma membrane of the endodermal cells to enter the symplasm. The efficiency of this process, relative to water, is measured by a transpiration stream concentration factor (TSCF)

TSCF = (concentration in xylem sap/concentration in the external solution bathing the roots)

$$\text{TSCF} = 0.784 \exp(-[(\log K_{ow} - 1.78)^2]/2.44)$$

The equation gives an optimum value for membrane permeability for log K_{ow} of about equal to 2. TSCF is a measure of the potential of a compound to be systemic in xylem when uptake is via the roots. Very lipophilic and very polar non-ionized compounds are not systemic.

However, once polar compounds (log K_{ow} < 2) have entered the xylem they are efficiently translocated into the transpiration stream, unlike lipophilic compounds (log K_{ow} > 5) such as pyrethroids, which are immobile. This is important for polar compounds such as chlormequat chloride applied to foliage.

The TSCF can be calculated for bases as well as for non-ionized compounds by allowing for any protonation that might occur in the xylem.

Translocation of acids in phloem and xylem is indicated graphically and quantified numerically by dynamic modelling of the plant's vascular system [6].

3.3. Bioconcentration factors (BCFs)

A range of BCFs (namely, for protein tissue, lipids, fish muscle, earthworms, whole organisms (e.g. fish) and *Daphnia pulex*) are calculated using equations, given below, obtained from the literature [7–10]. The BCF is an organism/water partition coefficent.

Whole organisms (including fish)	log BCF = 0.76 log K_{ow} − 0.5
Protein tissue	log BCF = 0.62 log K_{ow} + 0.46
Earthworms (dry weight basis)	log BCF = 0.48 log K_{ow} + 1.04
Fish muscle	log BCF = 0.56 log K_{ow} + 0.1
Lipids	log BCF = log K_{ow}

(generally the slope coefficient of the equation is a function of the lipid/protein mass rates of an organism)

Daphnia pulex	log BCF = 0.75 log K_{ow} − 0.43

The relationships may not hold for ultra lipophilic compounds (log $K_{ow} > 7$) nor for those with molecular masses greater than 650. Values are also calculated for acids and bases using the Henderson–Hasselbach equation and assuming that the log K_{ow} for the ion is 3.4 units less than that of the undissociated form.

The proportions and concentrations of the compound in a mesocosm at equilibrium are calculated. The components of the mesocosm are water, sediment, fish and *Daphnia*. For ionizable compounds, the pH of the water in the mesocosm can be set to any appropriate value.

In general, toxicity cannot be predicted from the physical properties of a compound, so toxicological endpoints are not given.

3.4. Volatility

Several properties that influence the volatilization of the compound, including Henry's constant and an air/soil partition coefficient, are calculated. The rates of volatilization from soil, water and foliage are predicted. An example is the method used to approximate the rates of volatilization from foliage in still air conditions at 20°C (summer). The method was derived from a method of Briggs [11], which in turn was derived from reliable data by Guckel et al. [12]. However, upward flux of pesticide is proportional to the gradient of the vapour density of the compound. Guckel et al. [12] found that a plot of log (vapour pressure) against the logarithm of the rate of volatilization (mol·ha^{-1}·d^{-1}) was linear. Assuming that the molecular weights of pesticides are similar, the approximation given below can be derived from Guckel's data

$$\text{AmtVol} = 10^{2.303 \log (\text{vapour pressure}) + 5.1}$$

where AmtVol is the amount volatilized (g·ha^{-1}·d^{-1}) under the conditions of Guckel's experiment (20°C and an air flow of 50 L/h; the apparatus volume was not given; the units of vapour pressure are Pa).

The rate of volatilization is the weight, not the proportion, of compound volatilized in unit time. Hence, it is not a first order but a zero order process, being a function of the amount of compound applied. Theoretically, the loss is linear and the DT_{50} value is half the time of complete disappearance. In practice, other processes will reduce such linearity and another function will operate as a unimolecular layer is approached (about 2 g/ha).

A practical example of the influence of the application rate was seen at Rothamsted when a relatively involatile fungicide (with high activity against *Erysiphe graminis*), triadimefon (vapour pressure = 0.00002 Pa), applied at moderately high rates, reduced mildew in untreated control plots adjacent to treated plots. The evaporative flux was just enough to affect cereal mildew.

TABLE III. QUANTITIES OF HERBICIDE PREDICTED TO LEACH THROUGH A LYSIMETER

	Isoproturon	Pendimethalin	Metsulturon-methyl	Mecoprop
Input				
Log K_{ow}/pK_a	2.24/–	5.18/–	1.64/3.64	3.2/3.18
Half-life (d)	25	50	60	4
Application rate (g/ha)	2500	100	12	2500
Output				
Amount leached in 1 year below 1 m (% of applied)				
Structured soil[a]	0.05	<0.005	0.99	0.02
Chromatographic flow[a]	<0.005	<0.005	0.01	<0.005
Concentration				
Average concentration in leachate over 1 year (μg/L)				
Structured soil[a]	0.3	<0.005	0.04	0.13
Chromatographic flow[a]	<0.005	<0.005	<0.005	<0.005

[a] Predictions in a non-structured soil.

3.5. Rates of degradation in soil

It is not usually possible to predict the rates or routes of degradation of compounds in soil, so a half-life measured at a well defined temperature and soil–water content needs to be entered. Then the program can calculate half-lives at other temperatures and soil–water contents using the methods of Walker [13]. Different units of soil–water content tend to be used by scientists from different disciplines, so routines are available to make conversions between different units. Soil texture diagrams can be displayed in the Windows Help system.

Some compounds such as DDT are degraded because they are strongly sorbed; these interactions can be investigated by calculating the rates of degradation for the soil–water phase.

The half-lives of metabolisms are calculated for the compound in mammals, plants and stored grain. However, the method assumes that the rates of metabolism in the different media are correlated with the rates of respiration in those media.

3.6. Leaching through lysimeters

The potential of a compound to leach (assessed by its sorption and persistence alone) is indicated by a graphic display of the groundwater ubiquity score index [14].

Such indicators of leaching potential do not take account of the rate of application of the compound, which is an important factor when considering the fate of highly active compounds such as pyrethroid insecticides and sulfonylurea herbicides. In an empirical routine, the concentration of the compound in the lysimeter leachate is estimated when the rate of application is entered. Predictions are for 1 year of exposure to average weather conditions in the United Kingdom.

The influence of lipophilicity, acidity, half-life and rate of application on the amounts leached is illustrated by the predictions given in the Table III. The results are consistent with UK experimental results from lysimeters and from field monitoring studies. Further study of experimental data will be required before the method can be applied to other climates.

4. CONCLUSIONS

The system is designed to be of interest to those working on registration data packages for the fate and behaviour of compounds in the environment. The system includes a full Windows Help file containing sources, methods and references, and for comparison purposes data on a range of benchmark compounds. A decision-support system such as that described allows the results of research to be very rapidly transferred from the bench to the user.

REFERENCES

[1] NICHOLLS, P.H., "Physicochemical Evaluation — The Environment, an expert system for pesticide preregistration assessment", Pests and Diseases (Proc. Conf. Brighton, 1994), British Crop Protection Council, Farnham, UK (1994) 1337–1342.

[2] BRIGGS, G.G., Theoretical and experimental relationships between soil adsorption, octanol–water partition coefficient, water solubilities, bioconcentration factors and the parachor, J. Agric. Food Chem. **29** (1981) 1050–1059.

[3] BROUWER, W.W.M., BOESTEN, J.J.T.I., SIEGERS, W.G., Adsorption of transformation products of atrazine by soil, Weed Res. **30** (1990) 123–128.

[4] NICHOLLS, P.H., EVANS, A.E., Sorption of ionisable organic compounds by field soils, Pestic. Sci. **33** (1991) 319–345.

[5] BRIGGS, G.G., BROMILOW, R.H., EVANS, A.A., Relationships between lipophilicity and root uptake and translocation of non-ionised chemicals by barley, Pestic. Sci. **13** (1982) 495–504.

[6] KLEIER, D.A., Phloem mobility of xenobiotics. V. Structural requirements for phloem-systemic pesticides, Pestic. Sci. **42** (1994) 1–11.

[7] LORD, K.A., BRIGGS, G.G., EVANS, A.A., Uptake of pesticides from water and soil by earthworms, Pestic. Sci. **11** (1980) 401–408.

[8] SOUTHWORTH, G.R., BEAUCHAMP, J.J., SCHMIEDER, P.K., Bioaccumulation potential of polycyclic aromatic hydrocarbons in *Daphnia pulex*, Water Res. **12** (1978) 973–977.

[9] ISNARD, P., LAMBERT, S., Estimating bioconcentration factors from octanol–water partition coefficient and aqueous solubility, Chemosphere **17** 1 (1988) 21–34.

[10] NEELY, W.B., BRANSON, D.R., BLAU, G.E., Partition coefficients to measure bioconcentration potential of organic chemicals in fish, Environ. Sci. Technol. **8** (1974) 1113–1115.

[11] BRIGGS, G.G., AgrEvo UK Ltd, Saffron Walden, unpublished data, 1995.

[12] GUCKEL, W., SYNNATSCHKE, G., RITTIG, R., A method for determining the volatility of active ingredients used in plant protection, Pestic. Sci. **4** (1973) 137–147.

[13] WALKER, A., Simulation of herbicide persistence in soil, Pestic. Sci. **7** (1976) 41–49.

[14] GUNSTAFSON, D.I., Groundwater ubiquity score, Environ. Toxicol. Chem. **8** (1989) 339–357.

IAEA-SM-343/4

EFFECTIVE USE OF PESTICIDE DATA FOR SUSTAINABLE FARMING IN AUSTRALIA AND CHINA*

I.G. FERRIS**,***, R.S. KOOKANA[+],
Xianfang WEN[++], B.M. HAIGH***, Jiarong PAN[++]

***Tamworth Centre for Crop Improvement,
NSW Agriculture,
Tamworth, New South Wales,
Australia

[+]Division of Soils,
Co-operative Research Centre for Soil
and Land Management,
Commonwealth Scientific and Industrial
Research Organization,
Glen Osmond, South Australia,
Australia

[++]Institute for the Application
of Atomic Energy,
Chinese Academy of Agricultural Science,
Beijing, China

Abstract

EFFECTIVE USE OF PESTICIDE DATA FOR SUSTAINABLE FARMING IN AUSTRALIA AND CHINA.

A strategy towards sustainable farming in Australia and China by making more efficient use of pesticide data is outlined. Specifically: (1) the pesticide dilemma — why the problem arose; (2) integrated pest management (IPM) — what strategies are needed for the future; (3) solutions — how multidisciplinary research can help solve the pesticide dilemma; and (4) implementation — how to reconcile the economic and environmental objectives. The Physicochemical Evaluation of the Environment (PETE) program predicted the organic carbon partition coefficient (K_{oc}) and the half-lives of atrazine and pirimicarb within an order of magnitude. Random samples of 247 pesticides drawn from a total of 585 were classified

---

* Research carried out in co-operation with the IAEA under Technical Co-operation Project No. CPR/5/008.
** Present address: Agrochemicals and Residues Section, Joint FAO/IAEA Division of Nuclear Techniques in Food and Agriculture, International Atomic Energy Agency, Vienna.

by oral LD_{50} to the rat, sorption (K_{oc}) and the half-life in soil. The pesticides fell into four categories: (a) 38% in a green zone that can be used with caution in sustainable farming systems — at issue is a lack of support for the agrochemical industry, including positive incentives to invest in the synthesis of pesticides that fit into a narrow band of physicochemical and toxicological properties; (b) 22% in an orange zone — at issue is IPM compatibility, conservation of useful pesticides and post-registration monitoring; (c) 24% in a red zone that is incompatible with the goals of sustainable farming, and phasing out is suggested — at issue is an acceptable phasing out period and research funding for alternative pest management strategies; and (d) 16% that remained unclassified because of missing data — at issue is a lack of data, primarily the pesticide half-life in soils. Although atrazine and pirimicarb fell in the orange zone, hazard assessment and long term simulations identified strategies to reduce risk. It was concluded that pesticides have a role to play in meeting the challenges of 21st century agriculture provided that: (i) public pressure is reduced for their elimination from farming systems; (ii) common goals are found with environmental groups; (iii) IPM is adopted; (iv) communication is improved between farmers (decision makers) and researchers (information providers); (v) farmers and support institutions respond to the global economic, social and environmental changes now in train; (vi) information technologies are made more affordable for rural communities; (vii) farmers receive a better return for their products; and (viii) sustainable farming practices are adopted that are economically and socially profitable, yet preserve environmental quality and biodiversity. Quality assurance programmes such as the International Standards Organization report ISO 14 000 offer a framework within which these objectives can be achieved.

1. INTRODUCTION

Pesticides are one of the great achievements of 20th century science. The so called Green Revolution (new plant varieties, fertilizers and pesticides) freed billions of people from the toil and misery of hunger and disease. Today, pesticides invoke fear and are often cited as a cause of environmental degradation. The simplest explanation for the change in perception on pesticides is that the magnitude of the problem confronting farmers and research organizations is not appreciated by the general community, particularly those with abundant food supplies [1]. Admittedly, the decline of the family farm has isolated the urban community from direct contact with food production systems. Yet public concern about pesticides also stems from more deep rooted causes. Controversy tends to shape their 'image'. Silent Spring [2] first raised the spectre of an environmental catastrophe from pesticide use. Herbicide controversy following the Viet Nam war, off-target drift, management of toxic waste and the banning of Australian beef by overseas markets have eroded public confidence in existing regulatory safeguards and reinforced fear about the adverse impact of pesticides. Scientists initially disregarded public concern because some of the assertions on pesticide risks were wrong, without acknowledging the genuine

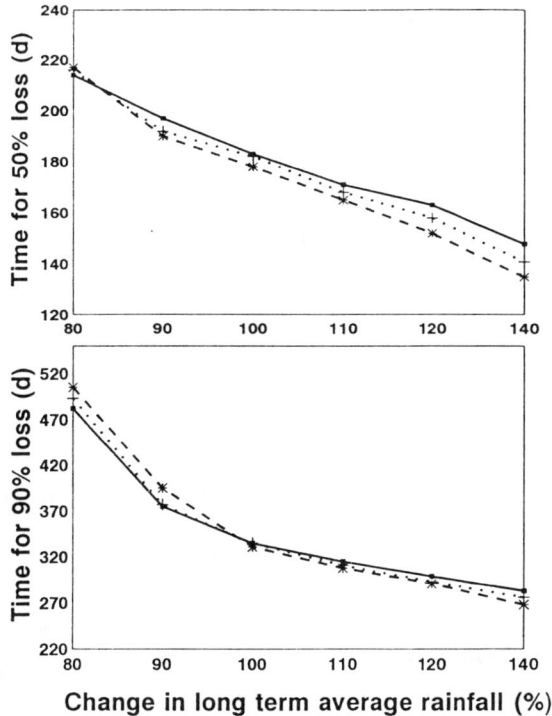

FIG. 1. *How climate changes could affect atrazine degradation: 100 year simulations with atrazine applied at 2.5 kg/ha during mid-autumn using the CALF (Varleach) model. At Tamworth, increasing rainfall is the dominant greenhouse factor affecting atrazine loss. The upper estimate for a greenhouse induced temperature increase of 4°C had only a minor effect on atrazine loss (see Haigh and Ferris [5] for model parameters) (temperature increase: ■—■ = 0°C; +··+ = 2°C; and *— —* = 4°C).*

underlying public concern about pesticide hazards (potential to cause harm). Today, litigation on pesticides diverts scarce technical resources away from the key issues confronting agriculture.

The United Nations Conference on Environment and Development held in Rio de Janeiro in 1992 committed Member States to the sound management of toxic compounds. Subsequently, EEC Directive 91/414 outlined unequivocal trigger and cut-off values for pesticides to protect the environment [3]. Currently, government agencies in Australia and China are considering pesticide risk reduction policies [4]. However, indiscriminate banning of pesticides may have long term repercussions (Fig. 1) [5], given the spectre of 'eco-political disasters' [6].

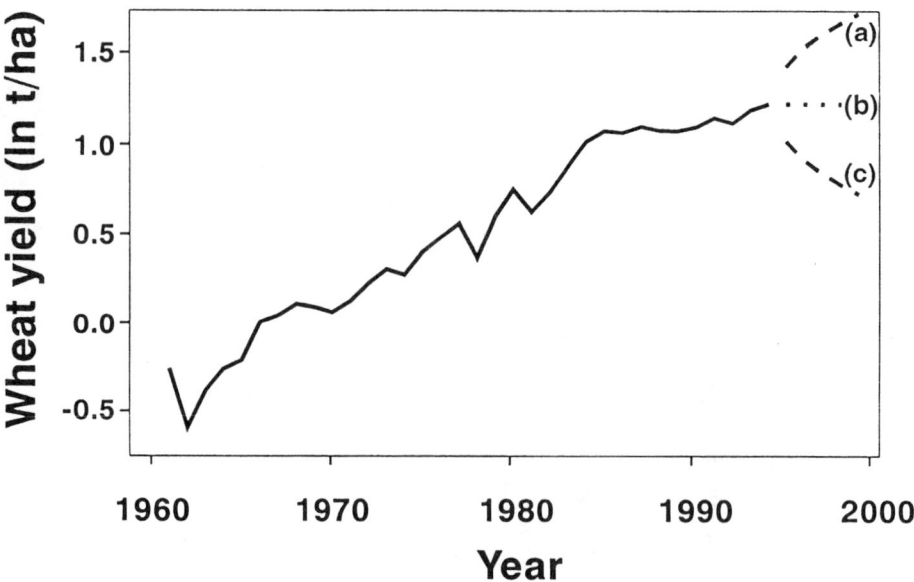

FIG. 2. Chinese wheat yields and forecasts to 2000. Data from the United States Department of Agriculture PSD program and forecasts from Alston [9] ((a) upper confidence limit; (b) forecast yield; and (c) lower confidence limit).

International agreements and adverse public perception of pesticides now threaten to overshadow the opportunities they provide to meet the unprecedented economic, social and environmental challenges of the 21st century [7]. According to Cassman [8], cereal production must increase by 60% within 25 years, but yield on prime agricultural land will have to show an even greater increase if the pressure on the remaining wilderness areas is to be reduced. Yet forecasts suggest that meeting cereal targets is becoming unlikely (Fig. 2) [9] as the 'easy ground has been conquered' [7]. Expansion of organic farming or introduction of genetically engineered crops could reduce the dependency on pesticides, but these measures are untested on a global scale. Reduced yields [10] can place additional pressure on forests and wetlands. In the case of genetically engineered organisms, regulatory issues are formidable obstacles [11, 12]. The issues are complex, covering scientific, economic and social dimensions. To date, public concern about pesticides has been addressed on an ad hoc basis. Most often, remedies are sought only after problems occur. In Australia, the response is often a ban on the offending pesticide. Most recently, the herbicide atrazine has come under scrutiny by the National Registration Authority. A ban on atrazine could jeopardize no tillage systems [13] that reduce carbon dioxide emissions [14] and soil erosion [15]. Combellack [16, 17] proposed an alternative to ad hoc banning of pesticides, namely, a phased reduction in herbicide use combined with targeted research to improve the efficiency of herbicide application and 'non-chemical' weed control strategies. Such a proposal addresses a fundamental need: equitable and sustainable farming systems [18, 19].

If an agreement is reached with environmental groups on common goals, the focus of the pesticide debate becomes meeting the need of farmers for short term survival and community expectations for sustainable land use [19]. Integrated pest management (IPM) bridges these needs [18, 20] and offers solutions for dealing with the major environmental issues — soil erosion [15, 21], and water and air pollution [22]. IPM demands new management skills and access to relevant knowledge, specifically the impact and fate of pesticides at the farm level. The latter include three unresolved issues: (1) how to obtain relevant information/technology; (2) which pesticides are compatible with sustainable farming systems; and (3) how to apply pesticides more efficiently. Can traditional agricultural research provide these data? Typically, replicated field trials are conducted over many sites and seasons. In Australia and China such an approach cannot succeed in the short term given that: (1) there are 600 known pesticide active ingredients and an unknown number of metabolites; (2) Australia and China represent less than 5% of the world's pesticide market and have large areas of 'problem' sodic and alkaline soils; (3) there is immense variation in climate and soil type; and (4) there are limited technical and financial resources for environmental toxicology. Nor is it appropriate to rely entirely on experience from abroad, without some testing under Australian and Chinese conditions. This paper outlines a proposal for more efficient use of pesticide data to develop more sustainable farming systems in Australia and China, including: (a) highlighting the utility of the Physicochemical Evaluation of the Environment (PETE) program for preliminary environmental assessments; (b) classifying pesticide hazards according to the Bernson [23] scheme; (c) identifying the pesticide risk reduction strategies; and (d) outlining the implementation options.

2. THE PETE PROGRAM

2.1. Background

The PETE program was developed to predict the fate of pesticides in soil, water and air [24]. It consists of a database of over 400 pesticides. Some properties are measured, such as the soil half-life, but where possible properties are predicted from simple physicochemical data, such as the solubility, melting point and octanol–water partition coefficients. This approach was advocated by Briggs [25], who realized that for non-ionic compounds partition coefficient (K) — pesticide in organic solvent ⟨=⟩ pesticide in water; bioconcentration factor (BCF) — pesticide in organism ⟨=⟩ pesticide in water; and

soil sorption K_d — chemical sorbed on soil ⟨=⟩ pesticide in water

are all interrelated. Any pair is related by a simple log–log relationship [25]

$$\text{Log } K_1 = \alpha \log K_2 - \beta$$

where K is the partition coefficient, while α and β are constants. The conceptual framework was extended to cover ionizable compounds [26, 27] and processes such as plant uptake [28].

Operation of PETE is straightforward under Microsoft Windows®. The common name of the pesticide is entered. If an entry exists, PETE indicates if the pesticide is a solid or liquid, and its $\log_{10} K_{oc}$. The K_d values are calculated for a soil with an organic carbon content of 1%. PETE calculates several indices, such as Helling's mobility class, to indicate the leaching potential. Users may enter their own organic carbon content or soil pH. K_{oc} is used subsequently to determine the plant uptake, bioconcentration and volatilization. Finally, leaching is estimated given a specific sorption, half-life and rate of application.

TABLE I. PREDICTED[a] AND EXPERIMENTAL K_{oc} (L/kg) AND HALF-LIFE (d) FOR ATRAZINE AND PIRIMICARB

	Observed	Predicted
	Atrazine (Australian soils)	
K_{oc}	217 ± 180	118
Half-life	69 ± 58	44
	Pirimicarb (Chinese soils)	
K_{oc}		
pH5.1	5350	6015
pH7.1	752	228
pH8.6	2550	69
Half-life	12 ± 4	14

[a] PETE program [24].

2.2. Assessment of atrazine and pirimicarb hazards

Atrazine and pirimicarb were selected to test the PETE program. The results, summarized in Table I [24], indicate reasonable agreement between pirimicarb sorption and degradation for Chinese soil types. The pseudo K_{oc} ranged from 69 to 6015 L/kg (Table I) [24]. The predictions agreed reasonably well with the values calculated by Pan and Wen [29], except in a highly alkaline soil. The large range in K_{oc} was expected, given the large variation in soil organic carbon (0.35–2.0) and pH (5.1–8.6). Pirimicarb is a base with a pK_a of 4.5 (i.e. 50% ionization occurs at pH4.5). Hence, the strongly sorbed cationic form would predominate in acid soils, while the more weakly sorbed non-ionic form is favoured in neutral and alkaline soils. Overall, the physicochemical properties placed pirimicarb in Helling's Class 1 or immobile pesticides. The half-life of pirimicarb averaged 12 days and agreed well with the PETE value of 14 days at 20°C. However, PETE gave a wide range of values, 7–275 days. The reason for this broad range needs to be determined. The bioaccumulation potential for pirimicarb was low, with greater than 95% remaining in the water phase, 4.8% in sediment and only 0.07% in fish. The half-life predicted for pirimicarb volatilization from leaf surfaces (27 days) was in agreement with the value of 16 days calculated from the initial loss of ^{14}C-pirimicarb from Chinese cabbage [30]. The low Henry's law constant (dimensionless 2.8×10^{-8}) indicated that pirimicarb is removed from the atmosphere by rain and is unlikely to be subject to long distance transport.

Atrazine is a weaker base than pirimicarb (pK_a of 1.7), and the overall environmental profile of atrazine is different. The predicted K_{oc} of 118 L/kg suggested intermediate sorption of atrazine, Helling's Class 2. Kookana et al. [31] found an average K_{oc} of 217 L/kg for soils from *Pinus radiata* plantations near Adelaide, South Australia. The predicted K_{oc} was within the range of variation (Table I) [24] and confirmed that the universal sorption equation can predict K_{oc} within an order of magnitude [25]. The half-life of atrazine averaged 56 days and agreed reasonably well with the value of 44 days at 20°C. It was concluded that PETE proved to be a useful preliminary environmental assessment of pirimicarb and atrazine.

3. CLASSIFICATION OF PESTICIDE HAZARDS

Pesticide hazards are of direct concern to applicators and farmers. The most readily available index is oral LD_{50} (median dose in mg/kg to kill 50% of a test species, usually the rat). These data are available from pesticide handbooks, material data sheets and databases on the World Wide Web such as EXTOXNET (http://www.oes.orst.edu:70/1/ext/extoxnet). For example, the oral LD_{50} of pirimicarb to the rat is 147 mg/kg, while atrazine is less toxic at 3080 mg/kg [32]. The

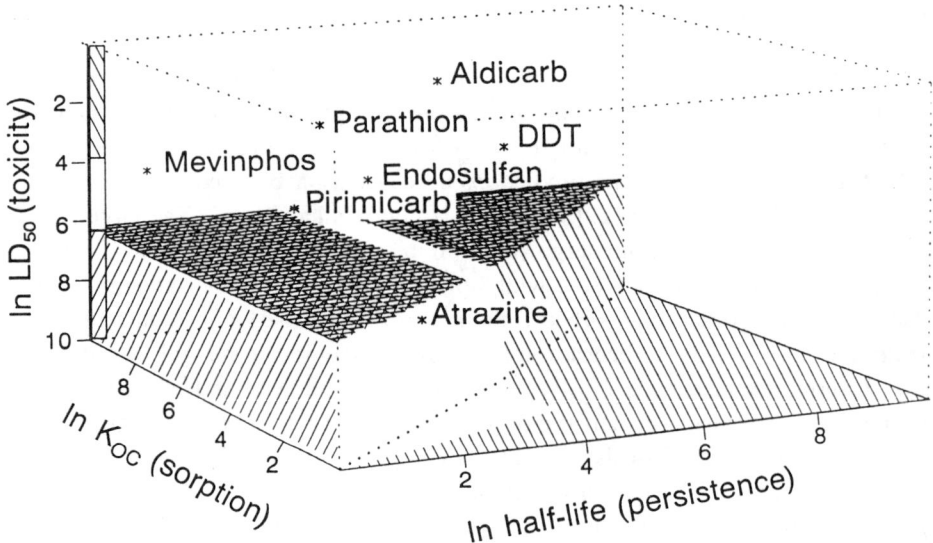

FIG. 3. *Classification of pesticides according to their toxicity and leaching potential. The four pesticides under review by the Australian National Registration Authority (atrazine, endosulfan, mevinphos and parathion) are shown with banned compounds such as DDT.*

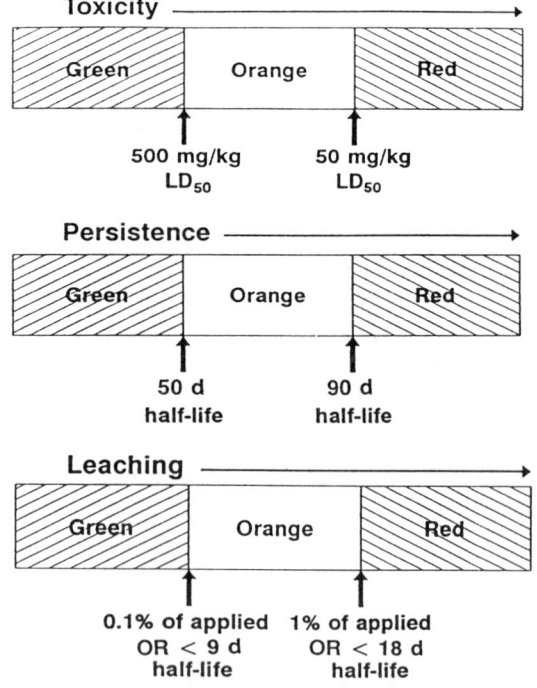

FIG. 4. *Toxicity, persistence and leaching.*

TABLE II. CLASSIFICATION OF PESTICIDES FOR SUSTAINABLE FARMING SYSTEMS
(After Bernson [23])

	Green zone[a]	Orange zone[b]	Red zone[c]	Unclassified[d]	Total[e]
Herbicides	48	17	19	9	93
Insecticides	19	25	25	12	81
Fungicides	21	7	12	12	52
Others	6	5	4	6	21
Total	94	54	60	39	247
%	38	22	24	16	100

[a] Use with caution.
[b] Subject to risk reduction.
[c] Incompatible.
[d] Lack of data to classify the selected pesticide, generally because of lack of soil half-life data.
[e] Random samples of 247 pesticides drawn from a total of 585, including 211 herbicides and 107 fungicides [32].

next step is to classify the toxicity and environmental hazards. Bernson [23] compared pesticide risks to the state of a traffic light: green, proceed with caution; orange, use determined by needs and alternatives; and red, reject or phase out. According to Ref. [23], regulatory and post-registration monitoring should focus on pesticides in the orange and red zones, where there is the greatest potential for risk reduction. This is presented graphically in Fig. 3. The toxicity, persistence and leaching criteria are shown in Fig. 4. The leaching criteria were superimposed to the relationships between sorption and half-life defined by Boesten and van der Linden [33].

These criteria do not represent sharp demarcation boundaries, since the pesticide parameters are continuous and not necessarily discrete. For example, unlike the half-life of an isotope, the half-life of a pesticide in soil depends on environmental conditions such as temperature and moisture [34]. With this proviso, pesticides were classified into three categories, as suggested by Bernson [23]. Pesticides in the red and orange zones comprised 24 and 22%, respectively, of random samples of 247 pesticides drawn from a total of 585 in The Agrochemicals Handbook [32] (Table II) [23]. The red zone included pesticides such as DDT and dieldrin, which are banned or restricted in Australia and China. This suggests that the scheme could identify problem pesticides. The orange zone included many pesticides used in no tillage and other conservation farming systems. Pirimicarb and atrazine fell in the orange zone, suggesting that a more detailed assessment is required.

4. PESTICIDE RISK REDUCTION

If there is no alternative to a hazardous pesticide, risk can be reduced by minimizing exposure. Here, dose is a crucial question. Control of insecticide resistant aphids and low toxicity to beneficial insects justifies use of pirimicarb, but it falls in the orange zone because of its toxicity (Fig. 3). Thus, the hazard is to applicators and farmers who may need to re-enter treated fields. Reducing exposure is an appropriate countermeasure. This could entail a range of options:

(1) Defining problems through a screening programme to determine the acetylcholinesterase (AChE) levels in order to define the magnitude of the problem and to help farmers with low AChE;
(2) Reducing pesticide toxicity by restricting Schedule 7 (World Health Organization Class 1) pesticides through eliminating small farmer packs and marketing safer and more readily disposable 'dissolve a bag' formulations;
(3) Minimizing farmer exposure and encouraging use of protective clothing, i.e. sale of a complete crop protection package, including gloves;
(4) Improving farmer education on the safe mixing and application of pesticides;
(5) Developing a cheap, 'no drift' knapsack applicator and formulations;
(6) Promoting spot spraying and better marking techniques to minimize overlaps;
(7) Marketing less toxic, slow release formulations;
(8) Improving diagnosis and treatment of affected farmers in remote areas.

Atrazine is also in the orange zone, but the primary risk is an environmental hazard. Thus, the aim is to identify the most appropriate IPM strategies for reducing leaching. Simulations conducted over long term historical meteorological records are useful for this task. The choice of model depends on the task and data available to drive the model. The BYPASS simulation model [35] met the essential key requirements: modest data requirement and more realistic water movement in structured soils.

Seasonal and annual trends have been identified in simulated atrazine leaching over the past 100 years. The seasonal trend reflects monthly rainfall before and immediately after atrazine application. On average, summer or winter applications leached more than autumn or spring applications. Annual trends were more complex, but followed a pattern of high atrazine leaching every 5–15 years (Fig. 5) [5]. Significantly, BYPASS predicts no atrazine leaching for 50% of the applications to a 'representative' grey clay soil in the Namoi Valley (New South Wales). A few extreme leaching events were found to bias average leaching. Hence, we are dealing

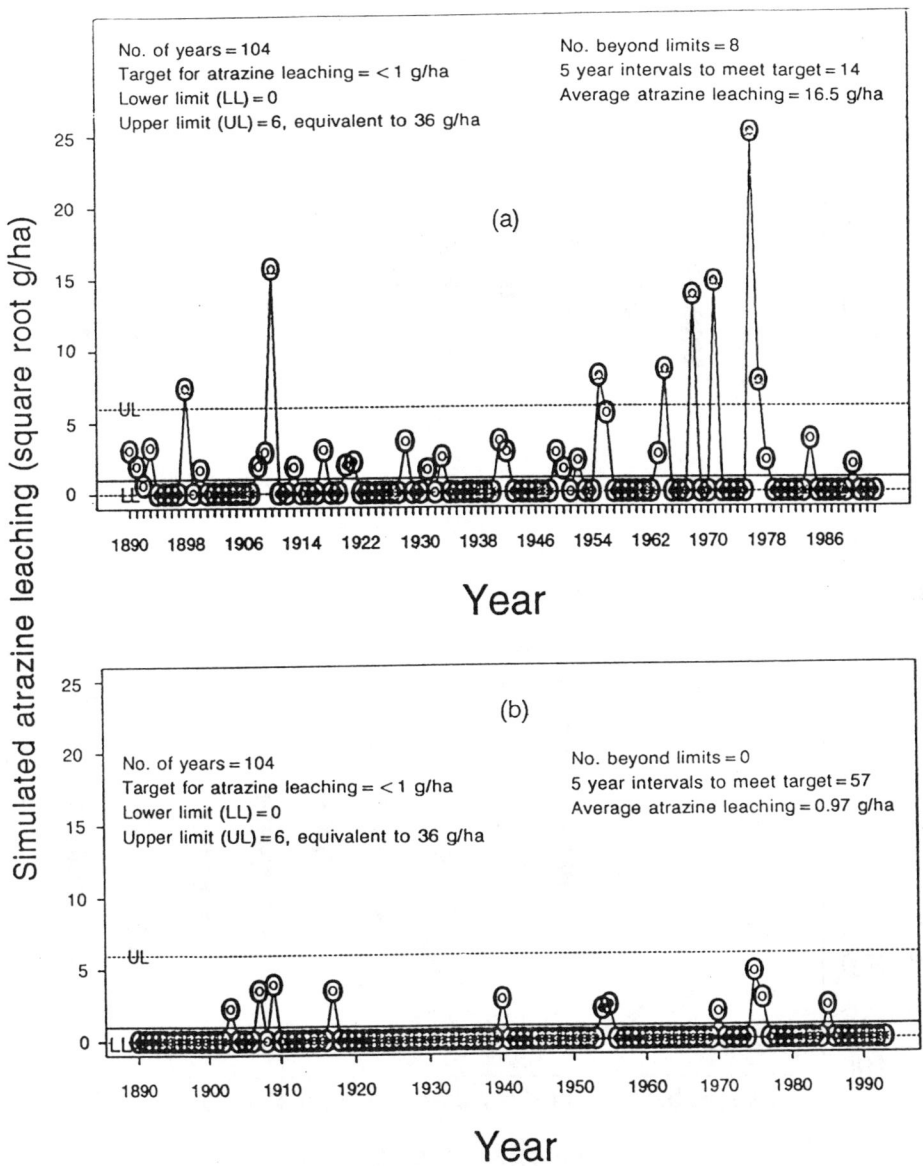

FIG. 5. *Time series (Shewhart quality control chart) of atrazine leaching below 1 m from annual 1.5 kg/ha treatments applied in (a) February or (b) September. The simulations were made with the BYPASS model (1890–1993) for a grey clay soil, with Tamworth rainfall data. The parameters are from Haigh and Ferris [5], with a BYPASS factor of 1 (i.e. maximum), and no change in sorption or loss with depth.*

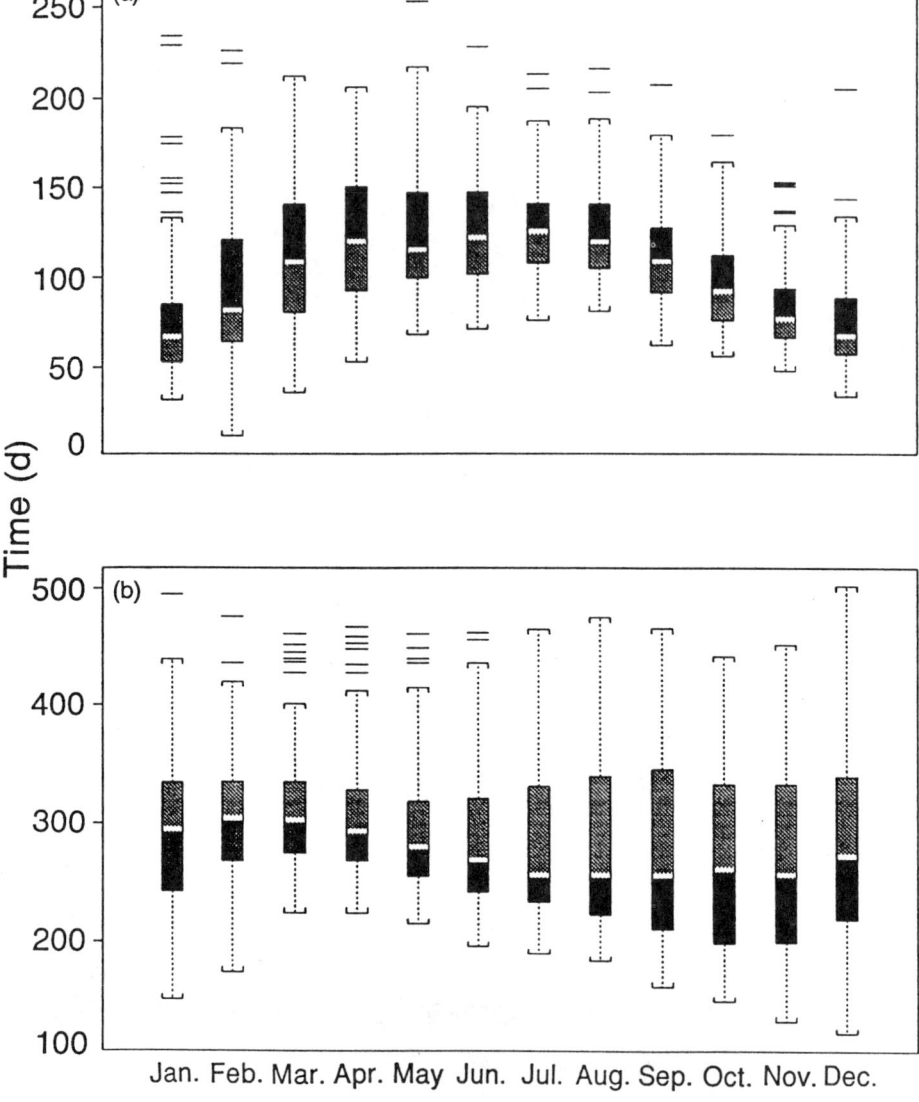

FIG. 6. Box plot showing the effect of application timing on (a) 50% and (b) 90% atrazine loss. The simulations were made with the BYPASS model (1890–1993) for a grey clay soil, with Tamworth rainfall data. The box plots were produced with SPLUS. The central box defines the median and interquartile range enclosing 50% of the values. The box 'whiskers' are 1.5 × the interquartile range, with extreme outliers beyond this range plotted individually.

with an episodic process such as soil erosion and need to consider individual events over long time-frames. Shewhart quality control charts were used to identify such events or 'control failures'. Consecutive years with leaching of less than 1 g·ha^{-1}·a^{-1} (our target for atrazine leaching) were more frequent for autumn or spring applications than for summer applications (Fig. 5) [5]. There were no extreme leaching events (greater than 36 g·ha^{-1}·a^{-1}) for mid-spring applications (Fig. 5(b)). In terms of the leaching potential, early spring applications were marginally better than mid-autumn applications. However, loss of atrazine to the atmosphere (not considered in the BYPASS model) would be lowest from mid-autumn applications, since cooler temperatures prevail during the first atrazine half-life. Although not yet an issue in Australia, this gives mid-autumn applications an advantage in terms of minimizing the overall off-target movement of atrazine.

From a historical perspective, the predicted atrazine leaching has been low since 1980. This suggests that leaching below 1 m, and subsequent movement to groundwater, are not the source of the atrazine residues at present in the Namoi River system [36]. A more likely source is surface runoff caused by excessive cultivation, since El Niño events have dominated the climate over the past 20 years. However, lower atrazine rates applied earlier in the cropping cycle would minimize leaching in periods of above average rainfall. Such a strategy is consistent with integrated weed control principles. For example, the time for 50% atrazine loss averaged better than 4 months for mid-autumn or early spring applications (Fig. 6(a)). There is also a 75% probability of sufficient rain for atrazine uptake by target weeds ($\geq$ 10 mm of rainfull per day, 1 month after application). This indicates mid-autumn or early spring applications would give reliable fallow or in-crop weed suppression. Recropping flexibility is another important issue for farmers. The time for 90% atrazine loss for mid-autumn or early spring applications averaged about 9 months (Fig. 6(b)). This suggests wheat could be sown safely in the following winter, i.e. a valuable double cropping option in the Namoi Valley. Unfortunately, the time for 90% atrazine loss for 40% of the early spring applications exceeded 9 months. Thus, some injury to wheat could occur every second or third year from a September application, consistent with anecdotal evidence. As a rough guide, injury could occur if the rainfall deficit (mm rain$_{\text{September to April}}$ — mm PanA evaporation$_{\text{September to April}}$) falls below −1100 mm. Often the damage to wheat is only cosmetic, but mid-autumn atrazine applications offer a better safety margin than early spring applications.

How do we interpret the significance of leachate concentrations for herbicide management over a catchment? The simplest approach is to consider leachate concentrations and to ignore the complexity of groundwater hydrology. Our target atrazine leaching is about 10 ppb (μg/L) in drainage water. This concentration is approximately the no observable effect level (NOEL) for wheat [5]. This is achievable with the adoption of best management practices. Meeting more stringent drinking water standards, such as the Health Advisory Level (USA), is also possible given that:

(a) Atrazine is only applied to a small proportion of the catchment;
(b) Atrazine degradation is rapid on neutral and acid soils;
(c) High risk alkaline soils comprise only about 40% of the Namoi Valley;
(d) Only areas with perched water tables are at immediate risk;
(e) Assumptions in the BYPASS mode are conservative;
(f) Buffer zones are highly efficient at preventing off-target movement of atrazine [37], which could minimize the risk of accumulation of atrazine in alkaline subsoils/groundwater.

To date, Department of Land and Water Conservation surveys have identified few situations where the atrazine levels have exceeded 10 ppb. Therefore, the issue is prevention of groundwater problems by devising practical countermeasures and by co-ordinating their implementation with farming groups such as LandCare. Banning atrazine will not stop nitrate and other agrochemicals from leaching to groundwater. On the contrary, appropriate use of atrazine and other herbicides could help solve pressing economic and environmental problems [38].

5. IMPLEMENTATION

Loss of pesticides in the red and orange zones has important implications:

(1) A further delay in restoring degraded land [21];
(2) A reduced number of pesticides with different modes of action, thus threatening the primary basis of the integrated pest [39] and weed management programmes [40];
(3) Loss of specific pest control strategies at a time of unprecedented economic and environmental demands on the farmer;
(4) A signal to the agrochemical industry to forego further investment in crop protection chemicals [20].

Care must be taken in deciding which pesticides to phase out of farming systems. An example highlights some of the difficulties involved. Spider mites (*Tetranychus* spp.) were injuring soyabean in the Liverpool Plains (NSW). Monocrotophos is a registered acaricide, but has an oral LD_{50} to the rat of 20 mg/kg and is classified in the red zone. An alternative strategy was to introduce predatory mites (*Phytoseiulus persimilis*). This appeared to be a good option, since spider mite resistance to monocrotophos is becoming more prevalent. Unfortunately, a supplier for the predatory mites took some time to locate. More important, the on-farm cost of predatory mites was US $125/ha versus US $25/ha for monocrotophos. The implication is inescapable — farmers must receive more money for produce in order to be in a position to adopt sustainable farming practices that benefit the whole community.

People in developing and industrialized countries perceived a decline in environmental quality and placed equal weight on causes such as soil erosion and pollution [41]. Significantly, the survey revealed that these people were willing to pay for environmental protection. Recent droughts and floods in Australia and China have demonstrated a tangible commitment to the plight of the farmer. Thus, product identification is one way of converting the goodwill of the urban consumer into tangible incentives for the farmer to adopt sustainable farming practices. The International Standards Organization report ISO 14000 is a quality assurance programme designed to achieve sustainable development and broad national and international support (http://www.dep.state.pa.us/dep/deputate/pollprev/ISO14000/standards.htm). ISO 14000 meets the call for sustainable development [18, 19]. It is global in intent, but is designed for local environmental conditions. Occupational health and safety issues, although not a formal part of an ISO 14000 audit, may be included under Codes of Conduct to encourage farmers' participation and involvement.

Adoption and compliance with ISO 14000 could:

(a) Establish quantifiable milestones for sustainable farming;
(b) Increase consumer confidence in the safety, quality and environmental impact of ISO 14000 products;
(c) Provide more accountability for the fees and charges involved in food distribution and marketing;
(d) Achieve higher economic and social benefits for those farmers who adopt sustainable farming practices;
(e) Bring an end to the factors that collectively contribute to the 'tragedy of the commons', in order to restore the world's degraded land [21];
(f) Address the perceived failure of countries to work together [41];
(g) Create profitable niche markets for agrochemical investment and precision farming technologies [20];
(h) Integrate local and global research approaches to benefit the decision makers, i.e. farmers and governments [42];
(i) Promote economic well being, environmental quality and cultural value, in order to contain population growth [43].

In Australia, farm produce is levied at about 1%, with matching funds from the Federal Government. These funds are distributed by rural industry funding bodies via competitive research grants. Thus, a mechanism exists to fund the public sector research and development needed to underpin higher yielding, more sustainable farming systems. Lacking are a commitment to implement research proposals at the national level [16, 17] and a new way of thinking about research and land management [44].

Efficient and sustainable use of natural resources is not a new development. To quote King [45], "China, Korea and Japan long ago struck the keynote of permanent agriculture but the time has now come when they can and will make great improvements and it remains for us and other nations to profit by their experience, to adopt and adapt what is good in their practice and help in a world movement for the introduction of new and improved methods". Our task is to use the knowledge gained on pesticide behaviour to implement his vision.

ACKNOWLEDGEMENTS

The project was undertaken in association with the IAEA project on Impact of Agrochemicals on Ecosystems. Formative ideas were contributed by C. Alston, G.G. Briggs, J. Messick and E.P. Lichtenstein. Special thanks are also due to P. Taylor and S. Harden for technical assistance.

REFERENCES

[1] EUE, L., World challenges in weed science, Weed Sci. **34** (1985) 155–160.
[2] CARSON, R., Silent Spring, Hamish Hamilton, London (1963) 304 pp.
[3] KLEIN, A.W., et al., Environmental assessment of pesticides under Directive 91/414/EEC, Chemosphere **26** 5 (1993) 979–1001.
[4] TOFFOLON, R., NSW Agriculture, Orange, personal communication, 1996.
[5] HAIGH, B.M., FERRIS, I.G., "Predicting herbicide persistence using the CALF model", Modelling the Fate of Chemicals in the Environment (Proc. Nat. Workshop Canberra, 1991) (MOORE, I.D., Ed.), Panther Publishing, Canberra, ACT (1991) 50–60.
[6] CRIBB, J., "Agronomy and the economy", Agronomy — Science with its Sleeves Rolled Up (Proc. 8th Aust. Agron. Conf. Toowoomba, 1996), The University of Southern Queensland, Toowoomba (1996) 9 pp.
[7] TRIBE, D.E., Feeding and Greening the World: The Role of International Agricultural Research, CAB International, Reading, UK (1994) 274 pp.
[8] CASSMAN, K.G., "Resource constraints, conservation of natural resources, and the need for further intensification of crop production systems", Integrated Resource Management for Sustainable Agriculture (Proc. Int. Conf. Beijing, 1993), Section 3, Beijing Agricultural University, Beijing (1993) 126.
[9] ALSTON, C., NSW Agriculture, Tamworth, personal communication, 1994.
[10] DERIA, A., et al., "Wheat production and soil chemical properties of organic and conventional paired sites in Western Australia", Agronomy — Science with its Sleeves Rolled Up (Proc. 8th Aust. Agron. Conf. Toowoomba, 1996), The University of Southern Queensland, Toowoomba (1996) 200–202.
[11] KESSLER, D.A., et al., The safety of foods developed by biotechnology, Science **256** (1992) 1747–1749, 1832.

[12] MIKKELSEN, T.R., et al., The risk of crop transgene spread, Nature (London) (1996) 31.

[13] HOLLAND, J.F., FELTON, W.L., Effect of tillage practice on grain sorghum production in northern New South Wales, Aust. J. Exp. Agric. **29** (1989) 843–848.

[14] FERRIS, I.G., et al., "Effect of tillage and herbicides on carbon cycling", No-Tillage Crop Production in Northern NSW (Proc. Project Team Meeting, Tamworth, 1983) (MARTIN, R.J., FELTON, W.L., Eds), Agdex 100/21, NSW Agriculture, Tamworth (1983) 64–73.

[15] MARSTON, D., Reduced tillage systems — Their potential for soil conservation in northern New South Wales, J. Soil Cons. Serv., NSW **34** (1978) 203–206.

[16] COMBELLACK, J.H., "Resource allocations for future weed control activities", Persistence of Herbicide Residues in Wheat Cropping Systems in Australia (Proc. Nat. Workshop Melbourne, 1988) (DURRANT, S., FERRIS, I.G., Eds), State Chemistry Laboratory, Department of Agriculture and Rural Affairs, Melbourne (1988) 136–163.

[17] COMBELLACK, J.H., "The importance of weeds and ways of reducing concerns about methods for their control", Weeds Control (Proc. 1st Int. Congr. Melbourne, 1992), Vol. 1 (COMBELLACK, J.H., et al., Eds), Weed Science Society of Victoria, Melbourne (1992) 43–63.

[18] PRETTY, J.N., "Sustainable agriculture in the 21st century: Challenges, contradictions and opportunities", Weeds (Proc. Conf. Brighton, 1995), British Crop Protection Council, Farnham, UK (1995) 111–120.

[19] HARE, W.L., et al., Ecologically sustainable development, Australia Conservation Foundation, Fitzroy (1990) 94 pp.

[20] DAVIES, D.H.K., "Technology developments in weed control and targeting for reducing environmental impact", Weeds (Proc. Conf. Brighton, 1995), British Crop Protection Council, Farnham, UK (1995) 613–622.

[21] DAILY, G.C., Restoring value to the world's degraded lands, Science **269** (1995) 350–354.

[22] CHAMEIDES, W.L., et al., Growth of continental-scale metro-agro-plexes, regional ozone pollution, and world food production, Science **264** (1994) 74–77.

[23] BERNSON, V., "Will regulatory pressure eliminate the need for new herbicides?", Weeds (Proc. Conf. Brighton, 1995), British Crop Protection Council, Farnham, UK (1995) 413–419.

[24] NICHOLLS, P.H., "Physicochemical Evaluation: The Environment, an expert system for pesticide preregistration assessment", Pests and Diseases (Proc. Conf. Brighton, 1994), British Crop Protection Council, Farnham, UK (1994) 1337–1342.

[25] BRIGGS, G.G., Theoretical and experimental relationships between soil adsorption, octanol–water partition coefficients, water solubilities, bioconcentration factors, and the parachor, J. Agric. Food Chem. **29** (1981) 1050–1059.

[26] NICHOLLS, P.H., EVANS, A.A., Sorption of ionisable organic compounds by field soils. Part 1. Acids, Pestic. Sci. **33** (1991) 319–330.

[27] NICHOLLS, P.H., EVANS, A.A., Sorption of ionisable organic compounds by field soils. Part 2. Cations, bases and zwitterions, Pestic. Sci. **33** (1991) 331–334.

[28] BROMILOW, R.H., et al., Physicochemical aspects of phloem translocation of herbicides, Weed Sci. **38** (1990) 305–314.

[29] PAN, J., WEN, X., "The persistence of ^{14}C-pirimicarb in three Chinese soils", Nuclear Agricultural Science for Young Scientists (Proc. Nat. Conf. Beijing, 1994) (LIANG, Q., Ed.), Chinese Agricultural Science Technology Press, Beijing (1994) 143–145.

[30] PAN, J., WEN, X., Institute for the Application of Atomic Energy, Beijing, personal communication, 1995.

[31] KOOKANA, R.S., et al., "Sorption and degradation of atrazine with depth in soils from a forest catchment of South Australia", Contaminants and the Soil Environment (Proc. 1st Int. Conf. Adelaide, 1996), Division of Soils, Commonwealth Scientific and Industrial Research Organization, Adelaide (1996) 171–172.

[32] HARTLEY, D., KIDD, H., The Agrochemicals Handbook, 2nd edn, Royal Society of Chemistry, Nottingham, UK (1989).

[33] BOESTEN, J.J.T.I., VAN DER LINDEN, A.M.A., Modeling the influence of sorption and transformation on pesticide leaching and persistence, J. Environ. Qual. **20** 2 (1991) 425–435.

[34] WALKER, A., Simulation of herbicide persistence in soil, Pestic. Sci. **7** (1976) 41–49.

[35] WALKER, A., National Horticultural Institute, Wellesbourne, UK, personal communication, 1993.

[36] COOPER, B., Central and Northwest Regions Water Quality Program: 1994/95 Report on Pesticide Monitoring, Technical Services Division, Department of Land and Water Conservation, Parramatta (1995) 77 pp.

[37] BAKER, J.L., et al., "Reducing herbicide runoff: Role of best management practices", Weeds (Proc. Conf. Brighton, 1995), British Crop Protection Council, Farnham, UK (1995) 479–487.

[38] FERRIS, I.G., A risk assessment of sulfonylurea herbicides leaching to groundwater, AGSO J. Aust. Geo. Geophys. **14** 2/3 (1993) 297–302.

[39] GUNNING, R.V., et al., Pyrethroid resistance mechanisms in Australian *Helicoverpa armigera*, Pestic. Sci. **33** (1991) 473–490.

[40] POWLES, S.B., MATTHEWS, J.M., "Multiple herbicide resistance in annual ryegrass (*Lolium rigidum*): A driving force for the adoption of integrated weed management", Resistance '91: Achievement and Developments in Combating Pesticide Resistance (DENHOLM, I., et al., Eds), Elsevier, London (1992) 75–87.

[41] BLOOM, D.E., International public opinion on the environment, Science **269** (1995) 354–358.

[42] ROOT, T.L., SCHNEIDER, S.H., Ecology and climate: Research strategies and implications, Science **269** (1995) 334–341.

[43] COHEN, J.E., Population growth and earth's human carrying capacity, Science **269** (1995) 341–346.

[44] HAMBLIN, A., "Agronomy and the environment", Agronomy — Science with its Sleeves Rolled Up (Proc. 8th Aust. Agron. Conf. Toowoomba, 1996), The University of Southern Queensland, Toowoomba (1996) 33–42.

[46] KING, F.H., Farmers of Forty Centuries, Rodale Press, Pittsburgh, PA (1911) 441 pp.

ENVIRONMENTAL FATE OF HERBICIDES IN HAWAII, PERU AND PANAMA

C.S. HELLING
Weed Science Laboratory,
Agricultural Research Service,
United States Department of Agriculture,
Beltsville, Maryland,
United States of America

Abstract

ENVIRONMENTAL FATE OF HERBICIDES IN HAWAII, PERU AND PANAMA.
Data on the fate and behaviour of herbicides in tropical soils are often unavailable. Here, the environmental fate of herbicides was compared in controlled field plots with the results derived from 'real world' pesticide use in tropical situations. As the intended herbicide use was to control illicit narcotic crops, sampling and verification are major problems because of the geographical remoteness, security concerns and political accessibility. Another fundamental problem is that exclusive use of either chemical analysis or biological monitoring may lead to an incomplete understanding of the environmental fate and ecological consequences of pesticide use. For hexazinone, tebuthiuron and imazapyr, the soil persistence and mobility, and the residual phytotoxicity to seven crops, were determined under controlled field conditions in Hawaii. This information was compared with field results in Peru and Panama for hexazinone, tebuthiuron and glyphosate. Herbicide dissipation was much faster under tropical than temperate zone conditions; also, herbicide persistence in Hawaii closely matched that found in the coca fields of Peru and Panama. Chemical and biological approaches together gave greater reliability to risk interpretation than would either approach alone; both showed low environmental risk for the pesticides tested.

1. INTRODUCTION

General interest in the environmental fate of pesticides, especially in soil and water, began in the early 1960s. Subsequent research on the behaviour and fate of these and other xenobiotics has, directly or indirectly, had major implications. Examples include the creation and/or expansion of government regulatory agencies to control the use of agrochemicals; restrictions on or the outright banning of many pesticides; increased efforts to find both reduced risk pesticides and alternative pest control methods; and expanded knowledge of basic processes such as solute transport through soils and groundwater, as well as pesticide degradation in soil, water and air.

While prediction of environmental behaviour on the basis of the chemical structure and properties is useful, such modelling typically has occurred following the acquisition of much evidence collected from laboratory and field experimentation. Such approaches are useful, but both mathematical modelling and laboratory tests must ultimately be supported by appropriate field observations for the validation of behaviour and/or risk assessment. This paper compares the results from two levels of such field observations, with research done at a relatively controlled, tropical field plot (in Hawaii) being compared with the monitoring of herbicides applied under less controlled, less accessible, actual use sites (in Peru and Panama).

The herbicides used in these studies were being tested for narcotic crop reduction/control, specifically coca (*Erythroxylum* spp.). Illicit coca is produced under tropical conditions of relatively high rainfall, mainly within three Andean region countries (Peru, Bolivia and Colombia) [1]; the net cultivation in 1995 was about 214 800 hectares, with additional, smaller production in Brazil, Venezuela, Panama and Ecuador [2]. Reducing acreage through narcotic crop destruction is a component of drug supply reduction strategy. However, herbicide use to accomplish this elicits the same fundamental concerns about toxicological and environmental safety as does pesticide use in conventional agriculture, forestry or other sectors. Among the common concerns are whether herbicides will cause long term soil 'sterilization', with an increased risk of soil erosion; preclude growing subsistence crops or other alternative crops on the treated coca fields; and contaminate drinking water.

Research at such drug production sites is perhaps the prime example of difficult environmental monitoring because of their remote locations, the high security risks involved and the need to co-ordinate activities across interagency and international boundaries. Compounding the logistical, social and political barriers are the technical difficulties inherent in field evaluation and sampling: on-site field visits are usually infrequent; on-ground time is usually very brief; and samples may be collected by non-scientific personnel or be delayed in transit. In lieu of utilizing agricultural research centres within coca producing areas, something as yet unachievable, we established long term experiments in Hawaii as a surrogate location for more controlled efficacy and soil environmental studies. (The same site was used concurrently for IAEA sponsored research on DDT dissipation under tropical conditions [3].)

This paper describes investigations of the environmental fate for several herbicides useful for coca control. Most published data for these herbicides — hexazinone, tebuthiuron, imazapyr and glyphosate — refer to temperate agroecosystems. Given that the afore-mentioned herbicides can be used for total vegetation control, and that some are known (from their use in temperate climates) to persist in soils, it was imperative to study their fate under more appropriate conditions. Two fundamental techniques were used in the field: (1) chemical analyses of soil or water samples; and (2) biological impact assessment, either by periodic recropping of the treated plots (in Hawaii), or by observation of plant growth (native vegetation or crop plants) at the illicit coca sites.

2. SITE LOCATIONS

2.1. Hawaii, United States of America

Controlled field experiments were conducted on the Island of Kauai, Hawaii (latitude 22°N, longitude 159°W), at an altitude of about 180 m above sea level. The mean annual rainfall is approximately 2500 mm; during the first 12 months of Experiment 1 (see Section 3.1.1), 2560 mm of rain was actually received. The mean air temperature range (minimum-maximum) is ca. 20-25.5°C, with little seasonal variation; and the soil temperature is 21-22°C. The soil at this field site is Halii sandy clay loam, a typic Gibbshumox (clayey, ferritic, isothermic), formed by the weathering of basic igneous rock and mixed with volcanic ejecta [4]. This reddish brown Oxisol contains a high proportion of iron and aluminium oxides, much as gravel sized concretions, and is highly permeable. Consequently, virtually no runoff occurs, despite the high rainfall. The soil organic matter content is relatively high, about $6.5 \pm 0.5\%$; the textural composition is sand (61%), silt (26%) and clay (13%); the pH is 5.4 ± 0.3; and the mineralogical composition of the surface soil clay is kaolinite (10%), gibbsite (30%) and goethite (60%). The field site (ca. 0.3 ha) is gently sloping ($\pm 5\%$).

2.2. Peru

Most illicit coca in Peru is grown on the eastern slopes of the Andes Mountains, especially in the large Upper Huallaga Valley (Alto Huallaga). In the Alto Huallaga, some coca is produced on flat land, near the Río Huallaga, a tributary of the Amazon River. (In this setting, the first herbicide tests on coca were conducted in 1987-1988; termed Phase I, they included environmental monitoring.) However, most cultivation within the Alto Huallaga watershed is on moderately sloping hillsides to the west; the Phase IV environmental tests reported on here were within the San Martín and Huanuco Provinces (near latitude 8°S, longitude 76°W). Much of the original cloud forest and montane jungle has been replaced by coca fields. The hilly zone of coca production, where the Phase IV tests were done, ranged from an altitude of about 1400 to 1600 m. The mean annual temperature in the Alto Huallaga is 22-26°C, and the rainfall about 2500-3500 mm [1]. Soils in five of the ten composited cores from the treated fields were clay loam, often grading to clay texture in the 15-30 cm subsoil; the subsurface clay content ranged from 40 to 47%. Two cores were loams throughout the 0-30 cm depth, while three were sandy loams at their surface. All the soils were acidic (pH of 4.4 ± 0.4); the organic matter in the surface 15 cm ranged from 2.4 to 7.8%.

TABLE 1. HERBICIDE TREATMENTS USED IN HAWAIIAN SOIL PERSISTENCE STUDIES[a]

Experiment	Treatment				Plot size (m²)	Plots per replicate	Total area treated (m²)	Recropped
	Herbicide	Formulation	Active ingredients (%)	Rate (kg/ha)				
1	Hexazinone	Pronone 10G	10	3.34	100	3	300	Yes
1	Hexazinone	Pronone 10G	10	10.1	100	3	300	Yes
1	Hexazinone	Velpar	90	10.1	123	2	246	No
1	Tebuthiuron	Spike 20P	20	3.34	100	3	300	Yes
1	Tebuthiuron	Spike 20P	20	10.1	100	3	300	Yes
1	Tebuthiuron	Spike 80W	80	10.1	123	2	246	No
2	Imazapyr	Arsenal 5G	5 (ae)	0.56	100	3	300	Yes
2	Imazapyr	Arsenal 5G	5 (ae)	2.24	100	3	300	Yes
2	Imazapyr	Arsenal	240 g/L (ae)	2.24	100	1	100	Yes

[a] ae = acid equivalent.

2.3. Panama

The herbicide treated areas were located in a very remote, essentially unpopulated region in the Darién Province, Panama, near the southeastern end of the Serranía del Darién mountain range. The approximate geographical co-ordinates were latitude 8°N, longitude 77°W, and the field elevations were about 400–600 m. The mean annual temperature is ca. 25°C, and the rainfall about 2500 mm [5]. Ecologically, the area is evergreen seasonal forest, grading to premontane forest [6], and many streams originate in this low mountainous region. Apart from illicit coca fields, created by slash and burn methods, most of the area is covered by rain forest. The coca fields ranged in size from ca. 0.5 to 3 ha, generally on moderately to steeply sloping (about 10–50%) sites.

The field selected for treatment with hexazinone was slightly sloping (± 5–10%) and about 30 m from a stream. Nearby, in the same watershed, tebuthiuron was applied to a steeper hillside (ca. 25–50% slope). The tebuthiuron treated coca field lay about 3–4 m from the stream referred to above; the buffer zone was heavily covered with natural grassy and broad leaf plants. The soils are unmapped in this region, but likely to be Oxisols.

3. METHODS

3.1. Herbicide treatments

3.1.1. Hawaii

All the plots were tilled and fallow at treatment; they were separated by 1.2 m alleys. The herbicide treatments for two experiments, conducted on the same site, are summarized in Table I. Experiment 1, using hexazinone and tebuthiuron, was initiated 2 years before Experiment 2. Not shown in the table are three and seven untreated control plots (100 m^2 each) for Experiments 1 and 2, respectively. Pronone 10G and Spike 20P are granule and pellet formulations. Sprayable formulations of Velpar (soluble powder) and Spike 80W (wettable powder) were included for comparison. Arsenal 5G was a granular form of imazapyr, while Arsenal was a liquid formulation; both were isopropylamine salts, but the rate applied is based on the acid equivalent (ae). To better assure accuracy and uniformity when hand broadcasting all the solid formulations, preweighed doses were applied to each of the 12 subplots per main plot.

3.1.2. Peru

The Phase IV tests in Peru used aerial applications of hexazinone (Pronone 10G) or tebuthiuron (Spike 20P). Hexazinone was applied to the illicit coca fields at two rates: 1.1 and 2.2 kg active ingredients per hectare; three rates of tebuthiuron were used: 1.1, 2.2 and 3.3 kg/ha.

3.1.3. Panama

Glyphosate, as Roundup diluted with water, was aerially applied to most of the coca fields at 4.5 kg ae/ha, but smaller scale testing was also done at 2.2, 6.7 and 9 kg/ha. Hexazinone (Pronone 5G) and tebuthiuron (Spike 40P) were hand applied at rates of 2.2 and 4.0 kg/ha, respectively. In the coca fields, the hexazinone treated area was 0.21 ha and that for tebuthiuron, 0.30 ha.

3.2. Sampling

3.2.1. Hawaii

At time 0, soil samples were collected from 10 × 10 cm areas at a depth of 4–5 cm. Thereafter, a stainless steel 'sampling sleeve', constructed to improve shallow soil sampling, was used. The 10 (diameter) × 25 cm cylinder was pressed 10 cm into the soil to obtain the first increment (0–10 cm), then the remaining 15 cm for the second sample (10–25 cm). A 10 cm collar around the upper end of the sampling sleeve helped to prevent the surface soil (which normally has the highest contaminant levels) accidentally contaminating the deeper cores. All such deeper soil samples were collected, at 25 cm increments, by using a 10 cm diameter bucket auger. Sampling at predetermined random locations ranged from one to two cores per plot (controls and solid formulations) to two to three cores per plot (spray formulations).

Sampling was carried out for 37 months in Experiment 1 (hexazinone-tebuthiuron), and for 15 months in Experiment 2 (imazapyr). On the basis of the residue analyses and the potential leaching, the sampling depth usually increased with the elapsed time. When terminated, sampling extended to a depth of 8 m for hexazinone and tebuthiuron, and 1.75 m for imazapyr. The soil samples collected in Hawaii were double bagged promptly and shipped within 2–5 days to Beltsville, Maryland, for processing (storing frozen until analysed). To calculate the pesticide dissipation rate, the elapsed time between the treatment and the sample receipt dates was usually used (depending on the conditions of sample storage prior to receipt).

3.2.2. Peru and Panama

In Peru, the sampling intervals were pretreatment, and then 1, 2, 3, 6 and 12 months post-treatment. Sampling extended to 0.5 m (0–15, 15–30, 30–50 cm), except at 12 months, when one core to 0.75 m and two cores to 1 m were collected. Given the limited time, and because they represented the greater environmental risks, only those fields treated with the higher herbicide rate(s) were sampled. Several stream water samples near the treated fields were also collected for analysis.

In Panama, no soil samples deeper than 0.3 m were collected; the last such sampling was 8.5 months after treatment. The proximate water samples were collected both upstream and downstream of the hexazinone and tebuthiuron treated fields at 0, 41 and 89 days after treatment. The water samples were similarly collected, 40 days after treatment, from a glyphosate area.

3.3. Herbicide residue analyses

3.3.1. Hexazinone and tebuthiuron

Hexazinone and tebuthiuron in soil or water matrices were analysed using the method of Lydon et al. [7], with two significant modifications: (1) the herbicide in water was first trapped by passing the samples through C_{18} solid phase extraction cartridges; and (2) either a phenyl or a C_{18} high performance liquid chromatographic column was used, with a ternary gradient mobile phase (methanol/H_2O/acetonitrile (15:70:15), changing to 50:0:50 after 10 minutes (at a flow rate of 2 mL/min)). The minimum detection limits for each herbicide were 4 ppb in soil and 0.5 ppb in water. The limit of quantification for soil was set at 20 ppb. Normally, recovery of spiked analytes from soil was quantitative. Recovery from spiked stream water (Panama) was 97 ± 1.5% for hexazinone and 95 ± 1% for tebuthiuron. Confirmation of identity by mass spectrometry was done for several samples using the method of Lusby et al. [8].

3.3.2. Imazapyr

A new procedure for analysing imazapyr in soils was developed [9] and used. The limit of detection was 5 ppb. Recovery of imazapyr from fortified samples, even after prolonged frozen storage, was 99 ± 3%.

3.3.3. Glyphosate

Glyphosate was analysed in the soil and water samples using the method of Glass [10]. The limits of detection were 50 ppm (soil) and 10 ppb (water).

3.4. Recropping

Beginning 1 week after herbicide treatment, seven species (bean, cassava, coffee, maize, okra, squash and wheat) were planted in each of the 15 main plots of Experiment 1 (four sprayed plots of hexazinone and tebuthiuron were not recropped). Since 25% of each plot was occupied by the two crop rows of Cycle 1, after the fourth cycle of recropping, the first recropping was subsequently replanted with Cycle 5 crops. Recropping continued periodically (11 cycles) up to 14 months post-treatment. The phytotoxicity ratings during the ≥ 10 week cycle indicated residual herbicide toxicity to each crop. For Experiment 2, with imazapyr, the same recropping design was used, except that peanut replaced wheat. Recropping began 3 weeks after the Arsenal 5G treatment, and immediately after Arsenal application.

4. RESULTS

The information presented here summarizes the major findings and conclusions from a series of related studies designed to understand the environmental fate of the herbicides considered or used for controlling narcotic crops in tropical climates. This information has either not been previously published, or else has appeared only in limited access technical reports or summaries.

4.1. Fate of hexazinone

4.1.1. Hawaii

Hexazinone and tebuthiuron were studied intensively in Hawaii because both appeared to be very effective in controlling *Erythroxylum coca* var. *coca*, based on controlled experiments in the USA and on Phase I (and later, Phase IV) testing in Peru. A likely rate for actual use (3.34 kg/ha, or 1×) was compared with a 'worst case' (10.1 kg/ha, or 3×) rate, bracketing possible extremes of environmental risk through deep leaching and/or long soil persistence. Although the preferred formulations of the two herbicides were solids (Pronone 10G and Spike 20P), spray formulations were also tested in case this altered the environmental fate of the chemicals.

Figure 1 summarizes persistence for up to 18 months of hexazinone and tebuthiuron at the Hawaiian site, as expressed by the dissipation half-life. It is derived from half-life estimates that were updated at each sampling time for each treatment. Two important adjustments were made: (1) curves were smoothed to better represent longer term trends; and (2) negative half-life values (which occurred only at early times for high rate, solid formulations) were disregarded. As expected, the variability in the pesticide residue levels during the first several months was lower for sprayed formulations because of their more uniform application to soil. In contrast, random sampling of several granules or pellets, prior to their disintegration and diffusion, can easily yield apparent residues far in excess of the theoretically applied dose, and therefore the derived half-life appears to be negative, 'forming' the pesticide with time.

Ideally, if pesticide loss actually follows first order or pseudo first order kinetics, the rate constant will be invariant under constant environmental conditions.

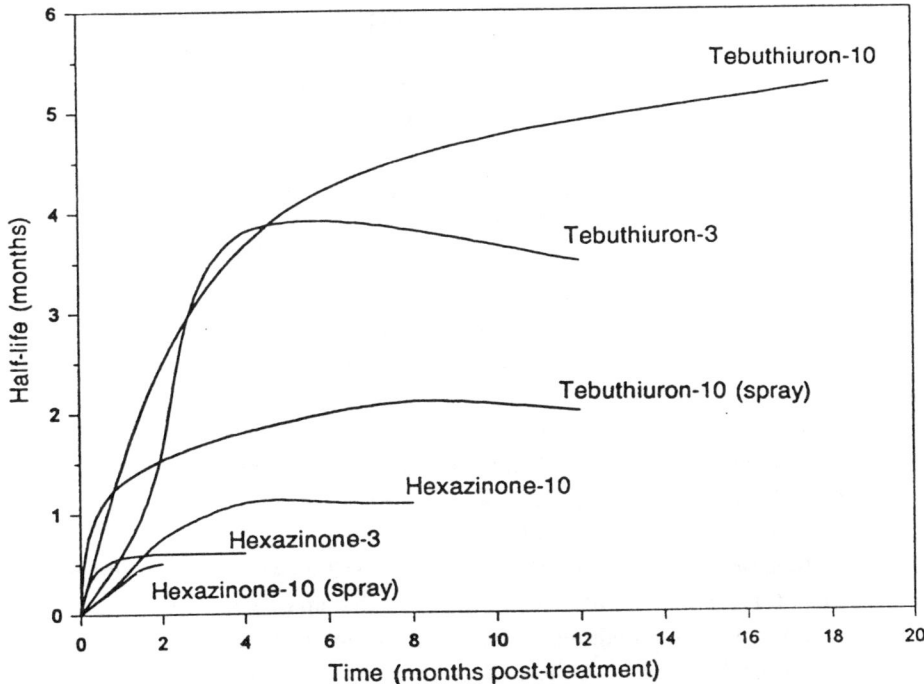

FIG. 1. Dissipation of hexazinone and tebuthiuron (3.4 and 10.1 kg/ha each) in Hawaiian soil, expressed as time dependent half-life.

According to the results shown in Fig. 1, the loss rate for hexazinone and tebuthiuron slows down, but there are at least three complicating factors: (a) sampling uncertainty is highest at early times; (b) leaching below the sampling depth, usually at later times, would lead to inaccurately low half-life estimates; and (c) residues near the limit of quantification would also provide less accurate half-life estimates. The latter is the reason why estimates of hexazinone persistence (Fig. 1) are shown only up to a maximum of 8 months. Combined evaluation of many samples indicated that the apparent half-life of hexazinone at the Hawaiian site was 0.5–0.6 month for 3.34 kg/ha of Pronone 10G; 1 month for 10.1 kg/ha of Pronone 10G; and 0.5 month for 10.1 kg/ha of Velpar. Estimation of the spray treatment, Velpar, is probably less accurate because of a shorter sampling period. An environmentally conservative statement of persistence for hexazinone would be that its half-life under these tropical conditions is 1 month or less. In Canada, use of hexazinone in forestry led to half-life estimates of 6 weeks when applied in summer [11], and to 6–7 months for an autumn application [12]. By regressing the percentage hexazinone remaining (from the 3.34 kg/ha rate) against the logarithm of time, hexazinone was projected to disappear from the Hawaiian field plots within 21 weeks.

The estimates of persistence on which Fig. 1 was derived were based on sampling for up to 18 months (tebuthiuron), or less for hexazinone. Usually, hexazinone was not found below 1.25 m, and the highest concentrations were in the upper 25 cm. However, sampling did proceed longer than indicated in Fig. 1. At 8 months, trace residues of hexazinone were detected to a depth of 1.5 m, and after 32 months to 5–6.5 m; the two cases were for the 3× rate. For both hexazinone and tebuthiuron, leaching and persistence increased as the rate of application increased. At the final, very deep (8 m) sampling 3 years after treatment, no detectable (i.e. <0.01 ppm) hexazinone was present. Because of its relatively rapid degradation in this tropical soil (and also, apparently, in Peru and Panama), the risk to groundwater from hexazinone leaching, when applied at normal rates effective for coca control, seems slight.

The recropping results tended to reflect the soil persistence data. No seeded crops (see Section 3.4) survived when planted 1 week after application of either the 1× (3.34 kg/ha) or 3× rates of hexazinone; only 8% of the coffee seedlings and 17% of the cassava cuttings survived the 1× rate. The delay in planting needed to ensure at least 50% survival of the crops was 5–8 weeks for 3.34 kg/ha of hexazinone, and 7–17 weeks for the 10.1 kg/ha rate. Cassava, coffee and bean were the most susceptible crops; by the 6 month replanting, all the crops survived except bean, which required at least 35 weeks. On bare plots sprayed with hexazinone (10.1 kg/ha), weed re-emergence began 2–3 months after treatment; by 12 months, the weeds were dense and varied.

TABLE II. HEXAZINONE AND TEBUTHIURON PERSISTENCE IN SOIL FOLLOWING AERIAL APPLICATION TO ILLICIT COCA IN PERU (PHASE IV TEST)[a]

Herbicide	Application rate (kg/ha)	Mean residual herbicide (±SD) (kg/ha) at time after treatment (months)				
		1	2	3	6	12
Hexazinone	2.2	0.23 ± 0.25	0.12 ± 0.07	0.017 ± 0.023	NS	NS
Tebuthiuron	2.2	0.11 ± 0.25	0.16 ± 0.26	0.18 ± 0.39	0 ± 0	NS
	3.3	NS	NS	1.54 ± 1.79	0.14 ± 0.21	0.071 ± 0.120

[a] NS = not sampled.

4.1.2. Peru

Residual hexazinone found in fields treated with 2.2 kg/ha is summarized in Table II (together with the tebuthiuron results). From the 15 cores collected post-treatment, only two suggested (on the basis of residue distribution) that some hexazinone may have leached below 50 cm. Dissipation, therefore, seems to occur via rapid degradation in the surface horizon, where soil organic matter and microbial activity are highest. The calculated half-life (assuming first order loss) is 0.3 month. Although this value may be low because of sampling problems inherent to granular or pelleted formulations (see Section 4.4.2), even doubling this value leads to an estimated dissipation half-life of <1 month for hexazinone in Peruvian soil. Hexazinone was undetected in nearby stream water (1 and 2 months post-treatment).

Natural vegetation in or near the coca fields was either unaffected or only temporarily affected by hexazinone application at 2.2 kg/ha. Minor leaf margin desiccation was observed in banana, but this was the only such phytotoxicity found for non-target crops growing near the coca fields. Artificial revegetation of the hexazinone treated fields would be unnecessary in this environment.

4.1.3. Panama

Surface soil (0–5 cm) was collected from the coca field treated 41 days earlier with hexazinone. The residue found, 0.060 ± 0.006 ppm, corresponded to ca. 98% herbicide dissipation. No trace of hexazinone was found in a 0–10 cm soil sample collected 8.5 months after treatment. Accurate evaluation of hexazinone dissipation (including with depth) would have required more sampling than was possible. Nevertheless, coca (in Panama, *E. novogranatense* var. *novogranatense*) had been almost entirely killed by the herbicide, and the field was heavily overgrown with grassy and broad leaf plants, indicating no further residual phytotoxicity.

Stream water collected 41 and 89 days after treatment contained no detectable hexazinone. Given the distance from the stream (about 30 m), the rapid soil degradation of hexazinone, and other attenuating factors (see Section 4.2.3), the lack of water contamination was not surprising.

4.2. Fate of tebuthiuron

4.2.1. Hawaii

The longer soil persistence of tebuthiuron compared with hexazinone is apparent from the half-life estimates (Fig. 1). For tebuthiuron applied at 3.34 kg/ha as Spike 20P, the estimated dissipation half-life in Hawaii was 3.5–4 months. After about 4 months, the 3× rate of Spike 20P gradually increased to an estimated half-life of 5 months when evaluated 18 months after treatment. When pesticides leach below the biologically most active upper soil horizon, degradation typically slows. Evidence here suggests that this had occurred with tebuthiuron, at least when the high rate was used. After 18 months, the tebuthiuron residues had disappeared from the surface soil, from the $1\times$ rate, and only a trace remained from the $3\times$ initial rate. However, small amounts of tebuthiuron leached below 1–2 m, persisting longer, so estimates of the overall half-life increased. At 23 months after treatment, deep sampling of both hexazinone and the lower, 'normal' rate of tebuthiuron failed to detect herbicide residues. However, at 27 months, sampling only the $3\times$ rate of tebuthiuron, small amounts of this chemical were found in one core at 3–5 m; on the basis of only this 27 month sampling, the half-life of tebuthiuron was recomputed to be 7.3 months. As sampling of the other two cores was limited (by rocks) to only 2.3 and 3.0 m, tebuthiuron might have been detected below 3 m as well, and the estimated half-life would again have had to be increased. Ultimately, plots were sampled to 8 m at 37 months after treatment, and very small residues (the maximum was 0.09 ppm) occurred deep in the subsoil from the high rate tebuthiuron application. To summarize these tebuthiuron results, the persistence of tebuthiuron in surface soil, from rates anticipated to control the coca, was about 3–4 months. Some tebuthiuron leached below the zone of maximum bioactivity, particularly when an excessive rate (10.1 kg/ha) was tested. Any such residue leaching below about 1 m appeared to dissipate more slowly, perhaps with a half-life of >1 year.

As expected, recropping demonstrated longer persistence of phytotoxic tebuthiuron residues from the $3\times$ rate (10.1 kg/ha) than from the 3.34 kg/ha application. For example, about 50% of the coffee plants would die if transplanted into this Hawaiian soil 10 weeks after treatment at the $1\times$ rate; the $3\times$ tebuthiuron rate would extend this period to >25 weeks. Although hexazinone caused faster injury than did tebuthiuron if the plantings were at 1 week post-treatment, tebuthiuron's residual phytotoxicity to all crops (except maize) was more prolonged. The estimated times for 50% crop survival if planted in soil that had received 3.3 kg/ha of tebuthiuron were: maize, 1.5 weeks; coffee, 8–12 weeks; cassava and squash, 12–15 weeks; and bean, 26–28 weeks. No phytotoxicity to any species was seen 17 months after replanting in the $3\times$ rate tebuthiuron plots, and only limited phytotoxicity occurred 12–14 months after treatment.

Natural revegetation by weed species was slower on the tebuthiuron sprayed plots (10.1 kg/ha) than on the adjacent hexazinone spray plots. The tebuthiuron plots remained largely weed free for 8–10 months. Similar trends were observed on the main plots: the order of weed re-emergence (earliest to latest) was hexazinone-3, hexazinone-10, tebuthiuron-3 and tebuthiuron-10.

4.2.2. Peru

Over 12 months, fields receiving 2.2 and 3.3 kg/ha of tebuthiuron were sampled (Table II). As in the Hawaiian tests, the sampling plots treated with a high load (in this case, 20% active ingredients) pelleted formulation present a major challenge in terms of obtaining *representative* soil samples. Not surprisingly, four of five cores at the first post-treatment sampling (1 month) and four of six cores at 2 months contained no quantifiable tebuthiuron. The calculated tebuthiuron dissipation rates may thus be artificially high. When data are fit to first order kinetics of loss, forcing the fit to the assumed initial treatment (i.e. 2.2 or 3.3 kg/ha), the estimated tebuthiuron half-life was 0.25 month and 2.0 months at the low and high application rates, respectively. Evaluation of the two cores with the highest residues from the 2.2 kg/ha treatment suggests that some tebuthiuron had leached below the sampling depth of 50 cm:

Second month: T2/46-48
   0-15 cm = 0.11 ppm; 15-30 cm = 0.10 ppm; 30-50 cm = 0.09 ppm.
Third month: T2/58-60
   0-15 cm = 0.15 ppm; 15-30 cm = 0.16 ppm; 30-50 cm = 0.15 ppm.

These residues are equivalent to 29 and 45% of the applied herbicide, so recovery of any chemical deeper than 50 cm would have increased the half-life estimate. However, a review of the 3.3 kg/ha samples, available from 3 to 12 months, did not suggest major leaching below 50 cm; by 12 months, the highest detectable residue was 0.17 ppm, and only in that core was residue detected in the 15-30 cm zone. The tebuthiuron half-life based on the 12 month samples was 2.2 months. In summary, tebuthiuron dissipated rapidly from Peruvian soil, although the half-life estimates of 0.25-2 months are probably too low because of inherent sampling problems; more realistically, tebuthiuron probably has a half-life of about 2-4 months in Peru. No tebuthiuron was found in the stream water samples collected 1, 3 and 12 months after treatment.

Grasses in the treated fields were unaffected by tebuthiuron. After coca was killed by the herbicide treatment, secondary plants (many grassy and broad leaf successional species) rapidly emerged in the abandoned fields. The subsistence crops (near to or within the treated fields) unaffected by tebuthiuron at 2.2 or 3.3 kg/ha included papaya, yuca (cassava) and pineapple.

4.2.3. Panama

Soil samples were collected at about 1.5, 3 and 8.5 months (41, 89 and 257 days) after treatment (Table III). Tebuthiuron present in the 0-5 cm samples represented 3% (upslope) and 30% (downslope) of the amount applied 41 days earlier. The composite sample taken at 3 months was equivalent to 36% of the theoretical dose; on the basis of tebuthiuron's distribution it seems likely that some leaching below 30 cm had occurred, so somewhat more than 30% actually remained. The absence of tebuthiuron after 8.5 months (the limit of detection was 0.010 ppm) must be viewed in the context of the few samples collected during the testing.

TABLE III. TEBUTHIURON PERSISTENCE IN SOIL FOLLOWING APPLICATION TO ILLICIT COCA IN PANAMA

Time after treatment (d)		Mean (±SD) tebuthiuron residue (ppm)				Notes
		Soil depth (cm)				
Sampled	Received	0–5	0–10	10–20	20–30	
41	61	0.17 ± 0.003	—	—	—	(a)
41	61	1.62 ± 0.13	—	—	—	(b)
89	97	—	0.65 ± 0.16	0.12 ± 0.03	0.21 ± 0.03	(c)
257	258	—	0	0	0	(d)

(a) Sampled upslope, from an area of about 0.1–0.15 m^2.
(b) Sampled downslope, about 0.1–0.15 m^2.
(c) Composite of two 0.22 m^2 areas, 1 m apart, midslope, near a dead coca plant.
(d) Sampled upslope, centre, 0.25 m^2.

Nevertheless, dissipation of tebuthiuron in Panama seems to be much faster than would have been expected from temperature zone soil conditions, and persistence is consistent with the estimates from Hawaii and Peru, i.e. a half-life of about 2–4 months.

Tebuthiuron was not detected in the adjacent stream, sampled 41 and 89 days after field treatment. On the basis of the stream flow (ca. 600 L/min) measured during the second sampling, the estimated recharge derived from the treated site (about 0.7% of the volume) and the analytical limits of detection (0.5 ppb), a stream loading of approximately 0.4 g/d of tebuthiuron (or hexazinone) would have been needed to detect either herbicide. Attenuation of both compounds through microbial degradation, adsorption and dilution, plus the long transit time that may occur for lateral subsurface transport, tend to minimize the likelihood of discovering these herbicides in stream water. In larger scale aerial application, some direct surface water contamination might occur, as well as surface runoff. For reference, the 0.5 ppb detection limit is far lower than any known level of toxicological concern for either hexazinone [13–16] or tebuthiuron [13].

The tebuthiuron treated coca field in Panama was relatively weedy. At all the observation periods after treatment, a vigorous, stoloniferous, sod forming grass continued to grow on the site. This served as an effective ground cover, reducing the risk of soil erosion on the steep hillside. At 8 and 22 months after treatment, other native broad leaf species were colonizing (some had remained uninjured) the abandoned field.

4.3. Fate of imazapyr in Hawaii

Imazapyr was lost relatively quickly from the Hawaiian soil plots, but this was not through leaching. Despite over 200 mm of rainfall per month, imazapyr was confined to the upper 25 cm of soil, with no more than traces found at 25–50 cm during the 15 months of sampling. Dissipation followed first order kinetics: using non-linear curve fitting techniques, the estimated half-life was 20 ± 4 days. Dissipation was unaffected by the formulation (granular versus spray) or the initial rate (0.56 versus 2.2 kg/ha).

In Hawaii, imazapyr seemed to be less phytotoxic to seven crops than hexazinone and tebuthiuron. This may reflect much stronger binding by the imidazolinone herbicide to this soil, something noted during early efforts to develop a satisfactory chemical extraction method [9]. In an extreme test, cropping occurred immediately after treatment with 2.2 kg/ha of Arsenal. When this recropping cycle was terminated after 1 month, all the seeded crops germinated poorly; cassava (begun from cuttings) growth was <10% that of the untreated control, and only coffee (which was transplanted) *may* have survived. In subsequent recropping, the most susceptible species included maize, cassava and, to a lesser degree, coffee; Arsenal 5G, at 2.2 kg/ha, continued to severely retard cassava replanted up to 9 weeks post-treatment; and the 0.56 kg/ha rate stunted cassava growth by 50–70% at this time. Maize, peanuts and possibly squash were also sensitive crops. Again, coffee showed symptoms of imazapyr phytotoxicity, but serious injury may be delayed as the roots proliferate into the contaminated soil. Crops grew normally when planted 4 months after treatment.

4.4. Fate of glyphosate in Panama

Soil was collected for glyphosate analysis at 40 (in a field receiving 6.7 kg ae/ha) and 89 (from a 4.5 kg/ha field) days after treatment. The latter was from a 900 cm^2 area, 1–2 cm deep, adjacent to a glyphosate killed coca plant, on a steep hillside. Glyphosate was not found, nor was glyphosate's major soil metabolite, aminomethylphosphonic acid (AMPA). AMPA, like glyphosate, is known to be degraded by soil microorganisms [17]. Glyphosate, a foliar herbicide, is inactivated in soils because of rapid, strong adsorption, as well as degradation. A fact that complicates analyses is that the glyphosate–soil method had much less sensitivity than the methods used for hexazinone, tebuthiuron or imazapyr in the present studies.

Analytical sensitivity was less of a problem for water samples, but this herbicide was not detected in water collected (at 40 days) near the treated fields. Transient contamination might have been detected during spraying, but sampling was impossible.

Eighteen days after glyphosate treatment at 4.5 kg/ha or higher, both the coca and perennial grasses growing on these fields were severely affected by foliar application. However, there was no soil residual activity; after 40 days, vigorous new weed seedlings were abundant. At 3 months, natural revegetation covered the soil; in the coca fields, large non-target shrubs defoliated by glyphosate were becoming refoliated. There was no visible damage to the adjacent forest from glyphosate aerially applied in Panama (nor from hexazinone or tebuthiuron, similarly applied in Peru). One unidentified, citrus like tree, between adjacent fields that received 9 kg/ha of glyphosate (far above the rate needed to kill coca), showed temporary crown damage, but was refoliating normally when seen at 3 months.

Fields in which the glyphosate treatments effectively controlled illicit coca were abandoned by growers; when observed at 8.5 and 22 months, natural revegetation was so extensive that it was often difficult to locate the original coca plants (which had been about 1–2 m tall). To summarize, glyphosate caused no obvious, long term ecological shift; in fact, treatment with any of the tested herbicides was reversing the damage done by coca production, as the very slow process of reforestation began.

5. CONCLUSIONS

Narcotic crop source reduction can be accomplished efficiently through use of herbicides, especially if applied aerially. However, just as toxicological and environmental safety issues must be satisfied before pesticides are approved for use in conventional agriculture, forestry and other applications, the same fundamental concerns exist when such chemicals are used against narcotic plants. In this paper, the comparable results for soil persistence and leaching across all the tropical locations demonstrated that Hawaii was an acceptable surrogate location for estimating the soil environmental fate of pesticides considered for use in coca control. The approximate half-lives in soil were: hexazinone (1 month), tebuthiuron (3–4 months) and imazapyr (3 weeks); glyphosate was undetected in Panama soil 3 months after application. In both Peru and Panama, hexazinone and tebuthiuron were not detected in streams near the treated coca fields; glyphosate, aerially applied in Panama, was similarly undetected when monitored 1.5 months after treatment.

Periodic recropping with seven legitimate crops in Hawaii demonstrated how long residual phytotoxicity can occur as a result of herbicide application. As expected, this is a function of the herbicide, the rate applied and the crop species, but these generally parallel the soil residue results. Several food crops growing in or near the coca fields in Peru had received hexazinone or tebuthiuron treatment, yet showed no apparent injury 3 months later. Similarly, natural vegetation in Peru and

Panama regrew within a few months after being treated with hexazinone, tebuthiuron or glyphosate (grasses were not injured by tebuthiuron). In Hawaii, natural weed recovery was prolific after only a few months on plots sprayed with a much higher application (10.1 kg/ha) of hexazinone than would ever be used for coca control. Regrowth of weeds on the corresponding high rate tebuthiuron plots was slower, but still substantial after 12 months. The combination of chemical and biological approaches gives greater reliability to risk interpretation than does either approach alone.

The results from tropical studies demonstrated the shorter soil persistence and lower environmental risk of these herbicides compared with their use in temperate zones. Clearly, this demonstrates that great caution must accompany any pesticide fate modelling — or risk assessment — that is based on inappropriate parameters or assumptions.

REFERENCES

[1] UNITED STATES CONGRESS, OFFICE OF TECHNOLOGY ASSESSMENT, Alternative Coca Reduction Strategies, Rep. OTA-F-556, US Government Printing Office, Washington, DC (1993) 214 pp.
[2] Major Coca and Opium Producing Nations: Cultivation and Production Estimates, 1991–95, Washington, DC (1996).
[3] HELLING, C.S., ENGELKE, B.F., DOHERTY, M.A., DDT dissipation in Hawaiian in-situ soil columns, J. Environ. Sci. Health **B29** (1994) 103–119.
[4] UNITED STATES DEPARTMENT OF AGRICULTURE, Soil Taxonomy: A Basic System of Soil Classification for Making and Interpreting Soil Surveys, Soil Conservation Service, Handbook No. 436, USDA, Washington, DC (1975) 323–326.
[5] INSTITUTO GEOGRAFICO NACIONAL "Tommy Guardia", Síntesis Geográfica, edn 2 (Map), Ministerio de Obras Públicas, Panama City (1991).
[6] DUKE, J.A., PORTER, D.M., "Darién phytosociological dictionary", Bioenvironmental and Radiological-Safety Feasibility Studies: Atlantic–Pacific Interoceanic Canal, Report of Contract No. AT(26-1)-171, Battelle Memorial Institute and United States Atomic Energy Commission, Columbus, OH (1970) 28 pp.
[7] LYDON, J., ENGELKE, B.F., HELLING, C.S., Simplified high-performance liquid chromatographic method for the simultaneous analysis of tebuthiuron and hexazinone, J. Chromatogr. **536** (1991) 223–228.
[8] LUSBY, W.R., HELLING, C.S., SOMICH, C.J., "Detection of tebuthiuron herbicide from soil by selected ion electron capture negative-ion mass spectrometry", Mass Spectrometry and Allied Topics (Proc. 38th Conf. Tucson, 1990), American Society for Mass Spectrometry, Santa Fe, NM (1990) 683–684.
[9] HELLING, C.S., DOHERTY, M.A., Improved method for the analysis of imazapyr in soil, Pestic. Sci. **45** (1995) 21–26.
[10] GLASS, R.L., Liquid chromatographic determination of glyphosate in fortified soil and water samples, J. Agric. Food Chem. **31** (1983) 280–282.

[11] ROY, D.N., et al., Determination of persistence, movement, and degradation of hexazinone in selected Canadian boreal forest soils, J. Agric. Food Chem. **37** (1989) 443–447.

[12] FENG, J.D., NAVRATIL, S., Sampling for zero-time hexazinone residues in forest soil dissipation study, Can. J. For. Res. **20** (1990) 1549–1552.

[13] Farm Chemicals Handbook, Vol. 78, Meister Publishing Co., Willoughby, OH (1995) C199-200, C362.

[14] ENVIRONONMENTAL PROTECTION AGENCY, Pesticide Fact Sheet: Hexazinone, Office of Toxic Substances, EPA, Washington, DC (1988) 9 pp.

[15] KRUETZWEISER, D.P., HOLMES, S.B., BEHMER, D.J., Effects of the herbicides hexazinone and triclopyr ester on aquatic insects, Ecotoxicol. Environ. Safety **23** (1992) 364–374.

[16] THOMPSON, D.G., HOLMES, S.B., THOMAS, D., MACDONALD, L., SOLOMON, K.R., Impact of hexazinone and metsulfuron methyl on the phytoplankton community of a mixed-wood/boreal forest lake, Environ. Toxicol. Chem. **12** (1993) 1695–1707.

[17] DUKE, S.O., "Glyphosate", Herbicides: Chemistry, Degradation, and Mode of Action, Vol. 3 (KEARNEY, P.C., KAUFMAN, D.D., Eds), Marcel Dekker, New York (1988) 1–70.

POLYCHLORINATED BIPHENYLS AND CYCLIC PESTICIDES IN SEDIMENTS AND MACROINVERTEBRATES FROM THE COASTAL REGIONS OF DIFFERENT CLIMATOLOGICAL ZONES*

J.M. EVERAARTS, E.M. VAN WEERLEE,
C.V. FISCHER, M.Th.J. HILLEBRAND
Nederlands Instituut voor Ondersoek der Zee,
Den Burg–Texel, Netherlands

Abstract

POLYCHLORINATED BIPHENYLS AND CYCLIC PESTICIDES IN SEDIMENTS AND MACROINVERTEBRATES FROM THE COASTAL REGIONS OF DIFFERENT CLIMATOLOGICAL ZONES.

The present study contributed towards establishing the status of a tropical and a moderate latitude marine ecosystem in terms of the concentration of anthropogenic, lipid associated, cyclic, halogenated hydrocarbons in surface sediments and benthic macroinvertebrates. Samples were obtained from the coastal region and the continental slope of Kenya (Indian Ocean), the Dutch coastal region and the continental shelf of the North Sea, and a bay area of Curaçao (Netherlands Antilles, Caribbean). In surface sediments (Kenya), congeners CB28, CB52, CB101, CB118, CB153, CB138 and CB180, and the pesticides alpha-HCH and gamma-HCH, dieldrin, endrin and members of the DDT family were not identified, except in the estuarine zone of the Sabaki River. In North Sea surface sediment, the characteristic PCB pattern was always determined, and concentrations were particularly enhanced in sediment from the river mouths. Also, alpha-HCH, p,p'-DDD and p,p'-DDE were quantified in most samples. Regarding benthic invertebrates, CBs and p,p'-DDE were quantified. Some samples showed particularly enhanced levels in bivalve molluscs and certain penaeid prawns from the Kenyan coastal region. The high levels present in the digestive (pyloric caeca) and reproductive (gonad) organs of seastar from the North Sea confirmed distinct concentration gradients. Sponges and tunicates from the bay area of Curaçao accumulated CBs and p,p'-DDE, but not at very high concentrations.

---

* Research carried out in association with the IAEA under Research Agreement No. 8013.

1. INTRODUCTION

Over 70% of the Earth's surface consists of oceans, coastal seas and estuarine zones. The importance of marine ecosystems is also illustrated by the fact that coastal regions are inhabited by about 60% of the world's population. The load of anthropogenic compounds induces the impairment of various physiological functions of individual organisms and affects the ecological quality of ecosystems. In particular, estuarine and coastal systems with high biological activity are exposed to a high degree of contamination. Polychlorinated biphenyls (PCBs) and chlorinated pesticides (e.g. DDT and dieldrin) show global distribution. They are ubiquitous toxic contaminants because of their bioaccumulative capacity and persistence, as well as their specific physicochemical properties. The technical PCB mixtures or the individual chlorinated biphenyl (CB) congeners and pesticides accumulate in biota at all trophic levels, and residues are reported in environmental compartments from all geographical latitudes [1, 2].

The present study contributed towards establishing the status of marine ecosystems from different latitudes in terms of the concentrations of anthropogenic, lipid associated, cyclic, halogenated hydrocarbons (PCBs, DDT, DDE and dieldrin) in sediment and selected benthic macroinvertebrates.

2. MATERIALS AND METHODS

Samples were obtained from the coastal region and the continental slope of Kenya (Indian Ocean), the Dutch coastal region and the continental shelf of the North Sea, and a bay area of Curaçao (Netherlands Antilles, Caribbean). The animal species in which the organochlorine (OC) contaminants were analysed represented the phyla Porifera (Demospongiae), Annelida (Polychaeta, bristle worms), Arthropoda (Crustacea, Natantia, shrimps) and Reptantia (hermit crabs), Mollusca (Bivalvia, clams), Chordata (Ascidiacea) and Pisces (Teleostei, bottom dwelling fish).

Two surveys with the research vessel, Tyro, from 18 June to 9 July 1992 (south-east monsoon) and from 19 November to 8 December 1992 (onset of the north-east monsoon) covered sampling locations in different areas of the coastal region and along transects radiating into the Indian Ocean, perpendicular to the Kenyan coast (Fig. 1).

In the North Sea, samples were collected during two surveys with the Dutch research vessel, Pelagia, from 26 August to 13 September 1991 and from 18 May to 6 June 1992. The surveys covered sampling locations in different areas of the coastal region and along transects radiating into the southern North Sea (Fig. 2).

At two sites in the industrialized bay area of Schottegat (south coast of Curaçao, Netherlands Antilles), samples of sponges and colonial tunicates were removed from the reef during scuba diving and transported at -4°C to the

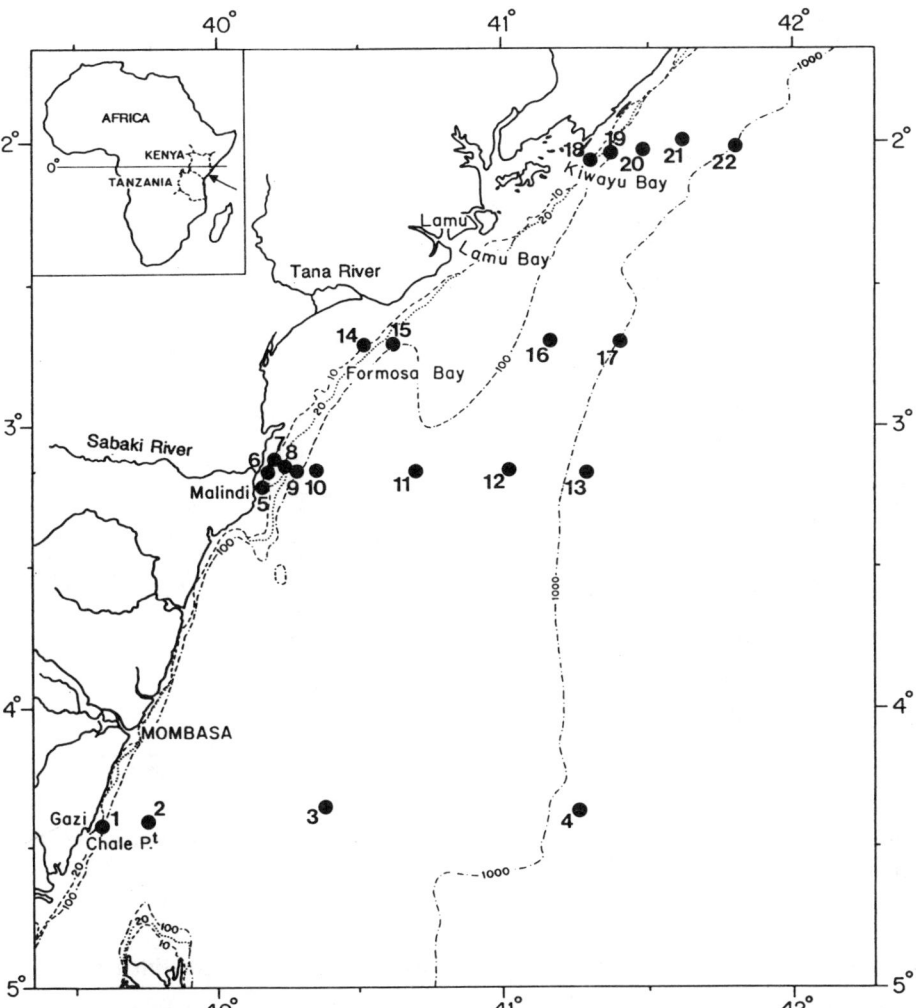

FIG. 1. Area of investigation in the estuarine, coastal region and continental slope of Kenya. The sampling stations are located along transects radiating into the Indian Ocean, perpendicular to the Kenyan coast: Gazi transect (stations 1–4); Sabaki transect (stations 5–13); Tana transect (stations 14–17); and Kiwayu transect (18–22).

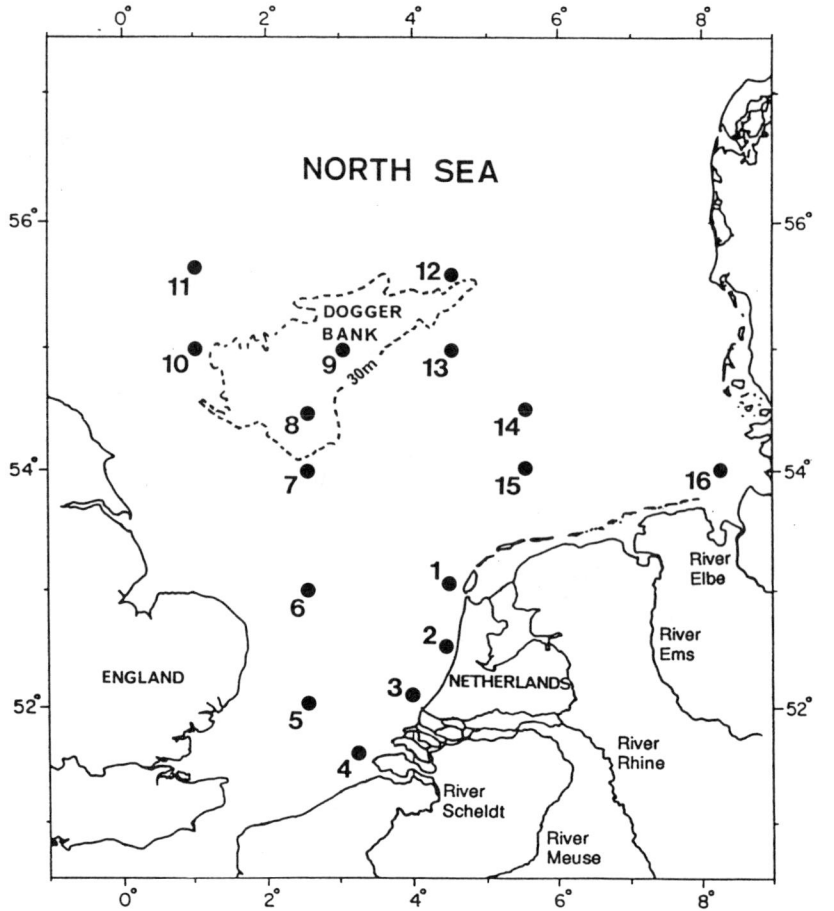

FIG. 2. *Sampling stations located in the coastal region of the Netherlands and along transects radiating into the southern North Sea.*

laboratory for analysis. All other epibenthic species were collected with a 5 m beam trawl, at a trawling time of 15 min. Sediment and infaunal invertebrates were collected with a 0.2 m^2 Van Veen bottom grab and a Reineck box corer; samples were stored at $-25°C$.

From the sediments, approximately 20 g wet weight samples were extracted by vigorously mixing (Ultra-Turrax) with isopropanol, hexane and water, modified according to the Blight and Dyer method descibed in Ref. [3]. About 3–5 g wet weight of whole body or specific tissue, measured to the nearest 0.01 g, was homogenized with as much anhydrous sodium sulphate (Na_2SO_4) as was necessary to obtain a dry running mixture. Subsequently, the samples were Soxhlet extracted for 8 h with 150 mL of a 1:1 mixture of dichloromethane and pentane.

The entire analytical procedure for polychlorinated hydrocarbons, including sample extraction, clean-up, fractional separation of the PCBs and most of the pesticides, and pretreatment of the chemicals used, has been described in Refs [1, 3–7].

Analyses of the samples were performed using a temperature programmed gas chromatograph (GC) (Hewlett-Packard 5880a) equipped with a 60 m fused silica capillary column (CPSil-19 (solid phase) and an inner diameter of 0.15 mm) and a ^{63}Ni electron capture detector (ECD). Verification of the chromatographic results on the CPSil-19 column were obtained by running selected samples through a temperature programmed GC (Carlo Erba HRGC 5160) equipped with a fused silica capillary column (CPSil-8, 50 m length, 0.15 mm inner diameter and 0.30 μm wall coating) and the same ECD. Hydrogen was used as the carrier gas at a pressure of 350 kPa, with a linear gas velocity of 30 cm/min. The temperatures of the injector and the detector were 230 and 340°C, respectively. Split/splitless injections of 1 μL aliquots of each sample dissolved in 2,2',4-trimethylpentane were made with an autosampler (Carlo Erba A200S). The temperature programme consisted of isothermal phases at 90°C (4.5 min), 215°C (25 min) and 270°C (20 min), with intermediate temperature increases of 10 and 5°C/min.

To identify and quantify the CB congeners, a synthetic mixture of 44 individual congeners was employed, with CB112 as the internal standard. The seven CB congeners reported on here were those that had been selected in several countries as indicators; quantification of these components can be used to determine whether the PCB levels in food products, waste mineral oil and environmental samples comply with the maximum levels permitted by legislation [8–11]. These congeners, given in sequence of elution from the GC CPSil-19 column, are: CB28 (2,4,4'-tri-CB), CB52 (2,2',5,5'-tetra-CB), CB101 (2,2',4,5,5'-penta-CB), CB118 (2,3',4,4',5-penta-CB), CB153 (2,2',4,4',5,5'-hexa-CB), CB138 (2,2',3,4,4',5'-hexa-CB) and CB180 (2,2',3,4,4',5,5'-hepta-CB).

They represent different levels of the overall chlorination of the molecule, and also differ in the site of the chloro-atom substitution to the biphenyl skeleton. Numbering of the CB congeners is according to International Union of Pure and Applied Chemistry (IUPAC) rules, as described by Ballschmiter and Zell [12].

The pesticides alpha-HCH and gamma-HCH, dieldrin, endrin and members of the DDT family (p,p'-DDT, o,p'-DDT, p,p'-DDD, o,p'-DDD and p,p'-DDE) were identified and quantified on the basis of a synthetic mixture of these compounds, using CB155 as the internal standard. The standards were used at every tenth injection to recalibrate the retention time and the response factor of each compound. The response factor of each single peak in the standard mixture, representing either a CB congener or a pesticide, did not differ more than 10% between successive standards. A peak in a chromatogram of a sample was only assigned to a compound in the standard mixture when its retention time did not differ more than 0.025 min from the compound in the standard. For positive identification, the peak pattern in the chromatogram was also taken into account. A compound was considered non-detectable when the peak height was less than the value of the average of the analyte blank signals, plus three times the standard deviation of the blank (SD_b). The peak height, with a value that was the average of the analyte blank signals, plus three times the SD_b, or plus nine times the SD_b, was considered as being detectable, or appropriate for quantification, respectively.

To control the procedural blank levels, each series of five samples was accompanied alternately by a blank run of pure Na_2SO_4 and a complete procedure with certified reference materials (CRMs) of the European Union Bureau of Standards (either sewage sludge (CRM 392) or cod-liver oil (CRM 349) of the Bureau for Certified Reference Materials). The certified and measured values for CRM 349 were 68 ± 7 and 69 for CB28, 149 ± 20 and 144 for CB52, 370 ± 17 and 353 for CB101, 456 ± 31 and 420 for CB153, and 282 ± 22 and 248 for CB180, respectively. For CRM 392, the certified and measured values were 100 ± 10 and 109 ± 4 for CB28, 79 ± 9 and 88 ± 4 for CB52, 134 ± 11 and 166 ± 7 for CB101, 97 ± 12 and 109 ± 7 for CB118, 320 ± 27 for DB153, and 313 ± 24 and 334 ± 27 for CB180, respectively. CB138 was not involved with CRMs.

All the analytical steps and procedures were performed according to formalized standard operating procedures and placed within the framework of the intercalibration exercises of the International Council for Exploration of the Sea and the EU program, Quality Assurance of Information for Marine Environmental Monitoring in Europe (QUASIMEME).

3. RESULTS AND DISCUSSION

3.1. Sediment

The concentrations of seven selected CB congeners and five cyclic OC pesticides in surface sediments, and the total particulate organic carbon contents are given in Table I [12]. In the tropical marine environment of the coastal region and the continental slope of Kenya, CB congeners could only be quantified in sediments from two shallow coastal stations (Fig. 1, stations 6 and 8) (chromatograms also showed the specific PCB pattern). In other samples, all the CBs were below the limit of detection (LOD).

Very different results were obtained in the North Sea. CB congeners could be quantified in surface sediments from all the sampling stations (also evidenced by the characteristic PCB pattern in the chromatograms). No significant concentration gradients were found, although two coastal stations located at the outflow of the Rivers Scheldt and Elbe (Fig. 2, stations 4 and 16, respectively) showed high CB levels. Enhanced levels were measured in an offshore area with a high percentage of silt (station 7) that is known to be a sedimentation zone and functions as a sink for particulate matter [13], and in an area (station 11) that might be influenced by

TABLE I. CONCENTRATIONS (ng/g OF ORGANIC CARBON) OF SEVEN SELECTED CB CONGENERS AND FIVE CYCLIC OC PESTICIDES IN SURFACE SEDIMENTS (0.5 cm TOP LAYER) FROM A TROPICAL (COASTAL AREA AND THE CONTINENTAL SLOPE OF KENYA) AND A TEMPERATE REGION (NORTH SEA). THE FRACTIONS OF ORGANIC CARBON (% OC) AND SILT (wt%, GRAIN SIZE: <63 µm) ARE SHOWN
(Numbering of the CB congeners is according to IUPAC rules [12]; they are listed in sequence of elution from the GC CPSil-19 column. The limit of detection (LOD) is the average of the analyte blank signals, plus three times its standard deviation; the limit of quantification (LOQ) is the average of the analyte blank signals, plus nine times its standard deviation. LOD and LOQ are considered on the basis of a 1 µL injection from a 1 mL extract)[a]

Station No./ depth (mm)	CB congener No.							Pesticides						Percentage	
	CB28	CB52	CB101	CB118	CB153	CB138	CB180	Alpha-HCH	Gamma-HCH	Dieldrin	p,p'-DDD	p,p'-DDE		OC	Silt

Coastal area and continental slope of Kenya (see Fig. 1)

Gazi transect

1/62	54.3	ND	ND	ND	ND	ND	ND	ND	ND	ND	ND	ND		0.46	NA
2/511	110.5	ND	ND	ND	ND	ND	ND	ND	ND	ND	ND	ND		1.05	NA
3/1004	200.0	ND	ND	ND	ND	ND	ND	ND	53.2	ND	ND	ND		0.47	NA
4/2053	78.5	ND	ND	ND	ND	ND	ND	ND	ND	ND	ND	ND		0.65	NA

Sabaki transect

5/18	ND	ND	ND	ND	ND	ND	ND	15.7	ND	37.3	90.2	508.8		0.51	NA
6/20	12.3	11.4	11.2	+	12.9	7.1	ND	3.6	10.0	ND	ND	32.1		1.40	NA
7/31	ND	ND	ND	ND	ND	ND	ND	23.9	ND	ND	ND	120.6		1.55	NA
8/40	47.5	12.4	18.8	ND	25.2	ND	ND	14.4	33.8	ND	ND	125.2		1.39	NA
9/64	62.2	ND	ND	ND	ND	ND	ND	86.7	ND	ND	ND	ND		0.45	NA
10/218	9.4	ND	ND	ND	ND	ND	ND	59.4	7.8	ND	ND	ND		0.64	NA
11/518	21.5	ND	ND	ND	ND	ND	ND	169.2	ND	ND	ND	ND		0.65	NA
12/1115	49.2	ND	ND	ND	ND	ND	ND	ND	ND	ND	ND	ND		0.65	NA
13/2063	17.9	ND	ND	ND	ND	ND	ND	15.8	ND	ND	ND	ND		0.95	NA

TABLE I (cont.)

Station No./ depth (m)	CB congener No.								Pesticides					Percentage	
	CB28	CB52	CB101	CB118	CB153	CB138	CB180	Alpha-HCH	Gamma-HCH	Dieldrin	p,p'-DDD	p,p'-DDE		OC	Silt
Tana transect															
14/19	ND	ND	ND	ND	ND	ND	ND	ND	ND	ND	ND	ND		0.11	NA
15/101	ND	ND	ND	ND	ND	ND	ND	ND	ND	ND	ND	ND		0.57	NA
16/992	*13.4*	ND	ND	ND	ND	ND	ND	10.3	ND	ND	ND	ND		0.97	NA
17/2145	*43.4*	ND	ND	ND	ND	ND	ND	ND	ND	ND	ND	ND		0.53	NA
Kiwayu transect															
18/24	*30.4*	ND	ND	ND	ND	ND	ND	ND	ND	ND	ND	ND		0.51	NA
19/54	*24.6*	ND	ND	ND	ND	ND	ND	8.7	ND	ND	ND	ND		1.38	NA
20/505	*14.9*	ND	ND	ND	ND	ND	ND	11.5	18.2	ND	ND	ND		1.48	NA
21/950	ND	ND	ND	ND	ND	ND	ND	7.4	ND	ND	ND	ND		1.08	NA
22/2023	*35.2*	ND	ND	ND	ND	ND	ND	ND	42.0	ND	ND	ND		0.88	NA
North Sea (see Fig. 2)															
1/28	3.8	2.3	4.6	3.8	9.2	7.7	5.4	3.1	ND	ND	36.2	1.5		1.30	2.27
2/17	3.5	+	3.5	3.7	7.0	5.3	3.5	ND	ND	ND	35.1	+		0.57	0.12
3/23	3.7	ND	3.8	3.8	7.5	5.7	+	1.9	ND	ND	143.4	ND		0.53	0.22
4/23	6.4	3.2	9.5	11.1	20.6	15.9	7.9	3.2	ND	ND	14.3	3.2		0.63	3.66
5/49	+	+	2.6	ND	5.2	+	+	ND	ND	ND	ND	+		0.77	0.06
6/34	+	+	+	+	5.0	+	+	ND	ND	ND	20.0	+		0.40	1.02

TABLE I (cont.)

7/66	7.0	ND	7.0	9.3	11.6	8.1	2.3	NA	NA	NA	NA	14.0	0.86	15.32
8/22	+	+	3.5	+	8.8	5.3	3.3	7.0	ND	ND	66.7	+	0.57	0.30
9/25	+	ND	+	ND	7.0	+	ND	4.7	ND	ND	139.5	+	0.43	0.50
10/62	2.7	ND	2.7	7.5	9.3	4.0	+	10.7	4.0	ND	238.7	4.1	0.75	2.08
11/80	2.6	ND	+	+	13.2	5.2	2.6	7.8	2.4	ND	13.0	2.6	0.77	1.56
12/33	+	ND	+	+	+	+	ND	3.3	ND	ND	123.3	ND	0.60	0.10
13/45	+	ND	+	+	+	+	ND	ND	ND	ND	147.6	+	0.42	3.78
14/43	+	+	+	+	3.8	+	ND	4.0	ND	ND	49.1	+	0.53	3.15
15/38	8.0	3.3	8.2	9.8	19.7	13.1	6.6	32.8	9.8	ND	159.0	13.1	0.61	10.12
16/19	68.0	56.0	136.0	116.0	260.0	222.0	112.0	48.0	20.0	32.0	268.0	76.0	0.25	31.09
LOD	0.08	0.04	0.05	0.08	0.04	0.08	0.04	0.04	0.05	0.07	0.08	0.07		
LOQ	0.20	0.10	0.11	0.20	0.10	0.20	0.10	0.11	0.14	0.19	0.20	0.19		

[a] ND = not detectable; NA = not analysed; and + = identified, but below the LOQ.

Notes: Endrin, o,p'-DDT and p,p'-DDT were not detected in any of the samples. In the sample GC runs for CB determination, the % recovery of the internal standard CB112 was 74.3 ± 6.8; the % recovery in the CRM and blank runs was 86.6 ± 5.1. Although identified and quantified as such, the values of CB28 indicated in italics do not refer to the actual concentration of CB28, but rather represent the compounds released by column bleeding or unknown components in the sample (e.g. phthalates).

the River Humber plume [14]. Thus, the data do not completely agree with the distinct concentration gradients along the transects perpendicular to the Dutch coast for PCBs in water and sediments, as described in Ref. [15].

Of the pesticides, p,p'-DDE could always be quantified in sediments from the shallow estuarine zone at the mouth of the Sabaki River, whereas the concentration of alpha-HCH increased across the continental shelf towards the deep sea (Table I). Also, alpha-HCH could be quantified at deeper stations along the transects across the continental shelf off Formosa Bay (at the mouth of the Tana River) and Kiwayu Bay. Gamma-HCH was analysed only in sediments from a few stations. Sediments from the shallowest sampling site (18 m) closest to the mouth of the Sabaki River showed high concentrations of dieldrin (37 ng/g of OC) and p,p'-DDE (510 ng/g of OC).

In surface sediments from the North Sea, p,p'-DDE and p,p'-DDD were quantified in all but one of the samples. In contrast, gamma-HCH and dieldrin were found in only a few samples. Alpha-HCH was found in the Dutch coastal region (stations 1–4), the western Dogger Bank area (stations 8–11) and in those areas with a very high percentage of silt (stations 15 and 16).

3.2. Macrobenthic organisms

The number of individuals of each species, the percentage lipid extracted, the amount of lipid extracted and the concentration of the CB congeners and p,p'-DDE are summarized in Table II [12]. A specific PCB pattern (technical mixture Clophen A50) was observed in the chromatograms, and the seven CB congeners selected were quantified, with a few exceptions that could only be identified. By far the highest levels of CB were found in bivalve molluscs from the mouth of the Sabaki River (30 and 65 ng/g of lipid for CB153) and Kiwayu Bay (40 ng/g of lipid for CB153), and in gastropod molluscs from Kiwayu Bay (25 ng/g of lipid for CB153). Only p,p'-DDE (representing the cyclic pesticides) was analysed and found to be present in all the samples, particularly at enhanced levels that varied between 15 and 48 ng/g of lipid in bivalve and gastropod molluscs. The concentrations were significantly enhanced compared with the same groups of species sampled in the coastal region and on the continental slope of the Banc d'Arguin (Mauritania), where only congeners CB153 and CB138 were identified [1]. In comparison, the levels found in the present study agreed with those found in macrobenthic fauna from the shallow Java Sea and the Strait of Madura, the Bali Basin (Indonesia) and the Java Trench (Indian Ocean), or were slightly lower, depending on the species considered [16]. In other tropical estuarine systems along the coast of the Malay Peninsula, the concentrations of dieldrin, DDT and in particular DDE were higher than those in samples from Kenyan marine waters [17].

Text cont. on p. 428.

TABLE II. SEVEN SELECTED CB CONGENERS AND p,p'-DDE IN MACROBENTHIC INVERTEBRATES AND TWO FISH SPECIES FROM A TROPICAL REGION (COASTAL AREA AND THE CONTINENTAL SLOPE OF KENYA) (SEE FIG. 1). THE NUMBER (No.) OF INDIVIDUALS OF EACH SPECIES, THE PERCENTAGE LIPID EXTRACTED, THE AMOUNT OF LIPID EXTRACTED (mg) AND THE CONCENTRATION (ng/g OF LIPID EXTRACTED) ARE SHOWN

(Numbering of the CB congeners is according to IUPAC rules [12]; they are listed in sequence of elution from the GC CPSil-19 column. The limit of detection (LOD) is the average of the analyte blank signals, plus three times its standard deviation; the limit of quantification (LOQ) is the average of the analyte blank signals, plus nine times its standard deviation. LOD and LOQ are considered on the basis of a 1 μL injection from a 1 mL extract)[a]

Station No./ depth (m)/ geographical position	Group/species	No.	% lipid extracted	Amount of lipid extracted (mg)	CB congener No.							Pesticide
					CB28	CB52	CB101	CB118	CB153	CB138	CB180	p,p'-DDE

Formosa Bay: mouth of the Sabaki River

South-east monsoon period — June/July 1992

Station No./ depth (m)/ geographical position	Group/species	No.	% lipid extracted	Amount of lipid extracted (mg)	CB28	CB52	CB101	CB118	CB153	CB138	CB180	p,p'-DDE
8–9/50 m/ 03°09'S 40°14'E	Echinodermata, Ophiuridae (brittle stars)	12	0.46	14.5	+	5.1	7.8	ND	6.1	3.3	3.6	34.5
	Crustacea, Penaeidae (tiger prawns)	6	1.28	37.7	+	1.8	2.8	0.8	4.1	3.4	ND	15.7
	Crustacea, Penaeidae (penaeid prawns)	5	1.04	28.5	+	1.5	+	ND	3.8	3.9	5.0	9.0
	Mollusca, Bivalvia	40	0.95	14.4	9.7	10.0	14.9	28.1	29.0	23.3	ND	18.5
	Mollusca, Bivalvia	6	0.94	12.0	7.5	39.5	49.2	54.5	65.6	64.4	7.5	47.8
10/220 m/ 03°10'S 40°18'E	Crustacea, Penaeidae (penaeid prawns)	10	1.34	31.2	3.2	1.7	10.0	2.3	6.7	5.1	ND	2.9
	Crustacea, Penaeidae (red prawns)	7	1.45	38.8	+	1.1	4.5	ND	4.8	2.1	1.3	3.0
	Pisces, Bothidae, flatfish	5	2.09	82.5	7.2	1.5	+	0.9	2.6	1.7	2.0	0.5

Formosa Bay: mouth of the Tana River

14/20 m/ 02°42'S 40°31'E	Echinodermata, Ophiuridae (brittle stars)	6	2.07	62.6	6.2	3.7	5.2	4.5	4.8	5.0	2.4	2.7
	Pisces, Tetraodonthidae (pufferfish, liver)	7	45.5	987.1	9.7	0.7	0.8	0.5	1.9	1.0	1.4	29.1

TABLE II (cont.)

Station No./ depth (m)/ geographical position	Group/species	No.	% lipid extracted	Amount of lipid extracted (mg)	CB congener No.								Pesticide
					CB28	CB52	CB101	CB118	CB153	CB138	CB180		p,p'-DDE

Kiwayu Bay

18/20 m/	Echinodermata, Ophiuridae (brittle stars)	15	0.22	32.0	12.6	7.4	4.1	1.7	2.5	2.0	1.3		2.3
02°04′S	Crustacea, Penaeidae (penaeid prawns)	10	1.12	35.1	+	+	+	ND	1.2	ND	0.8		4.8
41°18′E	Mollusca, Bivalvia	4	0.31	3.5	22.8	15.0	10.6	4.1	39.1	44.6	28.6		31.3
19/50 m/	Crustacea, Penaeidae (red prawns)	10	2.88	120.1	ND	1.0	+	ND	2.1	1.6	ND		4.6
02°01′S	Crustacea, Penaeidae (king prawns)	10	1.18	23.9	4.5	1.8	4.2	ND	7.6	3.9	ND		5.0
41°20′E	Crustacea, Penaeidae (penaeid prawns)	18	0.97	31.8	+	1.1	+	ND	2.8	2.4	ND		1.9
	Crustacea, Paguridae (*Pagurus* sp.)	10	1.08	63.4	9.3	1.5	1.3	1.6	2.0	1.8	0.8		1.9
	Mollusca, Bivalvia	7	3.45	17.0	*3.1*	ND	2.5	ND	4.1	ND	ND		2.4
	Mollusca, Gastropoda	25	0.92	18.1	7.9	11.8	30.7	27.9	24.2	36.4	5.8		23.1

North-east monsoon period — November/December 1992

Formosa Bay: mouth of the Sabaki River

8-9/50 m/	Echinodermata, Ophiuridae (brittle stars)	17	1.21	74.8	ND	ND	+	ND	2.1	1.1	1.8		7.7
03°09′S	Crustacea, Penaeidae (tiger prawns)	2	5.93	18.5	ND	2.4	5.5	ND	6.2	5.3	ND		21.2
40°14′E	Crustacea, Penaeidae (penaeid prawns)	10	1.18	30.9	+	1.5	4.0	4.3	5.7	5.8	3.1		6.1
	Mollusca, Bivalvia	17	0.75	38.1	+	0.5	+	ND	2.9	1.5	ND		15.1
	Mollusca, Bivalvia (*Perna viridis*)	9	0.49	28.8	+	ND	5.1	3.5	6.9	7.2	4.9		14.0
10/220 m/	Crustacea, Penaeidae (red prawns)	6	1.18	36.6	+	1.7	4.3	4.9	15.9	15.7	11.8		9.8
03°10′S	Crustacea, Paguridae (*Pagurus* sp.)	9	1.72	85.5	2.4	1.0	+	+	7.9	1.3	1.1		17.3
40°18′E	Crustacea, Paguridae (*Pagurus* sp. + eggs)	7	4.20	117.3	9.5	0.9	+	0.6	1.8	1.3	+		0.8
	Pagurus eggs	7	61.6	193.8	*11.1*	0.4	ND	0.5	2.6	2.1	1.7		ND
	Mollusca, Cephalopoda (*Sepia* sp.)	3	1.37	41.2	3.0	2.6	4.1	4.0	7.2	6.3	ND		8.6

TABLE II (cont.)

Kiwayu Bay													
18/20 m/ 02°04′S 41°18′E	Crustacea, Penaeidae (tiger prawns)	12	1.37	26.4	+	2.3	4.5	1.1	5.7	4.9	0.6	3.9
	Crustacea, Penaeidae (tiger prawns)	7	1.20	29.8	+	2.2	4.0	ND	3.5	2.6	ND	5.0
	Crustacea, Penaeidae (penaeid prawns)	7	0.90	53.2	+	0.9	3.0	1.9	1.4	ND	ND	2.2
	Crustacea, Penaeidae (shrimps)	5	1.36	32.7	+	1.7	3.5	2.6	3.4	5.9	ND	3.2
	Crustacea, Portunidae (swimming crabs)	6	0.62	32.2	+	1.0	+	ND	2.8	3.3	ND	2.3
	Crustacea, Majidae (spider crabs)	15	1.21	21.9	5.3	4.8	5.3	2.6	9.7	5.5	ND	4.8
	Mollusca, Bivalvia	10	0.62	38.2	+	ND	1.7	ND	2.0	1.8	ND	1.9
	Mollusca, Cephalopoda (*Sepia* sp.)	3	1.63	37.9	3.2	10.7	25.3	35.3	52.1	28.5	3.2	25.6
LOD					1.18	0.14	0.52	0.19	0.44	0.40	0.15	0.13
LOQ					3.04	0.32	2.60	0.49	1.10	1.06	0.39	0.31

[a] + = identified, but below the LOQ; and ND = not detectable.

Notes: In the sample GC runs for CB determination, the % recovery of the internal standard CB112 was 74.3 ± 6.8; and the % recovery in the CRM and blank runs was 93.7 ± 6.0. Although identified and quantified as such, the values of CB28 indicated in italics do not refer to the actual concentration of CB28, but rather represent the compounds released by column bleeding or unknown components in the sample (e.g. phthalates).

TABLE III. SEVEN SELECTED CB CONGENERS AND SOME CYCLIC OC PESTICIDES IN A HOMOGENATE OF PYLORIC CAECA FROM SIX SPECIMENS OF SEASTAR (*Asterias rubens*) FROM THE NORTH SEA (SEE FIG. 2). THE PERCENTAGE LIPID EXTRACTED, THE AMOUNT OF LIPID EXTRACTED (mg) AND THE CONCENTRATION (ng/g OF LIPID EXTRACTED) ARE SHOWN

(Numbering of the CB congeners is according to IUPAC rules [12]; they are listed in sequence of elution from the GC CPSil-19 column. The limit of detection (LOD) is the average of the analyte blank signals, plus three times its standard deviation; the limit of quantification (LOQ) is the average of the analyte blank signals, plus nine times its standard deviation. LOD and LOQ are considered on the basis of a 1 μL injection from a 1 mL extract)[a]

Station No./ depth (m)	%/amount (mg) of lipid extracted	CB congener No.							Pesticides					
		CB28	CB52	CB101	CB118	CB153	CB138	CB180	Alpha-HCH	Gamma-HCH	Dieldrin	p,p'-DDD	p,p'-DDE
1/28	3.95/38.7	26.3	50.5	138.4	188.2	456.2	295.8	20.3	69.7	38.8	73.6	397.5	109.7
2/17	4.37/41.9	74.3	188.0	417.4	351.6	1053.8	616.0	102.6	ND	+	+	ND	155.2
3/23	8.12/65.2	46.2	112.1	302.7	298.4	792.5	519.9	80.1	+	28.6	84.8	162.9	146.7
4/23	3.40/32.6	29.6	66.6	222.1	353.5	1009.9	599.1	102.4	+	68.5	+	ND	153.1
5/49	7.20/44.4	43.1	87.3	231.8	307.1	719.1	481.2	31.7	ND	ND	ND	ND	160.1
6/34	5.39/40.1	11.1	13.9	33.3	62.9	171.2	102.6	17.6	+	29.7	83.9	+	102.5
7/66	3.45/39.0	9.3	6.2	11.4	22.7	58.3	39.5	3.7	ND	+	+	ND	75.9
8/22	2.40/32.7	15.1	12.3	17.5	25.5	51.6	38.6	ND	18.8	+	45.1	ND	82.6
9/25	3.17/40.5	8.7	7.2	12.5	20.9	45.4	31.8	ND	+	ND	+	ND	56.4
10/62	6.29/64.3	28.0	58.6	170.6	220.8	527.7	336.3	29.4	ND	15.5	52.8	1093.7	146.7
11/80	3.26/29.9	8.4	6.3	11.0	12.2	41.1	25.8	ND	ND	ND	ND	ND	20.9
12/33	6.55/47.3	8.4	6.5	13.5	33.5	54.9	37.7	ND	28.1	29.8	59.2	75.8	64.1

TABLE III (cont.)

13/45	4.84/33.7	7.7	7.5	19.9	30.7	76.6	53.8	4.4	20.9	25.2	160.6	95.3	63.2
14/43	3.80/44.3	9.7	8.9	17.8	26.9	75.3	52.8	3.6	20.8	33.7	87.3	ND	59.4
15/38	5.95/37.6	11.7	14.9	42.1	55.9	125.4	80.9	4.7	ND	ND	69.9	+	115.4
16/19	3.37/25.4	15.8	22.8	69.3	59.5	232.0	126.7	26.0	+	37.9	45.1	+	58.9
LOD		0.02	0.12	0.12	0.04	0.19	0.17	0.06	0.12	0.14	0.24	0.43	0.14
LOQ		0.04	0.20	0.30	0.08	0.46	0.41	0.15	0.23	0.22	0.51	0.88	0.34

[a] ND = not detectable; and + = identified, but below the LOQ.

Notes: Endrin, o,p'-DDT and p,p'-DDT were not identified in any of the samples. In the sample GC runs for CB determination, the % recovery of the internal standard CB112 was 84.6 ± 13.1; and the % recovery in the CRM and blank runs was 99.4 ± 3.8. In the GC runs for pesticide determination, the % recovery of the internal standard CB155 was 68.7 ± 27.4; and the % recovery in the CRM and blank runs was 95.0 ± 3.5.

TABLE IV. SEVEN SELECTED CB CONGENERS AND SOME CYCLIC OC PESTICIDES IN A HOMOGENATE OF GONADS FROM SIX SPECIMENS OF SEASTAR (*Asterias rubens*) FROM THE NORTH SEA (SEE FIG. 2). THE PERCENTAGE LIPID EXTRACTED, THE AMOUNT OF LIPID EXTRACTED (mg) AND THE CONCENTRATION (ng/g OF LIPID EXTRACTED) ARE SHOWN

(Numbering of the CB congeners is according to IUPAC rules [12]; they are listed in sequence of elution from the GC CPSil-19 column. The limit of detection (LOD) is the average of the analyte blank signals, plus three times its standard deviation; the limit of quantification (LOQ) is the average of the analyte blank signals, plus nine times its standard deviation. LOD and LOQ are considered on the basis of a 1 μL injection from a 1 mL extract)[a]

Station No./ depth (m)	%/amount (mg) of lipid extracted	CB congener No.							Pesticides				
		CB28	CB52	CB101	CB118	CB153	CB138	CB180	Alpha-HCH	Gamma-HCH	Dieldrin	p,p'-DDT	p,p'-DDE
1/28	1.30/9.4	23.5	51.0	132.0	149.9	434.3	239.8	21.6	ND	60.9	135.8	ND	97.0
2/17	1.22/11.7	55.0	135.4	227.9	151.4	498.2	255.4	27.5	+	43.1	135.4	+	88.8
3/23	1.42/9.1	33.9	82.2	245.1	291.2	831.0	542.5	55.9	ND	45.3	135.8	ND	149.8
4/23	1.15/9.1	15.2	36.3	101.5	124.7	404.1	224.5	34.8	162.7	43.5	118.3	+	63.8
5/49	1.52/12.4	29.6	69.8	181.2	200.5	575.3	369.4	17.6	37.9	38.7	98.8	ND	111.8
6/34	1.27/9.3	9.8	11.4	21.6	38.7	122.0	69.1	8.5	+	43.5	121.7	ND	72.4
7/66	1.33/10.9	7.0	5.4	11.8	25.1	76.7	47.7	4.3	+	+	74.5	ND	81.7
8/22	1.97/17.5	5.0	4.5	8.7	19.9	40.8	33.5	3.6	+	13.0	52.3	+	58.7
9/25	1.48/11.3	7.2	6.5	10.2	13.9	27.2	18.8	ND	+	+	57.4	ND	42.7
10/62	1.09/6.8	24.9	52.0	138.2	161.6	431.5	239.5	20.1	ND	43.8	114.3	+	110.6
11/80	No gonads were available because of poor gonadal development												
12/33	1.60/12.1	6.4	5.9	11.9	19.1	43.5	31.4	2.6	28.2	26.1	81.0	ND	56.2

TABLE IV (cont.)

13/45	1.36/9.4	6.6	6.4	18.0	24.5	68.4	49.5	4.6	+	+	142.4	ND	58.4
14/43	1.29/12.4	8.2	7.6	14.9	25.6	77.7	52.4	4.5	24.6	37.2	140.3	+	62.8
15/38	0.90/7.7	6.6	9.3	26.3	33.5	87.6	59.1	3.9	ND	+	78.6	ND	79.9
16/19	0.88/7.1	12.8	23.8	80.4	68.7	315.9	178.0	29.9	+	54.4	111.8	+	87.5
LOD		0.14	0.08	0.05	0.02	0.05	0.06	0.02	0.12	0.14	0.24	2.18	0.06
LOQ		0.26	0.16	0.13	0.05	0.08	0.12	0.05	0.23	0.22	0.51	3.93	0.12

[a] ND = not detectable; and + = identified, but below the LOQ.

Notes: Endrin and p,p'-DDD were not identified in any of the samples; o,p'-DDT was identified in specimens from stations 2, 4, 10 and 12, but not quantified. In the sample GC runs for CB determination, the % recovery of the internal standard CB112 was 78.2 ± 13.3; and the % recovery in the CRM and blank runs was 88.2 ± 8.1. In the GC runs for pesticide determination, the % recovery of the internal standard CB155 was 69.9 ± 4.1; and the % recovery in the CRM and blank runs was 97.8 ± 0.1.

TABLE V. WHOLE BODY ANALYSES OF THE POLYCHAETE WORM, *Nephtys* sp., THE ECHINODERM ASTEROID, *Asterias rubens* (SEASTAR), AND TWO CRUSTACEANS, *Crangon crangon* AND *C. Allmanni* (SHRIMPS). THE NUMBER (No.) OF INDIVIDUALS OF EACH SPECIES, THE PERCENTAGE LIPID EXTRACTED AND THE CONCENTRATION (ng/g OF LIPID EXTRACTED) ARE SHOWN (*Data were taken from Ref. [21]*)[a]

Station No.	Depth (m)	Species	No.	% lipid extracted	CB28	CB52	CB101	CB118	CB153	CB138	CB180
1	28	*Nephtys* sp.	12	0.82	72	67	85	104	226	151	66
2	17	*Nephtys* sp.	7	0.91	ND	15	56	114	151	123	63
		Crangon crangon	12	1.57	43	100	79	237	271	239	175
3	23	*Nephtys* sp.	30	0.57	89	155	465	610	1160	736	600
		Crangon crangon	15	2.29	42	156	109	273	326	290	227
4	23	*Nephtys* sp.	15	0.45	54	86	263	445	977	704	505
		Asterias rubens	10	0.78	50	136	315	489	926	614	189
		Crangon crangon	12	2.05	51	166	118	260	283	260	262
5	49	*Nephtys* sp.	11	0.77	36	56	64	99	234	144	73
		Crangon allmanni	20	1.24	11	37	70	125	232	127	69
6	34	*Nephtys* sp.	15	0.78	92	74	92	89	129	90	66
		Crangon allmanni	25	1.33	ND	31	63	126	211	97	54
7	66	*Nephtys* sp.	10	0.77	+	ND	61	58	90	59	+
		Asterias rubens	6	1.57	19	27	58	149	235	121	30
		Crangon allmanni	40	1.20	6	21	38	58	112	59	28
8	22	*Nephtys* sp.	15	0.54	70	75	90	87	138	91	46

TABLE V (cont.)

9	25	Crangon allmanni	20	0.83	11	19	27	48	93	48	27
10	62	Asterias rubens	5	1.24	12	21	35	86	109	44	15
11	80	Nephtys sp.	18	0.50	+	ND	69	68	94	47	ND
		Asterias rubens	12	0.74	25	39	54	53	95	62	8
		Crangon allmanni	24	1.11	ND	24	25	42	77	43	23
12	33	Nephtys sp.	13	0.50	+	ND	92	96	148	94	ND
		Crangon allmanni	11	1.06	12	20	26	50	94	53	32
13	45	Nephtys sp.	10	0.70	60	38	103	34	119	79	ND
		Asterias rubens	10	1.29	17	23	53	82	144	89	24
		Crangon allmanni	30	1.25	15	22	29	44	87	47	24
14	43	Nephtys sp.	10	0.55	29	30	107	86	194	106	ND
		Crangon allmanni	45	0.78	15	25	39	61	139	78	33
15	38	Nephtys sp.	12	0.99	36	41	51	56	116	79	56
16	19	Asterias rubens	9	0.71	41	66	177	313	546	41	58

[a] ND = not detectable; and + = identified, but below the LOQ.

TABLE VI. TEMPORAL VARIATION IN THE RANGES OF CONCENTRATION (ng/g OF (PENTANE) LIPID EXTRACTED) IN WHOLE BODY SAMPLES AND THE PYLORIC CAECA OF THE SEASTAR *Asterias rubens* FROM THE DUTCH COASTAL REGION AND THE SOUTHERN NORTH SEA [19, 21][a]

CB congener No.	Concentration (ng/g) in whole body: Apr./May 1986 [21]		Concentration (ng/g) in pyloric caeca: Aug./Sep. 1991 [19]		Concentration (ng/g) in pyloric caeca: May/Jun. 1992 [19]		Concentration (ng/g) in pyloric caeca: Mar. 1995 (present study)	
	Dutch coastal region	Southern North Sea	Dutch coastal region	Southern North Sea	Dutch coastal region	Southern North Sea	Dutch coastal region	Southern North Sea
CB28	40–50	12–20	20–30	5–10	20–55	5–10	25–50	8–16
CB52	65–135	20–30	55–105	ND–40	60–40	20–40	50–190	5–15
CB101	175–315	35–60	115–180	15–85	110–270	10–80	135–420	10–40
CB118	310–500	50–150	95–145	15–70	100–230	10–80	180–360	10–60
CB135	550–940	100–235	200–290	25–160	220–670	20–180	450–1050	40–125
CB180	60–190	5–30	15–35	3–20	20–65	2–15	20–105	ND–5

[a] ND = not detectable.

TABLE VII. CONCENTRATION OF CB CONGENERS AND p,p'-DDE IN SPONGE *Terpios aurantiaca* (PORIFERA, DEMOSPONGIAE) AND IN COLONIAL TUNICATE SPECIES OF THE CLASS ASCIDIACEA (CHORDATA)

(Numbering of the congeners is according to IUPAC rules [12]; they are listed in sequence of elution from the GC CPSil-19 column. The limit of detection (LOD) is the average of the analyte blank signals, plus three times its standard deviation; and the limit of quantification (LOQ) is the average of the analyte blank signals, plus nine times its standard deviation. LOD and LOQ are considered on the basis of a 1 µL injection from a 1 mL extract)[a]

Species	% lipid extracted	CB congener No.							Pesticide
		CB28	CB52	CB101	CB118	CB153	CB138	CB180	p,p'-DDE
Terpios aurantiaca (sampling site 1)	4.73 ± 1.21	0.05 ± 0.01	0.09 ± 0.01	0.21 ± 0.01	0.16 ± 0.01	0.38 ± 0.02	0.34 ± 0.02	0.25 ± 0.02	+
Terpios aurantiaca (sampling site 2)	6.28 ± 0.33	0.05 ± 0.01	0.07 ± 0.01	+	0.09 ± 0.01	0.26 ± 0.02	0.24 ± 0.02	0.16 ± 0.01	ND
Tunicates (Ascidiacea) (sampling site 1)	0.77	0.25	0.36	0.66	0.47	1.10	1.07	0.68	0.20
Tunicates (Ascidiacea) (sampling site 2)	0.91	0.17	0.24	0.38	0.23	0.64	0.60	0.35	0.12
LOD		0.02	0.03	0.12	0.04	0.11	0.13	0.07	0.06
LOQ		0.04	0.06	0.18	0.08	0.24	0.22	0.16	0.11

[a] + = identified, but below the LOQ; and ND = not detectable.

All the CB congeners considered in the present study were present at high levels in the digestive (pyloric caeca) and reproductive (gonad) organs of the seastar, *Asterias rubens* (Tables III and IV, respectively). The data confirmed the distinct concentration gradients previously described for PCBs in benthic invertebrates and bottom dwelling fish [18–20]. In pyloric caeca, the concentrations of CB and pesticide were only slighly higher. The only remarkable difference between pyloric caeca and gonads was the absence of p,p'-DDT and o,p'-DDT in pyloric caeca, whereas in gonads p,p'-DDD was not detected, indicating a different capability for biotransformation and metabolic capacity.

Data derived from an earlier inventory in the North Sea [21] showed concentrations of CB that were one to two orders of magnitude higher in comparable species, e.g. echinoderms *(Asterias rubens)* and crustaceans (*Crangon crangon* and *C. allmanni)* (Table V) [21]. The results also showed well defined concentration gradients for PCBs in organisms. A comparison with another study on CB congeners in whole body analyses of *A. rubens* and the levels in the pyloric caeca and gonads indicates that these organs mainly contribute to the body burden of the specimens (Table VI) [19, 21]. The levels measured in pyloric caeca are somewhat higher, and thus the whole body values seem not to have been substantially 'diluted'. There is no evidence to suggest that the CB concentrations had not decreased over the past 10 years.

The concentrations of CB and p,p'-DDE in the sponge, *Terpios aurantiaca* (Porifera, Demospongiae), and in the colonial tunicate species of the Ascidiacea (Chordata) class are summarized in Table VII [12]. To the best knowledge of the author, there are not many studies that describe the uptake and distribution of PCBs and pesticides in tropical reef organisms such as sponges and tunicates. The CB levels were very low, if detectable at all, whereas the tunicate species showed somewhat higher levels. In tissue of the sponge, *Spongia officinalis*, from highly contaminated sites in the Mediterranean Sea, the PCB concentrations ranged from 10 to 25 ppm, and were about 1 ppm for p,p'-DDE, whereas at the reference sites, 0.1 ppm PCBs and no p,p'-DDE were detected [22, 23]. These levels are three to four orders of magnitude higher than the data presented here, which provide evidence of the uptake and accumulation of PCBs and p,p'-DDE; however, no speculation was allowed as to the extent to which the levels may affect reef integrity. The tunicate species did accumulate PCBs at a higher level than the sponges.

4. CONCLUSIONS

It can be concluded that the low degree of contamination of surface sediments from the estuarine and shallow coastal regions of Kenya is reflected in the presence of seven CB congeners (CB28, CB52, CB101, CB118, CB153, CB138 and CB180) and p,p'-DDE in samples from only a few sites in relation to the outflow of the

Sabaki River. Of the pesticides, alpha-HCH showed an increasing concentration gradient across the continental slope. In all species from the different animal groups, CB congeners and p,p'-DDE were always present, sometimes in substantial concentrations. In bivalve and gastropod molluscs in particular, high levels of CB and p,p'-DDE were established; enhanced levels were also found in edible penaeid prawns.

The CB congeners in surface sediments did not show a distinct gradient along transects radiating into the North Sea, although the concentrations were significantly higher in the nearby coastal region. Of the pesticides, p,p'-DDE and in particular p,p'-DDD were present in significantly enhanced concentrations. The CB and pesticide levels found in the digestive and reproductive organs of the seastar were high, up to about 1000 ng/g of lipid for CB153, which is always the most prominent congener present. Distinct concentration gradients for individual CB congeners were described, although certain 'hot spots' in offshore areas were found for CBs and for certain pesticides such as p,p'-DDE and p,p'-DDD. Also, alpha-HCH showed enhanced levels in seastar from the shallow offshore area of the North Sea (Dogger Bank).

In the marine environments discussed in the present study, the absence or occurrence of individual CB congeners and pesticides and their concentrations in surface sediment and biota could be explained by a number of mechanisms. Enhanced levels in estuarine and nearby coastal regions were due to direct river inputs. Declining concentration gradients across the continental slopes could have emanated from dilution. However, in the case of alpha-HCH, which showed increasing levels in the open sea and in the ocean, atmospheric transport must have played an important role. Owing to strong surface and deep ocean water currents, almost no sedimentation of any riverine material or oceanic particulate matter can take place; thus, the distribution of CBs and pesticides along the coast of Kenya was greatly affected. Apart from these hydrographical conditions, the chemical transition of compounds and their eventual biotransformation in organisms determined their environmental fate. The interspecific differences in the concentration of various CBs and pesticides reflect species specific uptake patterns in the different invertebrate organisms that are generally not related to sediment concentration or depth.

ACKNOWLEDGEMENTS

This research was carried out as a part of the Netherlands Indian Ocean Programme 1992/1995, Project A: Monsoons and Coastal Ecosystems in Kenya. The programme was organized and funded by the Netherlands Marine Research Foundation of the Netherlands Organization for Scientific Research. Research in the North Sea was carried out within the framework of the EU Indo-Dutch Joint

Research Project on the Distribution of Contaminants and Biomarkers. Special thanks are due to Captains J. de Jong, L.J. Blok and A. Souwer and the crews of their research vessels, Tyro and Pelagia, and to R. Kloosterhuis for analysing the total organic carbon in surface sediments. The research agreement with the IAEA is also acknowledged.

REFERENCES

[1] EVERAARTS, J.M., HEESTERS, R., FISCHER, C.V., HILLEBRAND, M.Th.J., Baseline levels of cyclic pesticides and PCBs in benthic invertebrates from the continental slope of the Banc d'Arguin (Mauritania), Mar. Pollut. Bull. **26** (1993) 515–521.

[2] PRESTON, M.R., "Marine pollution", Chemical Oceanography, Vol. 9 (RILEY, J.P., Ed.), Academic Press, London (1988) 54–196.

[3] HOLDEN, A.V., MARSDEN, K., Single-stage clean-up of animal tissue extracts for organochlorine residue analysis, J. Chromatogr. **44** (1996) 481–492.

[4] DUINKER, J.C., HILLEBRAND, M.Th.J., Minimizing blank values in chlorinated hydrocarbon analysis, J. Chromatogr. **150** (1978) 195–199.

[5] DUINKER, J.C., HILLEBRAND, M.Th.J., Characterization of PCB components in Clophen formulations by capillary GC–MS and GC–ECD techniques, Environ. Sci. Technol. **17** (1983) 449–456.

[6] BOON, J.P., et al., Concentrations of PAHs and PCBs in dab (*Limanda limanda*), Report 1994-12, Beleidsgericht Ondersoek Noordsee, The Hague (1994) 1–38.

[7] EVERAARTS, J.M., DNA integrity as a biomarker of marine pollution: Strand breaks in seastar (*Asterias rubens*) and dab (*Limanda limanda*), Mar. Pollut. Bull. **31** (1995) 431–438.

[8] PCB Regulation, Dutch Staatscourant, The Hague (1984).

[9] Verordnung über Höchstmengen an Schadstoffen in Lebensmitteln, Teil 1, Bundesgesetzblatt, Bonn (1988) 422–424.

[10] WELLS, D.E., DE BOER, J., TUINSTRA, L.G.M.T., REUTERGÅRDH, L., GRIEPINK, B., Improvements in the analysis of chlorobiphenyls prior to the certification of seven CBs in two fish oils, Fresenius' Z. Anal. Chem. **332** (1988) 591–597.

[11] DE BOER, J., DAO, Q.T., Analysis of seven chlorobiphenyl congeners by multidimensional gas chromatography, J. High Resolut. Chromatogr. **14** (1991) 593–596.

[12] BALLSCHMITER, K., ZELL, M., Analysis of polychlorinated biphenyls (PCBs) by glass capillary gas chromatography: Composition of technical Arochlor and Clophen–PCB mixtures, Fresenius' Z. Anal. Chem. **302** (1980) 20–31.

[13] EISMA, D., KALF, J., Dispersal, concentration and deposition of suspended matter in the North Sea, J. Geol. Soc., London **144** (1987) 161–178.

[14] KLAMER, H.J.C., FOMSGAARD, L., Geographical distribution of chlorinated biphenyls (CBs) and polycyclic aromatic hydrocarbons (PAHs) in surface sediments from the Humber plume, North Sea, Mar. Pollut. Bull. **26** (1993) 201–206.

[15] INTERDEPARTEMENTALE COMMISSIE VOOR NOORDSEE AANGELEGENHEDEN, North Sea Atlas for Netherlands Policy and Management, Stadsuitgeverij, Amsterdam (1992).

[16] BOON, J.P., et al., Cyclic organochlorines in epibenthic organisms from coastal waters around East Java, Neth. J. Sea Res. **23** (1989) 427–439.

[17] EVERAARTS, J.M., BANO, N., SWENNEN, C., HILLEBRAND, M.Th.J., Cyclic chlorinated hydrocarbons in benthic invertebrates from three coastal areas in Thailand and Malaysia, J. Sci. Soc. Thailand **17** (1991) 31–49.

[18] BOON, J.P., VAN ZANTVOORT, M.B., GOVAERT, M.J.M.A., DUINKER, J.C., Organochlorines in benthic polychaetes (*Nepthys* spp.) and sediments from the southern North Sea: Identification of individual PCB components, Neth. J. Sea Res. **19** 9 (1985) 93–109.

[19] EVERAARTS, J.M., et al., "DNA strand-breaks, cytochrome P450 dependent monooxygenase enzyme activity and levels of chlorinated biphenyl congeners in the pyloric caeca of the seastar (*Asterias rubens*) from the North Sea", The Integrated North Sea Programme 1991–1992 (VAN RAAPHORST, W., BOON, J.P., Eds), Report 1993-9, Nederlands Instituut voor Ondersoek der Zee, Den Burg–Texel (1993) 67–73.

[20] BOON, J.P., NIEUWENHUIZE, J.M., VAN LIERE, J., WILHELMSEN, S., KLUNGSØYR, J., "Concentrations of PCBs and PAHs in dab muscle", ibid., pp. 101–103.

[21] EVERAARTS, J.M., FISCHER, C.V., Micro Contaminants in Surface Sediments and Macrobenthic Invertebrates of the North Sea, Report 1989-6, Netherlands Instituut voor Ondersoek der Zee, Den Burg–Texel (1989) 1–44.

[22] VERNEDAL, B., ARNOUX, A., VACELET, J., "Pollutant levels in Mediterranean commercial sponges", New Perspectives in Sponge Biology: Biological Sponges (Proc. 3rd Int. Conf. Woods Hole, 1985), Woods Hole Oceanographic Institution, Woods Hole, MA (1985) 516–524.

[23] VERNEDAL, B., Spongiculture en Méditerranée nord-occidentale: Aspects culturaux, molysmologiques et économiques, PhD Thesis, Université de Provence, Marseille (1986).

EFFECT OF PESTICIDES ON NON-TARGET SPECIES

(Session 6)

Chairperson

J. ESPINOSA-GONZALEZ
Panama

Rapporteur

I.G. FERRIS
FAO/IAEA

IAEA-SM-343/3

USE OF SOIL MICROCOSMS IN ASSESSING THE EFFECT OF PESTICIDES ON SOIL ECOSYSTEMS*

C.A. EDWARDS**, T. KNACKER***,
A.A. POKARZHEVSKIJ[+],
S. SUBLER**, R. PARMELEE**

** Soil Ecology Program,
Department of Entomology,
The Ohio State University,
Columbus, Ohio,
United States of America

*** ECT Ökotoxikologie,
Florsheim, Germany

[+] Laboratory of Bioindication,
Institute of Evolutionary Morphology
   and Ecology,
Moscow, Russian Federation

Abstract

USE OF SOIL MICROCOSMS IN ASSESSING THE EFFECT OF PESTICIDES ON SOIL ECOSYSTEMS.

In recent years, considerable interest has been shown in model ecosystem techniques to predict the environmental impact of pollutants. The microcosms usually consist of soil units containing multiple biotic species that range in size from 0.05 dm^3 of soil to units of up to 1 m in diameter and that have more than 1 m^3 of soil. They include single species assays, mixed soil containing a range of organisms and intact soil cores from the field ecosystems they represent. Model ecosystem tests assess in an integrated form the impact of toxicants on soil organisms and soil processes at different levels of biological organization. At the same time, degradation of a pesticide and its leaching, volatilization and movement into plants and invertebrates can be followed. Such systems lend themselves readily to the use of radiolabelled pesticides, as well as stable isotopes such as ^{13}C and ^{15}N, to follow pesticide and nutrient transformations in soil. The Ohio State University (OSU) has developed an integrated, small scale soil microcosm to assess the effect of pollutants on soil ecosystems. This integrates the effect of chemicals on typical soil organisms (soil microbes, microarthropods, nematodes and earthworms) with data on the consequences of chemical contamination on soil ecosystem processes (nutrient cycling, enzyme activity and organic matter decomposition). It also

---

* Research carried out in association with the IAEA under Research Agreement No. 6755.

assesses the uptake of pollutants into plants and animals, as well as losses through degradation, volatilization and leaching. The microcosms are small, use sieved and well mixed soil treated with the pesticide, are easy to set up and handle, and are replicable. ECT Ökotoxikologie has designed a terrestrial model soil ecosystem to assess the potential fate and ecological effect of pesticides. This resembles the OSU microcosm, but it is larger and contains a soil core taken intact from the field, with the pollutant applied either to the soil surface or to the upper soil layer to mimic agricultural techniques. The measurements made include the nutrient and chemical concentrations in the soil/litter layers, the plants and leachates, the microbial activity, and the effect of pollutants on soil inhabiting invertebrates. Integrated soil microcosms, using radiolabelled and stable isotope labelled pesticides, which can be used in the laboratory or in the field, show great potential in assessing the environmental impact of toxicants. Such model ecosystems are ideal for providing environmental data to the manufacturers of pesticides and to the national registration authorities.

1. INTRODUCTION

The maintenance of soil quality, fertility and structure is essential to the protection and preservation of the biodiversity and integrity of terrestrial ecosystems. Critical to achieving this aim is an understanding of the potential effects of pesticides on the structure and function of soil ecosystems. Pesticides can affect soil organisms, species and community interactions, nutrient cycling and the functioning of soil food chains [1–3]. They can also be taken up into plants and animals, volatilize into the atmosphere, and leach into groundwater and waterways [4].

In recent years, interest has developed in the use of microcosm and model ecosystem techniques for ecotoxicological assays to predict the overall effect of contaminants on soil ecosystems [5–7]. Such systems usually consist of soil units, containing multiple biotic species, that range in size from a few grams of soil to units as large as 1 m in diameter. They include mixed cultures of organisms and artificial substrates [8], mixed soil containing endogenous or introduced organisms [9–11], and intact soil cores from the field ecosystems they represent [5, 7, 12, 13].

The attraction of microcosm and model ecosystem tests is that they can assess the impact of contaminants at different levels of biological and ecological organization, and can produce data that are much more relevant to the prediction of the overall effects of a pesticide in the field situation than single species assays. However, some currently used microcosm tests, although they include multiple organisms, can only assess the effects of the contaminant on one test species [9, 12].

Use of terrestrial microcosms or model ecosystems to assess the contaminants on soil ecosystems and the fate of these contaminants has been reviewed by Van Voris et al. [5], Sheppard [6], and Morgan and Knacker [7]. The latter authors defined a terrestrial model ecosystem as "a controlled reproducible system that attempts to simulate the field situation, i.e. the processes and interactions of components, in a portion of the terrestrial environment. It should have a boundary, although it may be an open or semi-enclosed system, and is subject to control of environmental factors, such as temperature, light, humidity, and fluxes of air and water".

Morgan and Knacker [7] reviewed 58 published techniques involving test organisms or processes to assess the impact of chemicals, but of these only seven involved truly multiple measurements on organisms, processes, and the fate and transport of pesticides. Nineteen assessed the effect of pesticides, and 16 used radiolabelled pesticides. Most of the techniques were not considered to be terrestrial model ecosystems or microcosms that access multiple pesticide impacts. These authors classified the systems into four main types.

(1) *Open intact terrestrial model ecosystems:* These are confined to the use of soil cores taken intact from the field and placed in containers. The pesticide is applied to the soil surface rather than mixed into the soil. Test organisms are not usually added, although plants may be grown. Various processes and end points are measured. These systems are excellent for leaching studies.

(2) *Closed intact terrestrial model ecosystems:* These systems, which involve intact soil cores in enclosed systems, with the pesticide applied to the surface soil, allow measurements to be made of the volatility of the pesticide, and are particularly suitable for studies on radiolabelled pesticides (to avoid their release into the environment), but not for leaching studies. Test organisms can be added, but plants do not grow well in such systems.

(3) *Open homogeneous terrestrial model ecosystems:* In these systems, soil taken from the field is mixed and sieved thoroughly into a homogeneous material before being placed in the model ecosystems. They allow the pesticide to be thoroughly mixed with the soil, so that it comes into close contact with the test organisms that can be introduced. They can also be used for surface applications or for mixing the pesticide with the surface layers of the soil. For leaching studies, they need to be large. Plants can be grown in these systems quite easily.

(4) *Closed homogeneous terrestrial model ecosystems:* These systems, which also use mixed soil, are particularly suitable for studies on radiolablled pesticides and for measurements of the volatilization of pesticides and the respiration from soil and plants. They are especially useful for confining test animals that are to be exposed to the test pesticide.

Sheppard [6] discussed at length the cost–benefit of microcosm tests compared with single species and process tests. He concluded that, although the former are cheaper, many multiple tests are needed and that the value of the information derived from integrated microcosm tests is usually greater than that obtained from a range of tests. If a microcosm test is fully standardized and sufficiently reproducible it can produce a large amount of much more relevant information at a relatively low cost, and with little need for additional single species tests. However, many national regulatory authorities still depend mainly on single species testing.

It is clear that no single test can assess the effects of a contaminant such as a pesticide in all settings, whether it is based on either single/multiple species or process testing, or on microcosm or model ecosystem assessment. However, use of an integrated microcosm allows better overall interpretation of the potential impacts of a pesticide on the individual soil inhabiting organisms, the interactions between them and the trophic structure of the soil communities, and provides useful information about the effect of pesticides on the ecological functions in soils; it can also assess losses of the pesticide from the system by various routes. There seems little doubt that extrapolation of the results of an integrated microcosm test to the field is much easier than trying to do this on the basis of the results of single or multiple species tests [14]. The question of the scale of microcosms has been addressed by Wright and Coleman [15], who monitored the effect of chemicals on nematode populations in replicated cores of 15, 49, 180 and 7600 cm^3. They concluded that, for such small organisms, there were serious ecological drawbacks in larger sized microcosms, where nematodes in the higher trophic levels were often lost. It is likely that similar conclusions hold for microarthropods and larger soil inhabiting invertebrates. Some workers have used samples as small as 4 g of soil to assess the effect of contaminants on soil organisms, but a slightly larger system was used to assess the effect of contaminants on communities of soil micro and meso-arthropods and nematodes [2]. The system described by Van Voris et al. [5], which was modified as the American Society for Testing and Materials method, Vol. 11.04, Section 11 [12], was considerably larger and involved intact cores of 17.5 cm diameter × 60 cm deep. Kuehle [16] described an even larger system (NATEC), which consisted of undisturbed soil cores of 72.0 cm diameter × 43.0 cm deep planted with four different crops.

Integrated microcosm or model ecosystem approaches, in which a suite of organismal and process level measurements are made involving different levels of biological and ecological organization, can provide a much broader understanding of the mechanisms by which pesticides affect the structure and function of soil ecosystems, resulting in a much more flexible and widely applicable predictive capability. Such a holistic system combines measurement of soil organism populations, soil processes and plant growth, as well as the bioaccumulation and persistence of chemical contaminants into a single testing system. This approach can offer great possibilities as a widely applicable and cost effective means of assessing the possible effect of pesticides on soil ecosystems and environmental quality, so providing data for environmental decisions.

A description is given of a small, inexpensive and cost effective integrated microcosm technique, using homogenized soil, that produces extensive and reproducible data. Because of the small size and relative simplicity of these soil microcosms, a large number of replicates can be handled and kept under environmental control in individual experiments. In contrast, a model ecosystem is reviewed that uses much larger intact soil cores. The benefits and drawbacks of the two systems are compared and discussed in some detail. No data on radiolabelled pesticides are presented, but experience has shown that such chemicals and both systems are well suited for assessing their effects.

2. THE OHIO STATE UNIVERSITY (OSU) INTEGRATED SOIL MICROCOSM FOR ECOTOXICOLOGICAL TESTING

A small, holistic/integrated, soil microcosm/model ecosystem assay methodology has been designed that can provide system level results on the effect of toxicants such as pesticides and on soil organisms and ecosystems, as well as data on losses of chemicals from the system [11], and that can be used in the laboratory or in the field. This microcosm, which is illustrated in Fig. 1, consists of well mixed, sieved, field collected soil containing both endogenous and introduced organisms packed gently into a plastic cylinder, with wheat seedlings planted in the soil. The cylinder of the microcosm is 5 cm (inside diameter) $\times$ 15 cm high, made of a commercially available high density polyethylene (HDPE) pipe. Fresh, field collected and homogenized soil, approximately 250–300 g dry weight, is used in the microcosm. A logarithmic series of four or five doses of the pesticide are applied to bulk batches of soil, thoroughly mixed and homogenized before placing in individual microcosms, or the pesticide is applied to the soil surface. Six replicate microcosms are used for each dose of each pesticide per sampling date.

A small quantity of organic material (chopped wheat straw), contained within a small cylinder of fibreglass screen material (1.6 $\times$ 1.8 mm mesh), is inserted into the surface of the microcosm soil to enable periodic measurement of the rates of decomposition of the organic matter, which can be a good indicator of the level of overall biological activity in the soil [11].

In addition to the soil microorganisms, microarthropods and nematodes that are present in the fresh, field collected soil, two small earthworms of the species *Aporrectodea tuberculata* (Eisen.) (total weight of 1 g), or similar species, are added to each microcosm. These represent the macrofaunal integrators of the bioaccumulation, excretion and toxicity of the pesticides. Upon termination of the microcosms, earthworm biomass (dry weight) and tissue concentrations of carbon, nitrogen and contaminants are determined.

Wheat seedlings are planted in the surface of each microcosm. The importance of including plants in soil microcosms is threefold: (1) the plant roots serve as a source of organic carbon in the soil, playing a fundamental role in below ground food chains; (2) the plant roots withdraw nutrients from the soil, altering the soil solution

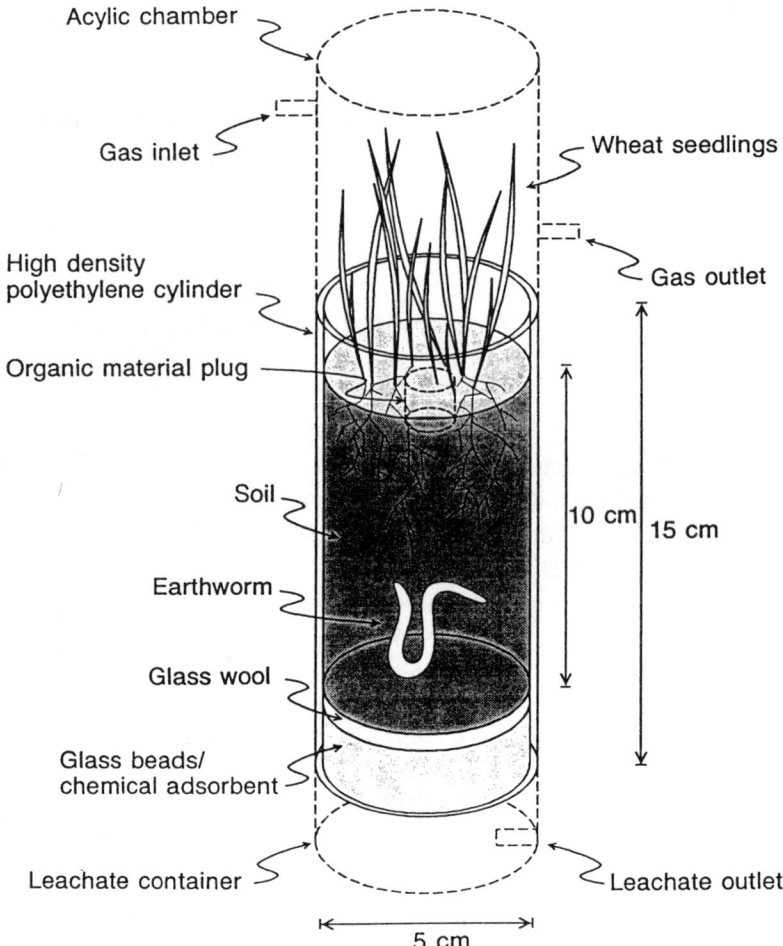

FIG. 1. *OSU soil microcosm.*

concentrations and affecting the processes controlled by positive or negative feedback; and (3) the plants can take up pesticides into their tissues. Ten wheat seeds are sown in the top 0.5 cm of the microcosm soil. After 1 and 2 weeks, the germination success is recorded for each microcosm, and the seedlings are thinned to four per microcosm.

At the bottom of each microcosm cylinder there is a layer of mixed bed, ion exchange resins separated from the soil core by a thin layer of glass wool. This allows free passage of the soil leachate from the microcosm, and acts as a partial barrier to prevent root growth out of the core bottom, while collecting nutrient ions leaching from the microcosm. Water is added two or three times a week to maintain a soil moisture content of 30–40% (dry weight basis). Excess water is added periodically to leach the soil core, as would occur under normal field conditions. Leachates passing out of the bottom of the microcosm are collected in a dish. Samples are cleaned up and analysed (to assess the pesticide and its degradation products) on a gas–liquid chromatograph (GLC) or by high performance liquid chromatography (HPLC).

At frequent intervals during the 2 month test period, a subset of microcosms is dismantled in order to sample the soil, plants and earthworms. Soil is removed from the microcosm, mixed gently and subsampled to determine the microbial biomass, the substrate induced respiration (SIR), the soil enzymatic activity and the concentrations of nitrogen in the inorganic and organic pools. Microbial biomass is measured as the carbon or nitrogen released from the soil following fumigation with chloroform [17]. Both the active and inactive (dormant) portions of the microbial community are measured. SIR, on the other hand, is indicative of the overall activity of the soil microbial community, and is measured as the evolution of CO_2 following the addition to the soil of a solution containing readily available carbon substrate (glucose). The activity of the dehydrogenase enzymes in the soil may also reflect the overall activity and biomass of the soil microorgansims, and is measured easily using calorimetric techniques. The activities of the other soil enzymes that are involved in nitrogen and phosphorus acquisition by microorganisms, such as chitinase, urease, and acid and alkaline phosphatase, can provide useful rates of mineralization and patterns of availability of nitrogen and phosphorus in the soil [18], and be measured with similar techniques. Extraction of the soil with a salt solution (e.g. 0.5N K_2SO_4) allows determination of the concentration of potentially mineralizable nitrogen, which is a pool of readily available organic nitrogen in the soil [19]. By repeating these measurements, at frequent intervals over a total of 56 days using separate replicated microcosms for each date, a dynamic picture of the effect of chemicals on soil organisms, soil processes and plant growth is obtained. Temporal patterns in the growth and activity of soil microorganisms may reveal both short and long term effects that are not readily discernable from single end point measurements. The net rate of processes such as ammonification, nitrification and total nitrogen mineralization can be calculated from the changes in pool size between sampling dates. Using this system it is also possible to measure the effect of pesticides on the gross rates of carbon and nitrogen mineralization using $^{13}CO_2$ evolution and ^{15}N pool dilution techniques. Because of the detailed dynamic information that can be obtained on the soil microbial activity, the carbon and nitrogen pools and transformations, and the plant nutrient uptake and growth, this microcosm methodology is ideal for developing and applying mathematical models of the short and immediate term effect of chemicals on soil ecosystems. If the system is enclosed it has great potential for tracing the fate and movement of radiolabelled pesticides.

These methods have been used to investigate the effect of the fungicides captan, benomyl and chlorothalonil on the soil organisms, dynamic processes and nitrogen dynamics [11]. For instance, captan had a more pronounced influence than benomyl on soil respiration and dehydrogenase activity, on soil microorganisms and on the patterns of nitrogen availability, increasing the net ammonification and nitrification significantly within 1 week of application to the soil. It is not clear if the increased nitrogen availability resulted from a flush of mineralization by the soil fungi, or from the degradation of the captan itself. Further studies using these methods, combined with use of radiolabelled fungicides, will help to resolve this question.

The effects of copper and zinc applied at 0, 50, 100, 200, 400 and 800 ppm to soil microorganisms were also studied. Earthworm populations were eliminated after 10 days at 800 ppm, and depressed significantly at 200 and 400 ppm. The microbial biomass and soil respiration were decreased significantly at doses above 200 ppm, and litter decomposition was inhibited significantly above 200 ppm. Doses greater than 200 ppm decreased nematode populations. Nitrogen mineralization increased dramatically above 800 ppm.

In a separate study, the mode of action of a soil biostimulant used commercially in extremely small quantities (applied at a rate of < 1 L of solution per hectare) was characterized. Using ^{15}N labelled fertilizer it was demonstrated conclusively that the biostimulant increased the rates of organic matter decomposition and the mineralization of nitrogen from the soil organic matter, leading ultimately to the enhanced uptake of nitrogen and the increased growth of plants.

3. GERMAN MODEL SOIL ECOSYSTEM FOR ASSESSING THE ENVIRONMENTAL IMPACT OF PESTICIDES

The German model soil ecosystem was designed originally by ECT Ökotoxikologie [20], and modified as the result of a workshop held in Germany in 1994 [13]. The model ecosystem differs from that used in Ohio in that it is considerably larger and consists of an intact core of soil (Fig. 2), and in the measurements and end points made. A single model ecosystem consists of a 60 cm deep HDPE tube (17.5 cm diameter), an 18.6 cm diameter porcelain Buchner funnel, a thin layer of inert gauze that fits between the Buchner funnel and the bottom of the soil core, and silicone tubing that connects the Buchner funnel to an Erlenmeyer flask, which acts as a collection vessel. A cap is required to cover the bottom of the soil core after extraction from the field to avoid loss of soil. If the test pesticide can be absorbed by HDPE it may be necessary to use aluminium tubing as a substitute. Because the model ecosystems are large, construction of carts to hold eight model ecosystems and adequate insulating material is necessary. Each cart has to have tubes running through it that are connected via a heat exchanger to a cooling unit in order to maintain the temperature at approximately the level that would be expected in the soil.

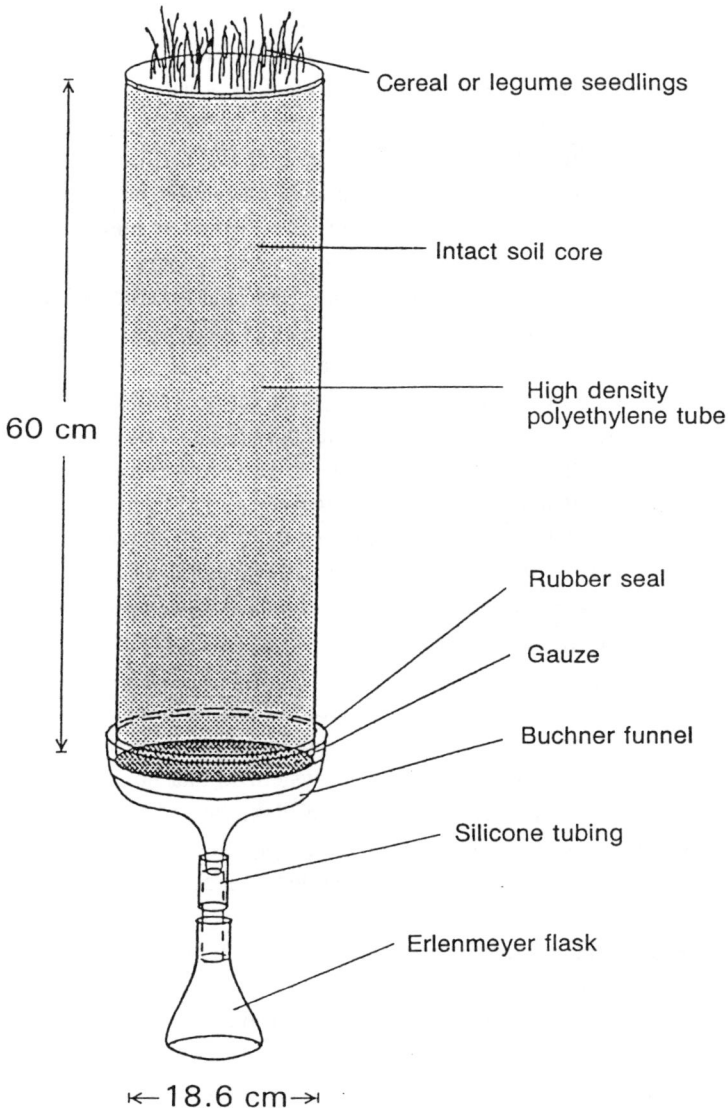

FIG. 2. German model ecosystem.

Cereal plants or legumes are included in the model system. The chosen species should be able to grow to relative maturity in the limited surface area of the model ecosystem. The seeds are sown in the extracted soil core, and the rate of seed application duplicates the standard field sowing rates. The seeds are sown evenly and covered to an appropriate depth with soil.

The soil cores are taken from the field at least 4–6 weeks before application of the test substance. To take the cores, a specially designed steel extraction tube and a hydraulic excavator or ram are used. Once the soil is cut by the leading edge of the driving tube, a core is forced up into the tube using the hydraulic excavator. A ram is used to grip the extraction tube and to pull the soil core slowly from the ground. The steel cap is removed from the steel extraction tube and the soil core lifted out. The cap is then fitted to the soil core, which is numbered or labelled appropriately. The steel extraction tube, with a new soil core tube, is then moved on to an adjacent section to take another core. The main method of application of the pesticide is to the surface as a liquid or as a dry powder or aerosol, or it is mixed with the upper soil layer to mimic agricultural application techniques.

At the beginning of the study, model ecosystems are treated with at least three concentrations of the test substance that bracket the predicted environmental concentration, plus a control. When selecting the lowest concentration, consideration should be given to the detection limits of the analytical method to be applied. Twelve replicate model ecosystems are used for each of the three concentrations and for the control, giving a total of 48 model ecosystems per pesticide assay. Six mobile carts hold the 48 model ecosystems. A randomized block design ensures that there is an equal number of model ecosystems, with the same test substance concentration in each cart.

Care is taken to provide sufficient water for normal plant functions, without overwatering. A method of adding water is used which prevents the water from finding channels down the sides of the model ecosystem so that no contact with the soil is made, e.g. by using a rain head (e.g. made of Plexiglas, 16.5 cm high $\times$ 14.0 cm diameter, ten evenly spaced holes that have micropipettes inserted) positioned above the model ecosystem. The volume of water added to each model ecosystem must be identical.

To determine the amount of pesticide taken up, plants are harvested at the end of the study period, or even once or twice during that time, and all the shoot material is separated from the soil and the roots. Whole roots can be removed and treated separately, or left in the soil and treated as part of these particular horizons. Normally, the harvested plant material is oven dried at 65°C and stored at -20°C until analysed.

Extraction of the test pesticide and its degradation products from the soil and the plant material, followed by clean-up, are necessary to bring the test substance into a form in which it can be analysed. Examples of the extraction procedures include acidification of the solvent extraction with various types of solvent, supercritical fluid extractions, alkaline or acid hydrolysis with or without heat, detergent extractions and protease digestion.

If the test substance is radiolabelled with ^{14}C, residues that cannot be extracted must be oxidized and analysed as $^{14}CO_2$, and reported as bound residues. The extracts and the oxidized or dissolved samples are then counted by ^{14}C liquid scintillation.

When a radiolabelled pesticide is not used, quantitative analysis of the parent compound is made using HPLC, gas chromatography (GC), or GC–mass spectrometry (GC–MS). Separation of the potential transformation products in the leachates is done using a suitable analytical technique, such as thin layer chromatography, HPLC or GC, and a suitable detector, such as a scintillation counter of mass spectrometry.

3.1. Nutrient loss

Nutrient loss from the model ecosystem, resulting from up to seven leaching episodes, is assessed. Nutrients that are analysed in the leachate include Ca^{2+}, Mg^{2+}, K^+, Na^+, NO_3^-, NH_4^+, SO_4^{2-}, dissolved organic carbon and PO_4^{3-}.

3.2. Substrate induced respiration

Substrate induced respiration can supply integrated information on the overall activity of microorganisms under optimized carbon source conditions (usually glucose) and, potentially, on the separate contributions of bacteria and fungi to microbial activity [21]. It can also be used to estimate microbial biomass [22].

The principle behind SIR is measurement of the maximum initial respiratory response of the soil sample after its amendment with glucose. Replicate samples (moist soil equivalent to 20–40 g dry weight) should be taken from each model ecosystem at the end of the experiment, homogenized and stored at 4–6°C until ready for testing. A glucose optimum must be determined on a soil/litter sample to check the microbial activity prior to carrying out the study. This amount is then added to all the replicates. The initial CO_2 production is measured using either semi- or fully automated systems, by Veep or on-line infrared gas analyis [23].

3.3. Plant litter decomposition

The effects of the pesticide on the rates of plant litter decomposition are assessed by the loss of weight from the litter bags. Litter bags of two mesh sizes are placed on top of the model ecosystem's surface before or after the test has started, depending on the type of situation being assessed. A litter bag with 20 μm mesh allows entry of microorganisms only, and will permit leaching of the soluble organic and inorganic materials, while a litter bag with a larger mesh (1–10 mm) also allows mesofauna and earthworms to enter [21, 24].

3.4. Extracellular enzymes

The microbial degradation of litter is regulated by extracellular enzymes, many of which are freely available in the soil. Cellulase deserves the most attention because the vast majority of organically bound carbon in plant material is present as cellulose. The dehydrogenase activity can also be measured and is taken to be indicative of the overall microbial activity [25].

Several specific enzymes involved in the breakdown of cellulose can be used, but it is recommended that the carboxymethylcellulase activity be examined. The activity of this enzyme is determined at the end of each test by sampling a proportion of the surface layer of the soil, by incubating the soil either directly [26] or after extracting the enzyme [27], and by measuring the degradation of carboxymethylcellulose to glucose, which can be coloured, and hence measured. The experimental set-up requires use of a bacterial inhibitor (e.g. tuluol), a buffer and a particular substrate concentration, each of which requires some degree of optimization before beginning the test. The dehydrogenase activity in the soil is determined by measuring the etriphenyltetrazolium chloride to orange–red formazon [28].

3.5. Microorganisms and invertebrates

The effects of the pesticide on the microorganisms and invertebrates are assessed according to the following criteria:

(1) Representatives of different ecological groups, e.g. primary decomposers or predators, should be selected;
(2) Representatives of different morphological and/or physiological groups with different potential exposure routes, e.g. food, soil or water, should be selected.

Having sampled, identified and quantified the species in the treated and control samples, the results should be presented as rank abundance curves and Renkonen coefficients. These methods highlight the changes in dominance, which are indicative of the overall changes in the community structure.

The fate of the test substance is assessed throughout the study period in the leachate and in the soil/litter layers, as well as in any plant material remaining at the end of the study. From these data, fate end points, such as the mass balance, the bioconcentration factors and the tendency of the pesticide to move through the model ecosystem, can be calculated.

A number of pesticides have been tested in these model ecosystems. When the insecticide lindane was applied at 0.04 and 0.2 g/m^2, 13% remained in the soil, and an equivalent half-life of 48 days was reported. Significant quantities of lindane were also found in the leachate (0.3%) and the plants (1.62%) after 12 weeks. It appears that most of the lindane was degraded over the 12 week assay period.

In other studies, the insecticide parathion and the herbicide Ustinex (active ingredients: amitrole and diuron) were tested in a model ecosystem, particular attention being paid to their effects on oligochaetes [29]. Parathion was applied at 610 and 3000 mg/kg of soil, and the herbicides at similar doses. Significant decreases in the number of earthworms were observed at the higher dose of parathion, while increases in the number of earthworms and enchytraeids were determined at the higher dose of Ustinex. In both oligochaete groups, the number of species was not changed by pesticide treatment, although a shift in the dominance spectrum occurred.

4. RUSSIAN WORK USING MICROCOSMS

Intact soil cores, covering a surface area of 400 cm^2 and at a depth of 40 cm, were used by Terytze and Terytze [30]. These were kept in plastic pots covered with polyethylene film. The heavy metals (zinc, calcium and cobalt) were applied as chlorides or nitrates to the surface of the cores. Carbon dioxide evolution and proteolytic activity were measured by the changes in weight of the dry granulated $Ca(OH)_2$. Proteolytic activity was assessed following the destruction of a gelatine layer of röntgen film, and cellulolytic activity by the loss of filter paper buried in the soil. The data generated were based on orthogonal experimental designs, using second order interpolation models for two and three factor experiments; the potential was extended to four or five factor data. The work was used for the coupled influence of pollutants [31] and extended to general application by Terytze and Pokarzhevskij [32] and Pokarzhevskij and Terytze [33].

5. KEY FACTORS IN THE DESIGN OF MODEL SOIL ECOSYSTEMS FOR SUITABLY ASSESSING THE OVERALL ENVIRONMENTAL IMPACT OF PESTICIDES

The types of model ecosystem or microcosm that can be used are closely linked to the required end points. No model ecosystem can fully mimic the field situation in all respects, and it is impossible to include all aspects of the ecosystem function in any model ecosystem. Some acceptance of a degree of compromise to achieve the desired aims is essential in choosing the type of model ecosystem.

A review is made of the main variables in model ecosystems or microcosms in terms of the results than can be obtained.

5.1. Use of intact soil cores versus homogenized soil

5.1.1. Intact soil cores

These allow realistic studies to be made of the leaching, volatilization and uptake of pesticides, and exposure of the soil organisms to the pesticide mimics the field situation much better. However, intact cores suffer considerable variability because of the horizontal variability in soils from the field; hence, numerous replicates are needed. In addition, it is difficult to work with small intact cores, particularly in terms of the leaching of pesticides, because the solution runs down the walls of the core container. Therefore, most studies involving intact cores use cores greater than 15 cm in diameter, and often much larger. This leads to large labour requirements (to take the cores from the field) and creates major problems of transport and storage, particularly under the constant environmental conditions found in the laboratory. Moreover, intact soil cores depend on endogenous populations of soil organisms, since it is difficult to introduce them.

In contrast, homogeneous soil cores can be of any size. They enable the pesticide to be mixed uniformly throughout the soil or the upper strata, or to be treated only on the soil surface. Because of the relatively uniform exposure of soil organisms to the pesticide, the results tend to be much more consistent and reproducible. Moreover, the variability in soil characteristics is minimized by thorough mixing. Such homogeneous cores are well suited to the introduction into the system of test animals of selected taxa. Homogeneous cores are particularly suitable for assessing the toxicity of pesticides to the smaller microorganisms, microarthropods and nematodes.

5.2. Open or closed model ecosystems

5.2.1. *Open ecosystems*

These have the advantage that plants in the system can respire and grow normally without any 'greenhouse' effects, and application of water to the system for leaching studies is facilitated, as is any periodic sampling of the soil or the plants that may be needed.

5.2.2. *Closed systems*

These allow studies on respiration to be made. They are also most suitable for studies involving radiolabelled pesticides that need to be contained. They are commonly used to contain test invertebrates such as crickets, centipedes, carabid and staphylinid beetles, as well as other macroinvertebrates. However, they are difficult to water, and plant growth may be abnormal because of the enclosures.

5.3. Size of the model ecosystems

5.3.1. *Small microcosms*

These are easily handled, have few transport problems and occupy a relatively small space in the environmental chambers. This allows numerous replicates to be processed, to account for the variability in soils and the susceptibility of soil organisms to pesticides, as well as a large number of subsamples with multiple end points.

They are best suited for tests involving microorganisms, microarthropods and nematodes, but are not usually suitable for testing the effect of pesticides on macroarthropods.

5.3.2. Large microcosms

These are particularly suitable for studies involving the leaching of pesticides, a large number of test plants and assessment of the toxicity of pesticides to macroarthropods. Their main drawback is the difficulty in taking and storing model ecosystems, since a large number of samples are needed to establish a series of end points, and most of the sampling is destructive.

5.4. Economics of the model ecosystems for testing the effect of pesticides

It is extremely difficult to compare the costs of a series of single or multiple species, tests of pesticide toxicity and tests on the effect of pesticides on the soil process, leaching, fate and degradation with those of a single, holistic, integrated model ecosystem. According to general principles, it is likely that the integrated system not only costs less but also produces interactive data that are much easier to interpret in terms of predicting the overall environmental impact of pesticides in the field.

6. CONCLUSIONS

On the basis of collaborative work between OSU, ECT Ökotoxikologie and the Moscow State University, significant progress has been made in designing and testing terrestrial model ecosystems and microcosms that can be used to predict the environmental effect of pesticides. These systems are ideally suited to the use of radiolabelled pesticides and have the potential to provide extensive, integrated data on the overall environmental impact of a pesticide on soil ecosystems, as well as its fate and transport into the atmosphere and aquatic systems. Such terrestrial model ecosystems provide a powerful tool for manufacturers of pesticides and the national registration authorities.

ACKNOWLEDGEMENTS

This work was supported by a North Atlantic Treaty Organization Environmental Linkage Grant. Additional collaborators in the project are: R. Kuperman, Argonne National Laboratory, Chicago, IL, USA; D. Krivolutskij and R. Butovskij, Moscow State University, Moscow, Russian Federation; and J. Römbke and B. Förster, ECT Ökotoxikologie, Florsheim, Germany.

REFERENCES

[1] TU, C.M., MILES, J.R.W., Interactions between insecticides and soil microbes, Res. Rev. **64** (1976) 17-65.

[2] PARMELEE, R.W., WENTSEL, R.S., PHILLIPS, C.T., SIMINI, M., CHECKAI, R.T., Soil microcosm for testing the effects of chemical pollutants on soil faunal communities and trophic structure, Environ. Toxicol. Chem. **12** (1993) 1477-1486.

[3] EDWARDS, C.A., BOHLEN, P.J., The effects of contaminants on the structure and function of soil communities, Acta Fenn. Zool. **194** (1995) 48-49.

[4] EDWARDS, C.A., "Pesticides as environmental pollutants", World Directory of Pesticide Control Organizations, 2nd edn (EKSTROM, G., Ed.), Royal Society of Chemistry, London (1994) 1-24.

[5] VAN VORIS, P., TOLLE, D.A., ARTHUR, M.F., CHESSON, J., "Terrestrial microcosms: Applications, validation and cost-benefit analysis", Multispecies Toxicity Testing (CAIRNS, J., Ed.), Pergamon Press, New York (1985) 117-142.

[6] SHEPPARD, S.C., "Toxicity testing using microcosms", Soil Ecotoxicology (BITTON, G., TARRADELLAS, Y., ROSSEL, D., Eds), Lewis, Boca Raton, FL (in press).

[7] MORGAN, E., KNACKER, T., The role of laboratory terrestrial model ecosystems in the testing of potentially harmful substances, Ecotoxicology **3** (1994) 213-233.

[8] AUSMUS, B.S., KIMBROUGH, S., JACKSON, D.R., LINDBERG, S., The behaviour of hexachlorobenzene organic soil pollutant in pine (*Pinus*) forest microcosms: Transport and effect on soil processes, Environ. Pollut. **20** 2 (1979) 103-111.

[9] LICHTENSTEIN, E.P., "Fate and behavior of pesticides in a compartmentalized microcosm", Microcosms in Ecological Research (GIEST, J.P., Ed.), Technical Information Center, United States Department of Energy, DOE Symposium Series 52, CONF-781101, USDOE, Washington, DC (1979) 954-970.

[10] EDWARDS, C.A., BOGLEN, P.J., The assessment of effects of toxic chemicals upon earthworms, Rev. Environ. Contam. Toxicol. **125** (1992) 23-99.

[11] EDWARDS, C.A., SUBLER, S., CHEN, Shu-Kang, A microcosm system for evaluating the effects of pesticides on dynamic soil processes and soil organisms, Proc. Brit. Crop Prot. Conf. 9C-4 (1994) 1301-1306.

[12] AMERICAN SOCIETY FOR TESTING AND MATERIALS, Standard Guide for Conducting a Terrestrial Soil Core Microcosm Test (Standard E1197-87), ASTM 1194, ASTM, Philadelphia, PA (1988) 743-755.

[13] Terrestrial Model Ecosystems (KNACKER, T., MORGAN, E., Eds), Umweltbundesamt, Berlin (1994) 1-35.

[14] CAIRNS, J., PRATT, J.R., Ecotoxicological effect indices: A rapidly evolving system, Water Sci. Technol. **19** (1988) 1-12.

[15] WRIGHT, D.H., COLEMAN, D.C., Patterns of survival and extinction of nematodes in isolated soil, Oikos **67** (1993) 563-572.

[16] KUEHLE, J.C., "Radioecological studies on earthworms and their value for ecotoxicological risk assessment", Earthworms in Waste and Environmental Management (EDWARDS, C.A., NEUHAUSER, E.F., Eds), SPB Academic Publishers, The Hague (1988) 377-388.

[17] BROOKES, P.C., LANDMAN, A., PRUDEN, G., JENKINSON, D.S., Chloroform fumigation and the release of soil nitrogen: A rapid direct extraction method to measure microbial biomass nitrogen in soil, Soil Biol. Biochem. **17** (1985) 837-842.
[18] FRANKENBERGER, W.T., Jr., DICK, W.A., Relationships between enzyme activities and microbial growth and activity indices in soil, Soil Sci. Soc. Am. J. **47** (1983) 945-951.
[19] KEENEY, D.R., BRENNER, J.M., Comparison and evaluation of laboratory methods for obtaining an index of soil nitrogen availability, Agron. J. **58** (1966) 498-503.
[20] KNACKER, T., SCHALLNASS, J.H., MARCINKOWSKI, A., FÖRSTER, B., VINCENA, R., Applicability of Semi-natural (Terrestrial) Systems for the Assessment of Environmental Hazard in Accordance with the German Chemical Act, Umweltbundesamt Project No. 10603069, UBA, Berlin (1990) 208 pp.
[21] ANDERSON, J.P.E., DOMSCH, K.H., A physiological method for the quantitative measurement of microbial biomass in soils, Soil Biol. Biochem. **10** (1978) 215-221.
[22] MARTENS, R., Estimation of microbial biomass in soil by the respiration method, Soil Biol. Biochem. **153** (1987) 77-81.
[23] HEINEMAYER, O., INSAM, H., KAUSER, E.A., WALENZIK, G., Soil microbial biomass and respiration measurements: An automated technique based on infra-red gas analysis, Plant Soil **116** (1989) 191-195.
[24] SWIFT, M.J., HEAL, O.W., ANDERSON, J.M., Decomposition in Terrestrial Ecosystems, Blackwell Scientific Publishers, Oxford (1979) 371 pp.
[25] TEUBEN, A., ROELFSMA, T.A.P.J., Dynamic interactions between functional groups of soil arthropods and micro-organisms during decomposition of coniferous litter in microcosm experiments, Biol. Fertil. Soils **9** (1990) 145-151.
[26] PANCHOLY, S.K., RICE, E.L., Soil enzymes in relation to old field succession: Amylase, cellulase, invertase, Soil Sci. Soc. Am. Proc. **37** (1973) 47-50.
[27] SPALDING, B.P., The effect of biocidal treatments on respiration and enzymatic activities on Douglas-fir needle litter, Soil Biol. Biochem. **10** (1978) 537-543.
[28] THALMANN, A., Methoden zur Bestimmung der Dehydrogenaseaktivität im Boden mittels Triphenyltetrazolium Chlorid (TTC), Landwirtsch. Forsch. **21** (1968) 249-258.
[29] RÖMBKE, J., KNACKER, T., FÖRSTER, B., MACRINOWSKI, A., "Comparison of the effects of two pesticides on soil organisms in laboratory tests, microcosms, and in the field", Ecotoxicology of Soil Organisms (LA PONT, T., GREIG-SMITH, P.W., Eds), Lewis, Boca Raton, FL (1994) 230-240.
[30] TERYTZE, H., TERYTZE, K., Mathematische Modellierung des Einflusses von Blei (Pb) und Zink (Zn) auf die Bodenatmung, Wiss. Z. Humboldt-Univ. Berl., Agrarwiss. **37** (1988) 182-188.
[31] POKARZHEVSKIJ, A.A., PERSSON, T., Effects of oxalic acid and lime on the enchytraeid *Cognettia sphagnetorum* (VEJD) in mor humus, Water Air Soil Pollut. **85** (1995) 1045-50.
[32] TERYTZE, K.V., POKARZHEVSKIJ, A.A., A Methodical Approach to an Estimation of Bioindication and Biomonitoring, Nauka, Moscow (1991) 247-263.
[33] POKARZHEVSKIJ, A.A., Laboratory of Bioindication, Institute of Evolutionary Morphology, Moscow, and TERYTZE, K.V., Institut für Geographie und Geoökologie, Berlin, personal communication, 1996.

IAEA-SM-343/39

ADVERSE EFFECTS ON FLORA AND FAUNA FROM USE OF ORGANOCHLORINE PESTICIDES IN A MAIZE AGROECOSYSTEM*

E.M. MINJA, R.A. MAKUSI, F. TESHA
Tropical Pesticides Research Institute,
Arusha, United Republic of Tanzania

Presented by A.S.M. Ijani

Abstract

ADVERSE EFFECTS ON FLORA AND FAUNA FROM USE OF ORGANOCHLORINE PESTICIDES IN A MAIZE AGROECOSYSTEM.
 Organochlorine pesticides have been used widely to manage insect pest problems in agriculture, forestry, and animal and human health. Most of these pesticides are also known to persist in the environment for long periods of time. It has been reported that this persistence is deleterious to animal and human life, and for this reason most of these pesticides have been banned in many countries, including DDT, gamma-HCH (lindane), dieldrin, endrin, toxaphene, heptachlor, chlordane and endosulfan. However, in other countries, e.g. the United Republic of Tanzania, some of these insecticides (endosulfan and lindane) are still registered for use in agriculture. Since no information is available locally on the effect of these chemicals on the environment in which they are used, this study was conducted to generate data on the effects of lindane spraying on the arthropod fauna of field maize. The results indicate that lindane caused significant adverse effects on most of the early season beneficial arthropod fauna, particularly Collembola, ants and spiders. Mid and late season beneficial arthropod populations, including ants, spiders and ladybird beetles, were only slightly affected. The buried leaf litter decomposition rate was slightly reduced in sprayed compared with unsprayed maize fields.

1. INTRODUCTION

 Synthetic pesticides have been used widely in agriculture, forestry, and animal and human health to manage pest problems. Some of these pesticides demonstrate excellent performance in the control of the target organism, but their effects on non-target organisms have not always been positive. This has particularly been so with first generation synthetic insecticides, the organochlorines, most of which are also known to persist in the environment for long periods of time. Organochlorine insecticides include DDT, heptachlor, toxaphene, methoxychlor, dieldrin, aldrin, endrin,

---

* Research carried out with the support of the IAEA under Research Contract No. 6680.

chlordane, gamma-HCH (lindane) and endosulfan. While most of these chemicals have been banned because of their persistence and accumulation in animal and human tissues, some are still registered for use in the United Republic of Tanzania [1].

Among the organochlorine insecticides registered for use in crop protection in Tanzania is lindane (Gammalin 20). In developed countries, the impact of organochlorine pesticides on the environment has been monitored and documented [2]. It has been calculated that as much as 50% of the pesticide sprays applied to crops miss their target and fall on the soil surface. There are also reports that the air can easily become contaminated with pesticides during spraying operations [3]. In most developing countries, including Tanzania, very little information is available on the pesticide effects on target and non-target organisms. The maize stem borer, *Busseola fusca* Fuller, is one of the major insect pests limiting maize production in Tanzania. Previously, DDT dust and sprays were recommended and commonly used by farmers. However, after the banning of DDT because of its persistence and hazardous effects on the environment, farmers switched to other insecticides, including lindane. The present work was therefore undertaken to investigate the effects of lindane spraying on the arthropod fauna of field maize.

2. MATERIALS AND METHODS

The experiments were conducted at the Tropical Pesticides Research Institute, Arusha, on a 0.7 ha field for four seasons, starting in the 1991/1992 short rainy season. Planting was always dependent on the onset of the short rains in November/December (the 1991/1992, 1992/1993 and 1994/1995 seasons), and of the long rains in March/April (1994 experiment). The short rainy season at Arusha is warm and humid, while the long rainy season is cool and wet. Insect pest populations on maize are larger during the warm season than during the cool season. The 1994 cool season was used for the experiment because of a failure in the 1993/1994 short rainy season crop after an unexpected dry period in January 1994. In each season, the field was ploughed and harrowed before planting. In the first two seasons, the field was ploughed, then 3 t/ha of animal manure (cow dung) were broadcasted and harrowed into the volcanic loamy soil, which is known to have low organic matter. Cow dung was applied to the soil 1 month before planting the maize. The field was later divided into four equal blocks, each of which was subdivided into two 25 × 25 m plots. The plots in each block were randomly assigned two treatments: sprayed and unsprayed.

The maize hybrid CG 4141 was sown at 90 × 30 cm (two seeds per hole), and the seedlings were thinned to one per hill 3 weeks after emergence (WAE). All the field operations, except tractor ploughing and harrowing, were carried out manually. Fertilizer top dressing (25 kg N/ha) was done twice (at 3 and 6 WAE).

In the two plots per block, 5 × 5 m subplots were marked out permanently for different investigations. Peripheral subplots acted as the guard rows. In one subplot, four standard pitfall traps were placed in the ground to monitor the population activity of the ground fauna. The traps were set for 3 days with a weak soap solution. They were then emptied and the arthropod fauna sorted out and recorded. In a second subplot used to monitor aerial fauna, the arthropods were swept from the plants using a sweep net (first three seasons) and a D-Vac (fourth season). A third subplot was used to assess visual fresh plant damage; all the plants with fresh borer damage were counted and the total damaged plants converted into a proportion of the total subplot plant population. A fourth subplot was used for organic matter decomposition studies. Four litter bags containing *Grevillea robusta* A. Cunn leaves were randomly buried in the soil at 5 cm below the soil surface, remaining there for 3 months. The contents of all the litter bags with fresh leaves were weighed and standardized before burial. Several bags with similar fresh leaf samples were dried in the laboratory for later comparison with the dry weight of the buried samples. After 3 months, the buried litter bags were carefully dug out of the soil and dried at room temperature. The difference between the dry weight of the buried litter and that of the unburied litter was converted to a proportion of the unburied litter weight.

The maize crop in the field was constantly inspected for stem borer eggs and first instar larvae to enable a decision to be made on the date for initial insecticide spraying. The first spraying took place at 4 WAE, when there were either eggs or first instar larvae on about 0.5–1.0% of the plants. Lindane (Gammalin 20 from Imperial Chemical Industries, United Kingdom) was sprayed once at a dose rate of 1.0 kg active ingredients/ha in the 1991/1992 and 1992/1993 seasons. Owing to severe leaf scorching, this application was divided into two, i.e. 0.5 kg active ingredients/ha were sprayed twice at 2 weekly intervals, starting at 4 WAE in 1994 and 1994/1995. Knapsack hand sprayers (CP15 and CP20), with a polyjet nozzle and operated at a pressure of 3 bar (2.0 kg/cm^2), were used at high volume spraying (300 L water/ha). The arthropod populations and plant damage were monitored just before the first spraying, 24 hours later, and then every 2 weeks until harvest. All the data were subjected to analysis of variance.

3. RESULTS

For the four seasons, the dominant maize stem borer was *B. fusca*. Eggs and first instar larvae were observed from the end of 3 WAE. Towards the middle of the seasons, however, *Sesamia calamistis* Hamp. was observed on the crop, and two stem borers cohabited on the plants until harvest. Damage assessments were based on *B. fusca* only. The beneficial arthropods monitored during the studies were Formicidae (ants), Araneae (spiders), Collembola (springtails) and Coccinelidae (ladybird beetles). The soil characteristics of the field used for the trials in the four seasons are shown in Table I.

TABLE I. SOIL CHARACTERISTICS (%) FOR MAIZE AT THE TROPICAL PESTICIDES RESEARCH INSTITUTE, ARUSHA (SOIL CLASSIFICATION: LOAM)

Organic matter	Clay	Silt	Sand
3.9	15.8	46.1	34.2

TABLE II. MEAN NUMBER OF ANT POPULATIONS IN SPRAYED AND UNSPRAYED PLOTS

Season	Sprayed	Unsprayed	Difference
1991/1992	1.81 ± 0.30	5.63 ± 1.20	Significant
1992/1993	40.28 ± 5.62	43.53 ± 6.89	Not significant
1994	58.78 ± 7.54	107.70 ± 10.75	Significant
1994/1995	28.88 ± 5.22	35.15 ± 6.80	Not significant

TABLE III. MEAN NUMBER OF SPIDER POPULATIONS IN SPRAYED AND UNSPRAYED PLOTS

Season	Sprayed	Unsprayed	Difference
1991/1992	0.94 ± 0.21	1.22 ± 0.40	Not significant
1992/1993	3.59 ± 0.84	4.53 ± 0.67	Not significant
1994	4.75 ± 0.96	10.10 ± 2.90	Significant
1994/1995	4.58 ± 1.10	6.45 ± 1.21	Not significant

TABLE IV. MEAN NUMBER OF COLLEMBOLA POPULATIONS IN SPRAYED AND UNSPRAYED PLOTS

Season	Sprayed	Unsprayed	Difference
1991/1992	Not monitored	Not monitored	—
1992/1993	178.41 ± 12.75	220.47 ± 20.61	Significant
1994	170.05 ± 18.69	288.05 ± 40.63	Significant
1994/1995	237.89 ± 17.50	356.57 ± 20.62	Significant

TABLE V. MEAN PERCENTAGE BURIED LEAF LITTER WEIGHT LOSS IN SPRAYED AND UNSPRAYED PLOTS

Season	Sprayed	Unsprayed	Difference
1992/1993	34.4 ± 2.6	37.6 ± 3.2	Not significant
1994	25.8 ± 2.5	27.7 ± 1.9	Not significant
1994/1995[a]	9.4 ± 2.4	10.1 ± 1.6	Not significant

[a] Figures are low because the litter bags were removed from the field 4 weeks earlier than in previous seasons because of ploughing carried out for long rainy season planting in March 1995.

The results indicate that lindane spraying significantly ($P < 0.05$) reduced the activity of ants in all seasons except 1991/1992 (Table II). Similarly, spider activity was adversely affected in 1994, but not in the other three seasons (Table III). In contrast, Collembola activity was adversely affected by lindane spraying in all three seasons in which it was monitored (Table IV). The ladybird beetle populations were very small in all seasons, and their activity in the fields was extremely erratic. They were normally active on maize plants towards the end of the season, when there were light infestations of cereal aphids on the cobs and tassels. The sprays did not appear to have affected the activity of these beetles. The buried leaf litter decomposition rate was slightly reduced by the chemical sprays (Table V). These results indicate that some of the pesticide spray reached the soil surface, where it had some adverse effects on the organisms responsible for the breakdown of organic matter.

4. DISCUSSION

The results showed that lindane spraying was effective in reducing *B. fusca* damage on maize plants. However, the insecticide caused adverse effects on the activity of most of the beneficial arthropods monitored during the studies. Collembola were the most severely affected in all four seasons. Spiders and ants were adversely affected in some seasons. These results suggest that, although lindane can be useful in reducing maize stem borer damage, it has pronounced negative effects on other, beneficial organisms. Critchley et al. [4] have reported similar reductions in population activity for ants and spiders as a result of DDT treatments in the humid tropical zone of Nigeria.

Collembola were abundant and most active early in each season, but their numbers gradually declined, even in the unsprayed fields, as the soil moisture decreased towards crop maturity. Lindane was sprayed early in the season, therefore Collembola activity coincided with the time at which the sprays were most effective. The activities of Collembola, spiders and ants were drastically reduced by the initial sprays, and when these sprays were divided into two applications the activities remained low for longer periods. Although the population activities of these organisms recovered about 1 month after final spraying, their numbers did not reach the prespraying levels. The mid and late season beneficial arthropods, including spiders, ladybird beetles and ants, were not significantly affected by spraying. These results suggest that lindane could have short lived adverse effects on the maize environment when it was only sprayed twice. Continuous spraying, as might be applied to perennial crops, would probably prolong the period of adverse effects on the environment. On the other hand, the warm and semi-humid climatic conditions in which the trials were conducted could have facilitated the break down of the insecticide. Edwards [5] observed that, in a combination of wet and warm weather, pesticides are released and may break down and disperse into the atmosphere much more rapidly than from dry soils. He also reported that there was evidence that pesticides which persist for more than 10 years in temperature climates may disappear almost entirely within 1 year in the tropics. Yeadon and Perfect [6] estimated 15.2% disappearance of DDT from soil surfaces within a growing season in southern Nigeria.

The results indicated that lindane slightly reduced the rate of buried leaf litter decomposition in three seasons. Although there was no microarthropod extraction from the buried litter in these studies, reduced Collembola activity in the pitfall traps could explain the slow down in decomposition. Cook et al. [7] showed that Collembola and Acari account for 90–100% of the fauna responsible for buried leaf litter decomposition, and that their numbers were significantly reduced by DDT spraying. These results, although preliminary, highlighted the importance of monitoring the effects of pesticides on the environment in which they are being used. There is a need to monitor the fate of lindane in soils and its effect on the soil organisms that are responsible for organic matter break down.

ACKNOWLEDGEMENTS

The authors are grateful to the Swedish International Development Agency (SIDA) and the IAEA for financial support. They also thank the Director, Tropical Pesticides Research Institute, Arusha, Tanzania, for providing facilities for the field work, and appreciate the constant help and advice given by C. Wiktelius and P. Chiverton of the Swedish University of Agricultural Sciences, Uppsala, Sweden. The conscientious supervision of the field work by G. Kayange and G. Shayo proved valuable to our work.

REFERENCES

[1] Tanzania Gazette, 18 January (1991) 29-63.
[2] CAREY, A.E., GARVEN, J.A., TAI, H., MITCHELL, W.G., WIERSMAN, G.B., Pesticide residue levels in soils and crops from 37 states — National Soils Monitoring Program (IV), Pestic. Monit. J. **12** 4 (1979) 209-233.
[3] TABOR, E.C., Contamination of urban air through the use of insecticides, Trans. N.Y. Acad. Sci. **2.28** 5 (1966) 569.
[4] CRITCHLEY, B.R., COOK, A.G., CRITCHLEY, U., PERFECT, T.J., RUSSELL-SMITH, A., The effects of crop protection with DDT on some elements of the subterranean and surface active arthropod fauna of a cultivated forest soil in the humid tropics, Pedobiologia **20** (1980) 31-38.
[5] EDWARDS, C.A., Environmental aspects of the use of pesticides in developing countries, Med. Fac. Landbouw. Rijksuniv. Gent **42/2** (1977) 853-868.
[6] YEADON, R., PERFECT, T.J., DDT residues in crop and soil resulting from application to cowpea *Vigna unguiculata* (L.) Walp. in the sub-humid tropics, Environ. Pollut., Ser. B **2** (1981) 275-294.
[7] COOK, A.G., et al., The effects of soil treatment with DDT on the biology of a cultivated forest soil in the sub-humid tropics, Pedobiologia **19** (1979) 279-292.

IAEA-SM-343/38

EFFECT OF ORGANOCHLORINE PESTICIDES ON BIRDS IN THE UNITED REPUBLIC OF TANZANIA*

A.S.M. IJANI, J.M. KATONDO, J.M. MALULU
Tropical Pesticides Research Institute,
Arusha, United Republic of Tanzania

Abstract

EFFECT OF ORGANOCHLORINE PESTICIDES ON BIRDS IN THE UNITED REPUBLIC OF TANZANIA.
Preliminary studies to investigate the effect of organochlorine pesticides on birds in the United Republic of Tanzania were conducted in the Lower Moshi area, at the National Agriculture and Food Corporation in the west Kilimanjaro region, on the Arusha seed farm and Tropical Pesticides Research Institute (TPRI) farms, on the Manyara ranch and in areas around Lake Victoria, as well as in the TPRI laboratory. Large quantities of the pesticides, particularly DDT, endosulfan, dieldrin, lindane and toxaphene, are still being applied against pests of cotton, coffee, maize, beans and other crops, as well as against disease vectors. Several groups of birds, including waterbirds, African fish eagles, marabou storks, oxpeckers, ducks, etc., were found to feed, roost and swim in water and to be exposed to other substances contaminated with organochlorine pesticides. The analytical results of African fish eagle tissues collected from the Lake Victoria areas showed that the kidneys were contaminated with p,p'-DDE and o,p'-DDE at levels of 0.4 ng/g and 1.45 ng/g, respectively. These organochlorine pesticides, and beta-HCH, were also present in the brain and liver tissues. The levels of organochlorine residue were well below the lethal and sublethal levels for raptorial birds reported in the literature.

1. INTRODUCTION

Organochlorine pesticides, including DDT, dieldrin, endosulfan, lindane and toxaphene, are still being used to control agricultural pests and vectors of human and animal diseases in the tropical and subtropical regions of developing countries. Use is greatly restricted in temperate developed countries. This is mainly because of their high persistence in the environment and their tendency to accumulate in living tissue and to move through different trophic levels in the food chains. Scientists have indicated that, compared with temperate regions, the behaviour of chemicals may differ in tropical environments because of the high temperatures, humidity, ultraviolet

---

* Research carried out with the support of the IAEA under Research Contract No. 6851.

light, microbiological complexes, etc. Little information is available on the amount and effect of organochlorine pesticide residues in water, food, soil, animals and birds in tropical and subtropical countries. Talekar et al. [1] stated that organochlorine pesticides were banned in some tropical countries without sufficient information about their impact on the local environment.

A number of organochlorine pesticides have been used in the United Republic of Tanzania for many years, and are still registered for use against insect pests of coffee, especially in the Kilimanjaro and Arusha regions, against pests of cotton, particularly in the Lake Victoria areas, and of maize and beans in many other regions, as well as against pests of domesticated animals and humans. However, no research has been conducted on their effect on higher fauna, particularly birds, despite the fact that in Tanzania there is a vast number of species, some of which are endangered. Therefore, studies on both the short and long term effects of organochlorine pesticides on birds is very important. In the Lake Victoria areas, piscivorous and raptorial birds, particularly the African fish eagle (*Haliaectus vocifer*), seem to be the species most affected by the pesticides. It has been reported that some fishermen pour endosulfan and other pesticides into the lake to kill fish. These pesticides, together with those washed into the lake from crop fields during heavy rains, accumulate in the lake, resulting in their being taken up into the bodies of those fish, birds and animals that use the water. The African fish eagle feeds on fish and swims in the contaminated water, hence a high concentration of the pesticides is likely to accumulate in its tissues. The population of these birds around Lake Victoria has declined over the past few years [2]. The pesticides probably accumulate in the tissue of the birds, interfering with their reproductive and metabolic activities.

This study was conducted to identify the various organochlorine pesticides used in the Lake Victoria areas, and in the Kilimanjaro and Arusha regions; to assess the utilization of crop fields by avifauna in these areas; to determine the distribution and abundance of bird species, particularly the African fish eagle; to investigate its feeding and breeding habits, and its reproductive success; and to determine the level of organochlorine residues present in different tissues of its body.

2. MATERIALS AND METHODS

2.1. Experimental site

Surveys were conducted on rice and sugarcane farms in the lower Moshi area and at the National Agriculture and Food Corporation (NAFCO) in the west Kilimanjaro region, on the Arusha seed farm and Tropical Pesticides Research Institute (TPRI) farms, on the Manyara ranch, the Manyara National Park (Arusha region) and in areas around Lake Victoria, namely, Bunda, Magu, Igoma, Mwanza and Butimba (Mwanza region), with a view to investigating the pesticides used and the bird species present. The experiment was started in March 1992 and lasted for about 3 years.

2.2. Surveys and ecological study

Shopkeepers, cattle herdsmen, crop farmers and agricultural extension and health officers were interviewed in order to find out the types and quantities of organochlorine pesticide used against crop pests and disease vectors in their areas. The time and persistence of the various pesticides were recorded. The results obtained were then used to demarcate the areas into 'contaminated zones', i.e. those where organochlorine pesticides were regularly used, and 'control zones', i.e. those where none was used.

Observations were made according to the respective crop and pesticide spraying calendars, the number of dead birds, and their behavioural changes during and immediately after sprayings were carried out. The birds in the different study sites were monitored once a month, while those on the TPRI farms were monitored once a week. The birds found were sampled and their distribution, abundance, feeding and breeding habits, reproductive success and population dynamics investigated. Depending on their movement, the birds were divided into sedentary and migratory groups. The results obtained from ecological studies were used to determine suitable indicator species for further intensive studies.

2.3. Sampling and laboratory analyses

During the different surveys, a small number of birds was sampled for residue analyses. However, because certain technical problems arose with gas–liquid chromatography (GLC), only a few samples of the African fish eagle collected from the Lake Victoria areas could be analysed. These birds were selected from different locations around the lake and shot at two different periods: at the beginning of the short rains in October 1994, and in February 1995 during the period of heavy rainfall. The dead birds were taken to the TPRI laboratory, where their liver, brain, kidneys and fat tissues were removed. These tissues, and water samples taken from the same areas, were then analysed according to the protocol outlined by the IAEA, with slight modifications (as shown in the technique used by Watts [3]) to meet circumstantial requirements.

Some tissue samples were prepared for residue analyses without using the solid phase extraction (SPE) column. In this case, 20 g of tissue were mixed with anhydrous sodium sulphate (four times the weight of the tissue) and then ground. This mixture was tumbled in 200 mL of acetone/hexane (1:1 (vol./vol.)) and left to settle for 1 hour. The suspension was then filtered into a separating funnel containing 200 mL of sodium sulphate solution. The hexane extract was separated and washed three times with 50 mL of sodium sulphate solution (20 g/L).

The hexane extract was concentrated to 2.0 mL, passed through a microcolumn (containing 1 g of 60–80 mesh silica gel for chromatography, and 0.5 g of anhydrous sodium sulphate) previously washed with hexane (50 mL) followed by methanol (50 mL), and dried at 75°C overnight. The column was eluted with a further volume of hexane (10 mL). The hexane extract was concentrated to 10 mL, and 5 μL were analysed by GLC. The conditions were: column length, 1.5 m; internal diameter, 6.35 mm; packing, 50 g/kg of QF1 on Chromosorb WHP; temperatures: column, 220°C, injector, 260°C and detection, 260°C; carrier gas, nitrogen; flow rate, 60 mL/min; detection, electron capture detector (^{63}Ni).

When the sample was extracted without using the SPE column, broad, unresolved peaks were observed. Consequently, the extracts were again cleaned using a microcolumn (or pipette column), as mentioned previously, and resolved peaks were obtained. The peaks were then quantified using analytical standards, after determining the retention time of the individual peaks and the concentration of the corresponding compounds. The pesticide residue concentrations were calculated as follows:

$$R = \frac{A \times B \times C \times P}{D \times E \times F \times 100}$$

where A is the weight of the standard used (ng/μL); B is the area or height of the sample peak (mm); C is the final volume of the extract (μL); D is the microlitres injected (μL); E is the weight of the tissue sample taken (g); F is the area or height of the standard peak (mm); R is the residue concentration (ng/g); and P is the purity of the standard (%).

3. RESULTS AND DISCUSSION

Endosulfan, dieldrin, lindane, DDT and several cocktails of DDT and other pesticides are still recommended as cheap and effective pesticides for controlling cotton and coffee insect pests as well as disease vectors in Tanzania, and are being sold in large quantities in Tanganyika Farmers Association (TFA) and other shops. Farmers of vegetables, maize, beans, rice and sugarcane, as well as other crop producers, have been applying various organochlorine pesticides to control crop pests, since few pesticides have been specifically recommended for such use. The water being used to irrigate sugarcane in the Lower Moshi area contains high concentrations of dieldrin, endosulfan and DDT, presumably washed down from the coffee farms in the Kilimanjaro and Arusha regions. Endosulfan (sold under the trade name Thiodan 35 EC), dieldrin and DDT have been extensively used against pests of coffee, maize, vegetables and beans on the Arusha seed farm and TPRI farms. Toxaphene has been extensively used against ticks and other animal disease vectors on the Manyara ranch. Large quantities of endosulfan (in ULV and EC formulations) have been used to control pests of cotton and maize in most of the areas around Lake Victoria (Mwanza region). Large quantities of lindane and DDT have been used in the lake areas in previous years. Endosulfan and other pesticides have also been used to kill fish in Lake Victoria.

The bird species observed in all these areas are listed in Table I. The Lower Moshi rice irrigation scheme (1200 ha) and sugarcane plantations are inhabited by numerous species of bird, most of which are permanent residents and pests of rice. Queleas in their millions visit farms between July and February each year and can devastate the rice crop if not managed properly. Most of the waterbird species (egrets, herons, ibises and storks) eat the rice grains, and drink and swim in the stagnant water, which is probably contaminated with endosulfan, dieldrin, DDT and other pesticides. During a survey conducted in the area in May 1994, immediately after extensive spraying of the sugarcane and rice fields with endosulfan and other pesticides, it was found that some birds, particularly weavers and fiscal shrikes, had died and others were unable to fly for long distances. The dead birds were collected for laboratory analysis of the pesticide residues present in their tissues. Unfortunately, because certain technical problems arose with GLC, these analyses could not be done.

The NAFCO west Kilimanjaro region is an open woodland area inhabited by a number of bird species, including grey crowned cranes and cattle egrets. The birds are likely to be affected by the DDT, dieldrin, endosulfan and other pesticides frequently applied against pests of coffee, beans, maize, vegetables and other crops. The Arusha seed farm and TPRI farms have fewer species of bird, of which only a few, including abdimins and storks, are sedentary, hence they are likely to be affected by the organochlorine pesticides being used in these areas. Yellow weavers, which feed on the bollworms found in the maize crops produced here, are at risk. Fiscal shrikes may be exposed to these pesticides when feeding on the insects that survive spraying on the maize, coffee and bean farms. The Manyara ranch, a wooded grassland area, has a high diversity of bird species, sharing many of them with the Lower Moshi area. Several waterbirds drink and swim in the five ponds that are used to wash cattle after the water has been treated with DDT, dieldrin, endosulfan and, particularly, toxaphene. Scavengers, such as marabou storks, are permanent residents; birds of prey, especially vultures, also visit the ranch. These birds, as well as oxpeckers, doves, weavers and other species, drink the contaminated water and some may even eat contaminated cattle meat. These five ponds are the only sites in the area where the birds can find drinking water. Other birds found on the ranch include ducks, geese, ibises, etc., which appear to be local immigrants from the nearby Manyara National Park. Queleas also utilize the acacia trees in the area as nesting and breeding sites, and later migrate to feed on wheat, rice and barley on

TABLE I. SPECIES, DISTRIBUTION AND ABUNDANCE OF BIRDS

Species	Manyara ranch	Lower Moshi area	NAFCO west Kilimanjaro area	Arusha seed farm and TPRI farms	Manyara National Park
Ostrich	+			+	++
Little grebe	+				
Long tailed cormorant	+	+			+
Squacco heron	+	+			
Great white egret	+	+			+
Black heron	+	+			
Little egret					+
Black headed heron		+		+	+++
Purple heron		+			+
Cattle egret		++	++	+	+++
Yellow billed egret		++			+
Woolly becked stork		+			+
Marabou stork	+	+			
Yellow billed stork		+			
Hadada ibis		+		+	
Glossy ibis		+			
Sacred ibis	+	+		+	+
African spoonbill	+	+			+
White faced tree duck	+	+			+
Egyptian goose	+	+			+
Red billed duck	+	+			+
Knob billed duck	+	+			+
Macwa duck					+
Augur buzzard	+	+			+
Black shouldered kite		+			+
Hilderbrandt's prancolin	+				+
Helmeted guinea fowl	+				+
Button quail	+	+			+

TABLE I. (cont.)

Species	Manyara ranch	Lower Moshi area	NAFCO west Kilimanjaro area	Arusha seed farm and TPRI farms	Manyara National Park
Grey crowned crane	+	+	+ +		+
Blacksmith plover	+	+		+	
Black crake		+		+	
Two banded courser		+			
Chestnut bellied sandgrouse	+			+	
Namagua dove	+	+			
Ring necked dove	+	+		+	
Laughing dove	+ +	+		+	+
Red eyed dove	+	+			+
White bellied go away bird	+				+
White browed coucal	+	+			+
African jacana		+			+
Sandpiper		+			+
Speckled mousebird	+		+	+	+
Little bee eater	+			+	
African hoopoe	+	+			
Red billed hornbill	+	+			
Long tailed fiscal	+	+		+	
Fiscal shrike	+		+	+	+
White crowned shrike	+				
Hilderbrandt's starling	+ +		+		
Superb starling	+ +		+		
Ashy starling	+ +				
Yellow billed oxpecker	+ +				
Quelea	+ + +	+ + +	+ + +	+	
Yellow weaver (*Ploceus* ssp.)	+	+ + +		+ +	
Pelican					+
White stork				+	
Black kite				+	
Abdimin				+	

No. of birds: + = <100; + + = <100–1000; and + + + = >1000.

the nearby farms. Those birds that are permanent inhabitants of the Manyara National Park, and do not visit areas contaminated with organochlorine pesticides, are considered to be in the 'control zone'. The amount of organochlorine residues obtained from the tissues of these birds (if any) can be compared with that obtained from similar species of birds living in 'contaminated zones'. The marabou stork, a raptorial and sedentary species, is a good indicator for this type of study, since it is found in both the national park, where no pesticide is used, and on the ranch, where several pesticides are applied.

In the Mwanza region, the African fish eagle, a raptorial bird, was found in all the areas surveyed around Lake Victoria, including Butimba, Igoma, Nyamahanga, Tahamu near the Rubara River and Guta; the number of birds observed ranged between four and eight. The eagles were found here, together with the marabou stork, black kite, great egret, yellow billed duck, yellow billed stork, black headed heron and other species of bird. Possibly, these birds have managed to stay in the same location because their ecological niches do not quite overlap. The African fish eagle feeds on fish and small animals such as frogs living in the shallow water at the lake shore and in river streams. They feed during the day, especially between 10:00 and 16:00 hours. It appears that they build their nests and lay their eggs during the dry season or towards the end of the rainy season, because no nests with eggs were found during the survey, which was conducted at the beginning of the rainy season. Only two young birds were observed, both of which laid eggs that hatched. Their nests were 30 cm in diameter and built from small twigs or tree branches. These were found on the highest treetops and were not easily viewed. Such building of nests and the parental care shown by African fish eagles towards their young usually results in chick survival. However, because of the small number of eggs laid, their population is limited in most areas.

The analytical results from the different tissues of the African fish eagle collected from the Mwanza region showed that the kidneys probably contained o,p'-DDE and p,p'-DDE, and were more heavily contaminated with residues of the DDT type than the liver and brain tissues (Table II). Possible traces of lindane and beta-HCH were also observed in the brain and liver tissues. Work on organochlorine pesticides in Africa has generally concentrated on residue levels without investigating the associated effects, except for eggshell thinning [4]. Therefore, the side effects on birds from these pesticides were estimated by comparing them with the results of studies made in other countries. However, this has been difficult, mainly because the sensitivity to pesticides varies widely between bird species [5], even though diurnal raptors have broadly similar sensitivities [6]. Mortality of raptors from DDT and its residues is reported to occur when the DDT level exceeds 30 mg/kg [7]. In addition to raptors, DDT and its residues are reported to cause eggshell thinning and breakage, as well as addling and a reduction in hatchability [8]. Several studies of organochlorine pesticides in Africa have indicated that the residue levels are generally lower than those found in Europe and North America [5]. This is mainly because

TABLE II. RESIDUES OF ORGANOCHLORINE PESTICIDES IN TISSUES OF THE AFRICAN FISH EAGLE AND IN SAMPLES OF LAKE WATER

Sampling		Sample identity	Residues	
Time	Location		Concentration (ng/g)	Identity
October 1994	Magu	Liver	0.002	o,p'-DDE
		Brain	0.024	o,p'-DDE
		Kidneys	0.430	o,p'-DDE
			0.022	p,p'-DDE
	Igoma (Mwanza region)	Liver	0.009	o,p'-DDE
			0.007	p,p'-DDE
		Kidneys	0.200	o,p'-DDE
			0.460	p,p'-DDE
	Mwanza	Brain	0.010	o,p'-DDE
			0.006	p,p'-DDE
			0.019	Lindane
February 1995	Mwanza	Kidneys	1.45	o,p'-DDE
		Liver	0.001	Beta-HCH
		Brain	0.001	Beta-HCH
	Kayenze (Mwanza region)	Water	0.001	o,p'-DDE
		Water	0.114	o,p'-DDT
	Bunda	Water	0.004	o,p'-DDE
	Magu	Water	0.154	o,p'-DDT

of the lower proportion of agricultural land in Africa and the shorter persistence of organochlorine pesticides in tropical compared with temperate climates. Assuming that the amount of residues obtained from tissues of the African fish eagle during this study can be confirmed, the values are still well below the lethal and sublethal levels calculated by other scientists [7]. However, the data obtained cannot be considered representative of the Mwanza area. This is because other important tissues of the bird, including muscle and fat, have not yet been analysed. In addition, the birds were sampled at the beginning of the rainy season, when most of the pesticides applied to the cotton fields had not yet been washed into the lake where the birds feed and swim. Furthermore, as the number of samples analysed was small it was not possible to provide consistent and reliable results, and the identity of the peaks has not yet been confirmed by gas chromatography–mass spectrometry.

Tentative data on the African fish eagle indicate that it is necessary to determine the level of organochlorine residues that are present in the tissues and/or in the eggs of the exposed birds, and to investigate the associated effects. The preliminary results presented in this paper show that further studies are needed on the effect of organochlorine pesticides on birds such as the African fish eagle and the marabou stork found around the Lake Victoria areas, where lindane and endosulfan are extensively used against crop pests. Oxpeckers in the Manyara area are also at risk, since large quantities of toxaphene, which is known to be highly toxic to birds [8], are used in cattle dips.

ACKNOWLEDGEMENTS

The authors are grateful to the IAEA and the Director of the TPRI for the financial support that enabled this project to be carried out.

REFERENCES

[1] TALEKAR, N.S., CHEN, J.S., KAO, H.T., Long term persistence of selected insecticides in subtropical soil: Their absorption by crop plants, J. Econ. Entomol. **76** (1983) 207.

[2] MGALLAH, S., Tropical Pesticides Research Institute, personal communication, 1994.

[3] WATTS, R.R., Analysis of Pesticide Residues in Human and Environmental Samples: A Compilation of Methods Selected for Use in Pesticide Monitoring Programs, Health Effects Research Laboratory, Environmental Protection Agency, Research Triangle Park, NC (1979).

[4] CRICK, H.Q.P., "Organochlorine pesticides and birds of prey in Africa", Proc. VII Pan-Africa Ornithology Congress (1992) 171–189.

[5] DE KOCK, A.C., LORD, D.A., Chlorinated Hydrocarbon Residues in African Fish Eagle (*Haliaectus vocifer*) Eggs, Report No. 10, Institute for Coastal Research (1986).

[6] NEWTON, I., "Effects of organochlorine pesticides on birds", Proc. 2nd Symp. African Predatory Birds, Durban, 1984, Natal Bird Club, Durban (1984) 151–159.

[7] BLUS, L.J., HENNY, C.J., STAFFORD, C.J., GROVE, R.A., Persistence of DDT and metabolites in wildlife from Washington State orchards, Arch. Environ. Contam. Toxicol. **16** (1987) 467–476.

[8] NEWTON, I., Population Ecology of Raptors (1979).

LINDANE AND ENDOSULFAN RESIDUES IN WATER AND FISH IN THE ASHANTI REGION OF GHANA*

S. OSAFO-ACQUAAH
Department of Chemistry,
University of Science and Technology,
Kumasi, Ghana

Abstract

LINDANE AND ENDOSULFAN RESIDUES IN WATER AND FISH IN THE ASHANTI REGION OF GHANA.
Pesticide residue analyses were performed on water and fish samples from the River Oda in Besease, the River Aframso in Nobewam near Kumasi, the River Atwetwe in Akomadan, and the River Kowire at Agogo in the Ashanti region of Ghana. The sampling sites were in regions of intense farming activities in the closed forest zone of the country. The community at Akomadan is involved in large scale vegetable and maize farming. Tomato cultivation constitutes about 80% of the activity in the area. Besease and Nobewam produce mainly rice, and are the location of the Anum Valley Irrigation Project. Close to the rivers are over 200 rice fields covering more than 400 ha of land. Agogo is noted for its vegetable, rice and maize production, and other surrounding villages are involved in cocoa farming. The residue levels for lindane and endosulfan were determined in the following water and fish samples: *Oreochromis niloticus, Tilapia zillii, Barbus trispulis, Heterobranchus* sp., *Tilapia busumana, Ophiocephalus obscura* and *Chana obscura*. Lindane residues ranging from 4.5 to 32 ng/L were obtained for the water samples, and from 0.2 to 25 ng/g for the fish samples. Endosulfan residues ranging from 6.4 to 35.2 ng/L were obtained for the water samples, and from 0.8 to 265 ng/g for the fish samples. These values are well below the World Health Organization recommended safe values for water and fish samples. Analysis of the pesticide residues was done by gas chromatography, with a ^{63}Ni electron capture detector.

1. INTRODUCTION

Organochlorine pesticides have found wide application in agriculture for the control of pests, and in public health for vector control [1, 2]. There has been an increase in pesticide use in Ghana over the past 10 years as a result of the efforts made by the government to boost food production.

---

* Research carried out with the support of the IAEA under Research Contract No. 7481.

In agriculture, organochlorine pesticides are widely used by farmers because of their cost effectiveness and their broad spectrum activity, but their possible impacts on the environment and humans have been neglected. Research in temperate regions and in some parts of Africa has found that pesticides are very persistent in the environment and that they remain in the soil [3] for long periods of time, resulting in serious damage to soil flora and fauna [4]. Detailed information on the effect of organochlorine pesticide residues and on the amount present in water, food, soil, flora and fauna in Africa is, however, very limited. In some African countries, organochlorine pesticides have been banned on the basis of data obtained from temperate regions in developed countries, without assessing the effect of these pesticides in a tropical environment.

Lindane (Gammalin 20) is widely used in Ghana on cocoa plantations for the control of capsids, on vegetable farms, and for the control of stemborers in maize [5]. Endosulfan is used in cotton growing areas, on vegetable farms and for the control of pests in coffee plantations.

Most of the farms in the commercial vegetable producing areas in Ghana such as Akomadan and Agogo are situated along rivers that act as the water supply source for farming and for drinking purposes. The runoff of pesticides from farms due to rain and to the aerial spraying of pesticides is known to pollute rivers, in addition to the improper disposal and handling of the chemicals during and after spraying. Endosulfan is known to be toxic to fish (e.g. the LC_{50} after 96 hours of exposure is 1.2 µg/L for *Fundus heteroclitus* and 0.75 µg/L for *Oreochromis niloticus*) [6, 7].

2. MATERIALS AND METHODS

2.1. Experimental materials

Pesticide grade hexane was obtained from Mallinckrodt Special Chemicals, United States of America. High performance liquid chromatograph grade methanol and acetonitrile were obtained from the Aldrich Chemical Company, USA. The bonded phase C_{18} solid phase extraction (SPE) columns were obtained from Varian, USA.

2.2. Experimental site

This general study covered about 500 000 ha of land in the closed forest region of Ghana, which is about 34.5% of the entire land area of the country.

Three regions of intense farming activity were selected for residue analyses after a general survey had established the agricultural activities undertaken in the areas under consideration, and the extent of pesticide usage. The community at Akomadan is involved in large scale vegetable and maize farming. Tomato cultivation constitutes about 80% of the activity in the area. Besease and Nobewam produce mainly rice, and are the location of the Anum Valley Irrigation Project. Close to the rivers are over 200 rice fields covering more than 400 ha of land. Agogo is noted for its vegetable, rice and maize production, and other surrounding villages are involved in cocoa farming.

2.3. Sampling

The water samples were collected at a depth of 25 cm from the River Oda at Besease near Kumasi, the River Kowire at Agogo and the River Atwetwe at Akomadan in the Ashanti region in 1 L sterilized bottles sealed with a stopper. The samples were collected in duplicate between January 1993 and July 1994, and between June 1995 and October 1995. The rivers pass through regions of intense cocoa production, as well as other farming activities. The initial stage of work involved random sampling in the various rivers.

More systematic sampling was undertaken between June and October 1995, and involved the farming communities at Agogo and Besease. No samples were taken from Nobewam, because neither lindane nor endosulfan was being used for farming in this area. The River Oda at Besease flows through several areas of cocoa production, where lindane is actively used. Endosulfan is used in farming at Agogo, and cocoa farms are sprayed with lindane.

After collection, the water samples were taken to the laboratory for analysis. The fish samples for pesticide analyses were collected from the River Oda near the rice fields and from the River Aframso in Agogo, where there is large scale vegetable production. The fish samples were obtained using the hook and line method, as well as cages and nets. These samples were immediately washed with distilled water and stored in a freezer. They were later taken to the Ghana Atomic Energy Commission laboratory and the Institute of Aquatic Biology in Accra, Ghana, for pesticide residue analyses. The present work involved seven species of fish.

2.4. Sampling of other species

Although several varieties of birds, rodents (rats, squirrels and grasscutters) and reptiles (lizards) were identified, the project focused on residues in fish. The grasscutter samples collected from the area have not yet been analysed.

2.5. Gas chromatographic analysis

Samples were analysed with a Hewlett-Packard Series II model 5890 gas chromatograph equipped with a 30 m SPB-5 capillary column (with a 0.5 mm internal diameter). The carrier gas (nitrogen) had a flow rate of 30 mL/min. The injector operating temperature was 250°C, the oven temperature 200°C, and that for the detector, 300°C. The sample injection volume was 1 μL. The limit of detection was 1 pg/μL. Recovery of lindane and endosulfan was 90%.

2.6. Experimental procedures

2.6.1. Water residue analyses

A reservoir was mounted on top of the SPE column before conditioning. Two 10 mL volumes of methanol, followed by two 10 mL volumes of distilled water, were passed through the column. Slight pressure was applied to the top of the reservoir to help the liquid flow down the column at a rate of at least 2 mL/min; 800 mL of each water sample were passed through the column at a flow rate of 2 mL/min. The column was then washed with 1 mL of 30% methanol, followed by 1 mL of distilled water, and air dried under vacuum for 20 min. The analytes trapped in the column were eluted with five units of 0.5 mL hexane.

2.6.2. Fish residue analyses

Ten grams of each fish sample were homogenized using a pestle and mortar. The sample was mixed with 20 g of sodium sulphate and then transferred to a 250 mL separatory funnel. Twenty millilitres of hexane were added, followed by

TABLE I. PESTICIDE RESIDUE ANALYSES FOR WATER

| Sample No. | Date of collection | Sample | Residue analyses | | Source |
			Lindane (ng/L)	Endosulfan (ng/L)	
1	15 Jan. 1993	Water sample	14.2	Nil	River Oda
2	21 Dec. 1993	Water sample	15.0	Nil	River Oda
3	15 Jan. 1994	Water sample	1.3	Nil	River Atwetwe
4	12 Mar. 1994	Water sample	3.9	Nil	River Aframso
5	7 Jul. 1994	Water sample	3.0	Nil	River Kowire
6	22 Jul. 1994	Water sample	0.3	Nil	River Aframso

TABLE II. PESTICIDE RESIDUE ANALYSES FOR FISH

Sample No.	Date of collection	Sample	Residue analyses		Source
			Lindane (ng/g)	Endosulfan (ng/g)	
1	15 Jan. 1993	*Heterobranchus* sp.	14.2	Nil	River Oda
2	21 Dec. 1993	*Heterobranchus* sp.	24.3	Nil	River Oda
3	15 Jan. 1994	*Barbus trispulis*	23.2	Nil	River Atwetwe
		Chana obscura	0.8	Nil	
4	12 Mar. 1994	*Tilapia* sp.	0.2	Nil	River Aframso
		Barbus trispulis	0.2	Nil	
5	7 Jul. 1994	*Ophiocephalus obscura*	1.5	Nil	River Kowire

TABLE III. PESTICIDE RESIDUES IN WATER (1995)

Sample No.	Month	Sample	Residue analyses		Source
			Lindane (ng/L)	Endosulfan (ng/L)	
1	June	Water	25.0	35.2	River Oda
2	July	Water	24	30	River Oda
3	August	Water	22	15	River Oda
4	September	Water	32	14	River Oda
5	October	Water	8.7	6.4	River Oda
6	September	Water	4.5	6.8	River Oda
7	September	Water	10.0	15.4	River Kowire
8	October	Water	14.8	25.6	River Kowire
9	September	Water	12.5	28.2	River Atwetwe
10	October	Water	20.2	14.6	River Atwetwe

100 mL of acetonitrile, which had been saturated with hexane, and the mixture was shaken for about 1 min. The acetonitrile was drained into a 1 L separatory funnel. The mixture was extracted with 3 × 50 mL portions of acetonitrile, and the extracts combined in a flask. This was washed with 500 mL of water, 40 mL of saturated sodium chloride solution and 50 mL of hexane. The mixture was allowed to stand, and the aqueous layer was drained into another 1 L separatory funnel and extracted with another 50 mL of hexane. The combined organic extracts were passed through a plug of 10 g of sodium sulphate into a flask. The solvents were evaporated almost to dryness, and the residue dissolved in 10 mL of methanol and diluted with 25 mL of water; 3.5 mL of the solution were taken for pesticide residue analyses. This was equivalent to 1 g of the fish sample. The 3.5 mL of the final extract were passed through a previously conditioned column at a flow rate of 2 mL/min. The column

TABLE IV. PESTICIDE RESIDUES IN FISH (1995)

Sample No.	Month	Sample	Residue analyses		Source
			Lindane (ng/L)	Endosulfan (ng/L)	
1	June	*Oreochromis niloticus*	20.4	80.2	River Oda
		Barbus trispulis	120.4	158	
		Heterobranchus sp.	14.0	160	
2	July	*Oreochromis niloticus*	22.5	267.5	River Oda
		Tilapia zillii	34	142.5	
		Heterobranchus sp.	35.2	64	
3	August	*Oreochromis niloticus*	36.4	20.2	River Oda
		Barbus trispulis	16.8	11.4	
		Heterobranchus sp.	24.8	6.5	
4	September	*Oreochromis niloticus*	100	186	River Oda
		Tilapia zillii	72	77	
		Heterobranchus sp.	80	66	
5	October	*Oreochromis niloticus*	38.2	5.0	River Oda
		Barbus trispulis	22	10	
6	September	*Oreochromis niloticus*	8.4	12.4	River Kowire
7	October	*Oreochromis niloticus*	13.4	22.8	River Kowire

was washed and the sample eluted, as described previously. After the injection of 1 μL of extract, the retention times of the various peaks obtained on the gas–liquid chromatograph were compared with those of the pesticide standards. The procedure was repeated for the other fish samples. No independent method of confirmation was applied.

3. RESULTS

The results of the pesticide residue analyses for the water and fish samples are summarized in Tables I–IV. The pH of the stream water ranged between 7.0 and 8.2, and the average temperature of the stream water was 28°C.

The results assume that the chromatogram peaks corresponding to lindane and endosulfan did in fact represent these compounds, but confirmation has not been possible.

4. DISCUSSION

Seven species of fish caught in the various rivers were subjected to residue analyses. The species caught depended on the time of year the trapping was done. It appeared that because of the adverse effects of some of the material, perhaps agrochemicals, on the aquatic environment, the fish populations had decreased considerably in these rivers. There were days when our team could not trap any fish in some of the rivers using a combination of the hook and line method, and cages and nets.

Our initial survey work indicated that farms near the River Oda at Besease were sprayed up to six times during the year. Farms near the River Atwetwe and River Kowire (tomato producing areas) were also subjected to up to six pesticide applications, depending on whether the tomatoes were farmed over one or two seasons.

Although lindane is not used widely in the spraying of tomatoes and rice, endosulfan is one of the most commonly used chemicals in these regions.

Peaks corresponding to the retention times of lindane and endosulfan were found to be much lower in the water samples than in the fish samples, which may have been due to the accumulation of organochlorine pesticide residues in fish over a period of time.

The pesticide residues found in all the fish were below the lethal (LC_{50}) level. This was probably because these pesticides have been extensively introduced into farming over the past 5–10 years. On average, the size of the farms is about 1 ha, and spraying is done with different pesticides at different times of the year; farming is not practised all year round.

Residues whose retention times corresponded to DDT and other organochlorine pesticides were observed in the chromatogram, but were not analysed because the focus of this project was lindane and endosulfan.

ACKNOWLEDGEMENTS

The authors are grateful to the IAEA for its support and to C.A. Edwards for his comments and assistance. They also acknowledge with gratitude the assistance of the Ghana Atomic Energy Agency and the Institute of Aquatic Biology, Accra, in performing the residue analyses.

REFERENCES

[1] WALISZEWSKI, S.M., Residues of lindane, HCH isomers and HCB in the soil after lindane application, Environ. Pollut. **82** (1993) 289–293.

[2] JAMAL, A., Pesticides: Hazards and Alternatives, African Development Foundation, Washington, DC (1991) 3–4.

[3] EDWARDS, C.A., "Pesticide residue in soil and water", Environmental Pollution by Pesticides (EDWARDS, C.A., Ed.), Plenum Press, New York (1973) 409–458.

[4] EDWARDS, C.A., THOMPSON, A.R., Residue and the soil fauna, Res. Rev. **45** (1973) 1–79.

[5] MINISTRY OF AGRICULTURE, Technical Bulletin No. GDB/MI, Accra, Ghana (1990).

[6] TRIM, A.H., Acute toxicity of emulsifiable concentration of three insecticides commonly found in non-point source runoff in estuarine waters to the mummichog, *Fundus heteroclitus*, Bull. Environ. Contam. Toxicol. **30** (1987) 681–686.

[7] MACEK, K.J., BURTON, K.S., DERR, S.K., DEAN, I.W., SAUTERS, S., Chronic toxicity of lindane to selected aquatic invertebrates and fisheries, Rep. EPA-600/3076-046, Environmental Protection Agency, Duluth, MN (1976).

IAEA-SM-343/14

EXPOSURE OF TOAD EMBRYOS AND LARVAE TO PESTICIDES
Use of nuclear techniques to determine their effect on the reproduction, survival and potential risk to Bufo arenarum *populations*

A. CABALLERO DE CASTRO, A. VENTURINO, V. KIRS,
M. LOEWY, G. CARVAJAL, A.M. PECHEN DE D'ANGELO
Universidad Nacional del Comahue,
Neuquen, Buenos Aires,
Argentina

Abstract

EXPOSURE OF TOAD EMBRYOS AND LARVAE TO PESTICIDES: USE OF NUCLEAR TECHNIQUES TO DETERMINE THEIR EFFECT ON THE REPRODUCTION, SURVIVAL AND POTENTIAL RISK TO *Bufo arenarum* POPULATIONS.

Application of pesticides is currently the most common method used to control agricultural pests. However, undesired effects on non-target organisms and pollution of the soil, air and water are frequent consequences. Amphibians are good bioindicators of the presence of contaminants in the environment, because they are semi-aquatic animals located at the top of the food chain. In many parts of the world, amphibian populations are declining, and many reasons have been suggested for these losses. Although a link between widespread decline and pesticide residues has yet to be established, it is suspected that contamination of their breeding sites with pesticide residues has had a deleterious effect on the reproduction and development of amphibians. Recent experiments with a widely distributed toad, *Bufo arenarum* Hensel, in South America, particularly in Argentina, indicate that a variety of insecticides affects the fertilization process through the activation of an enzyme that degrades the source of second messengers and is involved in transducing the sperm signal to the oocyte. This harmful effect is not restricted to the fertilization process, since embryonic and larval development are also affected, producing severe morphological and behavioural abnormalities in embryos. Embryonic and larval development influence the timing of metamorphosis, the suceptibility to predation, survival in the terrestrial environment, and even the success of future reproduction, these being the most sensitive periods of a toad's life. Evaluation of the pesticides in our region showed their presence in many of the potential breeding sites, confirming that they may influence the survival of toad populations.

1. INTRODUCTION

Application of pesticides is currently the most common method used to control or eradicate agricultural pests. Most of the pesticides used for this purpose are non-specific, and affect not only the 'target' but also other organisms. The magnitude of the adverse ecological effects depends on multiple variables such as the type of pesticide and formulation, the rate and means of application, the landscape, and the composition of the food chain.

The implication of some insecticides on the decline of an amphibian (the toad, *Bufo arenarum* Hensel) widely distributed in South America, particularly in Argentina, is discussed.

Over the past few years, several authors have recognized that the decline in amphibian populations is a global phenomenon [1–3]. Amphibians are good bioindicators of the overall conditions of the environment. They are found in different ecosystems, from desert to forest, in different climates, and at different altitudes, from sea level to mountains. They live in water during their embryonic and larval periods, while as adults they live on land. They reproduce by oviposition in most fresh water ecosystems, exposing their unshelled eggs to soil and water. The bioindicators currently under investigation are mainly fish and birds, but results from recent investigations suggest the usefulness of including other biota, e.g. amphibians, as indicators of the presence of contaminants.

Many reasons have been suggested for the decline of these animals: destruction of their habitats by human beings, fungus infections, chemical pollution, acid rain, depletion of the stratospheric ozone, etc. We have evidence that several insecticides have had an adverse effect on the reproduction and early development of *B. arenarum*.

2. EFFECT OF INSECTICIDES ON THE FERTILIZATION PROCESS

We followed different periods of toad development, from oocyte to adulthood, and concluded that the most sensitive periods of development were during the fertilization process and during embryogenesis. These periods of the toad life-cycle take place in water ecosystems. The time of the year at which females lay their eggs, waiting for external fertilization, is spring, which coincides with the period in which intensive use is made of insecticides in agriculture. An important proportion of the mass of insecticides applied is transported by volatilization, wind, rainfall, leaching, runoff and groundwater to most waterways, where they come into contact with the laid oocytes. The lack of shell in this species (they only have a jelly coat) facilitates the absorption of different substances, mainly those of a lipophylic nature such as pesticides.

We performed several experiments in the laboratory using persistent and non-persistent insecticides by exposing oocytes obtained from the ovisac through surgery to different concentrations of insecticide and for different periods of time. The oocytes were inseminated in vitro using a homologous sperm suspension and then the number of successful fertilizations was recorded [4]. An important decrease in the fertilization rate was detected for several chlorinated hydrocarbon, organophosphorus and carbamate insecticides. Two were selected to investigate the molecular event that produced this deleterious effect.

Iwamatsu [5] and Miyazaki [6, 7] have reported the involvement of intracellular second messengers in the fertilization process. PIP_2 (phosphatidylinositol-4,5 bis phosphate) hydrolysis is involved in the response of amphibian oocytes to external stimulation through the generation of IP_3 (inositol 1,4,5 triphosphate) and DAG (1,2 diacylglycerol) [8]. The first is known to interact with the intracellular receptors that promote Ca^{2+} release from intracellular stores; the second is a well known PKC (protein kinase C) activator [9]. Both IP_3 and PKC activation produce cortical granule exocytosis in amphibians, indicating that these events are an integral component of the fertilization pathway.

We used oocytes endogenously labelled with ^{32}P-orthophosphate in order to follow the hydrolysis of PIP_2 and the generation of IP_3 and DAG in control oocytes and in oocytes pretreated with the insecticides dieldrin and azinphosmethyl. The low level of ^{32}P-PIP_2 labelling found in oocytes pretreated with the two insecticides (Fig. 1), without any accumulation in ^{32}P-PIP and PI, constitutes the first evidence of in vivo phospholipase activation produced by the insecticides. The second evidence was obtained in stimulation experiments. Muscarinic stimulation of oocytes produced transient hydrolysis of PIP_2 in the control cells; this is necessary to maintain elevated levels of IP_3 and to release Ca^{2+} from internal storage in subcellular compartments. In the oocytes pretreated with dieldrin and azinphosmethyl, these events, characteristic of the fertilization process, were absent and no hydrolysis of PIP_2 was demonstrated. The percentage successful fertilization for dieldrin and

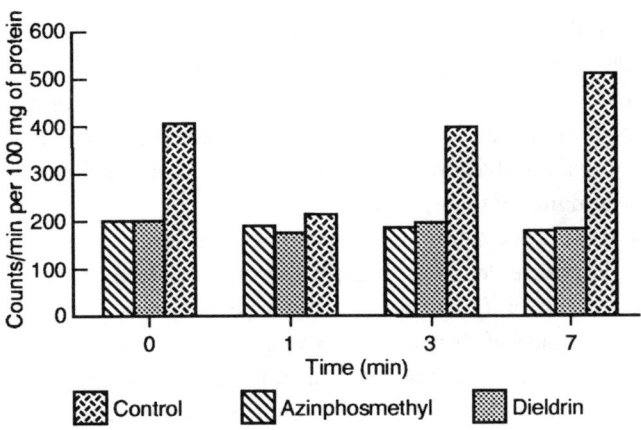

FIG. 1. PIP_2 labelling in control B. arenarum oocytes treated with dieldrin and azinphosmethyl: the effect of 1 mM carbachol stimulation.

azinphosmethyl treated eggs was 65 and 60%, respectively. We concluded that oocyte exposure to the two insecticides affected the reproduction of *B. arenarum* by phospholipase mediated depletion of PIP_2. Activation of such phospholipase is produced by intercalation of the insecticide in the oocyte membrane. The integrity of the PIP_2 pool is absolutely essential for transducing the sperm attachment to the oocyte membrane, for maintaining biochemical activities within the cell, and for completing the fertilization process.

Use of dieldrin is banned in Argentina, as in many other countries, but we included it in these experiments because of its resistance to biodegradation and its high persistence in different matrices. The amounts used were higher than those reported in most natural water ecosystems.

Oocytes prelabelled with ^{32}P-PI were exposed to a solution containing 4 mg/L of dieldrin or 0.05 mg/L of azinphosmethyl, or no insecticide, for 2 h. The oocytes were then washed and exposed to 1 mM of carbachol for different periods of time: 1, 3 and 7 min. At zero time and during the periods mentioned, samples were taken and polyphosphoinositides extracted; they were then purified by thin layer chromatography (TLC) and quantified by liquid scintillation counting (LSC) [4]. A marked decrease in PIP_2 labelling was observed at zero time in the oocytes treated with the two insecticides. A response to carbachol stimulation was only observed in the control oocytes, with a decrease in PIP_2 labelling at 1 min and further recuperation at 7 min. No significant changes were detected in the oocytes treated with the two insecticides. The results are the mean of three experiments.

3. EFFECT OF INSECTICIDES ON THE EARLY DEVELOPMENT OF *B. arenarum*

The harmful effect of pesticides on natural superficial water is not restricted to the fertilization process; embryos and larvae have the same potential for being adversely affected as oocytes. In fact, we observed that the rate of development, the length and weight of the larvae, the incidence of malformations and also the percentage survival are all influenced by insecticides such as dieldrin, lindane, DDT, parathion, malathion, azinphosmethyl and carbaryl.

Teratological information on insecticides has been based on macroscopic observations. Exposure to most chlorinated hydrocarbon insecticides produced severe abnormalities in *B. arenarum* embryos and larvae, e.g. body twisting, progressive appearance of dropsy, defective gills, hyperexcitability, erratic swimming and a shortening of the metamorphosis time. Organophosphorus and carbamate insecticides produced a shortening in the anterio-posterior axis, a bent notochord, abnormal pigmentation, defective gut and gills, and a circling swimming movement. Pyrethroid insecticides also cause excessive neural stimulation and neurotoxic symptoms that include loss of equilibrium and sudden bursts of swimming.

We tried to establish the relationship between the morphological and the biochemical data. The decrease in the rate of development observed in embryos and larvae treated with lindane, parathion or malathion is correlated with a depletion of the GSH (reduced glutathione) pool and an increase in the GSH-S transferase activity. The observation that elevated levels of this enzyme may be associated with tolerance to pesticides [10] indicates that the embryonic development of *B. arenarum* may be protected by this detoxifying mechanism. However, depletion of the GSH pool, which is associated with increased GSH-S transferase activity, is a dangerous combination that results in increased lipid peroxidation, DNA damage, alterations in polyamine metabolism and developmental arrest.

Drews [11] has demonstrated that the appearance of cholinesterase activity is closely linked to organ formation, and he postulated a functional relationship between cholinesterase activity and morphogenetic movement. This concept was extended by Schroder [12], who suggested that the cholinergic system is involved in the regulation of embryonic development. We have reported that cholinesterase inhibiting pesticides modify the pattern of expression and activity of acetyl, butyl and carboxylesterases [13], which produce neurophysiological, behavioural and functional alterations in larvae. We also found that dieldrin produced an alteration in acetylcholinesterase activity [14], and was associated with an increase in the rate of development, confirming the participation of the enzymes in regulating the proliferation rate and growth of embryos.

There are important differences in insecticide sensitivity between the different stages of toad development. It was recently found that fertilized oocytes and embryos are less sensitive to different insecticides than larvae. Mortality plots of bioassays with parathion on 5 day old embryos and 22 day old larvae indicate that young embryos are 4.4 times more tolerant to the acute lethal effect of parathion than larvae. The differences in toxicity were associated with the alteration in cholinesterase sensitivity to paraoxon and with the different metabolism of parathion in homogenates of early embryos and larvae [15]. In in vitro experiments with ^{14}C-parathion it was observed that cleavage of the arylphosphate linkage was done more efficiently by embryonic $16.000 \times g$ supernatants than by the corresponding larval preparation. In other experiments, in which we exposed embryos at different stages of development to ^{14}C-malathion, important differences in metabolite formation were detected. A net increase in the proportion of polar metabolites was observed in complete operculum and larvae, while in the earlier stages of neural fold and gill circulation a higher proportion of malathion and malaoxon was found (Table I). These results indicate a preponderance of the detoxifying versus the activation pathway in the advanced stages of toad embryonic development, but it fails to explain why early embryos are more tolerant to organophosphorus insecticides. We postulated that the presence of high lipid stores in early embryos facilitates the detoxification of malathion and malaoxon by binding to the lipidic pool, rendering no malaoxon to interact with cholinesterases. In the advanced stages of development,

TABLE I. METABOLISM OF ^{14}C-MALATHION IN DIFFERENT STAGES OF *B. arenarum* DEVELOPMENT[a]

% detected

Metabolites	Neural fold		Gill circulation		Complete operculum		Larvae	
	5 h	24 h	5 h	24 h	5 h	24 h	5 h	24 h
Malathion	91.10 ± 3.0	78.70 ± 1.8	82.34 ± 5.1	69.70 ± 1.6	85.50 ± 1.5	43.80 ± 4.4	10.30 ± 0.9	7.80 ± 0.7
Malaoxon	4.30 ± 0.5	7.83 ± 0.7	9.10 ± 0.6	7.54 ± 0.7	2.38 ± 0.7	0.73 ± 0.1	ND	ND
Polar metabolites	4.20 ± 0.8	14.10 ± 1.3	7.76 ± 0.6	22.60 ± 1.5	12.40 ± 1.7	55.30 ± 5.4	89.70 ± 1.8	92.90 ± 2.3
Non-extractable	0.01 ± 0.01	0.04 ± 0.01	0.04 ± 0.01	0.14 ± 0.01	0.08 ± 0.01	0.20 ± 0.1	ND	1.00 ± 0.2

[a] The distribution of radioactivity was calculated as the percentage total radioactivity recovered on the TLC plate. Global recovery was 83.5 ± 6.2%. Values are the mean of triplicates ± SD. ND = non-detected.

the lipid stores are depleted, so the minor amounts of malaoxon formed are free to inhibit cholinesterases, producing excessive neural stimulation, followed by paralysis and death.

The embryonic and larval growth rates are important ecological variables in amphibian populations, influencing the larval size, the metamorphosis time, the susceptibility to predation, competition between species, and even the success of future reproduction [16, 17]. Thus, considering the importance of these events, pesticides have a pronounced effect on the survival of amphibian populations through their effect on embryonic and larval growth.

4. DISTRIBUTION OF RESIDUES IN THE ECOSYSTEM

The temporal and spatial distribution of pesticide residues is important for characterizing the pesticide risk to non-target organisms. Spatial distribution of a pesticide is related to the horizontal and vertical profile of the residues after application of a pesticide. Temporal distribution is related to the half-life in different systems, and is very important for estimating the duration of exposure to target and non-target organisms.

In our region, the method most commonly employed for controlling or eradicating pests that affect fruit production is spraying with several organophosphorus, carbamate and pyrethroid insecticides for 4 months during the spring–summer period. A minor quantity of herbicides is used, and fungicides are mainly applied in fruit industries but not in the field. We measured the distribution of pesticide residues, including insecticides, herbicides and fungicides, in most surface and groundwater sources by gas–liquid chromatography (GLC). Most important, the rivers, lakes, streams, irrigation and drainage channels, and groundwater were sampled. The mean residue values were 10^3 times greater on drainages than in rivers or lakes, indicating that industrial activities are an important point source for the discharge of pesticides. The results of 2 years of groundwater quality monitoring also revealed that the most frequent residues found were azinphosmethyl, dimethoate, carbaryl and cypermethrin during the pesticide application season. Many of the wells showed chronic cases of contamination, where the residues fluctuate between high and low levels, or are above and below the detection limit. These residues mainly have vertical movement (leaching) in the soil, since runoff and erosion losses only account for a small percentage as the lands and fruit farms are flood irrigated.

Low residues of chlorinated hydrocarbon insecticides require special attention. They continue to be detected in water and soil, and remain troublesome in terms of their global mobility, persistence, toxicity and bioaccumulative potential. It is suspected that bioaccumulated chlorinated hydrocarbon pesticides impair reproduction in reptiles, birds and mammals [18]. We have previously described their deleterious effect on the fertilization and early embryonic development of the toad [4].

To understand and predict the relative mobility of pesticides in soil and the risk they pose of reaching surface and groundwaters, we used a soil leaching column with some representative insecticides. In experiments with non-radioactive insecticides, GLC was used to identify the extractable parent compounds and metabolites in the soil, while in experiments with radioactive insecticides, TLC separation and further LSC were employed. Analysis of their distribution in the soil after leaching showed that most of the organophosphorus pesticides were retained at a depth of 0.5 m, while some carbamates and pyrethroids had a higher mobility. We also determined their half-life values on soil in the laboratory to study the degradation pathway and the relative contribution of photochemical, chemical and microbial degradation to the permanence of the pesticides in the system.

This study reveals the need for care in the safe handling of pesticides and in the selection of the compounds to be used, taking into consideration not only their efficacy but also their fate in the environment, and their potential risk to non-target species. An on-site investigation showed that contamination was caused by pesticide spillage into or around waterways while mixing, by the loading or cleaning up of spray equipment; by pesticide drift due to inappropriate meteorological conditions at the time of application, and by pesticide transport through leaching on soil and runoff.

5. ASSESSMENT OF THE IMPACT OF PESTICIDES ON THE SURVIVAL OF *B. arenarum* POPULATIONS

Our studies indicate that *B. arenarum* are vulnerable to the toxic effect of several insecticides and that they are potentially affected by pesticide contamination of their habitats, mainly during the fertilization process and during the embryonic and larval periods.

Field studies carried out in the Alto Valley of Río Negro and Neuquen showed the presence of pesticide residues in many potential toad breeding sites. However, the level of residues found is not currently causing direct mortality, although sublethal effects may have occurred. Detailed chemical analyses of water samples failed to indicate any other source of contamination, e.g. heavy metals, hydrocarbons or a low pH. Thus, we conclude that pesticides influence the survival of toad populations in our region; however, the extent to which they have produced a decline in *B. arenarum* is still uncertain.

REFERENCES

[1] BARINAGA, M., Where have all the froggies gone? Science **247** (1990) 1033–1034.
[2] WAKE, D.B., Declining amphibian populations, Science **253** (1991) 860.

[3] BLAUSTEIN, A.R., WAKE, D.B., The puzzle of declining amphibian populations, Sci. Am. **272** (1995) 56–61.
[4] FONOVICH DE SCHROEDER, T.M., PECHEN DE D'ANGELO, A.M., Dieldrin modifies the hydrolysis of PIP_2 and decreases the fertilization rate in *Bufo arenarum* oocytes, Comp. Biochem. Physiol. **112C** (1995) 61–67.
[5] IWAMATSU, T., Exocytosis of cortical alveoli and its initiation time in medaka eggs induced by microinjection of various agents, Dev. Growth Differ. **31** (1989) 39–44.
[6] MIYAZAKI, S.I., Inositol 1,4,5-triphosphate-induced calcium release and guanine nucleotide-binding protein-mediated periodic calcium rises in golden hamster eggs, J. Cell. Biol. **106** (1988) 345–353.
[7] MIYAZAKI, S.I., Calcium wave in activating hamster eggs, Biol. Bull. **176** 5 (1989) 21–24.
[8] ORON, Y., DASCAL, N., NADLER, E., LUPU, M., Inositol 1,4,5-triphosphate mimics muscarinic response in Xenopus oocytes, Nature (London) **313** (1985) 141–143.
[9] NISHIZUKA, Y., The family of protein kinase C for signal transduction, J. Am. Med. Assoc. **262** (1989) 1826–1833.
[10] OPPERNOOTH, F.J., "Biochemistry and genetics of insecticide resistance", Comprehensive Insect Physiology, Biochemistry and Pharmacology, Vol. 12 (KERKUT, G.S., GILBERT, L.S., Eds), Pergamon Press, Oxford (1985) 731–773.
[11] DREWS, U., Cholinesterase in embryonic development, Prog. Histochem. Cytochem. **7** (1975) 1–52.
[12] SCHRODER, C., Characterization of embryonic cholinesterase in chick limb bud by colorimetry and disk electrophoresis, Histochemistry **69** (1980) 243–253.
[13] CABALLERO DE CASTRO, A., ROSENBAUM, E., PECHEN DE D'ANGELO, A.M., Effect of malathion on *Bufo arenarum* Hensel development. I. Esterase inhibition and recovery, Biochem. Pharmacol. **41** (1991) 491–495.
[14] CHIFFLET DE LLAMAS, M., CABALLERO DE CASTRO, A., PECHEN DE D'ANGELO, A.M., Cholinesterase activities in developing amphibian embryos following exposure to the insecticides dieldrin and malathion, Arch. Environ. Contam. Toxicol. **14** (1985) 161–166.
[15] ANGUIANO, O.L., MONTAGNA, C.M., CHIFFLET DE LLAMAS, M., GAUNA, L., PECHEN DE D'ANGELO, A.M., Comparative toxicity of parathion in early embryos and larvae of the toad, *Bufo arenarum* Hensel, Bull. Environ. Contam. Toxicol. **52** (1994) 649–655.
[16] WILBUR, H.M., COLLINS, J.P., Ecological aspects of amphibian metamorphosis, Science **182** (1973) 1305–1314.
[17] BRODIE, E.D., FOMANOWICZ, D.R., Prey size preference of predators: Differential vulnerability of larval amphibians, Herpentologica **39** (1983) 67–75.
[18] ENVIRONMENT CANADA, Toxic Chemicals in the Great Lakes and Associated Effects, Vol. 2, Effects, Toronto, ON (1991) 493–755.

EFFECTS OF METHAMIDOPHOS ON SOIL MICROBIAL ACTIVITY*

Bujin XU, Yongxi ZHANG

Institute of Nuclear Agricultural Sciences

Nanwen ZHU, Hong MING,
Meici CHEN, Yuhua ZAO

Section of Microbiology

Zhejiang Agricultural University,
Hangzhou, China

Abstract

EFFECTS OF METHAMIDOPHOS ON SOIL MICROBIAL ACTIVITY.
  The effects of methamidophos on soil microbial activity were studied. Soil was treated with methamidophos at doses of 0, 0.5, 2.5, 5 and 10 µg/g. The populations of bacteria, actinomyces, azotobacter and fungi were determined, as well as soil respiration, nitrogen fixation, nitrification, ammonification and Fe(III) reduction. The results showed that growth of bacteria, actinomyces and azotobacter was inhibited, as were soil nitrogen fixation and Fe(III) reduction, whereas fungi growth was stimulated. In general, soil respiration was also stimulated, but it showed a complex trend. The effects of methamidophos were stronger, and lasted longer, as the dose increased.

1. INTRODUCTION

Microorganisms are an important component of the soil ecosystem, and also play a significant role in soil fertility and material cycling. The effects of agrochemicals on soil microbial activity are of great concern to biologists. In the past, few studies on such effects related to organophosphorus insecticides, especially methamidophos, which is used extensively in China to control cotton pests. This study focuses on the effects of methamidophos on soil microbial activity.

---

  * Research carried out with the support of the IAEA under Research Contract No. 8079.

2. MATERIALS AND EQUIPMENT

2.1. Materials

(1) *Soil*: The top layer of fluvio, marine yellow, loamy soil from the cotton field of the Zhejiang Agricultural University experimental farm was used: organic matter content, 1.52%; nitrogen, 0.18%; K_2O, 1.87%; P_2O_5, 0.5%; and pH7.48.

(2) *Pesticide*: Crystalline methamidophos (99.3%) was produced by the Huzhou Ling Hu Chemistry Factory, Zhejiang.

2.2. Equipment

(1) *A gas chromatograph* (GC) (Model-102G), manufactured by the Shanghai Analytical Factory, Shanghai.

(2) *A colorimeter*.

3. METHODS

(1) *Soil treatment*: Methamidophos was applied at doses of 0 (control), 0.5, 2.5, 5 and 10 µg/g of soil (dry weight), with three replicates. The treated soils were placed in glass bottles with perforated caps, and the moisture content was adjusted to 60% of the maximum field capacity.

(2) *Incubation*: Indoors, under room temperature.

(3) *Sampling*: Samples were taken at 1, 4, 8, 16, 24 and 36 days after application.

(4) *Microbial activity*: The population of microorganisms in the soil samples was determined using the plate dilution method outlined in Ref. [1]. The respiration method described by Zelles et al. [2] was used. The nitrogen fixation capacity was determined by GC using the acetylene reduction technique described in Ref. [3]. The colorimetric methods used for Fe(III) reduction and for ammonification and nitrification are outlined in Refs [1, 4], respectively.

4. RESULTS AND DISCUSSION

Initially, methamidophos inhibited the growth of bacteria, but later showed slight stimulation; the details are given in Table I. Recovery from inhibition at a dose of 0.5 µg/g took place within 4 days; at 2.5 µg/g, 8 days; at 5 µg/g, 16 days; and at 10 µg/g, 24 days. This stimulation may have been caused by the phosphorus in methamidophos or by the proliferation of more methamidophos tolerant species.

TABLE I. EFFECTS OF METHAMIDOPHOS ON BACTERIA IN SOIL (10^7/g DRY WEIGHT)

Concentration (μg/g)/time (d)	0	0.5	2.5	5	10
1	1.84	1.82	1.75	1.73	1.54
4	1.54	1.59	1.52	1.47	1.34
8	3.28	3.31	3.34	3.01	2.62
16	2.12	1.98	2.16	2.37	1.93
24	6.95	6.41	6.37	7.08	7.08
36	6.52	6.35	6.70	6.58	6.61

TABLE II. EFFECTS OF METHAMIDOPHOS ON FUNGI IN SOIL (10^4/g DRY WEIGHT)

Concentration (μg/g)/time (d)	0	0.5	2.5	5	10
1	4.88	4.46	4.59	6.16	6.76
4	4.29	4.66	4.34	5.25	5.34
8	6.18	5.70	6.30	6.25	5.98
16	6.26	6.27	5.80	5.75	6.21
24	5.53	5.43	4.60	5.57	5.49
36	6.97	7.16	7.03	7.05	7.11

TABLE III. EFFECTS OF METHAMIDOPHOS ON Fe(III) REDUCTION IN SOIL (mg/100 g DRY WEIGHT)

Concentration (μg/g)/time (d)	0	0.5	2.5	5	10
1	29.37	22.31	18.55	17.01	18.28
4	34.89	30.57	26.22	22.23	23.89
8	30.48	29.15	29.09	27.86	28.36
16	30.44	32.87	31.97	33.53	33.21
24	32.78	—	32.78	33.53	33.21
36	29.71	29.37	28.94	29.49	29.93

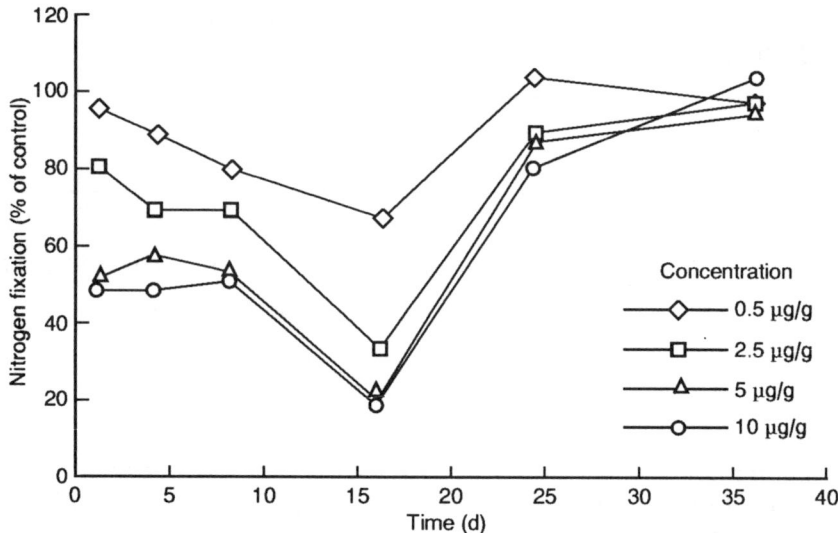

FIG. 1. *Effects of methamidophos on nitrogen fixation (control = 100%).*

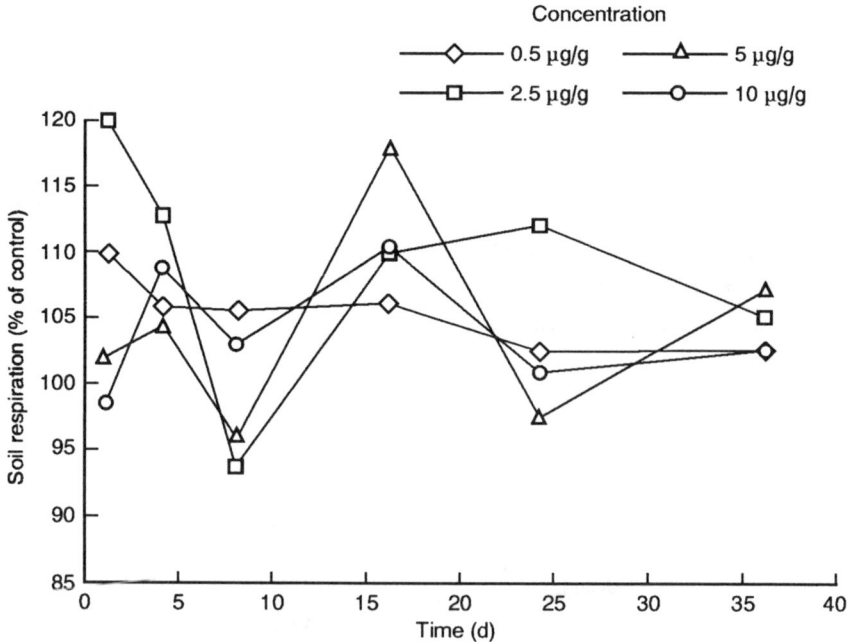

FIG. 2. *Effects of methamidophos on soil respiration (control = 100%).*

Actinomyces responded rapidly to methamidophos application. Within 24 hours, inhibition had reached 9.3–38.2%, and recovery took about 16 days.

Methamidophos stimulation of fungi started within 24 hours of application at doses of 5 and 10 μg/g, and within 4 days at doses of 0.5 and 2.5 μg/g; after about 16 days, the effects had disappeared (Table II). The effects were directly related to the doses and inversely related to the bacteria counts up to 4 days after methamidophos application.

Table III shows the effects of methamidophos on Fe(III) reduction. Inhibition was significant 24 hours after application, decreasing with time; the effects had disappeared after 16 days.

Soil nitrogen fixation was inhibited by methamidophos. The effects were observed 24 hours after application, increasing with time. Inhibition then increased, with the most significant influence observed between 8 and 16 days; about 36 days were needed for recovery. Figure 1 shows these effects (control = 100%). The decrease in nitrogen fixation correlated with the decrease in the azotobacter populations, which is in agreement with a previous review [5].

The effects of methamidophos on soil respiration (CO_2 evolution) are shown in Fig. 2 (control = 100%). In general, methamidophos promoted soil respiration, but it showed a complex trend. At all doses, except 10 μg/g, stimulation was shown within 24 hours. Later, stimulation at 0.5 and 2.5 μg/g showed a decrease, and at 5 and 10 μg/g, an increase. After 36 days, soil respiration for all the applications was slightly higher than that of the control. As soil respiration is a collective function of the soil microorganisms, the effects of methamidophos on the soil microbial population are complex, but the trend is understandable.

The effects of methamidophos on ammonification and nitrification in soil were weak, only showing a slight effect between 1 and 4 days.

5. CONCLUSIONS

The results showed that the activity and population of soil microorganisms were influenced by methamidophos, producing either inhibition or stimulation, depending on the type of microorganism. These effects were related to the methamidophos dose in soil. Soil bacteria, actinomyces and azotobacter showed inhibition, while fungi were only slightly stimulated. Because bacteria made up about 70–90% of the soil microbial population, this activity had a significant influence on the total effects, which differs from a previous review [6].

ACKNOWLEDGEMENT

The authors wish to thank the IAEA for the support received.

REFERENCES

[1] MICROBIOLOGY LABORATORY OF THE SOIL SCIENCE INSTITUTE, ACADEMIA SINICA, Study Methods for Soil Microorganisms, Scientific Press, Beijing (1986) 353 pp (in Chinese).

[2] ZELLES, L., SCHEUNERT, I., KORTE, F., Comparison of methods to test chemicals for side effects on soil microorganisms, Ecotoxicol. Environ. Saf. **12** (1986) 53-89.

[3] MING, Hang, ZHAO, Yuhua, LU, Yitong, Effects of trifluralin on soil microbes and earthworms, J. Rural Eco-Environ. **3** (1993) 40-43 (in Chinese).

[4] XU, Guanhui, ZHENG, Hongyuan, Analytical Manual of Soil Microbes, Agricultural Press, Beijing (1986) 314 pp (in Chinese).

[5] WAINWRIGHT, M., A review of the effects of pesticides on microbial activity in soil, J. Soil. Sci. **29** (1978) 287-298.

[6] MOORMAN, J.B., A review of pesticide effects on microorganisms and microbial processes related to soil fertility, J. Prod. Agric. **2** 1 (1989) 14-23.

__IAEA-SM-343/9__

EXTRACTION OF BETA-ENDOSULFAN AND ENDOSULFAN SULPHATE IN SOILS WITH SUPERCRITICAL CO_2

R.M. LOPEZ-ROMERO
Instituto de Recursos Naturales,
Colegio de Postgraduados,
Montecillo, Edo. de México

S. CAPELLA-VIZCAINO
Departamento de Química Analítica,
Facultad de Química,
Universidad Nacional Autónoma de México,
Mexico City

Mexico

Abstract

EXTRACTION OF BETA-ENDOSULFAN AND ENDOSULFAN SULPHATE IN SOILS WITH SUPERCRITICAL CO_2.
　Extraction of organochlorine pesticides from indigenous soils with supercritical CO_2 was optimized using Soxhlet extraction as the reference method to evaluate the recoveries. The extraction cell temperature and pressure, and the restrictor temperature, were studied. The best conditions were 400 atm at 50°C for the extraction cell, and a 50 μm (inner diameter), 40 cm fused silica restrictor heated to 180°C. Extraction was carried out for 10 min in the static mode and for 20 min in the dynamic mode, with methanol as the modifier. Collection of the extracted analytes was performed in CH_2Cl_2. The recoveries with supercritical CO_2, relative to those obtained with solvent extraction, were 175 and 148% (10 and 39% relative standard deviation) for beta-endosulfan and endosulfan sulphate, respectively.

1. INTRODUCTION

　Supercritical fluid extraction (SFE) with CO_2 is an interesting alternative to traditional solvent extraction methods for pesticide analysis in soils, with a significant reduction in time and hazardous solvent use. SFE is affected by a number of operation variables that have to be optimized to obtain acceptable recoveries. Such variables are the pressure and temperature for extraction, the addition of modifiers to the extraction fluid, and the restrictor and collection system. Most published reports deal with spiked samples, and data on field soil samples are scarce.

Brady et al. [1] used SFE–CO_2 to extract polychlorinated biphenyls, DDT (1000 ppm) and toxaphene (400 ppm) from indigenous soils, and reported recoveries in the range of 70%. For the same soils, Dooley et al. [2] reported recoveries as high as 95% using methanol as the modifier for SFE. With spiked samples of different soils, Snyder et al. [3] found mean recoveries of 84% for several organochlorine and organophosphate pesticides using SFE–CO_2 with methanol as the modifier. In the same work, SFE was compared with solvent extraction (sonication) for field samples of indigenous soils, and similar recoveries were obtained.

In this work, we present the results of optimizing SFE for the extraction of samples from three different soils cropped with potatoes at León, Guanajuato (Mexico), where application of pesticides has been intensive.

2. EXPERIMENTS

2.1. Samples

The soil samples were obtained at a depth of 0.1 m from three replicates registered in the laboratory as 201, 206 and Rangel. The samples were air dried in the laboratory and sieved (0.8 mm). The characteristics of soil 206 are given in Table I [4, 5].

2.2. Soxhlet extraction

Soxhlet extraction was used as the reference method [6] to compare the recoveries from SFE. Ten grams of the soil sample were extracted for 24 h with hexane/acetone (1:1). All the solvents used in this work were of analytical grade (Baker, United States of America). Extracts were concentrated to 0.01 L in a Kuderna–Danish evaporator. Further concentration to 0.001 L was carried out under N_2 flow. No further clean-up was undertaken.

TABLE I. CHARACTERISTICS OF SOIL 206

Soil	pH (1:2 H_2O)	OM[a] (%)	Sand (%)	Silt (%)	Clay (%)
Sandy clay loam[b]	7.1	1.7	49	20	31

[a] Organic matter: Walkley and Black method, described in Ref. [4].
[b] Bouyoucos method, described in Ref. [5].

2.3. SFE

The SFE extractions were performed with a Suprex instrument (SFC/200A) modified to use the chromatographic oven for the extraction cell (4.6 mm (inner diameter), 0.1 m stainless steel tubing) and to allow solvent trapping (5 mL of CH_2Cl_2 at 4°C) outside the oven. The fused silica tubing (50 μm (inner diameter), 0.4 m) was used as the restrictor, and the temperature of the restrictor was controlled independently with a heating tape. To switch from the static to the dynamic extraction mode, an on/off valve was installed between the cell and the restrictor. Two grams of the sample were thoroughly mixed with anhydrous Na_2SO_4; this mixture was then placed in the extraction cell between two (lower and upper ends) layers of Chromosorb W80/100 mesh (Supelco, USA). To the upper layer were added 200 μL of methanol and 10 μL of internal standard solution (aldrin (Polyscience Corp., USA), 20 ppm in iso-octane). Extraction was performed in the static mode for 10 min and in the dynamic mode for 20 min. After extraction, the volume of the trapping solvent was adjusted to 0.001 L. The effect of the extraction temperature and pressure, and the restrictor temperature, on recoveries was studied. Supercritical fluid grade CO_2 (Air Products, USA) was used as the extraction fluid.

2.4. Extract analysis

The extract solutions were analysed by capillary gas chromatography (Hewlett-Packard 5890GC) coupled to a mass spectrometric detector (Hewlett-Packard 5971A). The chromatographic conditions were as follows: DB1 column (30 m, 0.25 mm (inner diameter), 0.25 μm film (Supelco, USA)); the temperature was held at 60°C for 2 min; the temperature was increased at a rate of 20°C/min, until it reached 240°C; then the rate was changed to 10°C/min until the temperature reached 290°C, which was held for 7 min. The injector temperature was 250°C, and the transference line, 280°C. The injection volume of 1 μL was splitless. Helium was used as the carrier gas, with a 10 psi inlet pressure.[1] Chromatograms of each sample extract were run in both the SCAN and the SIM modes. For SIM, the most significant ions were monitored (four to six single ions). To identify the pesticides, the retention times and the mass spectra of the standard solutions (Chem Service, USA) were obtained.

---

[1] Pound force per square inch (psi) is 1 lbf/in^2 = 6.895 × 10^3 Pa.

3. RESULTS AND DISCUSSION

3.1. Soxhlet extraction

Endosulfan sulphate and beta-endosulfan were found to be the most important pesticides present in soil 206. In Rangel soil, only endosulfan sulphate was detected, and in soil 201 none of the pesticides studied was present. The reproducibility of the Soxhlet extractions was determined as 1.9 and 1.4% for endosulfan sulphate and beta-endosulfan, respectively, with soil 206.

3.2. SFE

Only soil 206 was used for the SFE experiments. With 350 atm and 50°C as the extraction parameters, the effect of the restrictor temperature was studied, experiments being performed at 100, 180 and 220°C. The best results were found at 180°C; at 100 and 220°C, obstructions occurred because of the low volatility or solubility of the analytes as they passed through the restrictor, where CO_2 is depressurized. These results agree with those reported by Reindl and Hofler [7].

To evaluate the effect of the pressure and temperature in the extraction cell, experiments were carried out at 300 and 350 atm, and at 40, 50, 67 and 97°C. Figure 1 gives the results for endosulfan sulphate as relative recoveries (100% for

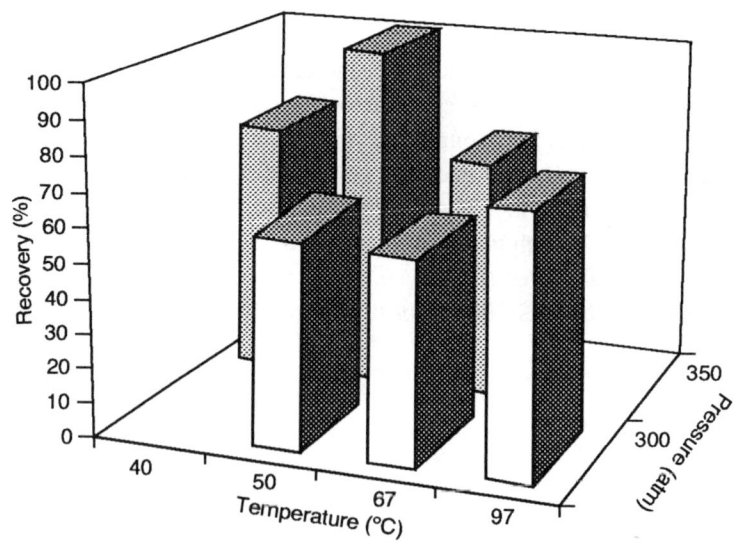

FIG. 1. Average extraction recovery of endosulfan sulphate versus pressure and temperature, relative to the 350 atm at 50°C extraction treatment of 100%.

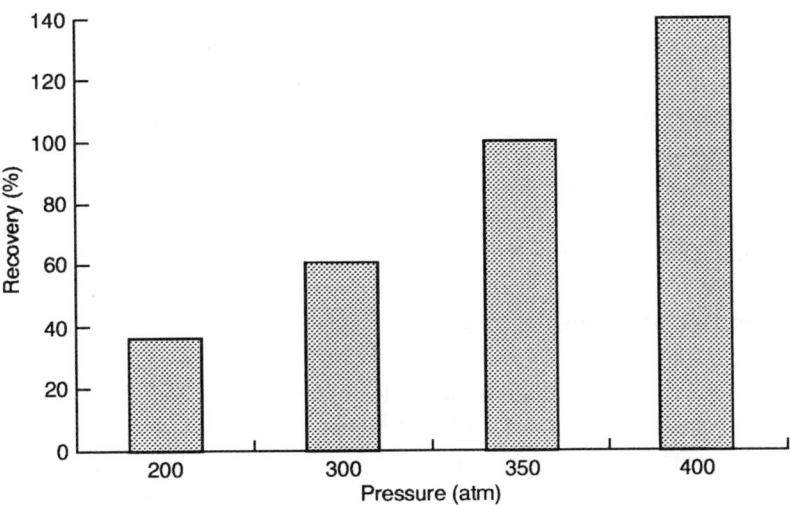

FIG. 2. Average extraction recovery of endosulfan sulphate versus pressure at 50°C, relative to the Soxhlet extraction treatment of 100%.

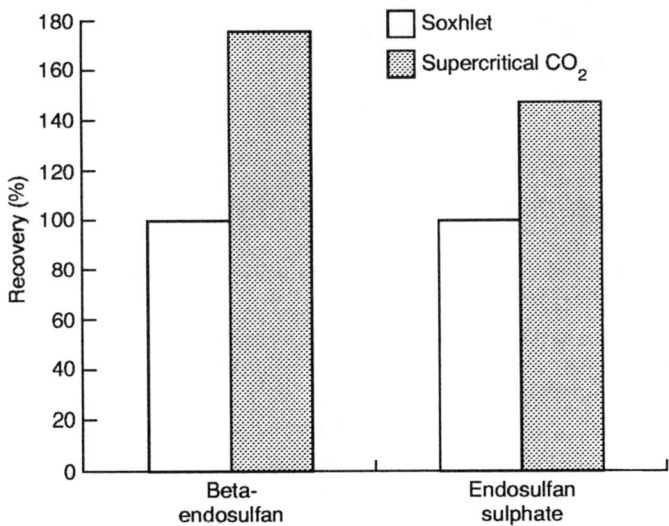

FIG. 3. Comparison of the mean recoveries of beta-endosulfan and endosulfan sulphate using Soxhlet and supercritical CO_2 (n = 6).

350 atm at 50°C). Two conditions, volatility and solubility, are needed to extract the analytes with supercritical CO_2. This is in agreement with the work of Snyder et al. [3]. According to Taylor [8], both the temperature and the pressure have to be tuned for each analyte–matrix system, since the effect of these variables depends on the specific interaction of the analyte with the matrix by which it is sorbed. In Fig. 2, the results of changing the pressure and maintaining the temperature at 50°C are shown. Larger recoveries are obtained at higher densities (pressures), which is the most important parameter of SFE.

3.3. Comparison of Soxhlet and SFE

Figure 3 shows the relative recoveries obtained with both procedures (100% assigned for Soxhlet extraction). SFE had higher recoveries than Soxhlet extraction, although the reproducibility was not as good (39% for endosulfan sulphate and 10% for beta-endosulfan). The recoveries obtained with a solvent as weak as CO_2 can be explained by the low amount of organic matter in the soil. With a higher organic content, extraction can be more difficult, and the extracts may not be as clean as those obtained in this work.

4. CONCLUSIONS

The best conditions for extracting endosulfan sulphate and beta-endosulfan from soil 206 were 400 atm and 50°C. SFE extraction was more efficient and less time consuming than Soxhlet extraction. Caution is recommended in extrapolating these results to other pesticides and soils, since each sample must be studied as a specific problem.

REFERENCES

[1] BRADY, B.O., et al., Supercritical extraction of toxic organics from soils, Ind. Eng. Chem. Res. **26** (1987) 261–268.

[2] DOOLEY, K., KAO, C., GAMBRELL, R.P., KNOPF, F.C., The use of entrainers in the supercritical extraction of soils contaminated with hazardous organics, Ind. Eng. Chem. Res. **26** (1987) 2058–2062.

[3] SNYDER, J.L., GROB, R.L., McNALLY, M.E., OSOTYK, T.S., Comparison of supercritical fluid extraction with classical sonication and extractions for selected pesticides, Anal. Chem. **64** (1992) 1940–1946.

[4] JACKSON, M.L., Soil Chemical Analysis, Prentice Hall, Englewood Cliffs, NJ (1958).

[5] DAY, P.R., Experimental confirmation of the hydrometer theory, Soil Sci. **74** (1965) 181–186.

[6] ENVIRONMENTAL PROTECTION AGENCY, Method 3540 B, Soxhlet Extraction, Revision 2, EPA, Washington, DC (1990) 3540B-1–3540B-8.

[7] REINDL, S., HOFLER, F., Optimization of the parameters in supercritical fluid extraction of polynuclear aromatic hydrocarbons from soil samples, Anal. Chem. **66** (1994) 1808–1816.

[8] TAYLOR, L.T., Course of Supercritical Fluid, Universidad Autónoma de Querétaro, Mexico (Jun. 1993).

MAXIMIZING THE USE OF ENVIRONMENTAL DATA
(Session 7)

Chairperson

R.J. HANCE
FAO/IAEA

CONCLUSIONS AND RECOMMENDATIONS

The final session of the Symposium was devoted to a discussion on Maximizing the Use of Environmental Data in order to evaluate the fate and behaviour of pesticides in the environment for regulatory purposes. As a logical corollary, this included consideration of the need for additional information. Several conclusions/recommendations were reached/agreed upon:

(1) It is essential that laboratories producing data on the interactions of pesticides with the environment should have quality assurance and control procedures that meet the criteria of International Organization for Standardization Guide ISO 25 and, where relevant, the Organization for Economic Co-operation and Development (OECD) Principles of Good Laboratory Practice. Only where these standards are achieved is there any likelihood that such data will be accepted internationally, which is essential if unnecessary duplication is to be avoided.

(2) Standard laboratory physicochemical data obtained in accordance with OECD or other internationally accepted guidelines, together with information on the metabolism and persistence obtained under temperate conditions, can be used to assess behaviour under tropical conditions, since data on persistence represent the 'worst case' situation. However, mobility and volatility, in particular, may not be well predicted.

(3) Standard data require supplementation with the available local data, taking note of the rainfall patterns, light quality and soil types (particularly if these are lateritic, volcanic or of high pH). Work to generate additional data that are likely to be required should also take into account the formulation, pattern of use and cropping system.

(4) Existing toxicological data may be inadequate for predicting effects in the tropics. Therefore, it is important that supporting toxicological studies are conducted with tropical species, especially aquatic organisms.

(5) Although the fate and behaviour of pesticides in lateritic and, to some extent, volcanic soils can be predicted in a general way from basic principles, experimental evidence is needed to support such predictions.

(6) Model systems provide useful additional data, but currently they cannot substitute entirely for field studies.

(7) There is a need to develop experimental systems in which temperature, light and irrigation can be varied in a controlled way in order to simulate natural fluctuations, since extrapolation of the data obtained under steady state conditions can be misleading.

(8) Runoff and erosion are important processes of dissipation in the tropics, even in areas that are arid for much of the year. There is a need to incorporate knowledge of these processes, obtained in the context of soil erosion, into strategies and models that assess the environmental fate of pesticides.

(9) Eroded soil transported into aquatic habitats may be more damaging than any pesticide residues carried simultaneously. Therefore, it is important that efforts are made to segregate these two components in assessing effects.

(10) Detailed information on impurities in commercial pesticide products is as important for environmental considerations as it is for human health.

CHAIRPERSONS OF SESSIONS

Opening Session	R.J. HANCE	FAO/IAEA
Session 1	R.J. HANCE	FAO/IAEA
Session 2	F. GONZALEZ-FARIAS	Mexico
Session 3	M.A. MATIN	Bangladesh
	M.F. ZARANYIKA	Zimbabwe
Session 4	A.M. PECHEN DE D'ANGELO	Argentina
	K. RAGHU	India
Session 5	E. CARAZO	Costa Rica
Session 6	J. ESPINOSA-GONZALEZ	Panama
Session 7	R.J. HANCE	FAO/IAEA

SECRETARIAT OF THE SYMPOSIUM

R.J. HANCE	Scientific Secretary (FAO/IAEA)
F.P. CARVALHO	Scientific Secretary (IAEA)
H. SCHMID	Symposium Organizer (IAEA)
P. HOWARD KITTO	Proceedings Editor (IAEA)

LIST OF PARTICIPANTS

Abtahi, M.
Liaison Office in Vienna,
United Nations Educational, Scientific
  and Cultural Organization,
Wagramerstrasse 5, A-1400 Vienna,
Austria

Al-Mashhadi, A.S.
National Agriculture and Water Research Centre,
Ministry of Agriculture and Water,
P.O. Box 17285,
Riyadh 11484, Saudi Arabia

Ambrus, A.
Pesticides Management Group,
Plant Protection Service,
Food and Agriculture Organization
  of the United Nations,
Viale delle Terme di Caracalla,
I-00100 Rome, Italy

Andreescu, L.
Agrochemicals and Residues Section,
Joint FAO/IAEA Division of Nuclear Techniques
  in Food and Agriculture,
International Atomic Energy Agency,
Wagramerstrasse 5, A-1400 Vienna,
Austria

Bahig, M.
Nuclear Research Centre,
Atomic Energy Authority,
Atomic Energy Post Office,
Postal Code 13759,
101 Kasr El-Ainy Street,
Cairo, Egypt

Balasubramaniam, A.
81 Sylvan Crescent,
Leeming, Western Australia 6149,
Australia

Barceló, D.
Department of Environmental Chemistry,
Centro de Investigación y Desarrollo,
Consejo Superior de Investigaciones Científicas,
Jordi Girona 18–26,
E-08034 Barcelona, Spain

LIST OF PARTICIPANTS

Cai, Fulong	Laboratory of Marine Radioactivity, Third Institute of Oceanography, State Oceanic Administration, P.O. Box 0570, Xiamen 361005, Fujian, China
Carazo, E.	Centro de Investigación en Contaminación Ambiental, Universidad de Costa Rica, San José, Costa Rica
Chandler, L.R.	Ministry of Agriculture and Rural Development, Graeme Hall, Christ Church, Barbados
Dang, Duc Nhan	Institute for Nuclear Science and Technique, Viet Nam Atomic Energy Commission, P.O. Box 5T, 160 Nghia do, Tu Liem, Hanoi, Viet Nam
Dekek, A.M.A.	National Agriculture and Water Research Centre, Ministry of Agriculture and Water, P.O. Box 17285, Riyadh 11484, Saudi Arabia
Dierksmeier, G.	Instituto de Investigaciones de Sanidad Vegetal, Calle 110 esq. 5ta B, No. 514, Municipio Playa, Havana, Cuba
Edwards, C.A.	Soil Ecology Program, Department of Entomology, The Ohio State University, 103 Botany and Zoology Building, 1735 Neil Avenue, Columbus, OH 43210, United States of America
Espinosa-González, J.	Laboratorio de Agroecotoxicología y Plaguicidas, Instituto de Investigación Agropecuria de Panamá, Apartado Postal 6A-4391, El Dorado, Panama City, Panama
Everaarts, J.M.	Nederlands Instituut voor Ondersoek der Zee, P.O. Box 59, NL-1790 AB Den Burg-Texel, Netherlands

LIST OF PARTICIPANTS

Ferris, I.G.	Agrochemicals and Residues Section, Joint FAO/IAEA Division of Nuclear Techniques in Food and Agriculture, International Atomic Energy Agency, Wagramerstrasse 5, A-1400 Vienna, Austria
Führ, F.	Institut für Radioagronomie, Forschungszentrum Jülich, D-52428 Jülich, Germany
Getoff, N.	Institut für theoretische Chemie und Strahlenchemie, Universität Wien, Währingerstrasse 38, A-1090 Vienna, Austria
Ghanem, I.	Department of Agricultural Applications, Atomic Energy Commission of Syria, P.O. Box 6091, Damascus, Syrian Arab Republic
Ghods-Esphahani, A.	Agrochemicals and Residues Section, Joint FAO/IAEA Division of Nuclear Techniques in Food and Agriculture, International Atomic Energy Agency, Wagramerstrasse 5, A-1400 Vienna, Austria
González-Farias, F.	Mazatlán Marine Station, Institute of Marine Sciences and Limnology, National Autonomous University of Mexico, P.O. Box 811, Mazatlán, Sinaloa 82000, Mexico
Haberhauer, G.F.	Österreichisches Forschungszentrum Seibersdorf, A-2444 Seibersdorf, Austria
Helling, C.S.	Weed Science Laboratory, Agricultural Research Service, United States Department of Agriculture, Beltsville, MD 20705, United States of America

LIST OF PARTICIPANTS

Hussain, M.
Agrochemicals and Residues Section,
Joint FAO/IAEA Division of Nuclear Techniques
   in Food and Agriculture,
International Atomic Energy Agency,
Wagramerstrasse 5, A-1400 Vienna,
Austria

Ijani, A.S.M.
Tropical Pesticides Research Institute,
P.O. Box 3024,
Arusha, United Republic of Tanzania

Ilim, M.
Ankara Nuclear Research and Training Center,
Turkish Atomic Energy Authority,
Saray, Ankara 06105, Turkey

Khan, S.U.
Centre for Land and Biological Resources Research,
Research Branch,
Agriculture and Agri-Food Canada,
960 Carling Avenue,
Ottawa, ON K1A 0C6, Canada

Kim, Y.
United Nations Industrial Development Organization,
Vienna International Centre,
Wagramerstrasse 5, A-1400 Vienna,
Austria

Klaine, S.J.
Department of Environmental Toxicology,
The Institute of Wildlife and
   Environmental Toxicology,
P.O. Box 709,
Pendleton, SC 29670,
United States of America

Krenn, A.
Österreichisches Forschungszentrum Seibersdorf,
A-2444 Seibersdorf, Austria

Lawani, C.
Service de protection des végétaux,
Direction de l'agriculture,
Porto-Novo, Benin

Lazoumar, H.
Direction de protection des végétaux,
B.P. 323,
Niamey, Niger

LIST OF PARTICIPANTS

Lee, Chin Min	World Health Organization, Vienna International Centre, Wagramerstrasse 5, A-1400 Vienna, Austria
López-Romero, R.M.	Instituto de Recursos Naturales, Colegio de Postgraduados, Carretera México Texcoco, km. 35.5, CP 56230, Montecillo, Edo. de México, Mexico
Luchini, L.C.	Instituto Biológico, Centro de Radioisótopos, P.O. Box 7119-01064-970, Avenida de Conselheiro Rodrigues Alves 1252, São Paulo 04014-002, Brazil
Mansingh, A.	Pesticide and Pest Research Group, The University of the West Indies, Mona, Kingston 7, Jamaica
Mashayekhi, S.	Nuclear Research Centre for Agriculture and Medicine, P.O. Box 31585-4395, Karaj, Islamic Republic of Iran
Matin, M.A.	Institute of Food and Radiation Biology, Atomic Energy Research Establishment, P.O. Box 3787, Dhaka, Bangladesh
Mirzoyan, V.S.	Armenian Plant Protection Research Institute, Toumanian Street 40/25, Yerevan 2, Armenia
Mustafa, T.M.	Faculty of Agriculture, University of Jordan, Amman, Jordan
Nicholls, P.H.	Integrated Approach to Crop Research (IACR)-Rothamsted, Harpenden, Hertfordshire AL5 2JQ, United Kingdom
Osafo-Acquaah, S.	Department of Chemistry, University of Science and Technology, University Post Office, Kumasi, Ghana

LIST OF PARTICIPANTS

Pak, L.	Control and Toxicological Laboratory, Republic Station for Plant Protection, Ryskulov 59, Almaty 480026, Kazakstan
Pancas, M.N.R.	Plant Protection Department, National Directorate of Agriculture, Ministry of Agriculture and Fisheries, Av. das FPLM, Recinto do Inia, C.P. 3658, Maputo, Mozambique
Pechen de D'Angelo, A.M.	Universidad Nacional del Comahue, 8300 Neuquén, Buenos Aires 1400, Argentina
Raghu, K.	Nuclear Agriculture Division, Bhabha Atomic Research Centre, Mumbai 400 085, India
Rathor, M.N.	Agrochemicals and Residues Section, Joint FAO/IAEA Division of Nuclear Techniques in Food and Agriculture, International Atomic Energy Agency, Wagramerstrasse 5, A-1400 Vienna, Austria
Scheunert, I.	GSF-Institut für Bodenökologie, Neuherberg, D-85758 Oberschleissheim, Germany
Skenderi, H.	Directorate of Plant Protection Service, Ministry of Agriculture and Food, Tirana, Albania
Sugavanam, B.	United Nations Industrial Development Organization, Vienna International Centre, Wagramerstrasse 5, A-1400 Vienna, Austria
Tavares, T.M.	Instituto de Química, Campus Universitário da Federação, Universidade Federal da Bahia, 40-210-340 Salvador, Bahia, Brazil

LIST OF PARTICIPANTS

Tayaputch, N.	Division of Agricultural Toxic Substances, Department of Agriculture, Bangkok 10900, Thailand
Tejada, A.W.	National Crop Protection Center, University of the Philippines at Los Baños, College, Laguna 4031, Philippines
Than, A.	Plant Protection Division, Myanmar Agriculture Service, Agricultural Lane, P.O. Box 11081, Yangon, Myanmar
Tiryaki, O.	Ankara Nuclear Research and Training Center, Turkish Atomic Energy Authority, Saray, Ankara 06105, Turkey
Tungguldihardjo, M.S.	Centre for the Application of Isotopes and Radiation, National Atomic Energy Agency, Jl. Cinere Ps. Jumat, P.O. Box 7002, Jakarta 12070, Indonesia
Vollner, L.	Institut für ökologische Chemie, GSF-Forschungszentrum für Umwelt und Gesundheit, Ingolstädter Landstrasse 1, D-85758 Neuherberg, Germany
Xu, Bujin	Institute of Nuclear Agricultural Sciences, Zhejiang Agricultural University, 268 Kai Xuan Road, Hangzhou 310029, China
Yeboah, P.O.	Department of Chemistry, National Nuclear Research Institute, P.O. Box 80, Legon, Ghana
Yücel, Ü.	Ankara Nuclear Research and Training Center, Turkish Atomic Energy Authority, Saray, Ankara 06105, Turkey
Zaranyika, M.F.	Chemistry Department, University of Zimbabwe, P.O. Box 167, Mount Pleasant, Harare, Zimbabwe

AUTHOR INDEX

Ambrus, A.: 11
Anwar, E.: 145
Appoh, F.E.: 163
Bajet, C.M.: 265
Baker, D.: 247
Balasubramaniam, A.: 61
Bali, S.: 127
Barceló, D.: 331
Barquero, M.: 215
Benjamin, R.: 247
Beretta, M.: 321
Brown, T.: 247
Caballero de Castro, A.: 479
Cai, Feng: 349
Cai, Fulong: 349
Calumpang, S.M.F.: 265
Cao, G.: 181
Capella-Vizcaíno, S.: 495
Carazo, E.: 215
Carvajal, G.: 479
Carvalho, F.P.: 35, 289
Casey, R.: 247
Chen, Ying: 349
Chen, Meici: 489
Chen, Shumei: 349
Costa, M.A.: 321
Dang, Duc Nhan: 313
Dargie, J.D.: 5
Dasgupta, T.P.: 301
Dierksmeier, G.: 343
Dodoo, D.K.: 163
Dust, M.: 179
Edwards, C.A.: 435
Espinosa-González, J.: 93
Everaarts, J.M.: 407
Fernández, D.: 247
Ferris, I.G.: 371
Fischer, C.V.: 407
Fowler, S.W.: 35, 289

Führ, F,: 179
Ghanem, I.: 127
González-Farias, F.: 289
Gözek, K.: 171, 223
Haigh, B.M.: 371
Hatough, A.: 79
Helling, C.S.: 389
Hernández, R.: 343
Hillebrand, M.Th.J.: 407
Hoque, E.: 279
Horvat, M.: 35
Hossain, M.M.: 279
Ijani, A.S.M.: 461
Ilim, M.: 171
Joab, B.: 247
Kale, S.P.: 205
Katondo, J.M.: 461
Khan, S.U.: 111
Khan, Y.S.A.: 279
Khatoon, J.: 279
Khattari, S.: 79
Kirs, V.: 479
Klaine, S.J.: 247
Klotz, D.: 187
Knacker, T.: 435
Kookana, R.S.: 371
Kulkarni, M.G.: 205
Lin, Zhifeng: 349
Loewy, M.: 479
López-Romero, R.M.: 495
Luchini, L.C.: 235
Machi, S.: 3
Makusi, R.A.: 453
Malulu, J.M.: 461
Mansingh, A.: 301
Martínez, K.: 343
Masha'al, K.: 79
Mashayekhi, S.: 73
Matin, M.A.: 279

Medina, M.J.B.: 265
Mee, L.D.: 289
Mian, A.J.: 279
Ming, Hong: 489
Minja, E.M.: 453
Mittelstaedt, W.: 179
Mohamad, F.: 127
Montford, K.G.: 163
Moreno, P.: 343
Mugari, P.: 151
Murthy, N.B.K.: 205
Mustafa, T.M.: 79
Naddy, R.: 247
Nguyen, Manh Am: 313
Nicholls, P.H.: 361
Oliveira de Rezende, M.O.: 235
Osafo-Acquaah, S.: 471
Overmeyer, J.: 247
Pak, L.: 91
Pan, Jiarong: 135, 371
Parmelee, R.: 435
Pechen de D'Angelo, A.M.: 479
Pokarzhevskij, A.A.: 435
Pütz, T.: 179
Qian, Lumin: 349
Raghu, K.: 205
Ricardo, C.: 343
Richards, P.: 247
Robinson, D.E.: 301
Rodríguez, O.M.: 215

Sa'adeh, A.: 79
Scheunert, I.: 181
Schroll, R.: 181
Sheikhigorgan, A.: 73
Stork, A.: 179
Subler, S.: 435
Talebi, K.: 73
Tavares, T.M.: 321
Tejada, A.W.: 265
Tesha, F.: 453
Tiryaki, O.: 223
Tungguldihardjo, M.S.: 145
Valverde, B.E.: 215
van Weerlee, E.M.: 407
Varca, L.M.: 265
Venturino, A.: 479
Villeneuve, J.P.: 35
Vo, Van Thuan: 313
Vollner, L.: 187
Wen, Xianfang: 135, 371
Xu, Bujin: 489
Yeboah, P.O.: 163
Yiang, Jiadong: 349
Yücel, Ü.: 171
Zao, Yuhua: 489
Zaranyika, M.F.: 151
Zhang, Yongxi: 489
Zhu, Nanwen: 489
Zolfagharieh, H.R.: 73

INDEX OF PAPERS BY NUMBER

IAEA-SM-343/	Page	IAEA-SM-343/	Page
1	331	23	163
2	61	24	205
3	435	26	145
4	371	27	73
5	179	28	301
6	111	29	79
7	389	30	91
8	361	31	289
9	495	32	93
10	11	33	265
12	35	34	127
13	247	35	171
14	479	36	223
15	279	38	461
16	235	39	453
17	321	40	313
18	135	41	151
19	489	42	343
20	215	43	349
21	181	44	471
22	187	45	407

HOW TO ORDER IAEA PUBLICATIONS

No. 4, October 1996

☆ ☆ **In the United States of America and Canada**, the exclusive sales agent for IAEA publications, to whom all orders and inquiries should be addressed, is:

Bernan Associates, 4611-F Assembly Drive, Lanham, MD 20706-4391, USA

☆ ☆ **In the following countries** IAEA publications may be purchased from the sources listed below, or from major local booksellers. Payment may be made in local currency or with UNESCO coupons.

AUSTRALIA	Hunter Publications, 58A Gipps Street, Collingwood, Victoria 3066
BELGIUM	Jean de Lannoy, 202 Avenue du Roi, B-1060 Brussels
BRUNEI	Parry's Book Center Sdn. Bhd., P.O. Box 10960, 50730 Kuala Lumpur, Malaysia
CHINA	IAEA Publications in Chinese: China Nuclear Energy Industry Corporation, Translation Section, P.O. Box 2103, Beijing
CZECH REPUBLIC	Artia Pegas Press Ltd., Palác Metro, Narodni tř. 25, P.O. Box 825, CZ-111 21 Prague 1
DENMARK	Munksgaard International Publishers Ltd., P.O. Box 2148, DK-1016 Copenhagen K
EGYPT	The Middle East Observer, 41 Sherif Street, Cairo
FRANCE	Office International de Documentation et Librairie, 48, rue Gay-Lussac, F-75240 Paris Cedex 05
GERMANY	UNO-Verlag, Vertriebs- und Verlags GmbH, Dag Hammarskjöld-Haus, Poppelsdorfer Allee 55, D-53115 Bonn
HUNGARY	Librotrade Ltd., Book Import, P.O. Box 126, H-1656 Budapest
INDIA	Viva Books Private Limited, 4325/3, Ansari Road, Darya Ganj, New Delhi-110002
ISRAEL	YOZMOT Literature Ltd., P.O. Box 56055, IL-61560 Tel Aviv
ITALY	Libreria Scientifica Dott. Lucio di Biasio "AEIOU", Via Coronelli 6, I-20146 Milan
JAPAN	Maruzen Company, Ltd., P.O. Box 5050, 100-31 Tokyo International
MALAYSIA	Parry's Book Center Sdn. Bhd., P. O. Box 10960, 50730 Kuala Lumpur
NETHERLANDS	Martinus Nijhoff International, P.O. Box 269, NL-2501 AX The Hague Swets and Zeitlinger b.v., P.O. Box 830, NL-2610 SZ Lisse
POLAND	Ars Polona, Foreign Trade Enterprise, Krakowskie Przedmieście 7, PL-00-068 Warsaw
SINGAPORE	Parry's Book Center Pte. Ltd., P.O. Box 1165, Singapore 913415
SLOVAKIA	Alfa Press Publishers, Hurbanovo námestie 3, SQ-815 89 Bratislava
SPAIN	Díaz de Santos, Lagasca 95, E-28006 Madrid Díaz de Santos, Balmes 417, E-08022 Barcelona
SWEDEN	Fritzes Customer Service, S-106 47 Stockholm
UNITED KINGDOM	The Stationery Office Books, Publications Centre, 51 Nine Elms Lane, London SW8 5DR

☆ ☆ Orders (except for customers in Canada and the USA) and requests for information may also be addressed directly to:

Sales and Promotion Unit
International Atomic Energy Agency
Wagramerstrasse 5, P.O. Box 100, A-1400 Vienna, Austria

Telephone: +43 1 2060 22529 (or 22530)
Facsimile: +43 1 2060 29302
Electronic mail: SALESPUB@ADPO1.IAEA.OR.AT